AN INTRODUCTION TO ELEMENTARY PARTICLES

Second Edition

This is Volume 12 in
PURE AND APPLIED PHYSICS
A Series of Monographs and Textbooks
Consulting Editors: H. S. W. MASSEY AND KEITH A. BRUECKNER
A complete list of titles in this series appears at the end of this volume.

AN INTRODUCTION TO ELEMENTARY PARTICLES

Second Edition

W. S. C. WILLIAMS

NUCLEAR PHYSICS LABORATORY
UNIVERSITY OF OXFORD

AND

ST. EDMUND HALL
OXFORD, ENGLAND

 *1971*

ACADEMIC PRESS   New York and London

Copyright © 1971, by Academic Press, Inc.
ALL RIGHTS RESERVED
NO PART OF THIS BOOK MAY BE REPRODUCED IN ANY FORM,
BY PHOTOSTAT, MICROFILM, RETRIEVAL SYSTEM, OR ANY
OTHER MEANS, WITHOUT WRITTEN PERMISSION FROM
THE PUBLISHERS.

ACADEMIC PRESS, INC.
111 Fifth Avenue, New York, New York 10003

United Kingdom Edition published by
ACADEMIC PRESS, INC. (LONDON) LTD.
Berkeley Square House, London W1X 6BA

LIBRARY OF CONGRESS CATALOG CARD NUMBER: 73-84251

PRINTED IN THE UNITED STATES OF AMERICA

CONTENTS

Preface   xi

I. Quantum Mechanics

1.1	Introduction	1
1.2	Bras, Kets, Vectors, and Linear Operators	1
1.3	Quantum Mechanics	5
1.4	Time Development of Vectors	9
1.5	The Lorentz Transformations	11
1.6	Transformations	17
1.7	Parity, the Parity Transformation, and Parity Conservation	20
1.8	Center-of-Mass and Laboratory Coordinates	22
1.9	Conclusions	23
	References	23

II. Angular Momentum

2.1	Introduction	24
2.2	Orbital Angular Momentum	24
2.3	Rotations (I)	26
2.4	Spin and Total Angular Momentum	28
2.5	The Eigenvalues of Angular Momentum	30
2.6	The Matrix Elements of Angular Momentum	32
2.7	Vector Addition of Angular Momentum	34
2.8	The Eigenfunctions of Orbital Angular Momentum	40
2.9	The Pauli Spin Matrices	42
2.10	Rotations (II)	43
2.11	Decay of Pure States	49
2.12	Tensor Operators	50
2.13	Polarization	54

	2.14 The Density Matrix	56
	2.15 Decay of Mixed Spin States	61
	2.16 Rotation of the Density Matrix	66
	References	66

III. Scattering and Reaction Theory

3.1	Introduction	67
3.2	The Partial-Wave Analysis	67
3.3	Scattering of Spin-0 by Spin-$\frac{1}{2}$ Particles	75
3.4	Polarization in Spin-0–Spin-$\frac{1}{2}$ Scattering	80
3.5	Angular Distributions in Spin-0–Spin-$\frac{1}{2}$ Scattering	86
3.6	The Ambiguities of Spin-0–Spin-$\frac{1}{2}$ Scattering	87
3.7	The Scattering of Spin-$\frac{1}{2}$ by Spin-$\frac{1}{2}$ Particles	91
3.8	The Scattering of Identical Particles	91
3.9	The Scattering Matrix (I)	93
3.10	Binary Reactions	96
3.11	The Scattering Matrix (II)	98
3.12	Reciprocity	103
3.13	The Principle of Detailed Balance	106
	References	110

IV. Energy Dependence in Scattering

4.1	Introduction	111
4.2	Phase-Space Considerations	111
4.3	Phase Shifts at Low Energy	112
4.4	The Wigner Condition	116
4.5	Resonance and the Breit–Wigner Formula	118
4.6	Unitarity	124
	References	125

V. Symmetry, Isotopic Spin, and Hypercharge

5.1	Introduction	127
5.2	Symmetry and Antisymmetry	127
5.3	Two-Nucleon State Vectors	129
5.4	Isotopic Spin	130
5.5	Strangeness and Hypercharge	136
5.6	Conclusion	142
	References	142

VI. Parity, Time Reversal, Charge Conjugation, and G-Parity

6.1	Introduction	144
6.2	Parity	144

Contents

6.3 Parity Conservation	145
6.4 Parity Nonconservation	151
6.5 Time Reversal	156
6.6 The Consequences of Time-Reversal Invariance	160
6.7 Charge Conjugation	165
6.8 G-Parity	167
6.9 The CPT Theorem	170
9.10 Conclusion	174
References	175

VII. The Bosons

7.1 Introduction	177
7.2 The Pions	177
7.3 The Spin and Parity of the Pions	179
7.4 The K-Mesons	187
7.5 The Neutral K-Meson System	188
7.6 Meson Resonances	193
7.7 Meson Resonances Decaying into Two Mesons	196
7.8 Meson Resonances Decaying into Three Mesons (I)	203
7.9 Meson Resonances Decaying into Three Mesons (II)	212
7.10 Conclusion	215
References	216

VIII. The Baryons

8.1 Introduction	218
8.2 The Stable Baryons	218
8.3 Baryon Resonances	222
8.4 Isotopic Spin of the Pion-Nucleon System	226
8.5 Low-Energy Pion-Nucleon Scattering	230
8.6 Pion-Nucleon Scattering up to 2500 MeV	233
8.7 The Kaon-Nucleon System	238
8.8 Low-Energy Kaon-Nucleon Scattering	239
8.9 The Antikaon-Nucleon System	243
8.10 The Production of Baryon Resonances	247
8.11 The Ω^-	252
References	253

IX. Unitary Symmetry

9.1 Introduction	255
9.2 Symmetry and the Classification of States	255
9.3 The Theory of Continuous Groups	256
9.4 The Hadrons and $SU(3)$ Multiplets	268
9.5 Properties of Representation	270
9.6 Applications of $SU(3)$	275
9.7 Applications of Broken $SU(3)$	281

9.8	Quarks	288
9.9	Higher Symmetry Schemes	289
9.10	Conclusion	292
	References	292

X. Field Theory

10.1	Introduction	294
10.2	First Quantization	294
10.3	Units and Notation	296
10.4	The Lagrangian Formalism	297
10.5	The Electromagnetic Field	302
10.6	The Dirac Field	303
10.7	Second Quantization and the Commutation Relations	308
10.8	Interaction and the S-Matrix	309
10.9	Renormalization and the Radiative Corrections	314
10.10	QED at High Energies	318
10.11	Field Theory and Strong Interactions	323
10.12	Conclusion	326
	References	326

XI. Weak Interactions

11.1	Introduction	328
11.2	The Description and Theory of Beta Decay	328
11.3	The Classification of Beta Decays	332
11.4	Beta Decay: Pre-1956	334
11.5	Beta Decay: Post-1956	337
11.6	The Two-Component Theory of the Neutrino	339
11.7	Conservation of Leptons in Beta Decay	345
11.8	Muon and Pion Decay	346
11.9	The Universal Weak Interaction	352
11.10	The Conserved Vector Current	353
11.11	Muon Capture	361
11.12	The Leptonic Decays of Strange Particles	362
11.13	The Cabibbo Angle	369
11.14	The Nonleptonic Decay of Strange Particles	374
11.15	The Intermediate Vector Boson	382
11.16	Neutrino Interactions	384
11.17	Conclusion	385
	References	385

XII. Strong Interactions

12.1	Introduction	388
12.2	The Mandelstam Variables	388
12.3	The Analytic Properties of the S-Matrix	393
12.4	Pion–Nucleon Scattering Dispersion Relations	400

Contents

12.5	Other Singularities	404
12.6	Strong Interactions at High Energies	405
12.7	Asymptotic Relations	407
12.8	One-Particle-Exchange Mechanisms	408
12.9	Regge Poles	418
12.10	The Chew–Frautschi Plot	427
	References	429

XIII. The Electromagnetic Interaction of Hadrons

13.1	Introduction	431
13.2	The Electromagnetic Interaction	431
13.3	Isotopic-Spin Selection Rules	435
13.4	The Angular-Momentum Properties of the Electromagnetic Field	435
13.5	Photoproduction Processes	441
13.6	Electromagnetic Form Factors	444
13.7	The Vector-Dominance Model	453
	References	461

XIV. The Neutral Kaons and *CP* Conservation

14.1	Introduction	463
14.2	The Time Development of Neutral-Kaon Systems	463
14.3	The Neutral-Kaon Mass Difference	467
14.4	The Theory of Regeneration	468
14.5	The Sign of the Mass Difference	471
14.6	*CP* Nonconservation	472
14.7	*CP* Noninvariant Analysis	474
14.8	The Experimental Situation	480
14.9	The Source of *CP* Violation	481
	References	482

Appendix A Functions of a Complex Variable

A.1	Functions and Singularities	484
A.2	Integral Theorems	485
	References	487

Appendix B Relativistic Kinematics

B.1	The Properties of Four Vectors	488
B.2	Laboratory and Center-of-Mass Coordinates	490
B.3	Particle Reactions	491
B.4	Dalitz Plots	494
B.5	The Mandelstam Variables	495
B.6	Transformation of Differential Cross Sections	495

	B.7	Transformation of the Polarization Vector	497
		References	498

Appendix C Phase Space

C.1	The Density-of-States Factor	499
C.2	Phase Space and Dalitz Plots	502
	References	504

Appendix D Dirac Matrix Elements

D.1	The Structure of Matrix Elements for Dirac Particles	505
	Reference	508

Appendix E Tables of Clebsch–Gordan Coefficients and of Rotation Matrix Elements

E.1	Clebsch–Gordan Coefficients	509
E.2	Rotation Matrix Elements	510
	References	511

Author Index   513

Subject Index   521

PREFACE

This book has been written for research workers in the field of experimental elementary-particle physics and is intended to present an introduction to the theoretical methods and ideas that are used to describe the behavior of elementary particles.

In the last 20 years, there has been a great increase in the use of symmetry properties, and of their related conservation laws, in the field of elementary particles. These subjects are very important because they give considerable insight into the behavior of elementary particles without requiring an extensive knowledge of quantum mechanics for their understanding. At the time of the first edition there was an obvious need for a book that covered these aspects of elementary-particle physics in a manner that would appeal to experimental research workers. I hope that this second edition will have the same appeal.

The scope and structure of the book follow my own preferences on the subjects and the order of presentation. In all cases, the treatment is as elementary as is consistent with clarity and no claim is made to be rigorous.

Compiling references has proved to be a problem. The rate of publication in this subject is so great that it is impossible to be up to date and comprehensive. Readers will find that my references are not dated later than early 1969. If the reader wishes to start a search of later literature the best place is the "High Energy Physics Index" published by DESY, Hamburg or in the latest publication of the Particle Data Group of the Lawrence Radiation Laboratory.

I wish to record special thanks to Dr. F. Gilman, SLAC, who read almost all the manuscript and gave me a great deal of very valuable criticism and advice. I am also indebted to Dr. I. Corbett who helped read the proofs.

I

QUANTUM MECHANICS

1.1 Introduction

The primary objective of this book is to serve as an introduction to the interpretation of some of the phenomena associated with elementary particles. It would be useless to discuss the meaning of elementary in this respect: for us the term will cover the "particles" of high-energy physics, and we shall avoid discussing their bound states, which is the subject of nuclear physics.

We are particularly interested in those properties of elementary particles that are a consequence of the symmetry properties or lack of symmetry of nature. These properties allow a great amount to be learned about particles without recourse to the construction of models, and it is this aspect of these properties with which we shall deal. In order to do this, the first chapter is devoted to restating the principles of quantum mechanics so as to familiarize readers with the notation to be used and to give a reference point.

1.2 Bras, Kets, Vectors, and Linear Operators

Dirac has introduced a notation into quantum mechanics that is particularly useful and that we shall use frequently. We shall not discuss the notation in detail, but content ourselves with stating its most important features. Readers can find a full discussion in Dirac's book (1958).

All the information about a physical system is supposed to be contained in a state vector, which is represented by a ket, $|\psi\rangle$. The ket represents a vector in what is called Hilbert space, just as the position or velocity of a particle can be represented by a vector in ordinary space. The two vectors $|\psi\rangle$ and $c\,|\psi\rangle$, where c is a nonzero complex number, represent the same state, so that a state is specified not by the magnitude of the vector but by its direction in Hilbert space. We shall see that for the purposes of interpretation it is usual to make the vectors have unit length. It is possible to construct linear sums of vectors to produce other vectors that represent new states: Thus,

$$|\phi\rangle = c_1|\phi_1\rangle + c_2|\phi_2\rangle,$$

where c_1 and c_2 are complex numbers, represents a state different from either of the states represented by $|\phi_1\rangle$ or $|\phi_2\rangle$ as long as these two vectors do not point in the same direction in Hilbert space and neither c_1 nor c_2 is zero. This is the principle of superposition, which will be discussed fully in Section 1.3.

It is necessary to introduce a second kind of vector in order to make a complete theory. For every ket vector, $|\phi\rangle$ say, there exists a bra vector $\langle\phi|$ (sometimes called an adjoint vector) such that the scalar product of two vectors $|\phi_1\rangle$ and $|\phi_2\rangle$ is represented by the bra–ket $\langle\phi_1|\phi_2\rangle$. Although this is analogous to the scalar product of two three-dimensional vectors, it is not the same, as we shall see by defining further the connection between the bra corresponding to a ket. If

$$|\phi\rangle = c_1|\phi_1\rangle + c_2|\phi_2\rangle,$$

then the bra corresponding to the ket $|\phi\rangle$ is

$$\langle\phi| = c_1^*\langle\phi_1| + c_2^*\langle\phi_2|.$$

The connection between a ket and its bra is the same as that between a complex column matrix and its hermitian conjugate, so that if $\langle|\rangle$ corresponds to matrix multiplication, a bra–ket product is, in general, a complex number with the property that

$$\langle\phi_1|\phi_2\rangle = \langle\phi_2|\phi_1\rangle^*. \tag{1.1}$$

Evidently $\langle\phi_1|\phi_1\rangle$ is real and by analogy with ordinary vectors is taken to be the square of the length of the vector. If $\langle\phi_1|\phi_1\rangle = 1$ the state is said to be normalized. Two vectors $|\phi_1\rangle$ and $|\phi_2\rangle$ that have their scalar product $\langle\phi_1|\phi_2\rangle = 0$ are said to be orthogonal.

Notice that since there is a one-to-one correspondence between bras and kets, it is irrelevant which is chosen to specify a physical system. We shall use this notation, placing within the bra or ket the quantum numbers or other parameters that we wish to use to specify the state. The bra and ket vectors obey the laws of distribution and association.

The next step is to introduce linear operators. Consider a ket $|\phi_1'\rangle$ that is a function of the ket $|\phi_1\rangle$, so that to each $|\phi_1\rangle$, there corresponds only one $|\phi_1'\rangle$. This connection may be represented by the action of an operator R, say, on the ket:

1.2 Bras, Kets, Vectors, and Linear Operators

Suppose
$$|\phi_1'\rangle = R|\phi_1\rangle.$$

then if
$$|\phi_j'\rangle = R|\phi_j\rangle, \quad j = 1, 2, \ldots;$$

and
$$|\psi\rangle = \sum_j c_j |\phi_j\rangle$$

$$|\psi'\rangle = R|\psi\rangle = \sum_j c_j R|\phi_j\rangle, \quad (1.2)$$

the operator is said to be a linear operator. These operators obey the laws of association and distribution: that is,

and
$$P(Q|\psi\rangle) = (PQ)|\psi\rangle$$

$$P(Q + R)|\psi\rangle = PQ|\psi\rangle + PR|\psi\rangle.$$

However the operators do not, in general, obey the commutative law of multiplication: thus, in general,

$$PQ|\psi\rangle \neq QP|\psi\rangle.$$

This is written alternatively as

$$[P, Q] = PQ - QP \neq 0. \quad (1.3)$$

The associative law also holds for the quantity $\langle\psi_1|P|\psi_2\rangle$, so that we can interpret P as operating on $|\psi_2\rangle$ to produce a new ket $P|\psi_2\rangle$ or as P operating on the bra $\langle\psi_1|$ to produce a new bra $\langle\psi_1|P$. One important quantity that has the action of an operator is the ket–bra $|\phi_1\rangle\langle\phi_2|$. Consider this right-multiplied into the ket $|\psi_2\rangle$ to give $|\phi_1\rangle\langle\phi_2|\psi_2\rangle$. This is evidently the ket $|\phi_1\rangle$ multiplied by the complex number $\langle\phi_2|\psi_2\rangle$. Thus the ket–bra is an operator. Since the scalar product of two vectors $\langle\psi|\phi\rangle$ is in general a complex number and since an operator changes a vector into another vector, it follows that quantities such as $\langle\psi|P|\phi\rangle$ are also complex numbers, which we shall call matrix elements.

There are some particularly important types of operator. The adjoint or Hermitian conjugate ($P^\dagger$) of an operator P is defined by the equation

$$\langle\phi|P|\psi\rangle = \langle\psi|P^\dagger|\phi\rangle^*. \quad (1.4)$$

It follows that the bra corresponding to the ket $P|\psi\rangle$ is $\langle\psi|P^\dagger$. An operator that is Hermitian has $P^\dagger = P$, while a unitary operator U has the property $U^\dagger = U^{-1}$, where $U^{-1}U = 1$.

If an operator P acting on a ket vector $|\psi\rangle$ produces the same ket times a constant p, then the ket is said to be an eigenvector of P with eigenvalue p. Thus

$$P|\psi\rangle = p|\psi\rangle. \quad (1.5)$$

The eigenvalues of an operator can be discrete or continuous. Consider the case of a discrete set of eigenvalues

$$p_1, p_2, p_3, \ldots, p_i, \ldots$$

with eigenvectors $|1\rangle, |2\rangle, |3\rangle, \ldots, |i\rangle, \ldots$ of the operator P. Then

$$P|i\rangle = p_i |i\rangle, \quad i = 1, 2, 3, \ldots,$$

and hence

$$\langle i| P |i\rangle = p_i \langle i|i\rangle.$$

From the definition of the Hermitian conjugate the left-hand side is

$$\langle i| P |i\rangle = \langle i| P^\dagger |i\rangle^*.$$

However, if P is Hermitian, $P = P^\dagger$ and the matrix element is real; and since $\langle i|i\rangle$ is real, it follows that p_i is real. Thus, the eigenvalues of a Hermitian operator are real. Similarly, the eigenvalues of a unitary operator are complex with modulus 1. In addition, all the eigenvectors of a Hermitian or or of a unitary operator form a complete set, which means that they are orthogonal,

$$\langle i|j\rangle = 0 \quad \text{if} \quad i \neq j,$$

and that they are linearly independent. Also, they span the vector space, which means that there are just a sufficient number, and no more, to permit the expansion of any vector in the space as a linear sum of these eigenvectors. If $\langle i|j\rangle = \delta_{ij}$, the vectors are said to be orthonormal.

We can now list some other properties of operators that are important for the development of quantum mechanics:

(a) If $[P, Q] = 0$, then $[P, f(Q)] = 0$, where $f(Q)$ is a function of Q that can be expanded as a power series in Q.

(b) If $[P, Q] = 0$, then it is possible to find vectors that are simultaneously eigenvectors of both P and Q.

(c) If $[P, Q] = 0$ and $[P, R] = 0$ but $[Q, R] \neq 0$, then it is not possible to find vectors that are simultaneously eigenvectors of P, Q, and R. However, it is possible to find vectors that are eigenvectors of P and one of Q and R.

(d) The matrix element of the product of two operators can be expanded thus:

$$\langle \phi | PQ | \phi \rangle = \sum_i \langle \phi | P | i \rangle \langle i | Q | \phi \rangle, \quad (1.6)$$

where the vectors $|i\rangle$ are a complete set in the sense above. We have mentioned that the ket-bra $|\phi\rangle\langle\phi|$ is an operator, so the ket-bra

$$\sum_i |i\rangle\langle i| = I, \quad (1.7)$$

1.3 Quantum Mechanics

the identity operator ($I|\phi\rangle = |\phi\rangle$ for all $|\phi\rangle$). The summation in Eq. (1.6) becomes an integration if the eigenvalues of $|i\rangle$ are continuous. The extension of Eq. (1.6) to products of more than two operators is obvious.

(e) If the eigenvalues of an operator are continuous, it is impossible to normalize the eigenvectors by having $\langle i|j\rangle = \delta_{ij}$: instead the normalization becomes

$$\langle i|j\rangle = \delta(i-j), \tag{1.8}$$

where $\delta(x)$ is the Dirac δ-function (Dirac, 1958).

1.3 Quantum Mechanics

The apparatus of vectors and operators is the one appropriate to quantum mechanics, but it is necessary to have postulates by which predictions may be made from quantum mechanics. These postulates are

(1) To every physically observable quantity there corresponds an operator.

(2) The result of one measurement of the observable is one of the eigenvalues of the corresponding operator.

(3) The average value $\langle P \rangle$ of an observable P to be expected from repeated measurements on states all represented by the vector $|\phi\rangle$ is

$$\langle P \rangle = \frac{\langle \phi | P | \phi \rangle}{\langle \phi | \phi \rangle}. \tag{1.9}$$

$\langle P \rangle$ is often called the expectation value.

The immediate consequence of (1) is that observables must be represented by Hermitian (or in special cases unitary) operators, since these are the only ones that have real eigenvalues. If the state is represented by an eigenvector of the observable, then every measurement yields the eigenvalue and the state is said to be an eigenstate of the operator. In situations where the state is not an eigenstate of the observable, the measurement will yield one of the eigenvalues: it is impossible to say which eigenvalue will be obtained in a measurement, and the best that can be done is to predict the probability that a certain result is obtained.

Let us consider the observable P, which must be a Hermitian operator. The eigenvectors of P form a complete set, and this means that any state $|\phi\rangle$ can be expressed as a linear superposition the eigenstates of P:

$$|\phi\rangle = \sum_i c_i |i\rangle, \tag{1.10}$$

where

$$P|i\rangle = p_i |i\rangle, \quad i = 1, 2, 3, \ldots,$$

and the eigenvectors $|i\rangle$ are orthonormal. The coefficients c_i are easily found by multiplying Eq. (1.9) by the bra $\langle j|$.

$$\langle j|\phi\rangle = \sum_i c_i \langle j|i\rangle = c_j ;$$

hence

$$|\phi\rangle = \sum_i |i\rangle \langle i|\phi\rangle . \tag{1.11}$$

The physical interpretation is that the *probability amplitude* for finding the state $|\phi\rangle$ in the eigenstate $|i\rangle$ by a measurement of the observable is the complex quantity $\langle i|\phi\rangle$. The probability of finding the state in this eigenstate will be the amplitude squared, viz.,

$$|\langle i|\phi\rangle|^2 = \langle \phi|i\rangle \langle i|\phi\rangle . \tag{1.12}$$

Since the eigenstates $|i\rangle$, $i = 1, 2, \ldots$, are a complete set, we must find the original normalized state in one of these, and hence must have

$$\sum_i \langle \phi|i\rangle \langle i|\phi\rangle = 1 ; \tag{1.13}$$

but from Eq. (1.6) the left-hand side is $\langle \phi|\phi\rangle$, which is consistent with the normalization. To extend the physical interpretation, we can consider the measurement that must be made to find the probability of finding our original state in one of the eigenstates of the observable. Since this probability is $|\langle i|\phi\rangle|^2$, it follows that the average value $\langle P\rangle$ of the observable will be given by

$$\langle P\rangle = \sum_i p_i |\langle i|\phi\rangle|^2$$
$$= \sum_i \langle \phi|i\rangle p_i \langle i|\phi\rangle .$$

Now

$$p_i = \langle i|P|i\rangle ;$$

hence

$$\langle P\rangle = \sum_i \langle \phi|i\rangle \langle i|P|i\rangle \langle i|\phi\rangle$$
$$= \langle \phi|P|\phi\rangle . \tag{1.14}$$

Considering the normalization of the state $|\phi\rangle$, this is consistent with postulate (3).

The complete set of states chosen for the resolution of $|\phi\rangle$ in Eq. (1.10) is called the base vectors. There may be more than one set of base vectors from which to choose, and the most convenient depends upon the problem. This ability to express one state as a linear sum of other states is called the principle of superposition and is central to the construction of quantum mechanics. However, a particularly simple and illuminating

1.3 Quantum Mechanics

example of its application comes from physical optics and its description of polarized light. Considering only fully polarized light, we know that there are two possible descriptions: one describes a polarized beam by a sum, with correct amplitudes and phase, of two orthogonal plane-polarized beams; the other describes the beam by a sum, with correct amplitudes and phase, of two opposite circularly polarized beams. This kind of summation is exactly like that of Eq. (1.10), and it is possible to construct state vectors for all the states of polarization and to find the correct linear sum, in either description, which describes the state of polarization of photons in the beam. Similarly operators can be defined that will predict the probability of finding a certain result if a particular measurement of the state of polarization is made on the beam of photons: for example, that a measurement of the state of circular polarization of a plane-polarized photon finds the photon having left- or right-hand circular polarization with equal probability. Evidently where classical physical optics gave amplitudes and intensities, quantum mechanics gives probability amplitudes and probabilities.

There are probability amplitudes in quantum mechanics that are of considerable importance. Consider the operator of the observable that is positioned as specified by Cartesian coordinates, for example. The eigenvalues of this operator $\mathbf{X}$ are continuous so that the amplitudes give probability densities rather than probabilities. Formally, we can write a ket $|\mathbf{x}\rangle$ that represents the state containing a particle at the position $\mathbf{x}$. The orthogonal properties and normalization are expressed by [see Eq. (1.8)]

$$\langle \mathbf{x}' | \mathbf{x} \rangle = \delta(\mathbf{x}' - \mathbf{x}),$$

while Eq. (1.7) becomes

$$\int |\mathbf{x}\rangle \langle \mathbf{x}| \, d\mathbf{x} = 1 \qquad (1.15)$$

where $d\mathbf{x}$ is a volume element and the integration is over all the available space. The average value of $\mathbf{x}$ for a normalized state $|\psi\rangle$ is given by

$$\langle \mathbf{X} \rangle = \langle \psi | \mathbf{X} | \psi \rangle = \int \langle \psi | \mathbf{x} \rangle \mathbf{x} \langle \mathbf{x} | \psi \rangle \, d\mathbf{x}$$

$$= \int \mathbf{x} \, |\langle \mathbf{x} | \psi \rangle|^2 \, d\mathbf{x}. \qquad (1.16)$$

Thus we can interpret $|\langle \mathbf{x} | \psi \rangle|^2$ as the probability of finding the particle at point $\mathbf{x}$, per unit volume. $\langle \mathbf{x} | \psi \rangle$ is the corresponding amplitude, and thus it is the more familiar Schrödinger wavefunction in the bra–ket notation. Another probability amplitude of interest is $\langle \mathbf{p} | \psi \rangle$ which as $|\langle \mathbf{p} | \psi \rangle|^2$ is the probability of finding a state $|\psi\rangle$ with momentum $\mathbf{p}$, per unit volume of momentum space.

The resolution of a state vector into a complete set of base states, Eq. (1.10), is not necessarily unique. There may exist a large choice of sets of base vectors into which a state can be expanded, just as we have an infinite choice of Cartesian coordinate systems upon which to base the resolution of a vector in ordinary space into its three components. Suppose $|\phi\rangle$ can be expanded into a second set of n base states $|j\rangle$:

$$|\phi\rangle = \sum_j d_j |j\rangle, \qquad j = 1, 2, \ldots, n, \qquad (1.17)$$

where $d_j = \langle j|\phi\rangle$. Just as $|\phi\rangle$ can be expanded, so can $|j\rangle$ in terms of states $|i\rangle$:

$$|j\rangle = \sum_i |i\rangle \langle i|j\rangle.$$

Substitute into Eq. (1.17):

$$|\phi\rangle = \sum_i \sum_j d_j \langle i|j\rangle |i\rangle. \qquad (1.18)$$

Comparing Eqs. (1.10) and (1.18) we find

$$c_i = \sum_j \langle i|j\rangle d_j,$$

that is,

$$\langle i|\phi\rangle = \sum_j \langle i|j\rangle \langle j|\phi\rangle. \qquad (1.19)$$

This expresses the transformation that changes the amplitudes d_j which are the coefficients of the expansion of $|\phi\rangle$ into the base states $|j\rangle$ into the coefficients of the expansion into the base states $|i\rangle$. Notice that the transformation is given by the n^2 amplitudes $\langle i|j\rangle$, which are independent of $|\phi\rangle$. In addition, Eq. (1.19) can be expressed readily in matrix form: if we have the column matrix

$$\phi = \begin{bmatrix} \langle j=1|\phi\rangle \\ \langle j=2|\phi\rangle \\ \vdots \end{bmatrix}, \qquad (1.20)$$

i.e., the jth row is $\phi_j = \langle j|\phi\rangle$, and in a similar way we define the matrices $\phi_i' = \langle i|\phi\rangle$, $U_{ij} = \langle i|j\rangle$, then Eq. (1.19) becomes

$$\phi' = U\phi. \qquad (1.21)$$

ϕ' and ϕ are two different matrix representations of the state $|\phi\rangle$. The transformation between them is given by the matrix U.

This matrix has elements $\langle i|j\rangle = U_{ij}$ with the following properties:

$$\langle i|j\rangle = \langle j|i\rangle^* \qquad \text{or} \qquad U_{ij} = U_{ji}^*$$

$$\sum_j \langle i|j\rangle \langle j|i'\rangle = \delta_{ii'} \qquad \text{or} \qquad \sum_j U_{ij} U_{i'j}^* = \delta_{ii'} \qquad (1.22)$$

These are the properties of the elements of a unitary matrix.

This change in base states is equivalent to a passive coordinate transformation in Hilbert space: that is, a transformation in which the vector $|\phi\rangle$ remains fixed and the coordinate system is rotated. However, this view is no different in its formalism from that which regards the transformation as active and moves the vector in Hilbert space, so that it changes from $|\phi\rangle$ to $|\phi'\rangle$. In this case, its matrix representation in terms of the unchanged base states $|j\rangle$ is

$$|\phi\rangle = \begin{bmatrix} \langle j=1|\phi\rangle \\ \langle j=2|\phi\rangle \\ \vdots \end{bmatrix} \longrightarrow |\phi'\rangle = \begin{bmatrix} \langle j=1|\phi'\rangle \\ \langle j=2|\phi'\rangle \\ \vdots \end{bmatrix}.$$

If we let ϕ and ϕ' stand for these matrix representations, then

$$\phi = U\phi'.$$

Since $|\phi'\rangle$ and $|\phi\rangle$ continue to represent physical states of the system, it can be shown (Wigner, 1959; Wightman and Schweber, 1955) that the operator U that transforms one into the other is a unitary operator; this agrees with the fact, which has been shown, that the matrix representations are connected by a unitary matrix. In addition it is consistent with the conservation of probabilities, as can be seen by the following. If

$$|\phi\rangle \rightarrow |\phi'\rangle = U|\phi\rangle,$$

then from the definition of the Hermitian conjugate,

$$\langle\phi| \rightarrow \langle\phi'| = \langle\phi|U^\dagger.$$

Hence

$$\langle\phi'|\phi'\rangle = \langle\phi|U^\dagger U|\phi\rangle, \qquad (1.23)$$

and, if the operator U is unitary, $U^\dagger = U^{-1}$, then

$$\langle\phi'|\phi'\rangle = \langle\phi|U^{-1}U|\phi\rangle = \langle\phi|\phi\rangle \qquad (1.24)$$

as required.

1.4 Time Development of Vectors

We now wish to consider a physical system and how it changes with time. Let us consider a system represented at the time t_1 by the vector $|\phi(t_1)\rangle$ and at time t_2 by the vector $|\phi(t_2)\rangle$. Since these vectors represent the same physical system, they will be connected by a unitary operator

$$|\phi(t_1)\rangle \rightarrow |\phi(t_2)\rangle = U(t_2, t_1)|\phi(t_1)\rangle. \qquad (1.25)$$

Suppose $t_1 = t$ and $t_2 = t + \Delta t$; then if $\Delta t = 0$ we must have $U(t, t) = 1$. Also $U(t + \Delta t, t)$ will depend only upon Δt, and by use of Eq. (1.47), can be expressed in terms of a Hermitian operator H,

$$U(t + \Delta t, t) = \exp(-iH\Delta t/\hbar). \qquad (1.26)$$

For infinitesimal Δt this becomes

$$U(t + \Delta t, t) = 1 - (i/\hbar)H \Delta t \, .$$

Hence, from Eq. (1.25) it follows that

$$i\hbar \frac{\partial}{\partial t} |\phi(t)\rangle = H |\phi(t)\rangle \, . \tag{1.27}$$

This is the Schrödinger equation. Here, H is called the Hamiltonian operator. By considering the development of bra vectors instead of kets it is possible to find the corresponding equation

$$-i\hbar \frac{\partial}{\partial t} \langle\phi(t)| = \langle\phi(t)| H \, . \tag{1.28}$$

Consider now an observable Q. At time t its expectation value is given by

$$\langle Q \rangle = \langle\phi(t)| Q |\phi(t)\rangle \, .$$

In the Schrödinger picture all the development in time is contained in the state vectors, and the operators are time independent. It follows from Eqs. (1.27) and (1.28) that

$$i\hbar \frac{\partial}{\partial t} \langle Q \rangle = \langle\phi(t)| QH - HQ |\phi(t)\rangle \, . \tag{1.29}$$

The alternative view, called the Heisenberg picture, is that the state vectors are constant and it is the operators that vary with time. In that case the operator Q satisfies the equation (Schweber, 1961)

$$i\hbar \frac{\partial Q}{\partial t} = QH - HQ = [Q, H] \, . \tag{1.30}$$

Many observable operators commute with the Hamiltonian, and it follows that where this is the case they have expectation values that are conserved.

The Hamiltonian is taken to be the operator of the observable that is the total energy of the system. Thus, an isolated system not experiencing any external forces will be in a state of constant total energy E and hence in an eigenstate of H with eigenvalue equal to E:

$$H|\phi\rangle = E|\phi\rangle \, . \tag{1.31}$$

Using Eq. (1.26), we can put for such an isolated system

$$\begin{aligned} |\phi(t_2)\rangle &= U(t_2, t_1) |\phi(t_1)\rangle \\ &= \exp\bigl(-iE(t_2 - t_1)/\hbar\bigr) |\phi(t_1)\rangle \, . \end{aligned} \tag{1.32}$$

Thus displacement in time has the effect of changing the phase of the ket vector.

1.5 The Lorentz Transformations

If we consider the time development of the Schrödinger wavefunction $\psi(\mathbf{x}, t) = \langle \mathbf{x} | \psi(t) \rangle$, it follows from Eq. (1.27) that

$$i\hbar \frac{\partial}{\partial t} \psi(\mathbf{x}, t) = H \psi(\mathbf{x}, t), \qquad (1.33)$$

where H is now an algebraic operator [for example, $-(\hbar^2/2m)\nabla^2 + V$ for a single particle moving in a potential V]. For an isolated system of constant energy, $\psi(\mathbf{x}, t)$ can be factored into a time-independent function $\psi(\mathbf{x})$ and a function describing the time variation $T(t)$, where

$$i\hbar \frac{\partial T(t)}{\partial t} = E\, T(t);$$

hence,
$$T(t) = \exp(-iEt/\hbar). \qquad (1.34)$$
Thus
$$\psi(\mathbf{x}, t) = \psi(\mathbf{x}) \exp(-iEt/\hbar),$$

where $\psi(\mathbf{x})$ satisfies the equation

$$H \psi(\mathbf{x}) = E\, \psi(\mathbf{x}). \qquad (1.35)$$

This is the time-independent Schrödinger equation.

1.5 The Lorentz Transformations

All modern quantum mechanics can proceed meaningfully from its assumptions only if these satisfy the requirements of the special theory of relativity (the only exception is the original Schrödinger wave equation; for this reason it can be used only in restricted cases). We will now discuss those postulates of relativity that will affect our subsequent material.

The special theory of relativity postulates that

(1) The laws of physics formulated by an observer are independent of the state of uniform motion or position of the observer and his apparatus.
(2) The velocity of light *in vacuo* is the same for all such observers.

Let us see how this works by considering two observers who have agreed to make identical observations on identical phenomena. We must distinguish carefully between the adjectives "same" and "identical." We take the words "same phenomenon" to mean a phenomenon that has a common source for both observers; for example, the diffraction of light from a specified star. The words "same observation" would mean the measurement of a quantity associated with the common phenomenon; for example, the wavelength of a particular hydrogen spectral line in the light from our specified star. Obviously the same quantities will depend on the relative motion of our observers: in our example the Doppler effect alters the wavelength. By "identical phenomena" we mean phenomena that are alike in all respects but do not have a common source; the "identical observations" are the separate measurements of a quantity associated with both these phenomena. Two observers who measure separately the wavelength of a cadmium spectral line from their own local lamp would be measuring identical quantities but not the same quantity. So long as the measurements are not performed in a gravitational field or under conditions of acceleration, two such observations of identical quantities must yield identical results, which will be independent of the two observers' state of relative motion. It follows that if our two observers set up separate equations of motion that describe the observed phenomena, the two equations will be formally identical; that is, form invariant.

The system of coordinates that each observer sets up (normally stationary to himself) is called an inertial frame. The coordinate axes of two intertial frames can be connected by one or more of the following: rotation about the origin, displacement of one origin from the other, uniform relative velocity. As observers we are restricted to one inertial frame, but we do know how to transform observables and equations to the values and forms they would have if observed from another inertial frame (here we mean an observable quantity that is the same in the sense defined earlier). This system of transformations is called the Lorentz transformations, after their originator. We have already stated that in formulating a theory for any physical phenomenon it is essential that the transformation of our equations to the inertial frame of a second observer does not change the consequence of the theory; when this requirement is satisfied, the theory is said to be Lorentz covariant. Maxwell's electromagnetic theory, Dirac's equation for the electron, and the Klein–Gordon equation for bosons are all Lorentz covariant. Schrödinger's equation is not.

We shall now consider the Lorentz transformations. We commence by defining two kinds of transformation:

(1) A "passive" or "coordinate" transformation is one in which the physical system is unchanged and we consider the relation between a quantity as observed from one intertial frame and the same quantity as observed from a second interial frame; geometrically, this corresponds to a displacement and rotation of a four-dimensional coordinate system.

(2) In contrast, an "active" transformation is one in which a Lorentz transformation is applied to the physical system while the observer's inertial frame is unchanged; geometrically, this corresponds to a rotation and dis-

1.5 The Lorentz Transformations

placement of the physical system within a fixed four-dimensional coordinate system.

In mathematical formalism, these two types of transformation are indistinguishable; however, for clarity in discussion we shall consider only coordinate transformations. We are therefore interested in the relation between the results of measurements made by two separate observers on the same quantity.

We must classify the Lorentz transformations. There is a general divisions into two types, defined as follows:

(a) Proper Lorentz transformations are those that can be reached by an integration of a large number of infinitesimal transformations: two inertial frames with a uniform relative velocity or two inertial frames related by a simple rotation are examples of frames connected by proper Lorentz transformations.

(b) Improper Lorentz transformations are those that involve a discontinuity such as reflection through a plane or in time: obviously, improper transformations cannot be reached by a sum of infinitesimal transformations.

We will notice a further division into homogeneous and inhomogeneous transformations:

(a) A homogeneous Lorentz transformation of coordinate axes corresponds to a rotation of the axes in four-dimensional space.

(b) An inhomogeneous Lorentz transformation of coordinate axes corresponds to a rotation of the axes and displacement of the origin in four-dimensional space.

We can clarify these definitions by considering two inertial frames having coordinate axes connected by a proper Lorentz transformation. We write quantities in the first frame unprimed and quantities in the second frame primed; in addition, the coordinates of a point ct, x, y, z ($c =$ velocity of light) are written x^λ ($\lambda = 0, 1, 2, 3$). Such a point will have different coordinates in the second system; if these coordinates are x'^κ, then

$$x'^\kappa = \sum_{\lambda=0}^{3} a^\kappa{}_\lambda x^\lambda + b^\kappa \qquad (\lambda, \kappa = 0, 1, 2, 3). \tag{1.36}$$

$a^\kappa{}_\lambda$ is a set of 16 appropriate numbers. If we compare this with a simple coordinate transformation in Euclidean space, we see that this equation corresponds to a rotation and a displacement of the axes in four dimensions: it is therefore an inhomogeneous transformation. If $b^\kappa = 0$ ($\kappa = 0, 1, 2, 3$), then the displacement is absent and the transformation is homogeneous. We shall not be concerned with inhomogeneous transformations, except briefly in Section 1.6. The determinant formed from the real quantities $a^\kappa{}_\lambda$ has the following properties:

$|a^\kappa{}_\lambda| = 1$ for all proper Lorentz transformations;

$\phantom{|a^\kappa{}_\lambda|} = -1$ for improper transformations that involve reversal of one or three axes;

$\phantom{|a^\kappa{}_\lambda|} = +1$ for improper transformations that involve reversal of two or four axes.

The Lorentz transformations lead to a classification of observable quantities that is widely used in physics; it arises from the connection between the value Q of an observable measured from one intertial frame and the value Q' of the same observable measured from a second frame. In general,

$$Q' = LQ,$$

where L is a number, a matrix, or an operator that expresses the connection. The nature of L determines the classification of Q.

(a) *Scalars:* A scalar S is a quantity that transforms according to

$$S' = S;$$

that is, L is the identity operator for scalars, or invariants, as they are sometimes called.

(b) *Pseudoscalars:* A quantity P is said to be pseudoscalar if it transforms according to

$$P' = |a^\kappa{}_\lambda| P.$$

The important test of a pseudoscalar is made by reversing the three space axes for which $|a^\kappa{}_\lambda| = -1$. For this so-called parity transformation, $P' = -P$.

(c) *Vectors:* In the theory of relativity, a vector can be transformed meaningfully only if it has four components; a four-vector V is defined by the transformation of its four components:

$$V'^\kappa = \sum_{\lambda=0}^{3} a^\kappa{}_\lambda V^\lambda \qquad (\lambda, \kappa = 0, 1, 2, 3). \qquad (1.37)$$

Equation (1.36) indicates that the space–time coordinates of a point are the components of a four-vector.

(d) *Pseudo- or axial vectors:* Four-vectors that transform according to the equation

$$A'^\kappa = |a^\kappa{}_\lambda| \sum_{\lambda=0}^{3} a^\kappa{}_\lambda A^\lambda$$

are called pseudo- or axial vectors. Under the parity transformation the

1.5 The Lorentz Transformations

space components of an axial vector do not change sign, whereas those of a vector do change sign.

(e) *Tensors:* A tensor T of rank n is a quantity of 4^n components, each component having n indices; the transformation of a tensor requires a summation over each subscript, thus a tensor (contravariant) of rank 3 transforms according to

$$T'^{\kappa\lambda\mu} = \sum_{\eta=0}^{3}\sum_{\theta=0}^{3}\sum_{i=0}^{3} a^{\kappa}{}_{\eta}\, a^{\lambda}{}_{\theta} a^{\mu}{}_{i} T^{\eta\theta i}.$$

Scalars and vectors are tensors of zero and unit rank, respectively.

(f) *Pseudotensors:* Tensors for which the transformation includes multiplication by the determinant $|a^{\kappa}{}_{\lambda}|$ are called pseudotensors. Pseudoscalars and axial vectors are pseudotensors of zero and unit rank respectively.

(g) *Spinors:* A spinor is a four-component quantity that transforms in a similar manner to a four-vector, but the coefficients of the transformation are functions of the $a^{\kappa}{}_{\lambda}$. Spinors are important as solutions of the Dirac equation.

There are other quantities, such as tensor densities, which are beyond the scope of this book.

It is useful at this point to make some extension in notation. If we consider the four-vector x (ct, **x**) its squared magnitude is the invariant interval

$$c^2 t^2 - \mathbf{x}\cdot\mathbf{x}.$$

This is not the same as $x \cdot x$ if that is formally defined in the same way as the usual three-vector scalar product. One way of avoiding this difficulty is to make the time component imaginary, but this has the result of leaving ugly minus signs in many places. We therefore define two kinds of four-vectors. A contravariant vector has an index in the upper position (for example, $x^0 = ct$, $x^1 = x$, $x^2 = y$, $x^3 = z$), which transforms as in Eq. (1.36). The covariant vector has an index in the lower position and has components that are the same as those of the contravariant vector but with the sign of the space components changed (for example, $x_0 = ct$, $x_1 = -x$, $x_2 = -y$, $x_3 = -z$). A contravariant vector transforms according to

$$x_{\kappa}' = \sum_{\lambda=0}^{3} a_{\kappa}{}^{\lambda} x_{\lambda},$$

where the coefficients $a_{\kappa}{}^{\lambda}$ are related to the coefficients $a^{\kappa}{}_{\lambda}$. The metric tensor $g_{\mu\nu}$ transforms a contravariant into a covariant vector, and $g^{\mu\nu}$ a covariant into contravariant vector:

$$x_{\mu} = \sum_{\nu=0}^{3} g_{\mu\nu} x^{\nu}, \qquad x^{\mu} = \sum_{\nu=0}^{3} g^{\mu\nu} x_{\nu}.$$

Evidently $g^{00} = g_{00} = +1$, $g^{ij} = g_{ij} = -\delta_{ij}$, and $g^{0i} = g_{0i} = 0$, etc., ($i, j = 1, 2, 3$). The relation between the contravariant and covariant transformation coefficients is

$$a_\kappa{}^\lambda = \sum_{\eta=0}^{3} \sum_{\theta=0}^{3} g_{\kappa\eta} a^\eta{}_\theta \, g^{\theta\lambda}.$$

It is obvious that tensors of rank 2 and greater can have mixed transformation properties, so that indices appear in the subscript and superscript positions. The metrics can be used to raise or lower indices.

The contravariant four-vector x^μ leads to the differential operators $\partial/\partial x^\mu$: applied to a scalar this operator produces a covariant vector, and to stress this the notation

$$\partial_\mu = \frac{\partial}{\partial x^\mu}$$

is used. Conversely $\partial/\partial x_\mu$ gives a contravariant vector and is sometimes written ∂^μ. Then

$$\partial_\mu = \sum_{\nu=0}^{3} g_{\mu\nu} \partial^\nu, \quad \text{etc.}$$

We shall now drop summation signs and use the usual notation in which summation over repeated indices is implied. Greek indices run from 0 to 3, roman indices from 1 to 3.

The scalar or inner-product of two four-vectors U and V is defined as follows:

$$U \cdot V = U_\lambda V^\lambda = U^0 U^0 - \mathbf{U} \cdot \mathbf{V}. \tag{1.38}$$

(1) The space–time coordinates of a point ct, x, y, z are the components of a four-vector x^λ ($\lambda = 0, 1, 2, 3$). The square of the scalar magnitude of this vector is given by

$$x \cdot x = x_\lambda x^\lambda = c^2 t^2 - x^2 - y^2 - z^2,$$

and this is a scalar. In actual fact, the true definition of a Lorentz transformation rests upon the required invariance of this interval. The scalar product of two different vectors x and x' is also invariant.

(2) The total energy and momentum of a particle ($E/c, p_x, p_y, p_z$) are the components of a four-vector p. Then it follows that

$$p \cdot p = E^2/c^2 - p_x^2 - p_y^2 - p_z^2 \tag{1.39}$$

is a scalar; it is actually $+m^2 c^2$, where m is the rest mass of the particle.

We shall be using the terms pseudoscalar and axial vector mainly in the context of discussions involving the parity transformation; in such cases the time coordinate plays no role, and the behavior of these quantities can be examined using only the three space coordinates. Angular momentum and magnetic field are examples of axial vectors. It follows that the scalar product of a vector and an axial vector is a pseudoscalar and that the scalar product of two axial vectors is a scalar.

In Appendix B of this book we have derived some of the relations between the quantities of nuclear and elementary particle reactions when observed from the two important inertial frames—namely, the laboratory and center-of-mass systems.

1.6 Transformations

We wish to consider a physical system that is under observation by two different observers in separate inertial frames connected by a Lorentz transformation L. Observer 1 will describe the system by a state vector $|\psi_1\rangle$, while observer 2 describes the same system by a state vector $|\psi_2\rangle$. The two vectors can be connected by an operator that depends upon L:

$$|\psi_2\rangle = U(L)|\psi_1\rangle. \tag{1.40}$$

This equation describes the effect on the state vectors of the passive transformation of the coordinate system from the inertial frame of observer 1 to that of observer 2. Suppose that there is a second possible state of the system described by the vectors $|\phi_1\rangle$ and $|\phi_2\rangle$ respectively, where $|\phi_2\rangle = U(L)|\phi_1\rangle$. The probability that observer 1 finds the original system in the state described by $|\phi_1\rangle$ must be the same as the probability that observer 2 finds it in the state $|\phi_2\rangle$. That is,

$$|\langle\phi_1|\psi_1\rangle|^2 = |\langle\phi_2|\psi_2\rangle|^2.$$

Once again it follows that the operator $U(L)$ is a unitary operator (or antiunitary; see Section 6.5). The reciprocal Lorentz transformation L^{-1} must change $|\psi_2\rangle$ into $|\psi_1\rangle$, so we have

$$|\psi_1\rangle = U(L^{-1})|\psi_2\rangle = U^{-1}(L)|\psi_2\rangle. \tag{1.41}$$

It is worth noting that an active transformation L of the physical system has the effect on the state vectors that is the reverse of the effect on the state vectors in the passive transformation. Thus, if the physical system is transformed from being at rest in the frame of observer 1 to being at rest in the frame of observer 2, to observer 1 the state vector will change according to

$$|\psi_1\rangle \rightarrow |\psi_1'\rangle = U(L^{-1})|\psi_1\rangle,$$

whereas if observer 1 transforms himself to the rest frame of observer 2, and thus becomes equal to him in observational result, the vector that describes the system to him will change from $|\phi_1\rangle$ to $|\phi_2\rangle$, where $|\phi_2\rangle$ is given by Eq. (1.40).

Consider now the operator R, which has the effect in inertial frame 1

$$|\phi_1'\rangle = R|\phi_1\rangle, \qquad (1.42)$$

where corresponding vectors in frame 2 are

$$|\phi_2\rangle = U|\phi_1\rangle \quad \text{and} \quad |\phi_2'\rangle = U|\phi_1'\rangle.$$

Then there is an operator R' such that

$$|\phi_2'\rangle = R'|\phi_2\rangle. \qquad (1.43)$$

What is the relation between R and R'? Substituting into Eq. (1.42) gives

$$U^{-1}|\phi_2'\rangle = RU^{-1}|\phi_2\rangle,$$

or

$$|\phi_2'\rangle = URU^{-1}|\phi_2\rangle.$$

Therefore,

$$R' = URU^{-1}. \qquad (1.44)$$

This is called a similarity transformation and gives the connection between the operators in different inertial frames that transform one state of the same physical system into a second state of the system.

We next consider the observable Q. We assume that the two observers in their different frames will use the same operator for the same observable. Let the physical system be in the state described by $|\phi_1\rangle$ and $|\phi_2\rangle$, respectively. Then the expectation values of the operator Q are given by

Observer 1: $\qquad \langle Q \rangle_1 = \langle \phi_1 | Q | \phi_1 \rangle,$

Observer 2: $\qquad \langle Q \rangle_2 = \langle \phi_2 | Q | \phi_2 \rangle.$

$\langle Q \rangle_1$ and $\langle Q \rangle_2$ may be different, as will be the case, for example, if Q is the momentum and one observer is moving with respect to the other. However, many observable quantities are scalars, and then

$$\langle Q \rangle_1 = \langle Q \rangle_2.$$

In this case, we find

$$\langle \phi_1 | Q | \phi_1 \rangle = \langle \phi_2 | Q | \phi_2 \rangle$$
$$= \langle \phi_1 | U^{-1}QU | \phi_1 \rangle.$$

Therefore,

$$Q = U^{-1}QU, \qquad (1.45)$$

that is,

$$[Q, U] = 0. \qquad (1.46)$$

1.6 Transformations

Thus, if the observable Q is scalar under the transformation caused by the unitary operator U, then Q commutes with U.

A unitary operator may be generated from an Hermitian operator by the following expansion (ϵ is a real number):

$$U = 1 + iP\epsilon + \frac{(iP\epsilon)^2}{2!} - \frac{(iP\epsilon)^3}{3!} + \cdots = \exp(iP\epsilon) \qquad (1.47)$$

Obviously U is unitary if P is Hermitian. It follows that for many transformations there will exist an Hermitian operator that generates the corresponding unitary operator. Such Hermitian operators exist for the proper Lorentz transformations, and it happens that they are the observables of quantum mechanics. Thus the generators of time and space displacements in the inhomogeneous Lorentz transformations are the operators of energy and linear momentum. The generators of space rotations are the operators of angular momentum. The improper transformations have no generators. These considerations are complicated in a fully relativistic language: for example, rotations in ordinary three-dimensional space are generated by the operator of angular momentum, whereas a complete description of rotations in four-dimensional space requires the concept of a total angular momentum that contains spin.

TABLE 1.1

TRANSFORMATIONS AND CONSERVED QUANTITIES

Transformation	Conserved quantity
Displacement	Linear momentum
Timewise displacement	Energy
Rotations	Total angular momentum
Parity transformation	Parity
Rotations in isotopic spin space	Isotopic spin

For many transformations there will exist operators and functions that remain invariant. In particular, we may expect that the behavior of the system is unchanged and consequently require that the Hamiltonian H of the system be invariant; that is,

$$H' = UHU^{-1} = H:$$

hence

$$[U, H] = 0.$$

For proper Lorentz transformations the result is also true for infinitesimal transformations for which $U = 1 + iP\epsilon$ and we have

$$[P, H] = 0.$$

The eigenvalues of an operator that commutes with the Hamiltonian are constants of the motion by virtue of the equation of motion [Eq. (1.30)]:

$$i\hbar \frac{dP}{dt} = [P, H]$$

Hence, the eigenvalues of P are conserved. We can state a general principle: for every transformation that leaves the Hamiltonian invariant, there is a corresponding operator whose eigenvalues are conserved.

We can make a list of the coordinate transformations that leave the Hamiltonian invariant alongside the corresponding conserved quantity. See Table 1.1.

1.7 Parity, the Parity Transformation, and Parity Conservation

Parity is an important aspect of elementary-particle physics. We shall defer a full discussion of its implications and context until Chapter VI, but since we need to use its simpler consequences before then, we shall describe briefly what is meant by parity in the context of the Schrödinger equation.

Parity is an important property of many functions. Suppose we have a function $f(\mathbf{x})$ of the space coordinates (x_k, $k = 1, 2, 3$). Then $f(\mathbf{x})$ is said to have even parity if

$$f(-\mathbf{x}) = f(\mathbf{x}),$$

and odd parity if

$$f(-\mathbf{x}) = -f(\mathbf{x}).$$

Many functions do not have a unique parity but are mixtures of functions of opposite parity.

We must now relate this property to the parity transformation, which, in the passive case, reverses the direction of the three spatial axes. Suppose a wavefunction $\psi(\mathbf{x})$ describes some physical system. Then the transformation defines a new function of $\mathbf{x}$, namely $\psi'(\mathbf{x})$, which describes the same physical system in reversed coordinates (or a mirror image of this system in the unreversed coordinates). The unitary operator P, responsible for this transformation, is defined by

$$\psi'(\mathbf{x}) = P\psi(\mathbf{x}).$$

If $\psi(\mathbf{x})$ has even or odd parity, then it is an eigenfunction of P with eigenvalues $+1$ or -1 respectively. We shall prove these statements in Chapter VI.

The Schrödinger Hamiltonian, Eq. (1.33), is invariant under the transformation $\mathbf{x} \rightarrow -\mathbf{x}$ if the potential V is invariant, and under these circumstances it follows from Eq. (1.46) that H and P commute. This has two consequences: first, that it is possible to find functions that are simultaneously

1.7 Parity, the Parity Transformation, and Parity Conservation

eigenfunctions of H and P; second, that the eigenvalues of P are conserved. This latter property is called the conservation of parity and is a feature of systems that obey the Schrödinger equation. Since the parity transformation defines a mirror image, we find that a second system, which is a mirror image of such a system, behaves so as to preserve the mirror relation. This is invariance under the parity transformation. This symmetry is so simple and, in our normal experience, so complete, that this invariance was assumed in all circumstances and, arguing back, the conservation of parity was assumed to be correct in all circumstances. The breakdown of this symmetry in weak interactions was discovered in 1956 and will be discussed in Chapter VI; until then we shall continue to assume that parity is conserved.

Parity has important application in elementary-particle physics. Let us consider a process in which a particle is produced; for example, neutral meson production in a proton–proton collision

$$p + p \to p + p + \pi^0.$$

The total wavefunctions of the initial and final states are required to have the same parity. When we make an analysis we find that it is sometimes necessary to attribute an odd intrinsic parity to the particle produced, and we then say particle has odd parity. If it is not necessary to give the particle odd parity, then we say the particle has even parity. Thus in the the above reaction, if it were possible to determine the orbital angular momentum states involved, we would find that parity is the same in initial and final states only if the π^0 carries odd parity.

In modern quantum mechanics, spinless particles that carry odd intrinsic parity are described by state vectors which transform like pseudoscalars. Thus, odd-parity spinless particles such as the pion are often referred to as pseudoscalar, and even-parity spinless particles as scalar particles. Odd or even particles of unit spin are referred to as vector or axial vector particles, respectively. Fermi particles (half-odd integer spin) such as electrons or nucleons do not have an observable intrinsic parity but do have odd parity with respect to their antiparticles. In addition, strongly interacting fermions have a definite parity with respect to other fermions of the same baryon number and strangeness (see Chaper VI). This does not cause any difficulties in checking parity in elementary-particle reactions because fermions are conserved and an undetermined parity cancels in the parity equation. This can be seen in the production reaction just mentioned. Some ambiguities arise: for example, if we consider the reaction

$$p + p \to n + p + \pi^+,$$

it is obvious that the π^+ could have even parity if the neutron had parity opposite to that of the proton. It is usual to assign the same parity to the charged as to the uncharged mesons, but there is no way of confirming this.

This qualification and others to the concept of parity are discussed by Wick *et al.* (1952). We have remarked that invariance under the parity transformation implies that a system and its mirror image behave in the same way; thus it is possible to test various reactions involving spin for equality of cross sections. We shall expand this point when we come to examine the properties of angular momentum under the parity transformation. At present we note that the active parity transformation of a system reverses the sign of all position and linear-momentum vectors.

1.8 Center-of-Mass and Laboratory Coordinates

The postulates of Newtonian mechanics indicate that the center of mass of an assembly of particles will continue in a state of uniform motion unless disturbed by an external force. Thus the movement of particles within a system uninfluenced by an external force can be separated from the movement of the whole system. This makes it convenient to describe the internal movements of the system by the use of coordinates that have their origin at the center of mass. This coordinate system is called the center-of-mass coordinates and may be in a state of uniform motion with respect to the observer. Coordinates that are stationary with respect to the observer are called laboratory coordinates. The nonrelativistic case of the Schrödinger equation is straightforward: Consider two particles of mass m_1 and m_2 experiencing a mutual potential $V(x)$, where x is their separation. In the center-of-mass system they behave like one particle of mass m, given by

$$\frac{1}{m} = \frac{1}{m_1} + \frac{1}{m_2}$$

(the reduced mass), moving in a potential $V(x)$, where x is the distance from the origin. In the case of collisions the incident plane wave is represented by the wave function $\exp(i\mathbf{p} \cdot \mathbf{x}/\hbar)$, where the momentum $\mathbf{p}$ is that which one of the particles has in the center-of-mass system (and which the other has equally and oppositely). The separation of the Schrödinger equation for two or more particles is discussed by Margenau and Murphy (1943).

The relativistic wave equations are covariant and hold in any coordinate system: however, they are single-particle equations and so separation is not applicable. The relativistic treatment of a many-particle system requires the full apparatus of field theory and second quantization. Then the results are often most conveniently calculated in a coordinate system in which the total three-momentum is zero, loosely called the center-of-mass system.

1.9 Conclusions

In this chapter, we have restated the principles of quantum mechanism and introduced the bra–ket notation of Dirac along with the ideas of vectors and linear operators. In our discussions of the representation of states of vectors we have assumed that the states were in fact "pure." This is frequently not the case, and the extension of these techniques to cope with "mixed" states is discussed in Section 2.14.

For more complete treatments of quantum mechanics readers are referred to the works, for example, Dirac (1958), Matthews (1968), Wigner (1959), Feynman *et al.* (1965), Messiah (1961), and others.

REFERENCES

Dirac, P. A. M. (1958). "Principles of Quantum Mechanics," 4th ed. Oxford Univ. Press, London and New York (1.2, 1.9).

Feynman, R. P., Leighton, R. B., and Sands, M. (1965). "The Feynman Lectures on Physics," Vol. III. Addison-Wesley, Reading, Massachusetts (1.9).

Margenau, H., and Murphy G. M. (1943). "The Mathematics of Physics and Chemistry." Van Nostrand, Princeton, New Jersey (1.8).

Matthews, P. T. (1968). "Introduction to Quantum Mechanics," 2nd ed. McGraw-Hill, New York (2.9).

Messiah, A. (1961). "Quantum Mechanics." North-Holland Publ., Amsterdam (1.9).

Schweber, S. S. (1961). "An Introduction to Relativistic Quantum Field Theory." Harper and Row, New York (1.4).

Wick, G. C., Wightman, A. S., and Wigner, E. P. (1952). *Phys. Rev.* **88,** 101 (1.7).

Wightman, A. S., and Schweber, S. S. (1955). *Phys. Rev.* **98,** 812 (1.3).

Wigner, E. P. (1959). "Group Theory and Its Application to the Quantum Mechanics of Atomic Spectra." Academic Press, New York (1.3, 1.9).

II

ANGULAR MOMENTUM

2.1 Introduction

In this chapter, we shall discuss the operators and observables of angular momentum; by considering the properties of physical systems under rotations we shall show that some of these operators have eigenvalues that are constants of the motion. We shall also consider the vector addition of angular momentum and introduce the spherical harmonics that are eigenfunctions of certain operators. We shall finish the chapter with a further discussion of rotation, and of the use of the density matrix to describe the polarization states of particles.

2.2 Orbital Angular Momentum

In Section 1.8 we indicated that the motion of a system of particles can be separated into two parts: the first part is the motion of the center of mass of the system in axes fixed by an external observer; the second part is the motion of the system in the coordinates in which the center of mass is stationary. In this chapter we are concerned with the angular momentum in this second part. Our starting point is the classical relation

$$\mathbf{L} = \sum_n \mathbf{x}_n \times \mathbf{p}_n.$$

The quantities are vectors, and the sum is taken over all the particles in the system, the nth particle having linear momentum $\mathbf{p}$ and coordinate $\mathbf{x}_n$ with respect to the center of mass. The transformation to quantum mechanics is performed by replacing the quantity $\mathbf{p}$ by $-i\hbar\nabla$ and this makes the relation an operator equation; consider the quantities operating on a wavefunction ψ thus:

2.2 Orbital Angular Momentum

$$\mathbf{L}\psi = -i\hbar \mathbf{x} \times \nabla\psi. \tag{2.1}$$

This can be decomposed into Cartesian components, and dropping the ψ for convenience, we have for the three components of **L**

$$\begin{aligned} L_x &= -i\hbar\left(y\frac{\partial}{\partial z} - z\frac{\partial}{\partial y}\right), \\ L_y &= -i\hbar\left(z\frac{\partial}{\partial x} - x\frac{\partial}{\partial z}\right), \\ L_z &= -i\hbar\left(x\frac{\partial}{\partial y} - y\frac{\partial}{\partial x}\right). \end{aligned} \tag{2.2}$$

The square of the magnitude of the angular momentum is defined thus:

$$L^2 = L_x^2 + L_y^2 + L_z^2. \tag{2.3}$$

As always in quantum mechanics, the commutation relations between the various operators are of interest. By manipulating the operators of (2.2) we find that

$$\begin{aligned}{} [L_x, L_y] &= +i\hbar L_z, \\ [L_y, L_z] &= +i\hbar L_x, \\ [L_z, L_x] &= +i\hbar L_y, \end{aligned} \tag{2.4}$$

$$[L_x, L^2] = [L_y, L^2] = [L_z, L^2] = 0. \tag{2.5}$$

These relations are a consequence of the classical definition of angular momentum and of the change to quantum-mechanical formalism. This process cannot be reversed; Eq. (2.1) is not derivable from Eqs. (2.4) and (2.5). In formal quantum theory, it is possible to start with relations (2.4) and to develop angular-momentum theory from that point, without introducing the more physical picture implied in Eq. (2.1). Such a course is necessary in the case of intrinsic spin, for which there is no classical starting point.

The commutation relations between angular-momentum operators, Eqs. (2.4) and (2.5), and the rules connecting observables and commuting operators, Section 1.2 (c), show that we can only find eigenfunctions that are simultaneously eigenfunctions of L^2 and one of the three operators L_x, L_y, L_z. It is usual to choose L_z, so that a state of pure orbital angular momentum requires two quantum numbers for its specification. These are l and l_z, where the eigenvalues of L^2 and L_z are $l(l+1)\hbar^2$ and $l_z\hbar$, respectively (see Section 2.5). The magnitude of the angular momentum is $\hbar[l(l+1)]^{1/2}$, but for conciseness we shall often speak of such a state having angular momentum l. There is one circumstance in which an eigenfunction can be found for the operator L^2 and two of the operators L_x, L_y, and L_z; this happens if the eigenvalue of one of these component operators is zero, in which case this eigenfunction can also be an eigenfunction of one of the other component operators.

2.3 Rotations (I)

We wish to show that the angular momentum operators defined by Eqs. (2.4) and (2.5) are the generators of rotations. Consider a scalar function $f(\mathbf{x})$ that gives some numerical value to a space point P, having coordinates $\mathbf{x}$ with respect to a right-handed Cartesian coordinate system F. We wish to examine the effect upon $f(\mathbf{x})$ of a rotation of this coordinate system. We make an infinitesimal rotation of the coordinates through an angle $d\phi$ around the z axis. The new coordinates of the same space point P in the new coordinate system F' are given by

$$x' = x + y\,d\phi,$$
$$y' = y - x\,d\phi,$$
$$z' = z.$$

There must exist a function $f'(\mathbf{x}')$ that has the same value as has $f(\mathbf{x})$ when $\mathbf{x}$ and $\mathbf{x}'$ describe the same space point from different coordinates. The existence of $f'(\mathbf{x}')$ defines a function $f'(\mathbf{x})$ that has the same value at the point P' (coordinate $\mathbf{x}$ referred to F') as $f(\mathbf{x} + \Delta\mathbf{x})$ has at this point ($\mathbf{x} + \Delta\mathbf{x}$ referred to F). We can make a Taylor expansion of $f(\mathbf{x})$ around $\mathbf{x}$ to give

$$f'(\mathbf{x}) = f(\mathbf{x} + \Delta\mathbf{x}) = f(\mathbf{x}) - \left(y\frac{\partial}{\partial x} - x\frac{\partial}{\partial y}\right)f(\mathbf{x})\,d\phi + \cdots$$

For infinitesimal rotations, this can be written

$$f'(\mathbf{x}) = \left(1 + \frac{i}{\hbar}L_z\,d\phi\right)f(\mathbf{x}).$$

For finite rotations, $d\phi \to \phi$ and we make a summation of the infinitesimal rotations to give

$$f'(\mathbf{x}) = \exp(+i\phi L_z/\hbar)\,f(\mathbf{x}).$$

This expression applies to a rotation through an angle ϕ about the z axis; it can be generalized for a rotation θ about a unit vector $\mathbf{n}$ to give

$$f'(\mathbf{x}) = \exp[+i\theta(\mathbf{L}\cdot\mathbf{n})/\hbar]\,f(\mathbf{x}).$$

So far in this section, we have developed the rotation operator as an algebraic operator that is operating on real functions. It is possible to develop these ideas in quantum mechanics so that the operator R acts on a state

2.3 Rotations (I)

vector $|\psi\rangle$ to transform it into the state vector $|\psi'\rangle$ that represents the same state in a rotated coordinate system:

$$|\psi\rangle \to |\psi'\rangle = R|\psi\rangle, \tag{2.6a}$$

where $R = \exp[+i\theta(\mathbf{L}\cdot\mathbf{n})/\hbar]$ and $\mathbf{L}$ is the quantum-mechanical operator with commutation relations given by Eqs. (2.4) and (2.5). If the system contains particles with spin, then it is necessary to replace $\mathbf{L}$ by $\mathbf{J}$, the operator of the total angular momentum; we put

$$D = \exp[+i\theta(\mathbf{J}\cdot\mathbf{n})/\hbar]. \tag{2.6b}$$

Consider now another operator Q that has the effect of changing one state vector into another

$$Q|\psi\rangle = |\phi\rangle \tag{2.7}$$

If we rotate the coordinates, then

$$|\psi\rangle \to |\psi'\rangle = D|\psi\rangle,$$

and

$$|\phi\rangle \to |\phi'\rangle = D|\phi\rangle.$$

Then there is an operator Q' that has the property

$$|\phi'\rangle = Q'|\psi'\rangle.$$

Substituting for $|\phi'\rangle$ and $|\psi'\rangle$ we find

$$D|\phi\rangle = Q'D|\psi\rangle \quad \text{or} \quad |\phi\rangle = D^{-1}Q'D|\psi\rangle.$$

Comparing this with Eq. (2.7) we find

$$Q' = DQD^{-1}. \tag{2.8}$$

For example, Q may be the operator that describes the results of displacing the coordinates system along the x axis. After a rotation of the coordinates the operator Q' describes the effect of a displacement of the new coordinates along the old x axis. We shall use Eq. (2.8) in Section 2.10 to transform some operators describing successive rotations around axes in different coordinate systems into operators describing an equivalent set of successive rotations around axes all in the same coordinate system.

We can now consider the development in time of a physical state. Let us suppose the state $|\psi_1\rangle$ becomes $|\psi_2\rangle$ after time t. Then the unitary operator S that describes the change is generated by the Hamiltonian H;

$$|\psi_2\rangle = S|\psi_1\rangle,$$

where

$$S = \exp[-iHt/\hbar].$$

This change is expected to be independent of the orientation of the physical system as long as there are no external fields, or of the coordinate system to which the state vectors are referred as long as there are no external fields associated with this coordinate system. That is, S is invariant under rotations of the coordinate system, and Eq. (2.8) applied to S reads

$$S = DSD^{-1},$$

that is,

$$[D, S] = 0.$$

We consider infinitesimal times, in which case

$$S = 1 - iH\,\Delta t/\hbar,$$

and it follows that

$$[D, H] = 0.$$

The rotations can also be made infinitesimal, and it follows that the generators of D also commute with H,

$$[\mathbf{J}, H] = 0,$$

and from property (a) of operators (Section 1.2) it follows that

$$[J^2, H] = 0.$$

Using the operator equation

$$i\hbar(dQ/dt) = [Q, M],$$

we see that the eigenvalues of J^2 and J_z (or of L^2 and L_z in the absence of spin) do not change with time; that is, they are conserved. Thus the symmetry of physical systems expressed as the invariance of S (i.e., of the Hamiltonian) under rotations leads to conservation laws.

This discussion merely verifies the material of Section 1.6 as applied to angular mometum.

2.4 Spin and Total Angular Momentum

We have already explicitly anticipated the existence of spin as a part of total angular momentum. In strict analogy with the orbital angular momentum, we can construct a space with state vectors and operators. Such operators are $\mathbf{S}$, corresponding to $\mathbf{L}$ (that is, S_x, S_y, S_z corresponding to L_x, L_y, L_z). Then if $|\chi\rangle$ is the vector describing the state of a particle in a pure spin state, it is an eigenvector of the operators S^2 and S_z with eigenvalues $s(s+1)\hbar$ and $s_z\hbar$, respectively, where s is the spin of the particle and s_z its z component.

2.4 Spin and Total Angular Momentum

The operators $\mathbf{S}$ (S_x, S_y, S_z) and S^2 are assumed to commute among themselves in a completely analogous way to the orbital angular-momentum operators. Since the spin operators operate on state vectors in "spin space," any spin operator commutes with any orbital operator.

Since spin is an angular momentum, it is necessary to construct a total angular-momentum operator; that is,

$$\mathbf{J} = \mathbf{L} + \mathbf{S}.$$

Then $\mathbf{J}$ also satisfies the usual angular-momentum relations; these can be written

$$[\mathbf{J} \times \mathbf{J}] = i\hbar \mathbf{J}.$$

In the absence of external fields, rotational symmetry ensures that the eigenvalues of J^2 and J_z are conserved. However, if J is made up of two or more angular momenta, it is necessary to investigate what other quantities are conserved. To do this, it is necessary to draw up a list of commuting operators, including the Hamiltonian. This Hamiltonian will now contain, for example, terms indicating interaction between the spins of particles or the interaction between the spin of a particle and the orbital angular momentum of the particle. The first gives terms like $\mathbf{S}_1 \cdot \mathbf{S}_2$, the second terms like $\mathbf{L} \cdot \mathbf{S}$. Let us consider a system containing only one particle with spin and only one orbital angular momentum, so that the Hamiltonian does not contain terms $\mathbf{S}_1 \cdot \mathbf{S}_2$ or $\mathbf{L}_1 \cdot \mathbf{L}_2$. We have that $\mathbf{L}$ does not commute with $\mathbf{L}$ and therefore does not commute with $\mathbf{L} \cdot \mathbf{S}$ or with the Hamiltonian. Similarly, $\mathbf{S}$ will not commute with H. However, $\mathbf{J} = \mathbf{L} + \mathbf{S}$ does commute with $\mathbf{L} \cdot \mathbf{S}$ and therefore with H. Operators L^2, S^2 commute with $\mathbf{L} \cdot \mathbf{S}$ and with H. Therefore the eigenvalues of the following operators are constants of the motion:

$$S^2, L^2, J^2, J_z.$$

Of the three operators in $\mathbf{J}$, only one, J_z, is allowed, since a state can be an eigenstate of only one of them. If the system contains two particles with spin, then the Hamiltonian can contain terms such as $\mathbf{S}_1 \cdot \mathbf{S}_2$ or

$$(3/r^2)(\mathbf{S}_1 \cdot \mathbf{r})(\mathbf{S}_2 \cdot \mathbf{r}) - \mathbf{S}_1 \cdot \mathbf{S}_2.$$

The former term commutes with all the operators we have just listed and does not alter the number of the constants of the motion. The latter term represents the noncentral tensor force and does not commute with L^2 or with S^2 ($\mathbf{S} = \mathbf{S}_1 + \mathbf{S}_2$). These operators must be removed from the list of commuting operators if the tensor force is present, and it follows that in these circumstances l and s are no longer constants of the motion.

Hereafter, we shall speak of conserved quantities by their quantum numbers; thus if J^2 commutes with H, then $j(j + 1)\hbar^2$ is a constant of the motion; for brevity we shall say j is the constant and refer to the total angular momentum as j; similarly, we shall use l to specify the eigenvalue of L^2, l_z for L_z, j_z for J_z, etc.

2.5 The Eigenvalues of Angular Momentum

We devote this section to developing the formalism of the angular-momentum operators. Let us construct a state vector that is simultaneously an eigenstate of H, J^2, J_z; let it be $|E, \alpha, \beta\rangle$. Then

$$H|E, \alpha, \beta\rangle = E|E, \alpha, \beta\rangle, \qquad (2.9)$$

$$J^2|E, \alpha, \beta\rangle = \alpha\hbar^2|E, \alpha, \beta\rangle, \qquad (2.10)$$

$$J_z|E, \alpha, \beta\rangle = \beta\hbar|E, \alpha, \beta\rangle. \qquad (2.11)$$

We introduce two further operators defined by

$$J_+ = J_x + iJ_y, \qquad (2.12a)$$

$$J_- = J_x - iJ_y. \qquad (2.12b)$$

Using Eq. (2.9) it can be shown that these operators satisfy the following commutation relations:

$$[J_z, J_+] = +J_+\hbar, \qquad [J_z, J_-] = -J_-\hbar, \qquad [J_+, J_-] = 2J_z\hbar.$$

We operate on Eq. (2.11) with J_+ and substitute the appropriate commutation relation: thus

$$J_+ J_z|E, \alpha, \beta\rangle = \beta\hbar J_+|E, \alpha, \beta\rangle$$

becomes

$$(-J_+\hbar + J_z J_+)|E, \alpha, \beta\rangle = \beta\hbar J_+|E, \alpha, \beta\rangle,$$

which can be rearranged to

$$J_z J_+|E, \alpha, \beta\rangle = (\beta + 1)\hbar J_+|E, \alpha, \beta\rangle. \qquad (2.13)$$

Equation (2.13) indicates that the state $J_+|E, \alpha, \beta\rangle$ is also an eigenstate of J_z with eigenvalue $(\beta + 1)\hbar$. In a similar way we can show that the state $J_-|E, \alpha, \beta\rangle$ is an eigenstate of J_z with eigenvalue $(\beta - 1)\hbar$:

$$J_z J_-|E, \alpha, \beta\rangle = (\beta - 1)\hbar J_-|E, \alpha, \beta\rangle \qquad (2.14)$$

The operators J_+ and J_- also commute with H and J^2, so that the new states given by Eqs. (2.13) and (2.14) are still eigenstates of H and J^2 with the same eigenvalues as the original eigenstate as in Eqs. (2.9) and (2.10).

Therefore J_+ and J_- have the power to generate, from one eigenstate of J_z, other eigenstates of J_z that differ in their eigenvalues by an amount $\pm\hbar$. However, there must be a limit to the repeated application of J_+

2.5 The Eigenvalues of Angular Momentum

or of J_-. This limit will occur for J_- if, when it is applied to the previously generated eigenstate $|E, \alpha, \beta_{\min}\rangle$, we have that

$$J_-\,|E, \alpha, \beta_{\min}\rangle = 0. \tag{2.15}$$

Operating on Eq. (2.15) with J_+, we must have

$$J_+J_-\,|E, \alpha, \beta_{\min}\rangle = 0. \tag{2.16}$$

Substituting into Eq. (2.16) yields the operator equality

$$J_+J_- = J^2 - J_z^2 + J_z\hbar.$$

Evaluating the eigenvalues shows us that

$$\hbar^2(\alpha - \beta_{\min}^2 + \beta_{\min}) = 0.$$

Thus the last eigenvalue $\beta_{\min}$ of J_z that can be generated by J_- is given by

$$\beta_{\min}(\beta_{\min} - 1) = \alpha. \tag{2.17}$$

Similarly, the limit to the repeated application of J_+ will come when

$$J_+\,|E, \alpha, \beta_{\max}\rangle = 0. \tag{2.18}$$

Operating on Eq. (2.18) with J_- and using the operator equality

$$J_-J_+ = J^2 - J_z^2 - J_z\hbar, \tag{2.19}$$

we find that

$$\beta_{\max}(\beta_{\max} + 1) = \alpha.$$

Thus

$$\beta_{\max} = -\beta_{\min}.$$

We see that $\hbar(\beta_{\max} - \beta_{\min})$ is an integral multiple of $\hbar$, since $\beta_{\max}$ and $\beta_{\min}$ are connected by an integral number of operations with J_+ or with J_-. Therefore we can put

$$\hbar\beta_{\max} = \hbar\beta_{\min} + n\hbar, \quad n \text{ a positive integer},$$

whence

$$\beta_{\max} = n/2.$$

Thus we see that the possible values for the eigenvalues of J_z can vary through the series

$$-\frac{n\hbar}{2}, \quad -\frac{n\hbar}{2} + \hbar, \quad \ldots, \quad \frac{n\hbar}{2} - 2\hbar, \quad \frac{n\hbar}{2} - \hbar, \quad \frac{n\hbar}{2}.$$

Substituting in Eq. (2.17), we see that

$$\alpha = \frac{n}{2}\left(\frac{n}{2} + 1\right).$$

This is usually expressed as $j(j+1)$, and our eigen equation becomes
$$J^2 |E, j, \beta\rangle = j(j+1)\hbar^2 |E, j, \beta\rangle,$$
while the possible eigenvalues of J_z are
$$-j, -j+1, -j+2, \ldots \quad \ldots, j-2, j-1, j.$$
The j can be zero or have positive integer or half-integer values.

We know from particular cases in elementary wave mechanics that L_z has eigenvalues that are integer multiples of $\hbar$, while the eigenvalues of S_z can be $\pm\hbar/2$. The total angular momentum and its permitted z components consequently may also have values that are half-integer multiples of $\hbar$. The relations among the eigenvalues show that this possibility can be accommodated.

Our eigenstate $|E, \alpha, \beta\rangle$ is now written $|E, j, j_z\rangle$, where

$$H|E, j, j_z\rangle = E|E, j, j_z\rangle, \tag{2.20}$$
$$J^2|E, j, j_z\rangle = j(j+1)\hbar^2|E, j, j_z\rangle, \tag{2.21}$$
$$J_z|E, j, j_z\rangle = j_z\hbar|E, j, j_z\rangle, \quad -j \leqslant j_z \leqslant j. \tag{2.22}$$

2.6 The Matrix Elements of Angular Momentum

In this section attention is devoted to some matrix elements of the form $\langle E, j_2, j_{2z} | K | E, j_1, j_{1z}\rangle$, where K is an operator made up of angular-momentum operators; the two states connected in this way are pure states of angular momentum, having the same properties as the eigenstate defined by Eqs. (2.20)–(2.22). As a trivial example, we put $K = J_z$, $j = j_1 = j_2$, and $j_z = j_{1z} = j_{2z}$; then we must have

$$\langle j, j_z | J_z | j, j_z\rangle = j_z\hbar.$$

We wish to consider the operator J_-J_+; by Eq. (2.19) this has eigenvalues $\{j(j+1) - j_z(j_z+1)\}\hbar^2$. Thus we have the matrix element

$$\langle j, j_z | J_-J_+ | j, j_z\rangle = \{j(j+1) - j_z(j_z+1)\}\hbar^2$$
$$= (j-j_z)(j+j_z+1)\hbar^2.$$

By the rule (d) of Section 1.2 we can write the left-hand side of this equation as

$$\sum \langle j, j_z | J_- | j', j_z'\rangle\langle j', j_z' | J_+ | j, j_z\rangle.$$

This sum must normally be taken over all possible intermediate states; however, since H and J^2 commute with J_+ and with J_-, all matrix elements are zero, except those connected to intermediate states having the same energy and total angular momentum. Thus the sum is taken over all possible values of j_z'; but we know that J_+ and J_- only generate states with

2.6 The Matrix Elements of Angular Momentum

eigenvalues of J_z differing by one unit. Therefore, from the orthonormality of states, we see that $J_+|j,j_z\rangle$ can only be matched with the eigenfunction $\langle j, j_z+1|$, and so on. Again all matrix elements are zero, except those having the two states such that their eigenvalues of J_z differ by one unit. Thus the sum reduces to one term:

$$\langle j,j_z|J_-J_+|j,j_z\rangle = \langle j,j_z|J_-|j,j_z+1\rangle\langle j,j_z+1|J_+|j,j_z\rangle .$$

The two terms on the right-hand side are the complex conjugates of one another. Therefore

$$(j-j_z)(j+j_z+1)\hbar^2 = |\langle j,j_z|J_-|j,j_z+1\rangle|^2 ,$$

or

$$\langle j,j_z|J_-|j,j_z+1\rangle = e^{i\delta}\hbar[(j-j_z)(j+j_z+1)]^{1/2} . \quad (2.23)$$

The $e^{i\delta}$ is a phase term appearing because the matrix elements are complex quantities; this phase is arbitrary, so we can chose $\delta = 0$. Similarly to Eq. (2.23) we have

$$\langle j,j_z+1|J_+|j,j_z\rangle = \hbar[(j-j_z)(j+j_z+1)]^{1/2} .$$

If we substitute Eqs. (2.12a) and (2.12b) into these matrix elements, we can solve for the matrix elements of J_x and of J_y. The nonzero matrix elements of J_x, J_y, J_z, J_+, and J_- are

$$\langle j,j_z+1|J_x|j,j_z\rangle = \tfrac{1}{2}\hbar[(j-j_z)(j+j_z+1)]^{1/2} , \quad (2.24)$$

$$\langle j,j_z|J_x|j,j_z+1\rangle = \tfrac{1}{2}\hbar[(j-j_z)(j+j_z+1)]^{1/2} , \quad (2.25)$$

$$\langle j,j_z+1|J_y|j,j_z\rangle = -\frac{i}{2}\hbar[(j-j_z)(j+j_z+1)]^{1/2} , \quad (2.26)$$

$$\langle j,j_z|J_y|j,j_z+1\rangle = +\frac{i}{2}\hbar[(j-j_z)(j+j_z+1)]^{1/2} , \quad (2.27)$$

$$\langle j,j_z|J_z|j,j_z\rangle = j_z\hbar , \quad (2.28)$$

$$\langle j,j_z+1|J_+|j,j_z\rangle = \hbar[(j-j_z)(j+j_z+1)]^{1/2} , \quad (2.29)$$

$$\langle j,j_z|J_-|j,j_z+1\rangle = \hbar[(j-j_z)(j+j_z+1)]^{1/2} . \quad (2.30)$$

The choice of phase that we made implies that we have chosen the relative phase between the eigenstate $|j,j_z\rangle$ and those generated by J_+ and J_-, that is,

$$J_+|j,j_z\rangle = [(j-j_z)(j+j_z+1)]^{1/2}|j,j_z+1\rangle , \quad (2.31)$$

$$J_-|j,j_z\rangle = [(j-j_z+1)(j+j_z)]^{1/2}|j,j_z-1\rangle . \quad (2.32)$$

This section effectively continues in Section 2.12, where we shall consider the matrix elements of tensor operators.

2.7 Vector Addition of Angular Momentum

In this section we are interested in a physical system containing two angular momenta that are coupled together. The interaction between the two parts A and B may be such that the individual total angular momenta (j_a, j_b) and the z components (j_{az}, j_{bz}) are constants of the motion, in which case the state is an eigenstate of the operators J_a, J_b, J_{az}, J_{bz} and can be represented by the state vector $|j_a, j_{az}, j_b, j_{bz}\rangle$. However, if there is an interaction between the two angular momenta, j_{az} and j_{bz} may no longer be good quantum numbers. Then it is important to be able to transform to a representation using base states that are eigenstates of $J^2 = (\mathbf{J}_a + \mathbf{J}_b)^2$ and of $J_z = J_{az} + J_{bz}$, these having eigenvalues that, in the absence of external fields, are expected to be conserved.

The state $|j_a, j_{az}, j_b, j_{bz}\rangle$ is already an eigenstate of J_z with eigenvalue $j_z = j_{az} + j_{bz}$; however, it is not necessarily an eigenstate of the operator J^2. It is in fact a sum of eigenstates of J^2, the sum being restricted to states with quantum numbers j that satisfy $-j \leqslant j_z \leqslant +j$ and $|j_a - j_b| \leqslant j \leqslant j_a + j_b$. The first condition arises from ordinary properties of angular momentum, the second from the fact that the sum of two vectors cannot be less than the magnitude of their difference or greater than the magnitude of their sum. To illustrate this we consider some simple examples of the addition of the two vectors $\mathbf{J}_a$ and $\mathbf{J}_b$. We pick out particular values of the total z component that can be reached in the addition; obviously the possible values must satisfy $-(j_a + j_b) \leqslant j_z \leqslant (j_a + j_b)$. We shall indicate eigenstates of J^2 and of J_z by $|j, j_z\rangle$.

(a) If $j_z = j_{az} + j_{bz} = j_a + j_b$, then the total state vector is

$$|j_a, j_a, j_b, j_b\rangle.$$

The z components of the total and of the separate vectors are at a maximum, and the state vector can only be an eigenstate of J^2 with quantum number $(j_a + j_b)$.

(b) If $j_z = j_a + j_b - 1$, then there are two ways of composing a total state vector with this J_z eigenvalue; they are

$$|j_a, j_a, j_b, j_b - 1\rangle, \qquad (2.33)$$

$$|j_a, j_a - 1, j_b, j_b\rangle. \qquad (2.34)$$

Neither of these states is an eigenstate of J^2, since there are two such possible states with J_z eigenvalues $(j_a + j_b - 1)$. They are

$$|j_a + j_b, j_a + j_b - 1\rangle, \qquad (2.35)$$

$$|j_a + j_b - 1, j_a + j_b - 1\rangle. \qquad (2.36)$$

2.7 Vector Addition of Angular Momentum

Either of the state vectors (2.33) or (2.34) can be expressed as a linear sum of the two functions (2.35) and (2.36), and vice versa. This follows from the principle of superposition, Section 1.3.

(c) If $j_z = j_a + j_b - 2$, then there are three ways of composing a total state vector with this J_z eigenvalue; we do not give them, as they are analogous to Eqs. (2.33) and (2.34). These three state vectors can each be expressed as a linear sum of the J^2 eigenstates that have J_z eigenvalues $(j_a + j_b - 2)$ and that are analogous to vectors (2.35) and (2.36).

This process continues; any value of j_z selected can have its state vector expressed as a linear sum of eigenstates of J^2. As we decrease the value of j_z, more J^2 eigenstates are required until $j_z = |j_a - j_b|$, after which no more are needed. When $j_z = -|j_a - j_b|$, the number of eigenstates required begins to decrease until only one is required when $j_z = -|j_a + j_b|$.

What we have said up to now can be reduced to the simple statement that

$$|j_a, j_{az}, j_b, j_{bz}\rangle = \sum_j C_j |j, j_z = j_{az} + j_{bz}\rangle. \qquad (2.37)$$

A set of such equations can be solved to give the reverse of an addition: thus, if we know that the state of angular momentum j, j_z is the vector sum of two angular momenta j_a and j_b with $j_z = j_{az} + j_{bz}$, then it is possible to put

$$|j, j_z\rangle = \sum_{j_{az}} G_{j_{az}} |j_a, j_{az}, j_b, j_{bz} = j_z - j_{az}\rangle. \qquad (2.38)$$

The coefficients of Eqs. (2.37) and (2.38) are the Clebsch–Gordan coefficients for the vector addition of the two angular momenta. In fact these equations are a part of a unitary transformation between two sets of base states, one consisting of the eigenstates ($|j, j_z, j_a, j_b\rangle$) of J^2, J_z, J_a^2, J_b^2 and the other of the eigenstates ($|j_a, j_{az}, j_b, j_{bz}\rangle$) of $J_a^2, J_{az}, J_b^2, J_{bz}$. Using the result of Eq. (1.11) we have

$$|j_a, j_{az}, j_b, j_{bz}\rangle = \sum_{j, j_z} \langle j, j_z, j_a, j_b | j_a, j_{az}, j_b, j_{bz}\rangle |j, j_z, j_a, j_b\rangle, \qquad (2.39)$$

and

$$|j, j_z, j_a, j_b\rangle = \sum_{j_{az}, j_{bz}} \langle j_a, j_{az}, j_b, j_{bz} | j, j_z, j_a, j_b\rangle |j_a, j_{az}, j_b, j_{bz}\rangle. \qquad (2.40)$$

The coefficients contain more quantum numbers than is essential to describe them fully, and we shall contract them to

$$\langle j, j_z | j_a, j_{az}, j_b, j_{bz}\rangle$$

and so on. The coefficients of Eq. (2.39) are the elements of a matrix with rows labeled by all pairs of j, j_z and columns by all pairs of j_{az}, j_{bz};

similarly for the coefficients of Eq. (2.40). The eigenstates involved are orthonormal, so that if we multiply Eq. (2.39) by $\langle j_a, j'_{az}, j_b, j'_{bz}|$ we find

$$\sum_{j,j_z} \langle j, j_z | j_a, j_{az}, j_b, j_{bz}\rangle \langle j_a, j'_{az}, j_b, j'_{bz} | j, j_z\rangle = \delta_{j_{az}j'_{az}}\delta_{j_{bz}j'_{bz}}. \quad (2.41)$$

Similarly

$$\sum_{j_{az}j_{bz}} \langle j_a, j_{az}, j_b, j_{bz} | j, j_z\rangle \langle j', j'_z | j_a, j_{az}, j_b, j_{bz}\rangle = \delta_{jj'}\delta_{j_z j'_z}. \quad (2.42)$$

Now

$$\langle j, j_z | j_a, j_{az}, j_b, j_{bz}\rangle = \langle j_a, j_{az}, j_b, j_{bz} | j, j_z\rangle^*,$$

so that Eqs. (2.41) and (2.42) confirm that the matrix of the coefficients is unitary, since this property requires

$$\sum_{\beta} U_{\alpha'\beta} U^*_{\alpha\beta} = \delta_{\alpha\alpha'}$$

for the matrix elements.

Since the Clebsch–Gordan coefficients are zero unless $j_z = j_{az} + j_{bz}$, the double sums in Eqs. (2.41) and (2.42) are purely formal and we can pick out from Eqs. (2.39) and (2.40) simpler sums corresponding to Eqs. (2.37) and (2.38):

$$|j_a, j_{az}, j_b, j_{bz}\rangle = \sum_j \langle j, j_z | j_a, j_{az}, j_b, j_{bz}\rangle | j, j_z = j_{az} + j_{bz}\rangle, \quad (2.43)$$

$$|j, j_z\rangle = \sum_{j_{az}} \langle j_a, j_{az}, j_b, j_{bz} | j, j_z\rangle | j_a, j_{az}, j_b, j_{bz}\rangle. \quad (2.44)$$

The Clebsch–Gordan coefficients can be determined by the use of the step operators on Eqs. (2.43) and (2.44). To illustrate this let us use the last equation to derive a recursion relation between the coefficients. We have

$$J_- = J_{a-} + J_{b-}.$$

We apply J_- to the left-hand side of Eq. (2.44) and expand the resulting eigenstate:

$$J_-|j, j_z\rangle = \hbar A(j, j_z) |j, j_z - 1\rangle$$
$$= \hbar A(j, j_z) \sum_{j'_{az}} \langle j_a, j'_{az}, j_b, j'_{bz} | j, j_z - 1\rangle | j_a, j'_{az}, j_b, j'_{bz}\rangle,$$

(2.45)

where

$$A(j, j_z) = [(j + j_z)(j - j_z + 1)]^{1/2}.$$

Operating with J_{a-} or J_{b-} on the right-hand side of Eq. (2.44) gives

2.7 Vector Addition of Angular Momentum

$$\sum_{j_{az}} \langle j_a, j_{az}, j_b, j_{bz} | j, j_z \rangle [J_{a-} | j_a, j_{az}, j_b, j_{bz} \rangle + J_{b-} | j_a, j_{az}, j_b, j_{bz} \rangle]$$

$$= \hbar \sum_{j_{az}} \langle j_a, j_{az}, j_b, j_{bz} | j, j_z \rangle$$
$$\times [A(j_a, j_{az}) | j_a, j_{az} - 1, j_b, j_{bz} \rangle$$
$$+ A(j_b, j_{bz}) | j_a, j_{az}, j_b, j_{bz} - 1 \rangle]. \qquad (2.46)$$

We equate (2.46) with (2.45) and compare the coefficients of $|j_a, j_{az}, j_b, j_{bz}\rangle$ to obtain

$$A(j, j_z) \langle j_a, j_{az}, j_b, j_{bz} | j, j_z - 1 \rangle$$
$$= A(j_a, j_{az} + 1) \langle j_a, j_{az} + 1, j_b, j_{bz} | j, j_z \rangle$$
$$+ A(j_b, j_{bz} + 1) \langle j_a, j_{az}, j_b, j_{bz} + 1 | j, j_z \rangle. \qquad (2.47)$$

Similarly application of $J_+ = J_{a+} + J_{b+}$ gives

$$A(j, -j_z) \langle j_a, j_{az}, j_b, j_{bz} | j, j_z + 1 \rangle$$
$$= A(j_a, -j_{az} + 1) \langle j_a, j_{az} - 1, j_b, j_{bz} | j, j_z \rangle$$
$$+ A(j_b, -j_{bz} + 1) \langle j_a, j_{az}, j_b, j_{bz} - 1 | j, j_z \rangle. \qquad (2.48)$$

Now $A(j, -j_z) = A(j, j_z + 1)$ and $A(j, -j_z + 1) = A(j, j_z)$, so that this last relation becomes

$$A(j, j_z + 1) \langle j_a, j_{az}, j_b, j_{bz} | j, j_z + 1 \rangle$$
$$= A(j_a, j_{az}) \langle j_a, j_{az} - 1, j_b, j_{bz} | j, j_z \rangle$$
$$+ A(j_b, j_{bz}) \langle j_a, j_{az}, j_b, j_{bz} - 1 | j, j_z \rangle. \qquad (2.49)$$

As an example of the use of these relations we shall evaluate the coefficients required for the vector addition of $j_a = 1$ and $j_b = \tfrac{1}{2}$. The eigenstate $|j_a, j_{az}, j_b, j_{bz}\rangle = |1, 1, \tfrac{1}{2}, \tfrac{1}{2}\rangle$ is already an eigenstate $|j, j_z\rangle$ of J^2 and J_z with quantum numbers $\tfrac{3}{2}, \tfrac{3}{2}$, and we choose the phase so that

$$|1, 1, \tfrac{1}{2}, \tfrac{1}{2}\rangle = |\tfrac{3}{2}, \tfrac{3}{2}\rangle, \qquad (2.50)$$

that is,

$$\langle 1, 1, \tfrac{1}{2}, \tfrac{1}{2} | \tfrac{3}{2}, \tfrac{3}{2} \rangle = 1.$$

From Eq. (2.49) we have

$$A(\tfrac{3}{2}, \tfrac{3}{2}) \langle 1, 1, \tfrac{1}{2}, \tfrac{1}{2} | \tfrac{3}{2}, \tfrac{3}{2} \rangle$$
$$= A(1, 1) \langle 1, 0, \tfrac{1}{2}, \tfrac{1}{2} | \tfrac{3}{2}, \tfrac{1}{2} \rangle + A(\tfrac{1}{2}, \tfrac{1}{2}) \langle 1, 1, \tfrac{1}{2}, -\tfrac{1}{2} | \tfrac{3}{2}, \tfrac{1}{2} \rangle,$$

that is,

$$\sqrt{3} = \sqrt{2} \langle 1, 0, \tfrac{1}{2}, \tfrac{1}{2} | \tfrac{3}{2}, \tfrac{1}{2} \rangle + \langle 1, 1, \tfrac{1}{2}, -\tfrac{1}{2} | \tfrac{3}{2}, \tfrac{1}{2} \rangle.$$

The condition (2.42) implies that the sum of the squares of these coefficients is 1, so we have

$$\langle 1, 0, \tfrac{1}{2}, \tfrac{1}{2} | \tfrac{3}{2}, \tfrac{1}{2} \rangle = \sqrt{\tfrac{2}{3}} \quad \text{and} \quad \langle 1, 1, \tfrac{1}{2}, -\tfrac{1}{2} | \tfrac{3}{2}, \tfrac{1}{2} \rangle = \sqrt{\tfrac{1}{3}}.$$

By putting $j = \tfrac{3}{2}, j_z = -\tfrac{1}{2}, j_{az} = 1, j_{bz} = -\tfrac{1}{2}$ in Eq. (2.49) we find that

$$\langle 1, 0, \tfrac{1}{2}, -\tfrac{1}{2} | \tfrac{3}{2}, -\tfrac{1}{2} \rangle = \sqrt{\tfrac{2}{3}}.$$

Similar substitution will give

$$\langle 1, -1, \tfrac{1}{2}, \tfrac{1}{2} | \tfrac{3}{2}, -\tfrac{1}{2} \rangle = \sqrt{\tfrac{1}{3}}, \quad \langle 1, -1, \tfrac{1}{2}, -\tfrac{1}{2} | \tfrac{3}{2}, -\tfrac{3}{2} \rangle = 1.$$

So far we have determined six out of ten coefficients to be found. The remainder are matrix elements of the kind $\langle 1, j_{az}, \tfrac{1}{2}, j_{az} | \tfrac{1}{2}, j_z \rangle$. If we put $j = j_z = \tfrac{1}{2}$ into Eq. (2.49), the left-hand side must be zero and we have

$$A(1, 1) \langle 1, 0, \tfrac{1}{2}, \tfrac{1}{2} | \tfrac{1}{2}, \tfrac{1}{2} \rangle + A(\tfrac{1}{2}, \tfrac{1}{2}) \langle 1, 1, \tfrac{1}{2}, -\tfrac{1}{2} | \tfrac{1}{2}, \tfrac{1}{2} \rangle = 0,$$

that is,

$$\sqrt{2} \langle 1, 0, \tfrac{1}{2}, \tfrac{1}{2} | \tfrac{1}{2}, \tfrac{1}{2} \rangle + \langle 1, 1, \tfrac{1}{2}, -\tfrac{1}{2} | \tfrac{1}{2}, \tfrac{1}{2} \rangle = 0. \quad (2.51)$$

A solution that satisfies this and the condition of Eq. (2.42) is

$$\langle 1, 0, \tfrac{1}{2}, \tfrac{1}{2} | \tfrac{1}{2}, \tfrac{1}{2} \rangle = -\sqrt{\tfrac{1}{3}}, \quad \langle 1, 1, \tfrac{1}{2}, -\tfrac{1}{2} | \tfrac{1}{2}, \tfrac{1}{2} \rangle = +\sqrt{\tfrac{2}{3}}. \quad (2.52)$$

Putting $j = \tfrac{1}{2}, j_z = -\tfrac{1}{2}, j_{az} = 0$, and $j_{bz} = \tfrac{1}{2}$ in Eq. (2.49) gives

$$\langle 1, -1, \tfrac{1}{2}, \tfrac{1}{2} | \tfrac{1}{2}, -\tfrac{1}{2} \rangle = -\sqrt{\tfrac{2}{3}}, \quad \langle 1, 0, \tfrac{1}{2}, -\tfrac{1}{2} | \tfrac{1}{2}, -\tfrac{1}{2} \rangle = +\sqrt{\tfrac{1}{3}}.$$

In the above derivation we have made the conventional (Condon and Shortley, 1951; Brink and Satchler, 1962) phase choices. The crucial places are Eqs. (2.31) and (2.32), where the phases of the matrix elements of J_+ and J_- are arbitrarily chosen and at Eqs. (2.50) and (2.52), which follow the convention that $\langle j_a, j_a, j_b, j - j_a | j, j \rangle$ should be real and positive. The recursion relation and normalization conditions then ensure that all the Clebsch–Gordan coefficients are real. Table 2.1 summarizes the results for $j_a = 1, j_b = \tfrac{1}{2}$. Down the left-hand side are the possible eigenstates of J_a^2, J_{az}, J_b^2, J_{bz}; across the top are the eigenfunctions of J^2, J_z, J_a^2, J_b^2 (as usual we drop the quantum numbers j_a and j_b). Reading along a row gives the superposition of total angular momentum states that is equal to the eigenstates on the left, as in Eq. (2.37); reading down a column gives the reverse superposition as in Eq. (2.38). The only spaces in the table that can contain entries are those for which $j_z = j_{az} + j_{bz}$.

2.7 Vector Addition of Angular Momentum

TABLE 2.1

Clebsch–Gordan Coefficients for $j_a = 1$, $j_b = \tfrac{1}{2}$

	$\|j, j_z\rangle$					
$\|j_a, j_{az}, j_b, j_{bz}\rangle$	$\|\tfrac{3}{2}, \tfrac{3}{2}\rangle$	$\|\tfrac{3}{2}, \tfrac{1}{2}\rangle$	$\|\tfrac{3}{2}, -\tfrac{1}{2}\rangle$	$\|\tfrac{3}{2}, -\tfrac{3}{2}\rangle$	$\|\tfrac{1}{2}, \tfrac{1}{2}\rangle$	$\|\tfrac{1}{2}, -\tfrac{1}{2}\rangle$
$\|1, 1, \tfrac{1}{2}, \tfrac{1}{2}\rangle$	1					
$\|1, 1, \tfrac{1}{2}, -\tfrac{1}{2}\rangle$		$\sqrt{\tfrac{1}{3}}$			$\sqrt{\tfrac{2}{3}}$	
$\|1, 0, \tfrac{1}{2}, \tfrac{1}{2}\rangle$		$\sqrt{\tfrac{2}{3}}$			$-\sqrt{\tfrac{1}{3}}$	
$\|1, 0, \tfrac{1}{2}, -\tfrac{1}{2}\rangle$			$\sqrt{\tfrac{2}{3}}$			$\sqrt{\tfrac{1}{3}}$
$\|1, -1, \tfrac{1}{2}, \tfrac{1}{2}\rangle$			$\sqrt{\tfrac{1}{3}}$			$-\sqrt{\tfrac{2}{3}}$
$\|1, -1, \tfrac{1}{2}, -\tfrac{1}{2}\rangle$				1		

As an example, we suppose that it is necessary to express $|1, 0, \tfrac{1}{2}, \tfrac{1}{2}\rangle$ in terms of eigenstates of J^2 and J_z. The table gives

$$|1, 0, \tfrac{1}{2}, \tfrac{1}{2}\rangle = \sqrt{\tfrac{2}{3}}\,|\tfrac{3}{2}, \tfrac{1}{2}\rangle - \sqrt{\tfrac{1}{3}}\,|\tfrac{1}{2}, \tfrac{1}{2}\rangle.$$

As an example of the opposite process, we find that

$$|\tfrac{1}{2}, -\tfrac{1}{2}\rangle = \sqrt{\tfrac{1}{3}}\,|1, 0, \tfrac{1}{2}, -\tfrac{1}{2}\rangle - \sqrt{\tfrac{2}{3}}\,|1, -1, \tfrac{1}{2}, \tfrac{1}{2}\rangle.$$

The physical interpretation of these equations is straightforward: the amplitude for finding the state $|1, 0, \tfrac{1}{2}, \tfrac{1}{2}\rangle$ in a total angular-momentum state $|\tfrac{3}{2}, \tfrac{1}{2}\rangle$ is the relevant coefficient, $\sqrt{\tfrac{2}{3}}$ in this case. Thus a measurement of total angular momentum will find $j = \tfrac{3}{2}$ with probability $\tfrac{2}{3}$ and $j = \tfrac{1}{2}$ with probability $\tfrac{1}{3}$.

It is frequently necessary to add j_a to $j_b = \tfrac{1}{2}$ or 1: the Clebsch–Gordan coefficients required are given in Appendix E.

The Wigner 3-j symbol is used frequently in place of the Clebsch–Gordan coefficient. The relation is

$$\langle j_a, j_{az}, j_b, j_{bz} | j, -j_z \rangle = (-1)^{j_a - j_b - j_z}(2j+1)^{1/2} \begin{pmatrix} j_a & j_b & j \\ j_{az} & j_{bz} & j_z \end{pmatrix}. \quad (2.53a)$$

Notice the change of sign in the z component of the total angular momentum: the condition on the z component is now $j_z + j_{az} + j_{bz} = 0$.

Edmonds (1960) gives the symmetry relations for the Clebsch–Gordan coefficients, and Brink and Satchler (1962) those for the Wigner 3-j symbol. An interesting property of the Clebsch–Gordan coefficients is that

$$\langle j_a, 0, j_b, 0 | j, 0 \rangle = 0 \quad \text{for} \quad j_a + j_b + j \text{ odd}. \quad (2.53b)$$

The coupling of two angular momenta is evidently only the simplest case of coupling. The cases of coupling three and four angular momenta are discussed by Brink and Satchler (1962).

2.8 The Eigenfunctions of Orbital Angular Momentum

So far in this chapter, the emphasis has been placed upon the operator properties of angular momentum; however, if the operators L^2 and L_z are expressed in polar coordinates θ and ϕ, then it is possible to find analytic solutions to the eigenequations

$$L^2 \psi(\theta, \phi) = l(l+1)\hbar^2 \psi(\theta, \phi), \qquad (2.54)$$

$$L_z \psi(\theta, \phi) = l_z \hbar \psi(\theta, \phi). \qquad (2.55)$$

A function that is simultaneously a solution of Eqs. (2.54) and (2.55) is

$$\psi(\theta, \phi) = \sin^m \theta \left(\frac{d}{d\cos\theta}\right)^m \{P_l(\cos\theta)\} \exp(\pm im\phi), \qquad (2.56)$$

where $P_l(\cos\theta)$ is the lth Legendre polynomial and $l_z = \pm m$. The parameters l and m must be positive integers, having $m \leqslant l$, in order that the solution be meaningful, single-valued, and well-behaved; such conditions are physically essential and thus lead naturally to quantum states. The functions

$$\sin^m \theta \left(\frac{d}{d\cos\theta}\right)^m P_l(\cos\theta)$$

are called the associated Legendre polynomials and are represented by $P_l^m(\cos\theta)$.

These solutions are unnormalized: each solution $P_l^m(\cos\theta) \exp(\pm im\phi)$ must be multiplied by a real numerical factor C_{lm} in order that the condition (e) of Section 1.2 be satisfied. We must have

$$\int \psi^* \psi \, d\mathbf{x} = 1.$$

Now $d\mathbf{x} = -d\phi \, d(\cos\theta)$ for angular functions only; thus

$$-\int_{+1}^{-1} \int_0^{2\pi} |C_{lm} P_l^m(\cos\theta)|^2 \, d(\cos\theta) \, d\phi = 1,$$

whence

$$C_{lm} = \left[\frac{2l+1}{4\pi} \frac{(l-m)!}{(l+m)!}\right]^{1/2}, \qquad m \geqslant 0.$$

Finally it is necssary to include a factor that makes the sign of each eigienfunction consistent with the choice of phases made in Section 2.6. The

2.8 The Eigenfunctions of Orbital Angular Momentum

result, which is correct for both positive and negative l_z, is the normalized eigenfunctions

$$i^{l_z+|l_z|} C_{l|l_z|} P_l^{|l_z|}(\cos\theta) \exp(il_z\phi),$$

which we write as $Y_l^{l_z}(\theta,\phi)$, or $Y(l,l_z)$ if it is unnecessary to display the independent variables explicitly. We shall refer to these functions as spherical harmonics. We tabulate these functions for the first three values of l:

$$\begin{aligned}
Y(0,0) &= 1/\sqrt{4\pi}, \\
Y(1,0) &= (3/4\pi)^{1/2} \cos\theta, \\
Y(1,\pm 1) &= \mp(3/8\pi)^{1/2} \sin\theta \exp(\pm i\phi), \\
Y(2,0) &= (5/16\pi)^{1/2}(3\cos^2\theta - 1), \\
Y(2,\pm 1) &= \mp(15/8\pi)^{1/2} \sin\theta \cos\theta \exp(\pm i\phi), \\
Y(2,\pm 2) &= (15/32\pi)^{1/2} \sin^2\theta \exp(\pm 2i\phi).
\end{aligned} \quad (2.57)$$

The parity of these angular-momentum functions is important. We recall that a test of parity is made by replacing **x** by $-$**x** (Section 1.7). For our spherical functions this corresponds to replacing θ and ϕ by $\pi - \theta$ and $\phi - \pi$. We have that

$$\cos(\pi - \theta) = -\cos\theta,$$

$$\left(\frac{d}{d\cos(\pi-\theta)}\right)^m = (-1)^m \left(\frac{d}{d\cos\theta}\right)^m,$$

and

$$\sin(\pi - \theta) = \sin\theta.$$

In addition, we have that

$$e^{im(\phi-\pi)} = (-1)^m e^{im\phi}.$$

Then we find on making the replacement in Eq. (2.56) that the parity of the differential operator and of the exponential cancel and we are left with the parity the Legendre polynomial, $P_l(\cos\theta)$. This polynomial contains the following powers of $\cos\theta$; $l, l-2, l-4, \ldots$, etc., and therefore has parity $(-1)^l$. Thus the parity of the eigenfunction $Y(l,l_z)$ is $(-1)^l$.

In quantum-mechanical language the spherical harmonic $Y(l,l_z)$ is the amplitude for finding a particle at a particular angle if it is in the state that is the eigenstate of L^2 and L_z with eigenvalues l, l_z. Suppose that this state is represented by the vector $|l,l_z\rangle$; then we use the following notation: the amplitude for finding the particle at the polar angles θ, ϕ is

$$\langle\theta,\phi|l,l_z\rangle \equiv Y_l^{l_z}(\theta,\phi). \quad (2.58)$$

A useful value is $\langle\theta=0, \phi=0|l,l_z\rangle = [(2l+1)/4\pi]^{1/2}$ if $l_z=0$, and 0 if $l_z \neq 0$.

Suppose we have a system containing both spin and orbital angular momentum l: this is represented by the vector $|l, l_z, s, s_z\rangle$. Then the amplitude for finding the system with a particle at the angle θ, ϕ with spin state s', s_z' is $\langle \theta, \phi, s', s_z' | l, l_z, s, s_z \rangle$. Since the state vectors for the two parts can be considered separately, this amplitude becomes

$$\langle \theta, \phi | l, l_z \rangle \langle s', s_z' | s, s_z \rangle = Y_l^{l_z}(\theta, \phi) \delta_{ss'} \delta_{s_z s_z'}. \tag{2.59}$$

2.9 The Pauli Spin Matrices

The matrix elements of the components of the angular-momentum vector for the case of angular momentum $\frac{1}{2}$ lead to the Pauli spin matrices. Thus we define

$$(\hbar/2)\sigma_x = \begin{bmatrix} \langle \tfrac{1}{2}, \tfrac{1}{2} | J_x | \tfrac{1}{2}, \tfrac{1}{2} \rangle & \langle \tfrac{1}{2}, \tfrac{1}{2} | J_x | \tfrac{1}{2}, -\tfrac{1}{2} \rangle \\ \langle \tfrac{1}{2}, -\tfrac{1}{2} | J_x | \tfrac{1}{2}, \tfrac{1}{2} \rangle & \langle \tfrac{1}{2}, -\tfrac{1}{2} | J_x | \tfrac{1}{2}, -\tfrac{1}{2} \rangle \end{bmatrix};$$

hence,

$$\sigma_x = \begin{bmatrix} 0 & 1 \\ 1 & 0 \end{bmatrix}. \tag{2.60a}$$

Similarly

$$\sigma_y = \begin{bmatrix} 0 & -i \\ i & 0 \end{bmatrix} \quad \text{and} \quad \sigma_z = \begin{bmatrix} 1 & 0 \\ 0 & -1 \end{bmatrix} \tag{2.60b}$$

Using the rules of matrix multiplication we find that these matrices have the properties

$$\sigma_x \sigma_y + \sigma_y \sigma_x = 2\delta_{xy} I \quad \text{and} \quad \sigma_x \sigma_y = i\sigma_z, \tag{2.60c}$$

where x, y, z can be cyclically permuted and I is the unit matrix. If we define

$$\mathbf{S} = \frac{\hbar}{2} \boldsymbol{\sigma},$$

the manner in which the Pauli matrices were derived ensures that the matrices $\mathbf{S}$ will satisfy the commutation relations of the angular-momentum operators [Eq. (2.4)] and the relations of Eq. (2.5) after defining $S^2 = (\mathbf{S})^2$.

The column vectors

$$\begin{bmatrix} 1 \\ 0 \end{bmatrix} \quad \text{and} \quad \begin{bmatrix} 0 \\ 1 \end{bmatrix}$$

are eigenvectors of S_z with eigenvalues $+\tfrac{1}{2}\hbar$ and $-\tfrac{1}{2}\hbar$ and of S^2 both with eigenvalue $\tfrac{3}{4}\hbar^2$, as required for a representation of spin $\tfrac{1}{2}$. They represent the spin-½ kets. The row vectors that are their Hermitian conjugates represent the bras.

This representation does not add anything to our knowledge of spin, but it is very convenient in various calculations and we shall use it frequently.

2.10 Rotations (II)

We have seen in Section 2.2 how the angular-momentum operators are responsible for generating the unitary operators for rotations; we wish to extend our consideration of these rotations. To do so, we begin by defining the Euler angles, which can specify a rotation completely. The orientation of a set of Cartesian coordinates with respect to another set requires three parameters, since there are three degrees of freedom; these parameters are the angular rotations made in turning one set of coordinates into the other. Consider a right-handed coordinate system S_1, with axes x_1, y_1, z_1; rotate the coodinates through an angle α about the z_1 axis to generate a new set of coordinates S_2 with axes $x_2, y_2, z_2(=z_1)$. (A positive rotation is in the direction that would move a right-handed screw in the positive direction along the axis of rotation.) The second step is to rotate S_2 through an angle β around y_2 to generate a third coordinate system S_3 with axes $x_3, y_3(=y_2), z_3$. Finally, rotate S_3 through an angle γ around z_3 to produce a fourth coordinate system S_4 with axes $x_4, y_4, z_4(=z_3)$. Summarizing the rotation from $S_1 \to S_4$:

$$S_1 \to S_2: \text{ angle } \alpha \text{ around } z_1,$$
$$S_2 \to S_3: \text{ angle } \beta \text{ around } y_2,$$
$$S_3 \to S_4: \text{ angle } \gamma \text{ around } z_3,$$

$\alpha, \beta,$ and γ are the Euler angles. The z_4 axis has conventional polar coordinates in S_1, $\theta = \beta$, and $\phi = \alpha$.

Consider a state vector $|\psi\rangle$ that describes some physical system in the coordinates S_1. The rotation defines a second vector $|\psi'\rangle$ that describes the same system in the transformed coordinate system S_4; then

$$|\psi'\rangle = D|\psi\rangle,$$

where D is the unitary operator introduced in Section 2.3 and for a rotation defined by the Euler angles is given by

$$D = \exp\left(+\frac{i}{\hbar}J_{z_3}\gamma\right)\exp\left(+\frac{i}{\hbar}J_{y_2}\beta\right)\exp\left(+\frac{i}{\hbar}J_{z_1}\alpha\right). \quad (2.61)$$

This is an inconvenient form for D, as each angular momentum operator is about an axis in different coordinate systems. We can transform these operators using Eq. (2.8). The operator for rotations about y_1, that is, $\exp[(+i/\hbar)J_{y_1}\beta]$, is transformed for rotations about y_2 by the operator that rotates $S_1 \to S_2$, that is, $\exp[(+i/\hbar)J_{z_1}\alpha]$; therefore,

$$\exp\left(+\frac{i}{\hbar}J_{y_2}\beta\right) = \exp\left(+\frac{i}{\hbar}J_{z_1}\alpha\right)\exp\left(+\frac{i}{\hbar}J_{y_1}\beta\right)\exp\left(-\frac{i}{\hbar}J_{z_1}\alpha\right).$$

Similarly,

$$\exp\left(+\frac{i}{\hbar}J_{z_3}\gamma\right) = \exp\left(+\frac{i}{\hbar}J_{y_2}\beta\right)\exp\left(+\frac{i}{\hbar}J_{z_2}\gamma\right)\exp\left(-\frac{i}{\hbar}J_{y_2}\beta\right).$$

Substituting in Eq. (2.61), we find

$$D = \exp\left(+\frac{i}{\hbar}J_{z_1}\alpha\right)\exp\left(+\frac{i}{\hbar}J_{y_1}\beta\right)\exp\left(+\frac{i}{\hbar}J_{z_1}\gamma\right). \quad (2.62)$$

This equation indicates that the rotation from coordinate system S_1 to S_4 could have been performed equally in the three stages

(1) angle γ about z_1,
(2) angle β about y_1,
(3) angle α about z_1,

where the directions z_1 and y_1 mean the original directions occupied by these axes in S_1 before rotations.

We are particularly interested in the case that occurs when $|\phi\rangle$ is an eigenstate of the angular-momentum operators J^2 and J_z, say $|j,j_z\rangle$. It is evident that $D|j,j_z\rangle$ will also be an eigenvector of J^2 with quantum number j, but it will not be an eigenvector of J_z. The eigenvectors of J_z form an orthonormal set; therefore it is possible to write

$$D|j,j_z\rangle = \sum_{j_z'=-j}^{+j} \mathscr{D}^j_{j_z j_z'} |j,j_z'\rangle. \quad (2.63)$$

As usual this change of basis means that the coefficients are the matrix elements of D between the states of the two bases: thus

$$\mathscr{D}^j_{j_z j_z'} = \langle j,j_z' | D | j,j_z\rangle. \quad (2.64)$$

To help keep the notation clear we shall use D to represent the rotation operator, without specifically giving the Euler angles as parameters unless it is necessary, in which case we shall put $D(\alpha,\beta,\gamma)$. Similarly with $\mathscr{D}^j_{j_z j_z'}$.

We will apply this immediately to the case of a particle having spin $\hbar/2$. Consider such a particle oriented so that every measurement of the

2.10 Rotations (II)

spin component S_{z_1} yields the value $\frac{1}{2}\hbar$; for brevity we say that the spin is "pointing" along the z_1 axis. This pure state can be represented by the Pauli column matrix

$$\begin{bmatrix} 1 \\ 0 \end{bmatrix}$$

and is an eigenvector of S_{z_1}. In a second coordinate system, a measurement of S_{z_2} will not always yield the value $\frac{1}{2}$ so long as z_2 is not parallel to z_1. Thus, in this second system, this pure state is not an eigenstate of S_{z_2} but is a linear sum of the eigenstates of S_{z_1}, namely

$$\begin{bmatrix} 1 \\ 0 \end{bmatrix} \quad \text{and} \quad \begin{bmatrix} 0 \\ 1 \end{bmatrix}.$$

The coefficients in the linear sum are the matrix elements

$$\begin{bmatrix} 1 & 0 \end{bmatrix} D \begin{bmatrix} 1 \\ 0 \end{bmatrix} \quad \text{and} \quad \begin{bmatrix} 0 & 1 \end{bmatrix} D \begin{bmatrix} 1 \\ 0 \end{bmatrix}$$

respectively. D is given by Eq. (2.62) with the total angular momentum operator $\mathbf{J}$ replaced by $\mathbf{S}$. We can evaluate these matrix elements using the matrix representation

$$S_{z_1} = \frac{\hbar}{2}\begin{bmatrix} 1 & 0 \\ 0 & -1 \end{bmatrix}, \quad S_{y_1} = \frac{\hbar}{2}\begin{bmatrix} 0 & -i \\ i & 0 \end{bmatrix}.$$

Certain parts of the matrix elements may be evaluated easily; for example,

$$\begin{bmatrix} 1 & 0 \end{bmatrix} \exp\left(+\frac{i}{2}\sigma_{z_1}\alpha\right) = \exp\left(+\frac{i\alpha}{2}\right)\begin{bmatrix} 1 & 0 \end{bmatrix},$$

$$\begin{bmatrix} 0 & 1 \end{bmatrix} \exp\left(+\frac{i}{2}\sigma_{z_1}\alpha\right) = \exp\left(-\frac{i\alpha}{2}\right)\begin{bmatrix} 0 & 1 \end{bmatrix},$$

$$\exp\left(+\frac{i}{2}\sigma_{z_1}\gamma\right)\begin{bmatrix} 1 \\ 0 \end{bmatrix} = \exp\left(+\frac{i\gamma}{2}\right)\begin{bmatrix} 1 \\ 0 \end{bmatrix}$$

With these equalities we can reduce the matrix elements to a simpler form involving the matrix σ_{y_1} alone, which may be evaluated by expanding the exponential and considering the matrix elements of σ_{y_1},

$$\begin{bmatrix} 1 & 0 \end{bmatrix} \exp\left(+\frac{i}{2}\sigma_{y_1}\beta\right)\begin{bmatrix} 1 \\ 0 \end{bmatrix} = \cos\frac{\beta}{2},$$

$$\begin{bmatrix} 0 & 1 \end{bmatrix} \exp\left(+\frac{i}{2}\sigma_{y_1}\beta\right)\begin{bmatrix} 1 \\ 0 \end{bmatrix} = -\sin\frac{\beta}{2}.$$

Hence we have that

$$D\begin{bmatrix}1\\0\end{bmatrix} = \exp\left(\frac{i\alpha}{2}\right)\exp\left(\frac{i\gamma}{2}\right)\cos\frac{\beta}{2}\begin{bmatrix}1\\0\end{bmatrix}$$
$$-\exp\left(\frac{-i\alpha}{2}\right)\exp\left(\frac{i\gamma}{2}\right)\sin\frac{\beta}{2}\begin{bmatrix}0\\1\end{bmatrix}$$
$$= \exp\left(\frac{i\gamma}{2}\right)\begin{bmatrix}\exp\left(\frac{i\alpha}{2}\right)\cos\frac{\beta}{2}\\-\exp\left(\frac{-i\alpha}{2}\right)\sin\frac{\beta}{2}\end{bmatrix}. \quad (2.65)$$

The state represented by Eq. (2.65) is the original state described in the new coordinate system. If we consider the z direction in this new system, the amplitude for finding the spin pointing along this direction is $\exp(i\gamma/2)\exp(i\alpha/2)\cos\beta/2$ $(= \langle \frac{1}{2}, \frac{1}{2} | D | \frac{1}{2}, \frac{1}{2}\rangle)$, that is, the probability that a measurement yields the result $\hbar/2$ for the component along z is $\cos^2\beta/2$. Similarly the probability that the measurement yields the result $-\hbar/2$ is $\sin^2\beta/2$ $(= |\langle \frac{1}{2}, -\frac{1}{2} | D | \frac{1}{2}, \frac{1}{2}\rangle|^2)$.

Sometimes it is more convenient to work in terms of an active transformation; in this case, the transformation rotates the physical system through the Euler angles α, β, γ leaving the coordinate system unchanged. The new state $|\psi'\rangle$ is described in terms of the original state $|\psi\rangle$ by

$$|\psi'\rangle = D^{-1}|\psi\rangle,$$

where D^{-1} is the reciprocal of D, the rotation operator for the passive transformation, and is obtained from D by changing $\alpha \to -\alpha, \beta \to -\beta$, and $\gamma \to -\gamma$. Thus rotating the system described by the vector $\begin{bmatrix}1\\0\end{bmatrix}$ through α, β, γ produces the state described by

$$\exp\left(\frac{-i\gamma}{2}\right)\begin{bmatrix}\exp\left(\frac{-i\alpha}{2}\right)\cos\frac{\beta}{2}\\\exp\left(\frac{+i\alpha}{2}\right)\sin'\frac{\beta}{2}\end{bmatrix}. \quad (2.66)$$

In more familiar terms, the new system has its spin pointing along polar coordinates $\theta = \beta, \phi = \alpha$, so that a system having spin $\frac{1}{2}$ pointing along the direction θ, ϕ is described by

$$\begin{bmatrix}\exp\left(\frac{-i\phi}{2}\right)\cos\frac{\theta}{2}\\\exp\left(\frac{+i\phi}{2}\right)\sin\frac{\theta}{2}\end{bmatrix}. \quad (2.67)$$

2.10 Rotations (II)

The common $\exp(-i\gamma/2)$ has been dropped, as it will be an unobservable phase factor. However, we must remember that it represents the transformation property of the state for rotations around the direction in which the spin points.

It is evident that the matrix elements of the rotation operator are important in angular momentum; generalizing from spin $\frac{1}{2}$ we have

$$D|j,j_z\rangle = \sum_{j_z'} \mathscr{D}^j_{j_z j_z'} |j,j_z'\rangle, \qquad (2.68)$$

where

$$\mathscr{D}^j_{j_z j_z'} = \langle j,j_z' | D | j,j_z \rangle.$$

Substituting for D we see

$$\begin{aligned}\mathscr{D}^j_{j_z j_z'} &= \exp(+i\alpha j_z')\langle j,j_z' | \exp(+i\beta J_y) | j,j_z \rangle \exp(+i\gamma j_z) \\ &= \exp i(\alpha j_z' + \gamma j_z)\, d^j_{j_z j_z'}(\beta).\end{aligned} \qquad (2.68\text{a})$$

With the choice of phase made in Eqs. (2.26)–(2.32) the matrix elements $d^j_{j_z j_z'}(\beta)$ are real.

If we express the state vectors as column vectors in Eq. (2.68), each having $2j+1$ rows, then this equation becomes equivalent to a matrix equation. A column matrix has the entry 1 in row j_z' (or j_z) and zeros elsewhere to repesent $|j,j_z'\rangle$ (or $|j,j_z\rangle$), where the rows are numbered from j down to $-j$. Thus

$$D \begin{bmatrix} 1 \\ 0 \\ 0 \\ \vdots \end{bmatrix} = \begin{bmatrix} \mathscr{D}^j_{jj} \\ \mathscr{D}^j_{jj-1} \\ \mathscr{D}^j_{jj-2} \\ \vdots \end{bmatrix}$$

so that the state represented by

$$\begin{bmatrix} 1 \\ 0 \\ 0 \\ \vdots \end{bmatrix}$$

is represented by

$$\begin{bmatrix} \mathscr{D}^j_{jj} \\ \mathscr{D}^j_{jj-1} \\ \mathscr{D}^j_{jj-2} \\ \vdots \end{bmatrix}$$

in the rotated frame.

Edmonds (1960) gives the derivation of a general formula for the matrices $d^j_{j_z j_z'}(\beta)$ (see also Feynman et al., 1965), but it is too clumsy for general use and most of the simple cases are tabulated (Brink and Satchler, 1962). In Appendix E we give the matrices for $j = \frac{1}{2}$ and 1.

The properties of the rotation matrices are given, for example, by Edmonds (1960) and Brink and Satchler (1962). We mention some of the simpler. For reasons of normalization we have a sum rule

$$\sum_{j_z'} |\mathscr{D}^j_{j_z j_z'}|^2 = 1 .$$

The matrix elements are orthogonal

$$(2j + 1) \int \mathscr{D}^j_{j_z j_z'} \mathscr{D}^J_{J_z J_z'} \sin \beta \, d\beta \, d\alpha \, d\gamma = 8\pi^2 \, \delta_{jJ} \delta_{j_z J_z} \delta_{j_z' J_z'} .$$

For integer j there are the following special cases:

$$d^j_{j_z 0}(\beta) = \left[\frac{(j - j_z)!}{(j + j_z)!} \right]^{1/2} P_j^{j_z}(\cos \beta) ,$$

and

$$d^j_{00}(\beta) = P_j(\cos \beta) . \tag{2.69}$$

Since D is a unitary operator the matrices $\mathscr{D}$ are also unitary.

The matrix elements for j half integral are double valued, changing sign for $\beta = 2\pi$. This gives no ambiguity as far as observable quantities are involved as long as there is a consistent choice of phase. There is a discussion of this difficulty by Werle (1966) and by Wigner (1959).

It is evident that the rotational properties of states are of great importance. If we consider a state $|\phi_1\rangle$, its rotational properties can be expressed by the fact that under a rotation of the coordinates the vector changes thus:

$$|\phi_1\rangle \to |\varphi_1\rangle = D|\phi_1\rangle .$$

If the state is developing so that after an interval of time t

$$|\phi_1\rangle \to |\phi_2\rangle = S|\phi_1\rangle ,$$

then it is interesting to ask about the rotational properties of $|\phi_2\rangle$. Since S commutes with D,

$$D|\phi_2\rangle = DS|\phi_1\rangle = SD|\phi_1\rangle .$$

The S in front has no effect on the rotational properties, which are therefore the same as those of the initial state. Thus when the system is invariant under coordinate rotations, the rotational properties of a state remain unchanged. This is just another way of expressing the conservation of angular momentum.

2.11 Decay of Pure States

Let us consider the decay of a pure state having total angular momentum j (integer), with $j_z = 0$, into two spinless particles. What is the angular distribution of the decay products? If angular momentum is conserved, the final state must have the same properties under rotation as the initial state; if $\langle \theta, \phi | j, 0 \rangle$ is the amplitude for finding a decay product at a polar angle θ, ϕ, then

$$\langle \theta, \phi | j, 0 \rangle = Y_j^0(\theta, \phi),$$

where Y_j^0 is the spherical harmonic having the same rotational properties as the initial state. Therefore the intensity of the decay product is

$$|Y_j^0(\theta, \phi)|^2.$$

This is proportional to $|P_j(\cos \theta)|^2$.

Another way of getting the same result is as follows. If we consider a decay product emitted in a direction (θ, ϕ), call it z', then for the final state there can be no component of the angular momentum along z', so only states that have $j_{z'} = 0$ can decay in that direction. Now the initial state has $j_z = 0$ so the probability the state decays with a product along z' is the probability the initial state is in a state with $j_{z'} = 0$. From the last section, this is just

$$|\langle j, 0 | D | j, 0 \rangle|^2,$$

where the rotation caused by D is that having Euler angles $\beta = \theta$, $\alpha = \phi$ (γ is irrelevant). Therefore the angular distribution as a function of θ and ϕ is

$$|\mathscr{D}_{00}^j|^2 = |\langle j, 0 | D | j, 0 \rangle|^2$$
$$= \{d_{00}^j(\theta)\}^2$$
$$= \{P_j(\cos \theta)\}^2.$$

For a more specific example let us consider the decay

$$\Sigma(1765) \to \pi + \Lambda(1520)$$
$$J^P \quad \tfrac{5}{2}^- \quad 0^- \quad \tfrac{3}{2}^-$$

This state is formed in K^-p interactions at about 970 MeV/c incident K^- momentum, and in these circumstances the $\Sigma(1765)$ is produced incoherently either in the state $j_z = \tfrac{1}{2}$ or in the state $j_z = -\tfrac{1}{2}$ (see Section 3.3). We will calculate the angular distribution of the outgoing π from the state $j_z = +\tfrac{1}{2}$ (it is the same for $j_z = -\tfrac{1}{2}$). The final state contains two angular momenta, $s = \tfrac{3}{2}$ and an orbital angular momentum l determined by the vector addition

$$\mathbf{s} + \mathbf{l} = \mathbf{j},$$

where $j = \frac{5}{2}$, and by the overall parity conservation, which requires l to be odd. Therefore $l = 1$ is the most likely, $l = 3$ being unfavored by the effect of the angular momentum barrier. Thus the final state must be the vector addition of $l = 1$ and $s = \frac{3}{2}$ to give $j = \frac{5}{2}$, $j_z = \frac{1}{2}$. From the Clebsch–Gordan coefficients, using Eq. (2.44)

$$|\tfrac{5}{2}, \tfrac{1}{2}\rangle = \sqrt{\tfrac{1}{10}} |1, -1, \tfrac{3}{2}, \tfrac{3}{2}\rangle + \sqrt{\tfrac{3}{5}} |1, 0, \tfrac{3}{2}, \tfrac{1}{2}\rangle + \sqrt{\tfrac{3}{10}} |1, 1, \tfrac{3}{2}, -\tfrac{1}{2}\rangle.$$

Therefore the amplitude for finding a pion in the direction θ, ϕ with the $\Lambda(1520)$ spin state having $s_z = \frac{3}{2}$ is [see Eq. (2.59)]

$$\langle \theta, \phi, \tfrac{3}{2}, \tfrac{3}{2} | \tfrac{5}{2}, \tfrac{1}{2}\rangle = \sqrt{\tfrac{1}{10}} \langle \theta, \phi | 1, -1 \rangle = \sqrt{\tfrac{1}{10}} \sqrt{\tfrac{3}{8\pi}} \sin\theta \, e^{-i\phi}.$$

Similarly the amplitude for the spin states $s_z = \frac{1}{2}$ and $-\frac{1}{2}$ are, respectively,

$$\sqrt{\tfrac{3}{5}} \sqrt{\tfrac{3}{4\pi}} \cos\theta \quad \text{and} \quad -\sqrt{\tfrac{3}{10}} \sqrt{\tfrac{3}{8\pi}} \sin\theta \, e^{+i\phi}.$$

In principle these spin states can be distinguished so that the outgoing intensity is given by the sum of the squares of amplitudes

$$I(\theta, \phi) = \frac{3}{8\pi} \left\{ \frac{1}{10} \sin^2\theta + \frac{6}{5} \cos^2\theta + \frac{3}{10} \sin^2\theta \right\}$$

$$= \frac{3}{20\pi} \{1 + 2\cos^2\theta\}.$$

Had the $\Sigma(1765)$ been of even parity, the orbital state would have $l = 2$ (or 4), in which case the angular distribution becomes, assuming $l = 2$,

$$(3/28\pi)\{1 + 10\cos^2\theta - 10\cos^4\theta\}.$$

The determination of the $\Sigma(1765)$ parity was made by the measurement of this angular distribution (Armenteros *et al.*, 1965). The results were in favor of odd parity.

2.12 Tensor Operators

We now introduce the concept of a tensor operator that is one of a set of $2k + 1$ operators $T(k, k_z)$, where $k_z = k, k - 1, \ldots, -k$. Each of these operators transforms under a change of the coordinate system according to Eq. (2.8):

$$T \to T' = DTD^{-1}.$$

These operators are defined by requiring that under rotations they transform like eigenstates of angular momentum; thus

2.12 Tensor Operators

$$D\, T(k, k_z)\, D^{-1} = \sum_{k_z'} \mathscr{D}^k_{k_z k_z'}\, T(k, k_z') \tag{2.70}$$

by analogy with Eq. (2.63).

If we make the rotation through infinitesimal angle α about the axis λ, Eq. (2.70) becomes

$$(1 + i\alpha J_\lambda/\hbar)\, T(k, k_z)\, (1 - i\alpha J_\lambda/\hbar)$$
$$= \sum_{k_z'} \langle k, k_z' | (1 + i\alpha J_\lambda/\hbar) | k, k_z \rangle\, T(k, k_z');$$

hence,

$$T(k, k_z) + \frac{i\alpha}{\hbar} \{J_\lambda\, T(k, k_z) - T(k, k_z)\, J_\lambda\}$$
$$= \sum_{k_z'} \Big\{ \langle k, k_z' | k, k_z \rangle\, T(k, k_z')$$
$$- \frac{i\alpha}{\hbar} \langle k, k_z' | J_\lambda | k, k_z \rangle\, T(k, k_z') \Big\},$$

or

$$J_\lambda\, T(k, k_z) - T(k, k_z)\, J_\lambda = \sum_{k_z'} \langle k, k_z' | J_\lambda | k, k_z \rangle\, T(k, k_z').$$

Substituting for the matrix elements on the right from Eqs. (2.26)-(2.32), we find

$$[J_\pm, T(k, k_z)] = \hbar[(k \mp k_z)(k \pm k_z + 1)]^{1/2}\, T(k, k_z \pm 1), \tag{2.71}$$

$$[J_z, T(k, k_z)] = \hbar k_z\, T(k, k_z). \tag{2.72}$$

If we apply the operator equation (2.72) to the state vector $|j, j_z\rangle$,

$$J_z\, T(k, k_z)|j, j_z\rangle - T(k, k_z)\, J_z |j, j_z\rangle = \hbar k_z\, T(k, k_z)|j, j_z\rangle.$$

It follows that

$$J_z\, T(k, k_z)|j, j_z\rangle = \hbar(k_z + j_z)\, T(k, k_z)|j, j_z\rangle;$$

therefore $T(k, k_z)$ generates from $|j, j_z\rangle$ an eigenstate of J_z with eigenvalue equal to $j_z + k_z$. Consider now the state vector

$$|\phi\rangle = T(k, k_z)|j_b, j_{bz}\rangle$$

and rotate the coordinates so that the rotation operator is D (we use j_b, not j, etc. to reach a consistent rotation at the end of this section). Then

$$|\phi\rangle \to D|\phi\rangle = D\, T(k, k_z)|j_b, j_{bz}\rangle$$
$$= D\, T(k, k_z) D^{-1} D |j_b, j_{bz}\rangle.$$

Substituting from Eq. (2.70) and recalling that

$$D|j_b, j_{bz}\rangle = \sum_{j'_{bz}} \mathscr{D}^{j_b}_{j_{bz}j'_{bz}}|j_b, j'_{bz}\rangle,$$

we get

$$\begin{aligned}D|\phi\rangle &= \sum_{k_z'} \mathscr{D}^{k}_{k_zk_z'} T(k, k_z') \sum_{j'_{bz}} \mathscr{D}^{j_b}_{j_{bz}j'_{bz}}|j_b, j'_{bz}\rangle \\ &= \sum_{k_z'} \mathscr{D}^{k}_{k_zk_z'} \sum_{j'_{bz}} \mathscr{D}^{j_b}_{j_{bz}j'_{bz}} T(k, k_z')|j_b, j'_{bz}\rangle. \end{aligned} \quad (2.73)$$

We shall now compare this with the effect of D upon a state containing two angular momenta, viz. $|j_a, j_{az}, j_b, j_{bz}\rangle$. D is given by, for example,

$$D = \exp\left(+\frac{i}{\hbar}(\mathbf{J}\cdot\mathbf{n})\alpha\right) = \exp\left(+\frac{i}{\hbar}(\mathbf{J}_a\cdot\mathbf{n})\alpha\right)\exp\left(+\frac{i}{\hbar}(\mathbf{J}_b\cdot\mathbf{n})\alpha\right).$$

$\mathbf{J}_a$ and $\mathbf{J}_b$ commute, so that the effect of the rotation can be applied separately to the two angular momenta to give

$$D|j_a, j_{az}, j_b, j_{bz}\rangle = \sum_{j'_{az}} \mathscr{D}^{j_a}_{j_{az}j'_{az}} \sum_{j'_{bz}} \mathscr{D}^{j_b}_{j_{bz}j'_b}|j_a, j'_{az}, j_b, j'_{bz}\rangle. \quad (2.74)$$

Comparing Eqs. (2.73) and (2.74) we see that the state $T(k, k_z)|j_b, j_{bz}\rangle$ transforms in the same way as $|j_a, j_{az}, j_b, j_{bz}\rangle$ if $j_a = k$ and $j_{az} = k_z$, etc. This means that the effect of $T(k, k_z)$ is to add into the system another angular momentum to give another state whose angular momentum is the vector sum of the two involved.

The next step is to examine the matrix elements of $T(k, k_z)$ between states of the system represented by state vectors $|\alpha, j, j_z\rangle$ and $|\alpha', j', j_z'\rangle$. The quantum number α is included and covers quantities other than angular momentum because T can connect states with other differing properties; energy, for example.

Now the states $T(k, k_z)|\alpha, j, j_z\rangle$ transform like two angular momenta, so by the use of Eq. (2.43) we can write

$$T(k, k_z)|\alpha, j, j_z\rangle = \sum_q \langle q, q_z|k, k_z, j, j_z\rangle|\beta, q, q_z\rangle, \quad (2.75)$$

where $q_z = k_z + j_z$ and the $|\beta, q, q_z\rangle$ are states given by the inverse sum

$$|\beta, q, q_z\rangle = \sum_{k_z} \langle q, q_z|k, k_z, j, j_z\rangle T(k, k_z)|\alpha, j, j_z\rangle.$$

These states are eigenstates of J^2 and J_z with quantum numbers q, q_z and are orthogonal but not orthonormal. It follows that the matrix element is

$$\langle \alpha', j', j_z'|T(k, k_z)|\alpha, j, j_z\rangle = \sum_q \langle q, q_z|k, k_z, j, j_z\rangle\langle \alpha', j', j_z'|\beta, q, q_z\rangle.$$

2.12 Tensor Operators

The matrix element on the far right is zero unless $q = j'$ and $q_z = j_z'$, but is not unity, as the $|\beta, q, q_z\rangle$ are not orthonormal. Then

$$\langle \alpha', j', j_z' | T(k, k_z) | \alpha, j, j_z \rangle = \langle j', j_z' | k, k_z, j, j_z \rangle \langle \alpha', j', j_z' | \beta, j', j_z' \rangle$$

expresses the fact that the amplitude for finding the state $\langle \alpha', j', j_z' |$ depends on a geometrical factor that is a Clebsch–Gordan coefficient and a matrix element $\langle \alpha', j', j_z' | \beta, j', j_z' \rangle$ that contains the dynamical features of the system. This separation of geometrical from other factors is emphasized by showing that $\langle \alpha', j', j_z' | \beta, j', j_z' \rangle$ is independent of j_z'. The state $|\beta, j', j_z'\rangle$ can be expanded

$$|\beta, j', j_z'\rangle = \sum_{\alpha'} |\alpha' j', j_z'\rangle \langle \alpha', j', j_z' | \beta, j', j_z' \rangle, \qquad (2.76)$$

and if $j_z' < j'$,

$$|\beta, j', j_z' + 1\rangle = \sum_{\alpha'} |\alpha', j', j_z' + 1\rangle \langle \alpha', j', j_z' + 1 | \beta, j', j_z' + 1 \rangle.$$

From Eq. (2.31)

$$\begin{aligned}
|\beta, j', j_z' + 1\rangle &= [(j' - j_z')(j' + j_z' + 1)]^{-1/2} J_+ |\beta, j', j_z'\rangle \\
&= [(j' - j_z')(j' + j_z' + 1)]^{-1/2} \sum_{\alpha'} J_+ |\alpha', j', j_z'\rangle \\
&\quad \times \langle \alpha', j', j_z' | \beta, j', j_z' \rangle \\
&= \sum_{\alpha'} |\alpha', j', j_z' + 1\rangle \langle \alpha', j', j_z' | \beta, j', j_z' \rangle. \qquad (2.77)
\end{aligned}$$

It follows from Eqs. (2.76) and (2.77) that

$$\langle \alpha', j', j_z' | \beta, j', j_z' \rangle = \langle \alpha', j', j_z' + 1 | \beta, j', j_z' + 1 \rangle.$$

It is usual to put these invariant matrix elements proportional to what is called a reduced matrix element $\langle \alpha', j' \| T(k) \| \alpha, j \rangle$; then one definition has

$$\langle \alpha', j', j_z' | T(k, k_z) | \alpha, j, j_z \rangle = \langle j', j_z' | k, k_z, j, j_z \rangle \frac{\langle \alpha', j' \| T(k) \| \alpha, j \rangle}{(2j' + 1)^{1/2}} \qquad (2.78)$$

This is the Wigner–Eckart theorem and provides a considerable simplification in the use of tensor operators.

We emphasize that $T(k, k_z)$ operating on $|j, j_z\rangle$, produces a state that is the vector sum of two angular momenta. It follows that $T(k, k_z)$ can have matrix elements only between states that satisfy the vector rules

$$j_z' = j_z + k_z,$$
$$|j - k| \leq j' \leq j + k.$$

Tensor operators are of considerable importance. For example, a state $|\phi_1\rangle$ may be able to emit a photon to give a final state $|\phi_2\rangle$ by an electric dipole transition. This process can be described by the use of a tensor operator T, and the amplitude for the transition is proportional to its matrix element between the states, $\langle\phi_2|T|\phi_1\rangle$. In this case the tensor operator has $k = 1$; i.e., it is a vector. It also has odd parity, and thus it connects states of opposite parity. An electric quadrupole transition requires a tensor operator with $k = 2$ and even parity. The reduced matrix elements $\langle\phi_2\|T\|\phi_1\rangle$ now contain the effect of the coupling strength of the electromagnetic field, the structure of the initial and final states, etc. The geometrical effects, that is, the effects due to rotational properties of the angular momenta involved, appear as a Clebsch–Gordan coefficient.

We shall use the Wigner–Eckart theorem as a method of comparing transition rates in strange-particle decays. (See Sections 11.10 and 11.11.)

2.13 Polarization

Polarization is an important phenomenon in elementary-particle physics. Targets of polarized protons are available, and the decay of hyperons and baryon resonances gives information on their polarization. The polarization of proton or neutron beams may be analyzed by scattering. Thus the ability to measure or produce polarization gives us further means of examining elementary-particle interactions.

The polarization of a system of particles in an eigenstate of linear momentum is defined in the center-of-mass frame of the particles. A statement that the system is polarized indicates that there exists a net angular momentum in this frame; such an angular momentum can only be due to spin and will be the result of a nonuniform and asymmetric population of the spin states. (These statements must be modified for photons, which have no center-of-mass frame.) It is self-evident that particles having zero spin cannot be polarized.

The porlarization of particles of spin $\hbar/2$ is an important phenomenon. The polarization of a spin state represented by the state vector $|\chi\rangle$ is the expectation value of the operator $\mathbf{S}$ (Section 2.4) in units of $\hbar/2$. This is identical to the expectation value of the Pauli spin operator $\boldsymbol{\sigma}$ (Section 2.9). Thus the polarization of a system of particles in a pure spin state $|\chi\rangle$ is

$$\langle\boldsymbol{\sigma}\rangle = \langle\chi|\boldsymbol{\sigma}|\chi\rangle. \qquad (2.79)$$

If the particles in the system are not all in the same spin state, then the polarization is the expectation value averaged over all the particles in the system. Since this is usually the case, we do not make any distinction in notation between polarization for a "pure" system and for a "mixed"

2.13 Polarization

system. The polarization is a vector, and the net angular momentum in any direction in units of $\hbar/2$ is the component of $\langle \boldsymbol{\sigma} \rangle$ in that direction; thus the z component of the spin polarization is $\langle \sigma_z \rangle$. An individual measurement on one particle of the spin angular-momentum component in a particular direction can only yield one of the two values, $+\hbar/2$ or $-\hbar/2$. Suppose the probability that such a measurement yields the result $+\hbar/2$ is N_+ and the probability it yields $-\hbar/2$ is N_-; then the component of $\langle \boldsymbol{\sigma} \rangle$ in that direction k is

$$\langle \sigma_k \rangle = \frac{N_+ - N_-}{N_+ + N_-}. \tag{2.80}$$

A statement about the magnitude of the polarization when the direction is not defined uses $|\langle \boldsymbol{\sigma} \rangle|$, where

$$|\langle \boldsymbol{\sigma} \rangle|^2 = \langle \chi | \boldsymbol{\sigma} | \chi \rangle \cdot \langle \chi | \boldsymbol{\sigma} | \chi \rangle.$$

The polarization of particles having spin greater than $\hbar/2$ presents a more complicated situation. The polarization is again the expectation value of the operator **S** in units of the spin, but the number of parameters required to describe the state of polarization of a beam of such particles is considerable (see Section 2.14). We shall deal later with techniques for describing the polarization, but at this point we wish to draw the reader's attention to the possibility of "alignment" that exists for particles of spin $\hbar$ and greater (except photons). Consider an assembly of particles, each having spin $\hbar$, that has no net angular momentum. This can happen in several ways; for example:

(1) All spin states are equally populated so that a measurement of spin component in a fixed direction yields the values $\hbar$, 0, and $-\hbar$ with equal probability; this is an entirely unpolarized assembly.

(2) The zero-spin-component state is unoccupied, so that measurements yield spin components $+\hbar$ and $-\hbar$ with equal probability but never yield the 0 result; this is complete alignment. The alignment will be incomplete if a measurement can yield the zero result but with a probability less than the probability that it yields the result $+\hbar$ or $-\hbar$.

(3) The $+\hbar$ and $-\hbar$ spin-component states are unoccupied or have equal populations that are smaller than that in the zero spin component state.

It is obvious that alignment does not produce a net angular momentum.

The polarization of photons is defined in a manner depending on the problem. Although their spin is $\hbar$, their polarization properties are closer to those of particles of spin $\hbar/2$. This happens because there are only two possible orientations of the spin of a photon, and although polarization is possible, alignment is not.

We must stress one important point, which we do by discussing one example. The uncertainty principle does not allow a fully polarized assembly of electrons to have all the spin angular-momentum vectors pointing the same way; the magnitude of the spin is $\hbar[s(s+1)]^{1/2} = \sqrt{3}\,\hbar/2$, whereas the maximum value of spin that can be observed is a component $\hbar/2$. Thus the direction of the polarization in this case is not the direction in which all the spins point but the direction in which every measurement of spin component yields the value $\hbar/2$.

2.14 The Density Matrix

It is evident that we require a technique capable of describing fully the state of polarization or alignment of a system of particles. The technique is that of the density matrix (Fano, 1957).

We examine this technique by returning to some of the basic concepts of quantum mechanics, in particular to the concept of a state of a system. Since the results of measurements are the most important aspects, we consider the result of making a measurement on one state. If there exists a measurement that always yields the same result, then the state is said to be a "pure state." All the particles in a fully polarized system of particles are in a pure spin state, since repeated measurement of the component of spin angular momentum of a particle in the direction of polarization always yields the same result. In addition, the same result is obtained for all the particles on which the measurement is performed. The measurement that always yields this unique result is called a "complete experiment." To such a complete experiment there will correspond an operator, and the pure state is represented by an eigenstate of this operator; this eigenstate is called a state vector, or sometimes a wave function. However, it is not always convenient to represent the pure state by such an eigenstate, but is preferable to invoke the principle of superposition and to represent the pure state as a linear sum of the eigenstates of a complete set. The choice of the complete set is a matter of convenience; thus in Section 2.10 a pure state of spin $\frac{1}{2}$, fully polarized along the direction having polar angles θ, ϕ, was represented by a linear superposition of two pure states, one having spin pointing along the z axis, the other having spin along the $-z$ axis; that is [Eq. (2.67)],

$$\exp\left(-\frac{i\phi}{2}\right)\cos\frac{\theta}{2}\begin{bmatrix}1\\0\end{bmatrix} + \exp\left(+\frac{i\phi}{2}\right)\sin\frac{\theta}{2}\begin{bmatrix}0\\1\end{bmatrix}.$$

Such a procedure does not affect the purity of the state.

2.14 The Density Matrix

There can exist systems for which no measurement yields the same result for all members of the system. Thus there is no orientation of a perfect Nicol prism that will permit the transmission of every photon of a partially transverse polarized beam of photons. Such systems are said to be in a "mixed state." These mixed states can be described by data that are sufficient to predict the probability of a particular result to any measurement. Mixed states can be represented as an incoherent sum of pure states. This must be compared with the representation of a pure state as a linear sum of other pure states; the sum implied is a coherent one.

We consider a pure state represented by the eigenvector $|\phi_n\rangle$. We expand this vector as a linear sum of the eigenvectors of some convenient basis

$$|\phi_n\rangle = \sum_i \langle i|\phi_n\rangle |i\rangle. \tag{2.81}$$

Then the expectation value of a Hermitian operator Q is

$$\langle Q\rangle_n = \langle\phi_n|Q|\phi_n\rangle = \sum_{ij} \langle\phi_n|j\rangle\langle j|Q|i\rangle\langle i|\phi_n\rangle.$$

If a mixed state is an incoherent sum of states $|\phi_n\rangle$ with weights ω_n, then

$$\langle Q\rangle = \sum_n \omega_n \sum_{i,j} \langle\phi_n|j\rangle\langle j|Q|i\rangle\langle i|\phi_n\rangle$$
$$= \sum_{i,j} \langle j|Q|i\rangle \sum_n \omega_n \langle\phi_n|j\rangle\langle i|\phi_n\rangle.$$

We define the density matrix ρ by

$$\rho_{ij} = \sum_n \omega_n \langle\phi_n|j\rangle\langle i|\phi_n\rangle. \tag{2.82}$$

If we put

$$\langle j|Q|i\rangle = Q_{ji}, \tag{2.82a}$$

then

$$\langle Q\rangle = \text{Tr}\, Q\rho, \tag{2.82b}$$

where Tr is a contraction for "trace," which means the sum of the diagonal elements of the following matrix, $Q\rho$ in this case. This Q in Eq. (2.82b) is the matrix representation of the operator Q; that is, the ji element of Q is $\langle j|Q|i\rangle$ of Eq. (2.82a).

The density matrix provides a convenient description of a mixed state, since the results of any measurement can be predicted by the use of Eq. (2.82b).

We must give the properties of the density matrix, and we do so with a view to its application to the polarization experiments. Under these circumstances the matrix refers to mixed spin states of a system of particles.

The important properties of the density matrix are as follows.

(1) The density matrix is Hermitian. This follows from the fact that $\langle Q \rangle$ is a real quantity and Q is a Hermitian operator; therefore

$$\rho_{lm} = \rho_{ml}^*.$$

(2) Since any particle of a system must be in one of the available states, we require

$$\operatorname{Tr} \rho = 1.$$

Later we shall be using two matrices ρ_1 and ρ_2, which represent the spin states of a beam of incident particles and a beam of scattered particles respectively. We shall see that there is a matrix M [Eq. (3.23a)] that tranforms ρ_1 into ρ_2, in which case we require

$$\operatorname{Tr} \rho_1 = 1,$$

which, in the nonrelativistic case, implies a beam of flux $\hbar k/m$ particles per cm² per sec, and

$$\operatorname{Tr} \rho_2 = \frac{d\sigma}{d\Omega}.$$

(3) In view of the normalization undertaken in (2) we have to modify the expectation value of Eq. (2.82b) to

$$\langle Q \rangle = \frac{\operatorname{Tr} Q\rho}{\operatorname{Tr} \rho} \tag{2.83}$$

(4) By a comparison of Eq. (2.81) and the statements made in Section 1.3, we see that the diagonal elements of the density matrix are the probability of finding any member of the system in a pure state described by one of the state vectors such as $|i\rangle$. Thus the diagonal elements of the matrix are positive definite or zero:

$$\rho_{nn} \geqslant 0.$$

(5) A matrix with N rows and N columns requires $2N^2$ real quantities; the Hermitian density matrix therefore requires N^2, and if $\operatorname{Tr} \rho = 1$, $N^2 - 1$ parameters suffice.

We cannot discuss fully the properties and applications of the density matrix; readers who are interested are referred to literature on the subject (Fano, 1957, and his bibliography). We can now examine the specification of mixed spin states of a system of spin-s particles. It is usual to quantize the z component of a single-particle spin, and there are $(2s + 1)$ such states to choose from. Any other pure spin state can be expressed as a coherent sum of these $(2s + 1)$ states, and thus there are that number of terms in a sum such as that of Eq. (2.81). The density matrix describing a system

2.14 The Density Matrix

of such particles in a mixed state will have $N = 2s + 1$ and requires $(2s + 1)^2$ real numbers to describe the state completely; if the intensity of a beam of such particles is of no consequence, then the state of polarization is described by $(2s + 1)^2 - 1$ real numbers.

Let us consider a pure spin-state represented by a vector $|\chi\rangle$. Its expansion in terms of eigenstates of S_z [Eqs. (1.10), (2.81)] will be

$$|\chi\rangle = \sum_{s_z} a_{s_z} |s, s_z\rangle = \sum_{s_z} \langle s, s_z | \chi \rangle | s, s_z \rangle;$$

for example, in Eq. (2.67) we have

$$a_{1/2} = \exp\left(-\frac{i\phi}{2}\right) \cos \frac{\theta}{2}, \quad a_{-1/2} = \exp\left(+\frac{i\phi}{2}\right) \sin \frac{\theta}{2}.$$

Thus we can write $|\chi\rangle$ as a column matrix χ which that has as elements the matrix elements $\langle s, s_z | \chi \rangle$. Suppose this spin state is that of an incident beam (indicated by subscript 1), with the state normalized so that $\langle \chi_1 | \chi_1 \rangle = 1$. The density matrix ρ_1 for this system is given by

$$\rho_{1 s_z s_z'} = \langle \chi_1 | s, s_z' \rangle \langle s, s_z | \chi_1 \rangle = \chi_1 \chi_1^\dagger, \quad (2.84)$$

which follows from the definition given by Eq. (2.82) and where χ_1 is the matrix representation of $|\chi_1\rangle$. The matrix representing the spin state of the scattered beam (subscript 2) is given by operating on χ_1 with a matrix M, and we normalize so that the differential scattering cross-section is given by

$$\frac{d\sigma}{d\Omega} = \chi_2^\dagger \chi_2.$$

The density matrix ρ_2 for the scattered wave spin states is

$$\rho_2 = \chi_2 \chi_2^\dagger = M \chi_1 \chi_1^\dagger M^\dagger = M \rho_1 M^\dagger.$$

Now

$$\frac{d\sigma}{d\Omega} = \chi_2^\dagger \chi_2 = \text{Tr } \rho_2 = \text{Tr } M \rho_1 M^\dagger.$$

If the initial state is a mixed state, then it is specified by the density matrix ρ_1, where

$$\rho_1 = \sum_r \omega_r \chi_{1r} \chi_{1r}^\dagger, \quad (2.85)$$

which follows from Eq. (2.82). The sum is performed over all the different pure spin states χ_{1r} that occur in the incident beam. The matrix M is independent of r; thus it follows that

$$\rho_2 = M \rho_1 M^\dagger, \quad (2.86)$$

$$\frac{d\sigma}{d\Omega} = \mathrm{Tr}\, M\rho_1 M^\dagger, \tag{2.87}$$

$$\langle Q \rangle_1 = \mathrm{Tr}\, Q\rho_1 / \mathrm{Tr}\, \rho_1, \tag{2.88}$$

as for the pure state.

The important case that we shall treat is that of $s = \frac{1}{2}$. This occurs in the analysis of the scattering of spin-0 particles by a target of spin-$\frac{1}{2}$ particles (protons), which can be polarized. Density matrices are required to describe the initial polarization of protons in the target and the polarization of protons recoiling at a particular angle. ρ_1, ρ_2, and M are 2×2 matrices, and it follows that we can write them as linear sums of the 2×2 unit matrix $\mathbf{1}$ and the Pauli matrices $\boldsymbol{\sigma}$; for example,

$$\rho_n = l_n \mathbf{1} + \mathbf{m}_n \cdot \boldsymbol{\sigma} \qquad (n = 1, 2). \tag{2.89}$$

We normalize so that $\mathrm{Tr}\, \rho_1 = 1$ and $\mathrm{Tr}\, \rho_2 = (d\sigma/d\Omega)|$. (We use $(d\sigma/d\Omega)|$ to represent the differential cross section for a polarized target as distinct from $d\sigma/d\Omega$, the cross section for an unpolarized target.) Using the Pauli matrices [Eqs. (2.60)] we have

$$l_1 = \frac{1}{2} \quad \text{and} \quad l_2 = \frac{1}{2} \frac{d\sigma}{d\Omega}\bigg|.$$

The polarization of either of these beams is the expectation value of $\boldsymbol{\sigma}$,

$$\langle \boldsymbol{\sigma} \rangle_n = \frac{\mathrm{Tr}\, \boldsymbol{\sigma} \rho_n}{\mathrm{Tr}\, \rho_n} = \frac{\mathrm{Tr}[l_n \boldsymbol{\sigma} + \boldsymbol{\sigma}(\mathbf{m}_n \cdot \boldsymbol{\sigma})]}{\mathrm{Tr}\, \rho_n},$$

whence

$$\langle \boldsymbol{\sigma} \rangle_n = \frac{2\mathbf{m}_n}{\mathrm{Tr}\, \rho_n}.$$

Thus for beam 1 having the density matrix ρ_1, normalized to $\mathrm{Tr}\, \rho_1 = 1$, and having polarization $\langle \boldsymbol{\sigma} \rangle_1$, we have

$$\rho_1 = \tfrac{1}{2}\{\mathbf{1} + \langle \boldsymbol{\sigma} \rangle_1 \cdot \boldsymbol{\sigma}\}. \tag{2.90}$$

Similarly, the scattered-beam density matrix is given by

$$\rho_2 = \frac{1}{2}\frac{d\sigma}{d\Omega}\bigg| \{\mathbf{1} + \langle \boldsymbol{\sigma} \rangle_2 \cdot \boldsymbol{\sigma}\} \tag{2.91}$$

and

$$\rho_2 = M\rho_1 M^\dagger. \tag{2.92a}$$

Evidently the polarization of the final state is given by ρ_2, which can be found from ρ_1 using M. Thus if the polarization of the initial state is known, we can obtain information about M by measuring differential cross sections and final-state polarization.

2.15 Decay of Mixed Spin States

In the above we have indicated how the density matrix can be applied to the spin state of a spin-½ particle involved in scattering; we shall do this in detail in Chapter III. In the next section we indicate how it can be used to obtain the angular distribution in the decay of mixed states.

2.15 Decay of Mixed Spin States

In Section 2.11 we considered the decay of pure spin states and showed how to obtain the angular distribution of decay products. In this section we consider the same problem for some simple cases of the decay of mixed spin states. First, let us consider a particle of spin s deaying into two dissimilar spinless particles: evidently the orbital angular momentum l of the final state has $l = s$. We define the density matrix for the initial states with respect to a set of base vectors that are eigenstates of the operator S^2 and S_z [corresponding to $|i\rangle \cdots |j\rangle$ in Eqs. (2.81) and (2.82)]:

$$\rho_{s_z s_z'} = \sum_n \omega_n \langle \phi_n | s, s_z' \rangle \langle s, s_z | \phi_n \rangle . \tag{2.92b}$$

Now consider Q, the matrix corresponding to the observable with which we are concerned, the intensity $I(\theta, \phi)$ of particles in direction θ, ϕ. It is therefore the matrix representation of the operator $|\theta, \phi\rangle\langle\theta, \phi|$. Thus

$$Q_{l_z' l_z} = \langle l, l_z' | \theta, \phi \rangle \langle \theta, \phi | l, l_z \rangle$$
$$= Y_l^{l_z'*}(\theta, \phi) \, Y_l^{l_z}(\theta, \phi) .$$

Then, if we put $l = s, l_z = s_z, l_z' = s_z'$,

$$I(\theta, \phi) = \text{Tr } Q\rho = \sum_{s_z'} \sum_{s_z} Q_{s_z' s_z} \rho_{s_z s_z'} . \tag{2.93}$$

An alternative form of this result can be obtained in the following way: consider the rotation of coordinates that transforms θ and $\phi \to 0$ and thus produces a new coordinate system having its z axis along the direction θ, ϕ referred to the old axes. This is caused by the operator $D(\alpha, \beta, \gamma)$, with $\alpha = \phi, \beta = \theta$ and $\gamma = 0$ (its value will be irrelevant). Then

$$\langle \theta, \phi | s, s_z \rangle = \langle \theta, \phi | D^{-1} D | s, s_z \rangle$$
$$= \langle 0, 0 | D | s, s_z \rangle$$
$$= \sum_{l_z} \langle 0, 0 | l = s, l_z \rangle \langle l = s, l_z | D | s, s_z \rangle$$
$$= \sum_{l_z} \langle 0, 0 | l = s, l_z \rangle \mathscr{D}^s_{s_z l_z} .$$

Using Eq. (2.58) we find that this gives

$$\langle \theta, \phi | s, s_z \rangle = \left(\frac{2s+1}{4\pi} \right)^{1/2} \mathscr{D}^s_{s_z 0} .$$

Similarly

$$\langle s, s_z' | \theta, \phi \rangle = \left(\frac{2s+1}{4\pi}\right)^{1/2} (\mathscr{D}^{s\dagger})_{0s_z'}.$$

Hence

$$Q_{s_z's_z} = \frac{2s+1}{4\pi} (\mathscr{D}^{s\dagger})_{0s_z'} (\mathscr{D}^s)_{s_z 0} \tag{2.94a}$$

Substituting into Eq. (2.93) gives

$$I(\theta, \phi) = \frac{2s+1}{4\pi} (\mathscr{D}^{s\dagger} \rho \mathscr{D}^s)_{00}, \tag{2.94b}$$

where $\mathscr{D}^s$ is the rotation matrix for the angular momentum s. Notice that we have not transformed Q, but have found an expression for it in terms of the rotation-matrix elements. The second coordinate system considered (z axis along direction of intensity measurement) is one in which its own matrix Q has a particularly simple form: every element is zero except the mid-diagonal element. Then the matrix Q of Eq. (2.94a) is this simple matrix rotated back into the original coordinate system. Equation (2.94b) suggests yet another view. The matrix ρ can be transformed to the matrix $\mathscr{D}^\dagger \rho \mathscr{D}$, which corresponds to changing the base states to which ρ is referred into base states of angular momentum in the second coordinate system, thus expressing ρ in a coordinate system in which Q has a particularly simple form.

As an example we give the result for $s = 1$. Using the Hermitian property of ρ, we get

$$I(\theta, \phi) = \frac{3}{4\pi} \left\{ \rho_{00} \cos^2\theta + \frac{1}{2}(\rho_{11} + \rho_{-1-1} - 2\,\text{Re}[\rho_{1-1}\exp(-2i\phi)]) \right.$$
$$\left. \times \sin^2\theta - \sqrt{2}\,\text{Re}[(\rho_{10} - \rho_{0-1})\exp(-i\phi)] \sin\theta \cos\theta \right\}. \tag{2.95}$$

We can pursue this example by considering the decay of a pure state $|\psi\rangle$ having $s_x = 1$; i.e., the spin points along the direction $\theta = 90°$, $\phi = 0°$. The decay intensity in a given direction should be proportional to $\sin^2\alpha$ ($|Y_1^1(\alpha, \beta)|^2$), where α is the angle between this direction and the x axis. From geometry

$$\sin^2\alpha = 1 - \cos^2\phi \sin^2\theta.$$

Equation (2.95) should give the same decay distribution if we substitute the correct density matrix. Since $|\psi\rangle = D(\alpha, \beta, \gamma)|1, 1\rangle$ with $\alpha = 0°$, $\beta = -90°$ and $\gamma = 0°$, we have

$$\langle 1, 1 | \psi \rangle = \frac{1}{2}, \qquad \langle 1, 0 | \psi \rangle = \frac{1}{\sqrt{2}}, \qquad \langle 1, -1 | \psi \rangle = \frac{1}{2},$$

2.15 Decay of Mixed Spin States

and using Eq. (2.84) the density matrix is

$$\rho = \begin{bmatrix} \frac{1}{4} & \frac{1}{2\sqrt{2}} & \frac{1}{4} \\ \frac{1}{2\sqrt{2}} & \frac{1}{2} & \frac{1}{2\sqrt{2}} \\ \frac{1}{4} & \frac{1}{2\sqrt{2}} & \frac{1}{4} \end{bmatrix};$$

hence

$$I(\theta, \phi) = \frac{3}{4\pi} \left\{ \frac{1}{2} \cos^2 \theta + \left(\frac{1}{4} - \frac{1}{4} \cos 2\phi \right) \sin^2 \theta \right\}$$

$$= \frac{3}{8\pi} \{ 1 - \cos^2 \phi \sin^2 \theta \},$$

as expected. As we have indicated, this result can be obtained easily from simple reasoning, but it does serve to illustrate the method.

The next problem is to consider the decay of a system of spin j into a system of two particles with relative orbital angular momentum l and a spin s. The density matrix ρ for the initial state is a square matrix with $(2j + 1)$ columns and rows. To get the angular distributions the matrix must be transformed to that it would be if represented with respect to the states $|l, l_z, s, s_z\rangle$. This requires a matrix having $(2l + 1)(2s + 1)$ rows and columns: in this case the rows and columns are labeled by all possible pairs [labeled $(l_z s_z)$] of the quantum numbers l_z and s_z. The matrices that do the transformation are made up of Clebsch–Gordan coefficients, since they are associated with the transformation between the two representations. Thus the new density matrix ρ' is given by

$$\rho'_{(l_z' s_z')(l_z s_z)} = \sum_{j_z'} \sum_{j_z} \langle l, l_z', s, s_z' | j, j_z' \rangle \rho_{j_z' j_z} \langle j, j_z | l, l_z, s, s_z \rangle.$$

The angular distribution $I(\theta, \phi)$ is given by the operator

$$Q = |\theta, \phi\rangle\langle\theta, \phi|,$$

so that

$$Q_{(l_z s_z)(l_z' s_z')} = \langle l, l_z, s, s_z | \theta, \phi \rangle \langle \theta, \phi | l, l_z', s, s_z' \rangle$$

$$= Y_l^{l_z *}(\theta, \phi) \, Y_l^{l_z'}(\theta, \phi) \, \langle s, s_z | s, s_z' \rangle$$

$$= Y_l^{l_z *}(\theta, \phi) \, Y_l^{l_z'}(\theta, \phi) \, \delta_{s_z s_z'},$$

and

$$I(\theta, \phi) = \text{Tr} \, Q\rho. \qquad (2.96)$$

To obtain the expectation values of the spin angular-momentum operators it is useful to extract the density matrix for the final-state spin. This is done by omitting the summations over s_z and s_z' in Eq. (2.96). To do this most conveniently we divide the matrix $\rho_{(l_z's_z')(l_zs_z)}$ into $(2s+1)^2$ submatrices, each having all its elements with the same value of s_z and s_z' and of size $(2l+1) \times (2l+1)$. Multiply each submatrix by the matrix

$$Q_{l_zl_z'} = \langle l, l_z | \theta, \phi \rangle \langle \theta, \phi | l, l_z' \rangle = Y_l^{l_z*}(\theta, \phi) \, Y_l^{l_z'}(\theta, \phi)$$

and take the traces. This leaves a $(2s+1) \times (2s+1)$ matrix, which is that required. It is normalized so that its trace is $I(\theta, \phi)$. The expectation values are found in the usual way.

As an example we consider the parity-conserving decay

$$\Delta(1236) \to \pi + N$$
$$J^p: \quad \tfrac{3}{2}^+ \quad\quad 0^- \quad \tfrac{1}{2}^+$$

hence $j = \tfrac{3}{2}, l = 1, s = \tfrac{1}{2}$. We take an initial pure state having $j_z = \tfrac{1}{2}$. The density matrix is

$$\begin{bmatrix} 0 & 0 & 0 & 0 \\ 0 & 1 & 0 & 0 \\ 0 & 0 & 0 & 0 \\ 0 & 0 & 0 & 0 \end{bmatrix}$$

(the rows and columns are labeled by $j_z = \tfrac{3}{2}, \tfrac{1}{2}, -\tfrac{1}{2}, -\tfrac{3}{2}$). The matrix of Clebsch–Gordan coefficients has its rows labeled by the pairs $l_z s_z$ in the order $(1, \tfrac{1}{2})(0, \tfrac{1}{2})(-1, \tfrac{1}{2})(1, -\tfrac{1}{2})(0, -\tfrac{1}{2})(-1, -\tfrac{1}{2})$ and its colums by j_z. It is

$$C = \begin{bmatrix} 1 & 0 & 0 & 0 \\ 0 & \sqrt{\tfrac{2}{3}} & 0 & 0 \\ 0 & 0 & \sqrt{\tfrac{1}{3}} & 0 \\ 0 & \sqrt{\tfrac{1}{3}} & 0 & 0 \\ 0 & 0 & \sqrt{\tfrac{2}{3}} & 0 \\ 0 & 0 & 0 & 1 \end{bmatrix}.$$

The transformed density matrix is

$$\rho = C\rho C^\dagger$$

						l_z	s_z
0	0	0	0	0	0	1	$\tfrac{1}{2}$
0	$\tfrac{2}{3}$	0	$\tfrac{\sqrt{2}}{3}$	0	0	0	$\tfrac{1}{2}$
0	0	0	0	0	0	-1	$\tfrac{1}{2}$
0	$\tfrac{\sqrt{2}}{3}$	0	$\tfrac{1}{3}$	0	0	1	$-\tfrac{1}{2}$
0	0	0	0	0	0	0	$-\tfrac{1}{2}$
0	0	0	0	0	0	-1	$-\tfrac{1}{2}$

2.15 Decay of Mixed Spin States

The labeling of the rows is shown and the columns are labeled in the same order; the submatrices of elements having common $s_z s_z'$ are bounded by the broken lines. The matrix $Q_{l_z l_z'}$ is

$$\frac{3}{8\pi} \begin{bmatrix} s^2 & -\sqrt{2}\, sce^{i\phi} & -s^2 e^{2i\phi} \\ -\sqrt{2}\, sce^{-i\phi} & 2c^2 & \sqrt{2}\, sce^{i\phi} \\ -s^2 e^{-2i\phi} & \sqrt{2}\, sce^{-i\phi} & s^2 \end{bmatrix}$$

where $s = \sin\theta$ and $c = \cos\theta$. The final spin density matix is then

$$\frac{1}{8\pi} \begin{bmatrix} 4c^2 & -2sce^{i\phi} \\ -2sce^{-i\phi} & s^2 \end{bmatrix},$$

which gives

$$I(\theta, \phi) = \frac{1}{8\pi}(1 + 3\cos^2\theta)$$

$$\langle s_z \rangle = \frac{\hbar}{2}\left(\frac{4\cos^2\theta - \sin^2\theta}{3\cos^2\theta + 1}\right),$$

$$\langle s_y \rangle = \frac{\hbar}{2}\left(\frac{4\cos\theta \sin\theta \sin\phi}{3\cos^2\theta + 1}\right),$$

$$\langle s_x \rangle = -\frac{\hbar}{2}\left(\frac{4\cos\theta \sin\theta \cos\phi}{3\cos^2\theta + 1}\right).$$

In this case, with a general density matrix ρ, the angular distribution is given by

$$I(\theta, \phi) = \frac{1}{8\pi}\{4(\rho_{11} + \rho_{-1-1}) + \sin^2\theta(3\{\rho_{33} + \rho_{-3-3} - \rho_{11} - \rho_{-1-1}\}$$

$$- 2\sqrt{3}\, \text{Re}[(\rho_{-13} + \rho_{-31})\, e^{2i\phi}])$$

$$+ 4\sqrt{3}\, \sin\theta \cos\theta\, \text{Re}[(\rho_{-3-1} - \rho_{13})\, e^{i\phi}]\}. \tag{2.97}$$

Analyses of this kind become more complicated if there is the possibility of two waves of different l in the final state; for example, the parity-conserving decay of a $\frac{5}{2}^+$ state into two particles 0^- and $\frac{3}{2}^+$ can have a final state of $l = 1$ or 3. In this case interference can take place between the orbital states. There also exists the possibility of parity-nonconserving decays; for example,

$$\Lambda \to \pi^- + p\,.$$
$$J^P: \quad \tfrac{1}{2}^+ \quad 0^- \quad \tfrac{1}{2}^+$$

In this case both $l = 0$ and $l = 1$ final orbital states occur (see Section 6.4). For a more general treatment than we have given, which includes these cases, see Werle (1966).

2.16 Rotation of the Density Matrix

We consider briefly the problem of transforming the density matrix from the form it has in one coordinate system to the form it has in a second coordinate system related to the first by a rotation. We assume that we are considering base states that are eigenstates of J_{z_1} or of J_{z_2}, where z_1 and z_2 are the z axes in the two coordinate-systems. In the first system

$$\rho_{j_{z_1} j'_{z_1}} = \sum_i \omega_i \langle \phi_i | j, j'_{z_1} \rangle \langle j, j_{z_1} | \phi_i \rangle .$$

Now

$$|j, j'_{z_1}\rangle = \mathscr{D}^j_{j'_{z_1} j'_{z_2}} |j, j'_{z_2}\rangle \quad \text{(sum over } j'_{z_2}\text{)}$$

and

$$\langle j, j_{z_1} | = (\mathscr{D}^j)^{-1}_{j_{z_2} j_{z_1}} \langle j, j_{z_2} | ;$$

hence,

$$\rho_{j_{z_1} j'_{z_1}} = \sum_i \omega_i \sum_{j_{z_2}} \sum_{j'_{z_2}} \mathscr{D}^j_{j'_{z_1} j'_{z_2}} \langle \phi_i | j, j'_{z_2} \rangle \langle j, j_{z_2} | \phi_i \rangle (\mathscr{D}^j)^{-1}_{j_{z_2} j_{z_1}}$$

$$= \sum_{j_{z_2}} \sum_{j'_{z_2}} \mathscr{D}^j_{j'_{z_1} j'_{z_2}} \rho'_{j_{z_2} j'_{z_2}} (\mathscr{D}^j)^{-1}_{j_{z_2} j_{z_1}} ,$$

where ρ' is the required density matrix; hence

$$\rho' = \mathscr{D}^{-1} \rho \mathscr{D} . \tag{2.98}$$

This confirms the interpretation following Eq. (2.94b) that this form transforms the density matrix [in comparing with Eq. (2.94b) recall that $\mathscr{D}$ is a unitary matrix; i.e., $\mathscr{D}^\dagger = \mathscr{D}^{-1}$]. The matrix $\mathscr{D}$ is, of course, a function of the Euler angles relating the two coordinate-systems.

REFERENCES

Armenteros, R. et al. (1965). *Phys. Lett.* **19**, 338 (2.11).

Brink, D. M., and Satchler, G. R. (1962). "Angular Momentum." Oxford Univ. Press, London and New York (2.7, 2.10).

Condon, E. U., and Shortley, G. H. (1951). "The Theory of Atomic Spectra." Cambridge Univ. Press, London and New York (2.7).

Edmonds, A. R. (1960). "Angular Momentum in Quantum Mechanics." Princeton Univ. Press, Princeton, New Jersey (2.7, 2.10).

Fano, U. (1957). *Rev. Mod. Phys.* **29**, 74 (2.14).

Feynman, R. P., Leighton, R. B., and Sands, M. (1965). "The Feynman Lectures in Physics," Vol. III. Addison-Wesley, Reading, Massachusetts (2.10).

Werle, J. (1966). "Relativistic Theory of Reactions." North-Holland Publ., Amsterdam (2.10, 2.15).

Wigner, E. P. (1959). "Group Theory and Its Application to the Quantum Mechanics of Atomic Spectra." Academic Press, New York (2.10).

III
SCATTERING AND REACTION THEORY

3.1 Introduction

In this chapter, we discuss some of the methods available for the analysis of the scattering of elementary particles where scattering includes both elastic scattering and reactions. Most important is the partial wave analysis which is considered for the scattering of spinless particles by spinless particles and for the scattering of spin-$\frac{1}{2}$ particles by spinless particles. The chapter also includes a discussion of the properties of the S-matrix and of reciprocity. We find it most convenient to discuss much of the material of this chapter using the wavefunction rather than the bra–ket vector method.

3.2 The Partial-Wave Analysis

We have seen in the previous chapter that angular momentum is conserved in many situations; if we consider the interaction of two spinless particles, then our conservation principle applies to the orbital momentum l, and its z component l_z. In quantum-mechanical terms we might have that initially the particles are in a state of relative motion described by an eigenfunction of L^2 and L_z. If these operators commute with the Hamiltonian that describes the system and the interaction which causes the scattering, then the final-state function of the system will also be an eigenfunction of L^2 and L_z, with the same eigenvalues. Then the probability distribution of the scattered particles as a function of angle is given by the square of the spherical harmonic that has these eigenvalues. Another scattering might take place in a state of different orbital angular momentum for which a different interaction may apply, and, in fact, a positive-energy scattering situation has an initial state that is not an eigenfunction of L^2

but which may be represented by a sum of such eigenfunctions; each term in the sum is called a partial wave. The final state is also represented by a sum of the angular-momentum eigenfunctions, each with an amplitude determined by the strength of the interaction. The value of the amplitude is given by a parameter that introduces the effect of the interaction on the partial wave to which the amplitude applies. The sum is the total amplitude of the scattered wave, and the differential cross section is the square of the total amplitude.

We shall not give the complete development of the analysis because readers can find rigorous presentations in several places (e.g., Blatt and Weisskopf, 1952). However, we summarize the results. We consider a beam of particles that may be scattered by a spherically symmetrical potential that has its center at the origin of coordinates and has an infinitely heavy source, which therefore does not recoil. Actual cases can be transformed so that the origin of coordinates is at the center of mass; the two situations are the same and are described by the same wave function. We take the Cartesian coordinates directed so that the z axis is in the direction of motion of the incident particle; the initial state is conveniently described by the plane-wave solution of the Schrödinger equation

$$\psi = e^{ikz}, \tag{3.1}$$

where $\hbar k$ is the momentum of the incoming particle. After scattering we assume that the unaffected part of the incident wave plus the relatively small scattered wave are represented by

$$\psi = e^{ikz} + e^{ikr} \frac{f(\theta)}{r}; \tag{3.2}$$

$f(\theta)$ is a function to be determined. The variable θ is the polar angle between the z axis and the scattered wave described by $f(\theta)$.

The incoming wave is already an eigenfunction of L_z with eigenvalue $l_z = 0$; application of the operator

$$L_z = -i\hbar \left(x \frac{\partial}{\partial y} - y \frac{\partial}{\partial x} \right)$$

to Eq. (3.1) will confirm this physically evident fact. It is only necessary to decompose the incoming wave into eigenfunctions of L^2, thus:

$$e^{ikz} = \sum_{l=0}^{\infty} (2l + 1) i^l P_l(\cos \theta) F_l(kr); \tag{3.3}$$

r is the distance between the particles, $F_0(kr) = (\sin kr)/kr$, and the other $F_l(kr)$ are connected with the Bessel functions. $P_l(\cos \theta)$ is the lth Legendre

3.2 The Partial-Wave Analysis

polynomial and is an eigenfunction of L^2 and L_z with $l_z = 0$ (Section 2.8). Then the $f(\theta)$ of the scattered wave is given by

$$f(\theta) = \frac{1}{2ik} \sum_l (2l + 1)[\exp(2i\delta_l) - 1] P_l(\cos \theta). \quad (3.4)$$

In Eq. (3.4) δ_l is the parameter required to describe the effect of the scattering potential in producing a scattered wave from the lth partial wave; these parameters are called phase shifts, since the potential actually changes the phase of the scattered wave with respect to the incoming wave by an angle δ_l.

The differential scattering cross section is then given by

$$(d\sigma/d\Omega) = |f(\theta)|^2 = (1/k^2) \left| \sum_l (2l + 1) \exp(i\delta_l) \sin \delta_l \, P_l(\cos \theta) \right|^2, \quad (3.5)$$

where Ω is the solid angle in steradians. The total cross section σ is given by

$$\sigma = \int_{-1}^{+1} (d\sigma/d\Omega) \, 2\pi \, d(\cos \theta) = (4\pi/k^2) \sum_l (2l + 1) \sin^2 \delta_l. \quad (3.6)$$

Equation (3.6) is simpler than (3.5) because of the orthogonality of the Legendre polynomials. The total cross section for the lth partial wave has a maximum value of $4\pi(2l + 1)/k^2$ when $\delta_l = (2n + 1)\pi/2$ (n an integer). In Eq. (3.6) the total cross section is shown to be a simple sum of the individual partial-wave cross sections, but this is not true of the differential cross section since terms belonging to different l in Eq. (3.5) contribute cross terms in the square of the magnitude; these are the interference terms.

The phase shifts are the parameters that contain the effect of the actual interaction that causes the scattering, but they are not determined by this analysis. To predict fully the scattering for a given potential, the phase shifts must be found by matching the wavefunctions inside and outside the potential; this is usually a complex calculation and can be done analytically only for the Coulomb and some other simply shaped potentials. The phase shifts are positive for attractive and negative for repulsive potentials.

The reverse procedure is sometimes valuable; the results of a scattering experiment are subjected to a partial-wave analysis which determines the phase shifts, and their behavior as a function of energy may yield considerable information about the interaction involved. The pion scattering on hydrogen is an outstanding example of the value of this procedure; we shall do this analysis to illustrate the method and to show the results obtained (Sections 3.5 and 8.6).

It is convenient to rewrite the expression for the differential cross section. First, we replace the Legendre polynomials by spherical harmonics, which have the properties enumerated in Section 2.8. In Eq. (3.3) we substitute

$$Y_l(\cos\theta) = \left(\frac{2l+1}{4\pi}\right)^{1/2} P_l(\cos\theta)$$

to obtain

$$e^{ikz} = \sqrt{4\pi} \sum_l (2l+1)^{1/2} i^l Y_l(\cos)\, F_l(kr). \tag{3.7}$$

Equation (3.4) becomes

$$f(\theta) = \frac{1}{k} \sum_l [4\pi(2l+1)]^{1/2} \left(\frac{\exp(2i\delta_l)-1}{2i}\right) Y_l(\cos\theta). \tag{3.8}$$

Equation (3.5) becomes

$$\frac{d\sigma}{d\Omega} = \frac{4\pi}{k^2} \left|\sum_l (2l+1)^{1/2} \frac{\exp(2i\delta_l)-1}{2i} Y_l(\cos\theta)\right|^2. \tag{3.9}$$

It is convenient to express the results in terms of the spherical harmonics, since these are normalized. Later we shall have to make the substitution of $Y_l^m(\cos\theta)$ for $Y_l(\cos\theta)$, which is less confusing with such normalized functions. The quantities

$$[\exp(2i\delta_l) - 1]/2ik \left(\equiv \exp(i\delta_l)\sin\delta_l/k \equiv 1/k(\cos\delta_l - i)\right)$$

are the partial-wave amplitudes. The quantity $f(\theta)$ is the total scattered-wave amplitude.

The differential cross section is proportional to the intensity of particles scattered in the direction implied; thus, this intensity is a square of a sum of partial-wave amplitudes. Amplitudes that add in this way can be said to be coherent. This coherence is analogous to that occurring in classical physical optics; in that science there are rules for connecting the intensities and amplitudes of light waves, and there are also stringent requirements on the coherence of two sources of waves that are to interfere, where interference implies that the total intensity is found by squaring the sum of amplitudes; waves that interfere are said to be coherent. In quantum mechanics two or more amplitudes are coherent if there is no possibility of determining by which "path" the system has reached the final state, where this implies that there is an amplitude associated with each path (cf. Young's slits experiment in optics). Thus, in scattering experiments, the observation of a particle scattered at a particular angle cannot determine the orbital angular-momentum state through which the scatter occurred, and it follows that the intensity of scattered particles is the square of the

3.2 The Partial-Wave Analysis

sum of the partial wave amplitudes. To extend this, consider proton–proton scattering: this scattering takes place in singlet or triplet spin states but, in principle, there exists the possibility of determining at a particular scattering angle the total spin state of the emerging proton–proton system, and then the scattered intensity is the sum of the square of two total amplitudes, one for scattering in the singlet and one for the triplet state.

As given so far the analysis covers only elastic scattering of nonidentical spinless particles. We can now remove the restriction to elastic scattering. Let us reformulate Eq. (3.7) by noting that $F_l(kr)$ is a solution of the radial wave equation and has asymptotic value as $r \to \infty$ given by

$$\lim_{r \to \infty} F_l(kr) = \frac{1}{kr} \sin\left(kr - \frac{l\pi}{2}\right)$$

$$= \frac{1}{kr} \frac{\exp[i(kr - \tfrac{1}{2}l\pi)] - \exp[-i(kr - \tfrac{1}{2}l\pi)]}{2i}$$

Therefore, as $r \to \infty$,

$$e^{ikz} = (\sqrt{\pi}/kr) \sum_l (2l+1)^{1/2} i^{l+1}$$
$$\times \{\exp[-i(kr - \tfrac{1}{2}l\pi)] - \exp[i(kr - \tfrac{1}{2}l\pi)]\} Y_l(\cos\theta).$$

Thus we have two spherical waves, one traveling toward $r = 0$, the other traveling outward from the same point. The existence of the scattering potential at $r = 0$ will alter the outgoing wave; therefore at large distances

$$e^{ikz} + (f(\theta)/r)e^{ikr} = (\sqrt{\pi}/kr) \sum_l (2l+1)^{1/2} i^{l+1}$$
$$\times \{\exp[-i(kr - \tfrac{1}{2}l\pi)] - S_l \exp[i(kr - \tfrac{1}{2}l\pi)]\} Y_l(\cos\theta), \quad (3.9a)$$

where S_l is a quantity that modifies the outgoing wave. Therefore

$$(f(\theta)/r) e^{ikr} = (\sqrt{\pi}/kr) \sum_l (2l+1)^{1/2} i^{l+1} (1 - S_l)$$
$$\times \exp[i(kr - \tfrac{1}{2}l\pi)] Y_l(\cos\theta)$$
$$= (\sqrt{4\pi}/kr) \sum_l (2l+1)^{1/2} \frac{(S_l - 1)}{2i} e^{ikr} Y_l(\cos\theta). \quad (3.10)$$

The flux of particles in direction r of any wave is given by

$$\frac{h}{2mi}\left(\phi^* \frac{\partial \phi}{\partial r} - \phi \frac{\partial \phi^*}{\partial r}\right)$$

particles per square centimeter per second. Applying this to an incoming wave e^{ikr}/r we find $4\pi v$ particles per second arriving at the scattering center, where $v \ (=\hbar k/m)$ is the particle velocity. An outgoing wave (Se^{ikr}/r) has $4\pi v S^*S$ particles per second leaving. Hence elastic scattering requires

$$S^*S = 1.$$

Conservation of angular momentum means that we can discuss each individual partial wave in the same way. The incoming flux associated with any partial wave must, in the case of elastic scattering, be associated with an equal outgoing flux of the same l value, so that

$$S_l^*S_l = 1.$$

Formally this is assured, on integrating over the sphere, by the orthonormal properties of the spherical harmonics. Comparing Eq. (3.4) with Eq. (3.10) we see that $S_l = \exp 2i\delta_l$ satisfies this condition. The presence of inelastic scattering in an angular-momentum state l requires

$$S_l^*S_l < 1.$$

This can be realized by a complex phase shift or by having

$$S_l = \eta_l \exp 2i\delta_l,$$

where η_l is the real absorption parameter ($0 \leqslant \eta_l \leqslant 1$). The differential elastic scattering cross section is now given by

$$\frac{d\sigma_s}{d\Omega} = \frac{4\pi}{k^2} \sum_l \left| (2l+1)^{1/2} \left(\frac{\eta_l \exp 2i\delta_l - 1}{2i} \right) Y_l(\cos\theta) \right|^2 \quad (3.11)$$

and the total elastic cross section by

$$\sigma_s = \int \frac{d\sigma_s}{d\Omega} d\Omega = \frac{\pi}{k^2} \sum_l (2l+1) |S_l - 1|^2. \quad (3.12)$$

The waves not elastically scattered go into reactions not specified. The total reaction cross section is

$$\sigma_r = (\pi/k^2) \sum_l (2l+1)(1 - |S_l|^2)$$

$$= (\pi/k^2) \sum_l (2l+1)(1 - \eta_l^2). \quad (3.13)$$

The total cross section σ is

$$\sigma = \sigma_r + \sigma_s = (2\pi/k^2) \sum_l (2l+1)(1 - \operatorname{Re} S)$$

$$= (2\pi/k^2) \sum_l (2l+1)(1 - \eta_l \cos 2\delta_l). \quad (3.14)$$

The allowed values of the contribution to the cross section due to individ-

3.2 The Partial-Wave Analysis

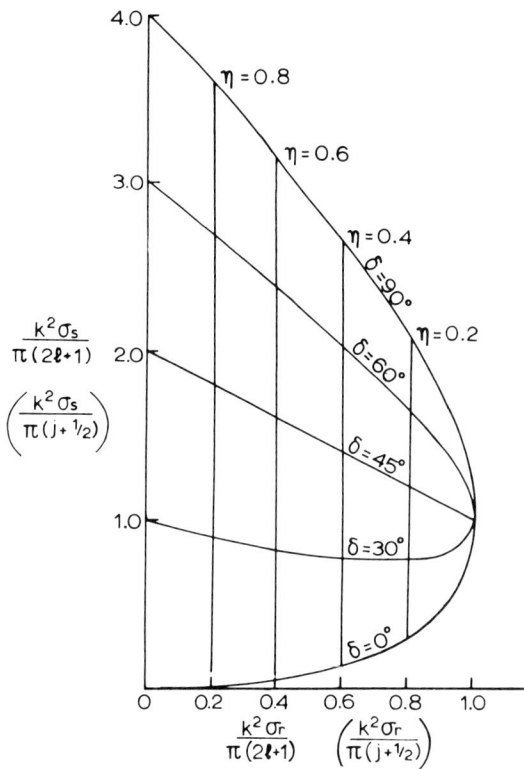

FIG. 3.1. Graph showing the allowed values of σ_s and σ_r in a single state of orbital angular momentum for various values of η and δ in the case of scattering without spin. The values in brackets are similar quantities for 0-½ scattering in a state of given j.

ual partial waves σ_{sl}, σ_{rl} are restricted by $\eta_l \leqslant 1$ in a manner shown in Fig. 3.1. We notice that if $\eta_l = 0$, which implies complete absorption, there is still elastic scattering with elastic and absorption reaction cross sections given by

$$\sigma_{sl} = \sigma_{rl} = (\pi/k^2)(2l + 1).$$

In fact there is always elastic scattering if $\eta < 1$, even though $\delta = 0$. This is equivalent to the diffraction of light by an absorbing object and is therefore called diffraction scattering.

There is another useful way of considering the elastic partial-wave amplitudes. Let us put

$$A_l = (\eta_l \exp 2i\delta_l - 1)/2ik$$

Then

$$A_l k = (i/2) - (\eta_l/2) \exp i[2\delta_l + (\pi/2)]. \tag{3.15}$$

Figure 3.2 shows the Argand diagram for $A_l k$ represented, for example, by the point P. Since $\eta_l \leqslant 1$, P must lie on or inside the circle of radius ½

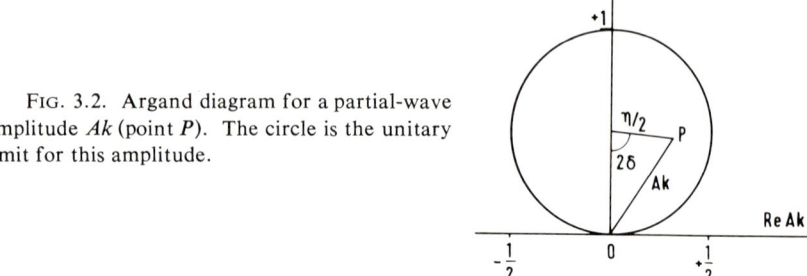

FIG. 3.2. Argand diagram for a partial-wave amplitude Ak (point P). The circle is the unitary limit for this amplitude.

centered at the point $i/2$. This is called the unitary circle, and the limit it places on the scattering amplitude is a consequence of the conservation of matter.

An important relation can be derived from these results. The forward scattering amplitude is

$$f(\theta = 0) = (1/k) \sum_l (2l + 1) [(\eta_l \exp 2i\delta_l - 1)/2i]$$
$$= (1/k) \sum_l (2l + 1) [\eta_l \cos 2\delta_l + i\eta_l \sin 2\delta_l - 1]/2i .$$

The imaginary part is given by

$$\operatorname{Im} f(0) = (1/k) \sum_l (2l + 1)(1 - \eta_l \cos 2\delta_l) . \tag{3.16a}$$

Comparing this with Eq. (3.14) we find

$$\operatorname{Im} f(0) = \sigma k/4\pi . \tag{3.16b}$$

Thus the total cross section is related to the imaginary part of the forward elastic scattering amplitude, which is, as a consequence, always greater than zero. This result is called the optical theorem and is, in fact, a general result for all scattering processes, with or without spin.

We can avoid the complexity of the infinite series (3.5) by general considerations on the partial waves that will be involved in any collision. We consider a collision classically: if the particle momentum in the center of mass is p and the impact parameter b, then the orbital angular momentum is pb, which we can put equal to $\hbar[l(l + 1)]^{1/2}$. If b is greater than the range R, there will be no scattering by the interaction, and since $b \simeq \hbar[l(l + 1)]^{1/2}/p$ we can express the condition for no scattering as $\hbar[l(l + 1)]^{1/2}/p > R$. Thus we need only consider the partial waves that have $[l(l + 1)]^{1/2} \leqslant (pR/\hbar)$. We can apply this, for example, to the scattering of pions by hydrogen; the interaction is expected to have a range of about the Compton wavelength of the pion, $\hbar/m_\pi c$; consequently, it is only necessary to consider partial waves for which $[l(l+1)]^{1/2} \leqslant (p/m_\pi c)$. For π-mesons, $m_\pi c^2 = 140$ MeV, and if the center-of-mass kinetic energy is 100 MeV, then $pc = 200$ MeV and we find that scattering only occurs if $[l(l+1)]^{1/2} \leqslant 1.4$, that is, if $l \leqslant 1$. Therefore the scattering is well described by an analysis that includes only the first two partial waves, $l = 0$ and $l = 1$, for kinetic energies up to 100 MeV. In fact, such an analysis is adequate up to about 200 MeV in the center of mass.

This argument can be made firmer by considering the properties of the $F_l(kr)$, which are radial solutions of the Schrödinger equation [see Eq. (3.3)]. The amplitude of these functions in a region near the origin decreases rapidly with increasing l, so that an interaction of limited range, i.e., near the origin, will have an appreciable effect only on a few low-order partial waves.

The unscreened Coulomb potential has an essentially infinite range, and the screened Coulomb potential has a range that is very large compared to nuclear forces; these properties mean that a partial-wave analysis of Coulomb scattering must continue up to very high-order partial waves (Mott and Massey, 1949). However, Coulomb scattering of high-energy particles is very small compared to nuclear-force scattering, except at very small scattering angles, and can normally be neglected unless we are interested in scattering at these small angles (see Section 3.6).

3.3 Scattering of Spin-0 by Spin-½ Particles

Let us now look formally at the scattering of spin-0 particles by spin-½ particles. We shall not include isotopic spin, so that the result will apply, for example, to K^+p, π^+p, or $n\alpha$ elastic scattering. Apart from Coulomb forces, we expect a symmetric short-range force and a spin–orbit force only. From the discussion of Section 2.4 we know that the constants of motion are the eigenvalues of J^2, J_z, L^2, and S^2. Although we shall include the effect of inelastic scattering, we are interested only in the elastic scattering, for which S^2 is automatically conserved. We shall consider first scattering from unpolarized spin-½ particles so that, taking the z axis along the direction of incidence, $j_z = s_z = \pm\frac{1}{2}$, depending on the spin orientation. These two states are, in this situation, incoherent, and give rise to incoherent scattering amplitudes. If parity is conserved, then by an argument given in Chapter X these two states give rise to the same differential scattering cross section, so that although we shall consider both states, we are not obliged to do so. If this were not the situation, it would be necessary to calculate both cases and average. We have $s = \frac{1}{2}, j_z = s_z = \pm\frac{1}{2}$. Every partial wave has two possible values of j, namely $l + \frac{1}{2}$ and $l - \frac{1}{2}$, and because the scattering can be different in the two and because j is conserved it is necessary to define a phase shift in each state, δ_{l+} and δ_{l-} respectively.

During the collision it is possible for the spin orientation to change, but since $\Delta j_z = 0$, this change can be made only at the expense of a reverse change in l_z. Thus this spin flip, $\Delta s_z = \pm 1$, is accompanied by the appearance of a scattered wave having a nonzero l_z; i.e., $l_z = \Delta l_z = \mp 1$. This outgoing partial wave is no longer represented by a Legendre polynomial in the equation for the scattered amplitude (3.4), but by an associated Legendre polynomial (see Section 2.8).

These functions have different angular dependence, and thus the presence of spin affects the distribution of scattered particles. In addition, the spin-flip states are orthogonal to the normally scattered waves, and thus these two waves do not interfere and the spin-flip contribution appears as an additional intensity of scattered particles. The incoming wave is

$$e^{ikz} \chi(\tfrac{1}{2}, \pm\tfrac{1}{2}) = (4\pi)^{1/2} \sum_l (2l+1)^{1/2} i^l F_l(kr) Y_l(\cos\theta) \chi(\tfrac{1}{2}, \pm\tfrac{1}{2}). \quad (3.17)$$

$\chi(s, s_z)$ is a wavefunction or spinor representing the spin state. Now $Y_l(\cos\theta)$ is an eigenfunction representing a state of orbital angular momentum l, with $l_z = 0$, which can be written $Y(l, 0)$. By the vector addition of angular momentum we can express the angular-momentum parts in terms of eigenfunctions of total angular momentum $\phi_l(j, j_z)$, where we put an l in subscript to remind us that each partial wave has well-defined and conserved l value (i.e., for elastic scattering with parity conservation, states of same j but different l do not mix). From Table E.1 (p. 509) we find

$$Y(l, 0) \chi(\tfrac{1}{2}, \pm\tfrac{1}{2}) = \left(\frac{l+1}{2l+1}\right)^{1/2} \phi_l(l+\tfrac{1}{2}, \pm\tfrac{1}{2})$$
$$\mp \left(\frac{l}{2l+1}\right)^{1/2} \phi_l(l-\tfrac{1}{2}, \pm\tfrac{1}{2}).$$

This gives an outgoing amplitude, using Eq. (3.8) with appropriate change, of

$$f_\pm(\theta) = \sqrt{4\pi} \sum_l \{A_{l+}(l+1)^{1/2} \phi_l(l+\tfrac{1}{2}, \pm\tfrac{1}{2})$$
$$\mp A_{l-} l^{1/2} \phi_l(l+\tfrac{1}{2}, \mp\tfrac{1}{2})\}, \quad (3.18a)$$

where

$$A_{l\pm} = \frac{\eta_{l\pm} \exp 2i\delta_{l\pm} - 1}{2i} \quad (3.18b)$$

are partial-wave amplitudes for each j state associated with the lth partial wave. Returning to orbital-spin wavefunctions, we have from Table E.1

3.3 Scattering of Spin-0 by Spin-½ Particles

$$\phi_l(l + \tfrac{1}{2}, \pm\tfrac{1}{2}) = \left(\frac{l+1}{2l+1}\right)^{1/2} Y(l, 0) \chi(\tfrac{1}{2}, \pm\tfrac{1}{2})$$

$$+ \left(\frac{l}{2l+1}\right)^{1/2} Y(l, \pm 1) \chi(\tfrac{1}{2}, \mp\tfrac{1}{2}),$$

$$\phi_l(l - \tfrac{1}{2}, \pm\tfrac{1}{2}) = \mp\left(\frac{l}{2l+1}\right)^{1/2} Y(l, 0) \chi(\tfrac{1}{2}, \pm\tfrac{1}{2})$$

$$\pm \left(\frac{l}{2l+1}\right)^{1/2} Y(l, \pm 1) \chi(\tfrac{1}{2}, \mp\tfrac{1}{2}),$$

so that Eq. (3.18a) becomes

$$\begin{aligned}
f_{\pm}(\theta) &= \sum_l \left(\frac{4\pi}{2l+1}\right)^{1/2} \{((l+1) A_{l+} + l A_{l-}) Y(l, 0) \chi(\tfrac{1}{2}, \pm\tfrac{1}{2}) \\
&\quad + [l(l+1)]^{1/2} (A_{l+} - A_{l-}) Y(l, \pm 1) \chi(\tfrac{1}{2}, \mp\tfrac{1}{2})\} \\
&= \sum_l \{[(l+1) A_{l+} + l A_{l-}] P_l(\cos\theta) \chi(\tfrac{1}{2}, \pm\tfrac{1}{2}) \\
&\quad \pm (A_{l+} - A_{l-}) P_l^1(\cos\theta) e^{i\phi} \chi(\tfrac{1}{2}, \mp\tfrac{1}{2})\}.
\end{aligned} \quad (3.19)$$

The first term in this equation corresponds to the spin-nonflip amplitude, since it goes with spin wavefunction the same as that of the initial state: the second term is called the spin-flip amplitude, since the outgoing wave has a spin that has been reversed.

Putting for the spin-nonflip amplitude

$$g(\theta) = \sum_{l=0} [(l+1) A_{l+} + l A_{l-}] P_l(\cos\theta) \quad (3.20)$$

and for the spin flip amplitude

$$h(\theta) = \sum_{l=1} (A_{l+} - A_{l-}) P_l^1(\cos\theta), \quad (3.21)$$

we have

$$d\sigma_s/d\Omega = |g(\theta)|^2 + |h(\theta)|^2, \quad (3.22)$$

since the spin wavefunctions are orthonormal. As expected, $d\sigma/d\Omega$ is independent of j_z.

An alternative way of writing the scattered-wave amplitude is

$$f(\theta) = g(\theta) - i h(\theta) \boldsymbol{\sigma} \cdot \mathbf{n}, \quad (3.23a)$$

where $\mathbf{n}$ is the unit vector normal to the plane of scattering:

$$\mathbf{n} = \frac{\mathbf{k}_i \times \mathbf{k}_f}{|\mathbf{k}_i \times \mathbf{k}_f|}.$$

$\mathbf{k}_i$ and $\mathbf{k}_f$ are the initial and final wave vectors of one of the particles. If the plane of scattering is at an azimuthal angle ϕ,

$$\mathbf{n} = -\mathbf{i}\sin\phi + \mathbf{j}\cos\phi,$$

so that

$$i\,\boldsymbol{\sigma}\cdot\mathbf{n} = i\cos\phi \begin{bmatrix} 0 & -i \\ +i & 0 \end{bmatrix} - i\sin\phi \begin{bmatrix} 0 & 1 \\ 1 & 0 \end{bmatrix}$$

$$= \begin{bmatrix} 0 & e^{-i\phi} \\ -e^{+i\phi} & 0 \end{bmatrix}.$$

In this representation of the Pauli spin matrices

$$\chi(\tfrac{1}{2}, \tfrac{1}{2}) = \begin{bmatrix} 1 \\ 0 \end{bmatrix} \quad \text{and} \quad \chi(\tfrac{1}{2}, -\tfrac{1}{2}) = \begin{bmatrix} 0 \\ 1 \end{bmatrix},$$

from which

$$i\,\boldsymbol{\sigma}\cdot\mathbf{n}\,\chi(\tfrac{1}{2}, +\tfrac{1}{2}) = -e^{+i\phi}\,\chi(\tfrac{1}{2}, -\tfrac{1}{2}),$$
$$i\,\boldsymbol{\sigma}\cdot\mathbf{n}\,\chi(\tfrac{1}{2}, -\tfrac{1}{2}) = +e^{-i\phi}\,\chi(\tfrac{1}{2}, +\tfrac{1}{2}).$$

If we now regard $f(\theta)$ of Eq. (3.23) as a matrix

$$M(\theta) = g(\theta)\,\mathbf{1} - i\,h(\theta)\,\boldsymbol{\sigma}\cdot\mathbf{n}, \tag{3.23b}$$

where $\mathbf{1}$ is the unit 2×2 matrix, then $M(\theta)$ operating on the initial spin states $\chi(\tfrac{1}{2}, \pm\tfrac{1}{2})$ gives the scattered wave:

$$f_\pm(\theta, \phi) = M(\theta)\,\chi(\tfrac{1}{2}, \pm\tfrac{1}{2}) = g(\theta)\,\chi(\tfrac{1}{2}, \pm\tfrac{1}{2}) \pm h(\theta)e^{\pm i\phi}\,\chi(\tfrac{1}{2}, \mp\tfrac{1}{2}). \tag{3.24}$$

This is the same as the scattered wave of Eq. (3.19), so that Eq. (3.24) represents a general way of writing the scattered wave. We shall make use of this in discussing polarization in $(0, \tfrac{1}{2})$ scattering (Section 3.4).

It is convenient to find expressions for the total elastic scattering cross-sections, σ_s in this scattering situation. Integrating Eq. (3.22) and extracting g and h from Eqs. (3.20) and (3.21) we find, remembering the orthonormal properties of $Y(l, l_z)$,

$$\sigma_s = 4\pi\left\{\sum_{l=0} \frac{|(l+1)A_{l+} + lA_{l-}|^2}{2l+1} + \sum_{l=1} \frac{l(l+1)}{2l+1} |A_{l+} - A_{l-}|^2\right\}$$

$$= 4\pi \sum_{l=0} \left((l+1)|A_{l+}|^2 + l|A_{l-}|^2\right). \tag{3.25}$$

Remembering that any j can have $l \pm \tfrac{1}{2}$, we can rearrange this as a sum over j:

$$\sigma_s = 4\pi \sum_{j=1/2} \sum_{l=j-1/2}^{l=j+1/2} (j+\tfrac{1}{2})|A_{jl}|^2. \tag{3.26}$$

3.3 Scattering of Spin-0 by Spin-½ Particles

If, as in Section 3.2 we put

$$A_{jl} = \frac{\eta_{jl}\exp(2i\delta_{jl}) - 1}{2ik} = \frac{S_{jl} - 1}{2ik},$$

then the scattering cross section

$$\sigma_s = (\pi/k^2) \sum_{j,l} (j + \tfrac{1}{2}) |S_{jl} - 1|^2, \qquad (3.27)$$

the reaction cross section

$$\sigma_r = (\pi/k^2) \sum_{j,l} (j + \tfrac{1}{2})(1 - |S_{jl}|^2), \qquad (3.28)$$

and the total cross section

$$\sigma = (2\pi/k^2) \sum_{j,l} (j + \tfrac{1}{2})(1 - \mathrm{Re}\, S_{jl}). \qquad (3.28a)$$

Then, as before, the contribution to the cross sections due to a state specified by total angular momentum j is restricted as in Fig. 3.1.

As an example let us consider the phase-shift analysis of $\pi^+ p$ scattering up to 200 MeV. The range of expected forces suggests that $l = 0$ and 1 will be adequate. The states involved are

$l = 0 \quad j = \tfrac{1}{2},$ phase shift δ;
$l = 1 \quad j = \tfrac{3}{2}$ or $\tfrac{1}{2}$ phase shifts δ_3 and δ_1, respectively.

We put

$$S_1 = \frac{\exp(2i\delta) - 1}{2i}, \quad P_3 = \frac{\exp(2i\delta_3) - 1}{2i}, \quad P_1 = \frac{\exp(2i\delta_1) - 1}{2i},$$

with all $\eta = 1$, since inelastic scattering is not appreciable below about 300 MeV (the notation is S, P for $l = 0$, 1 as in spectroscopy, with subscript $= 2j$). Then

$$g(\theta) = (1/k)\{S_1 + (2P_3 + P_1)\cos\theta\}, \qquad (3.29)$$

$$h(\theta) = (1/k)(P_3 - P_1)\sin\theta, \qquad (3.30)$$

and

$$(d\sigma/d\Omega) = (1/k^2)\{|S_1 + (2P_3 + P_1)\cos\theta|^2 + |(P_3 - P_1)\sin\theta|^2\}.$$

If we put

$$k^2(d\sigma/d\Omega) = B_0 + B_1\cos\theta + B_2\cos^2\theta, \qquad (3.31)$$

then

$$B_0 = |S_1|^2 + |P_3 - P_1|^2 = \sin^2 \delta + \sin^2(\delta_3 - \delta_1),$$
$$B_1 = 2 \operatorname{Re} S_1^*(P_1 + 2P_3)$$
$$= 2 \sin \delta [\sin \delta_1 \cos(\delta_1 - \delta) + 2 \sin \delta_3 \cos(\delta_3 - \delta_1)],$$
$$B_2 = |2P_3 + P_1|^2 - |P_3 - P_1|^2$$
$$= 3 \sin^2 \delta_3 + 6 \cos(\delta_3 - \delta_1) \sin \delta_1 \sin \delta_3,$$

and the total cross section is given by

$$\sigma = \frac{4\pi}{k^2}(B_0 + B_2/3) = \frac{4\pi}{k^2}(\sin^2 \delta + \sin^2 \delta_1 + 2 \sin^2 \delta_3). \quad (3.32)$$

The analysis has reduced the differential cross section to an expansion involving three phase shifts, but these can be determined only by experiment or from theoretical postulates. The former procedure has been used extensively, and the values of the phase shifts and their variation with energy indicate the form of the interaction between pions and protons. We shall defer full discussion of this point until we can include the analysis of the scattering of negative pions.

However, we can illustrate the possibility of obtaining some information by a simple examination of Eqs. (3.31) and (3.32):

(1) If there is no scattering in the $l = 1$ state, then there is no $\cos^2 \theta$ term.

(2) The presence of a $\cos \theta$ term indicates scattering in both the $l = 0$ and $l = 1$ states. (This is an interference term.)

(3) For small phase shifts, if the sign of the B_1 coefficient is negative, the signs of the phase shifts and the scattering potentials of the S state and of at least one of the P states are opposite (i.e., one of $\sin \delta$ and $\sin \delta_1$ or $\sin \delta_3$ is negative).

We shall defer more general discussion of the features of angular distributions in spin-0–spin-½ scattering until we can include polarization effects (see Section 3.5).

3.4 Polarization in Spin-0–Spin-½ Scattering

Before using the density-matrix formalism we shall obtain some results in a simple approach. First, consider the polarization of the spin-½ particles recoiling from the scattering in the situation in which they were unpolarized before scattering: as we shall find, this polarization can be directed only perpendicular to the plane of scattering (Section 6.3); to show this and to evaluate the polarization we must find the expectation value of σ_x, σ_y, and σ_z for the scattered wave. This must be done for each

3.4 Polarization in Spin-0–Spin-½ Scattering

of the two incoherent waves (incoherent because the initial state is unpolarized) $f_+(\theta, \phi)$ and $f_-(\theta, \phi)$ and the average taken. Then, since $\sigma_x = 2S_x/\hbar$,

$$\langle \sigma_x \rangle = \frac{1}{\hbar} \left\{ \frac{\langle f_+ | S_x | f_+ \rangle}{\langle f_+ | f_+ \rangle} + \frac{\langle f_- | S_x | f_- \rangle}{\langle f_- | f_- \rangle} \right\}, \quad (3.33)$$

and similarly for $\langle \sigma_y \rangle$. Note that these quantities are functions of θ and ϕ. Now $\langle f_+ | f_+ \rangle = \langle f_- | f_- \rangle = d\sigma/d\Omega$ [Eq. (3.5)], so that

$$\langle \sigma_x \rangle \frac{d\sigma}{d\Omega} = \frac{1}{\hbar} \{ \langle f_+ | S_x | f_+ \rangle + \langle f_- | S_x | f_- \rangle \}.$$

Substituting from Eqs. (3.19)–(3.21) and using the matrix elements from Eqs. (2.24) and (2.25) we find

$$\langle \sigma_x \rangle (d\sigma/d\Omega) = -2 \sin \phi \operatorname{Im} g^* h. \quad (3.34)$$

Similarly,

$$\langle \sigma_y \rangle (d\sigma/d\Omega) = +2 \cos \phi \operatorname{Im} g^* h \quad (3.35)$$

and

$$\langle \sigma_z \rangle = 0. \quad (3.36)$$

So if the scattering is in xz plane ($\phi = 0$), $\langle \sigma_x \rangle = 0$ and $\langle \sigma_y \rangle = \langle \sigma_n \rangle$, where **n** is the unit normal to the scattering plane, is given by

$$\langle \sigma_n \rangle (d\sigma/d\Omega) = 2 \operatorname{Im} g^* h. \quad (3.37)$$

There is nothing unique about this orientation of the axis, so this result tells us that the polarization is perpendicular to the plane of the scattering.

Returning to the case of $\pi^+ p$ scattering below 200 MeV, we can substitute from Eqs. (3.29) and (3.30) to find

$$k^2 \langle \sigma_n \rangle (d\sigma/d\Omega) = 2 \operatorname{Im}[S_1^* + (2P_3^* + P_1^*) \cos \theta][P_3 - P_1] \sin \theta. \quad (3.38)$$

Notice that the polarization varies as $\sin \theta$, which is to be expected as it must be zero at $\theta = 0$ and $180°$, where it is impossible to define a scattering plane.

A second case worth considering is scattering from a polarized target of spin-½ particles. Let us orient our axis so that the polarization has

$$\langle \sigma_x \rangle = 1, \quad \langle \sigma_y \rangle = 0, \quad \langle \sigma_z \rangle = 0,$$

i.e., the spin "points" in a direction having polar angles $\theta = 90°$, $\phi = 0°$. Thus the initial pure spin state is described by the spinor,

$$\begin{bmatrix} \cos \frac{\theta}{2} \\ \sin \frac{\theta}{2} \end{bmatrix} = \cos \frac{\theta}{2} \begin{bmatrix} 1 \\ 0 \end{bmatrix} + \sin \frac{\theta}{2} \begin{bmatrix} 0 \\ 1 \end{bmatrix}$$

$$= \sqrt{\tfrac{1}{2}} \, [\chi(\tfrac{1}{2}, +\tfrac{1}{2}) + \chi(\tfrac{1}{2}, -\tfrac{1}{2})].$$

Thus the outgoing amplitude is the same superposition of f_+ and f_- [Eq. (3.19)]:

$$f(\theta, \phi) = \sqrt{\tfrac{1}{2}}\,[f_+(\theta, \phi) + f_-(\theta, \phi)]$$

(θ is now the angle of scattering), and hence the differential cross section is

$$\begin{aligned}
\left.\frac{d\sigma}{d\Omega}\right| &= |f(\theta, \phi)|^2 \\
&= \frac{1}{2}|f_+(\theta, \phi)|^2 + \frac{1}{2}|f_-(\theta, \phi)|^2 + \mathrm{Re}\,\{f_+^*(\theta, \phi) f_-(\theta, \phi)\} \\
&= \frac{d\sigma}{d\Omega} - 2\sin\phi\,\mathrm{Im}\,g^*h \\
&= \frac{d\sigma}{d\Omega}\left(1 - \sin\phi\,\frac{2\,\mathrm{Im}\,g^*h}{|g|^2 + |h|^2}\right),
\end{aligned} \qquad (3.39)$$

where $(d\sigma/d\Omega)|$ is the cross section observed with the target polarized and $d\sigma/d\Omega$ is the cross section with the target unpolarized.

The $\sin\phi$ factor gives an azimuthal variation of intensity to the scattered beam that has its maximum and minimum at azimuthal directions perpendicular to the direction of the initial polarization. It is usual to call these directions left and right, and the scattering asymmetry ϵ is defined as

$$\epsilon = \frac{\left.\dfrac{d\sigma}{d\Omega}\right|_L - \left.\dfrac{d\sigma}{d\Omega}\right|_R}{\left.\dfrac{d\sigma}{d\Omega}\right|_L + \left.\dfrac{d\sigma}{d\Omega}\right|_R} \qquad (3.40)$$

where all cross sections refer to the same angle (θ) of scattering. If we evaluate Eq. (3.40), using Eq. (3.39), and compare the result with Eq. (3.37), we find that

$$\epsilon = \langle \sigma_n \rangle = \frac{2\,\mathrm{Im}\,g^*h}{|g|^2 + |h|^2}.$$

Thus, the scattering asymmetry in scattering from a fully transversely polarized proton-target is equal to the magnitude of the recoil-proton polarization generated at the same scattering angle from an unpolarized target.

These, and more general results, can be derived using the density-matrix formalism of Section 2.14. If $\langle \boldsymbol{\sigma} \rangle_1$ and $\langle \boldsymbol{\sigma} \rangle_2$ are the target and recoiling particle polarizations, respectively, then the density matrices for the target and recoiling-particle spin states are

$$\rho_1 = \frac{1}{2}(1 + \langle \boldsymbol{\sigma} \rangle_1 \cdot \boldsymbol{\sigma}), \qquad (3.41)$$

3.4 Polarization in Spin-0–Spin-½ Scattering

$$\rho_2 = \frac{1}{2} \frac{d\sigma}{d\Omega} \bigg| (1 + \langle \sigma \rangle_2 \cdot \sigma), \tag{3.42}$$

respectively.

The matrix M is now given by Eq. (3.23):

$$M = g - ih\,\mathbf{n} \cdot \boldsymbol{\sigma},$$

and we have

$$\rho_2 = M \rho_1 M^\dagger. \tag{3.43}$$

Therefore

$$\frac{1}{2} \frac{d\sigma}{d\Omega} \bigg| \{1 + \langle \sigma \rangle_2 \cdot \boldsymbol{\sigma}\} = \frac{1}{2}(g - ih\mathbf{n} \cdot \boldsymbol{\sigma}) \\ \times (1 + \langle \sigma \rangle \cdot \boldsymbol{\sigma})(g^* + ih^* \mathbf{n}_a \cdot \boldsymbol{\sigma}). \tag{3.44}$$

To evaluate Eq. (3.44) we use the commutation and anticommutation relations of the Pauli spin matrices [derived from Eq. 2.60c)] and the following vector relation:

$$(\mathbf{A} \cdot \boldsymbol{\sigma})(\mathbf{B} \cdot \boldsymbol{\sigma}) = \mathbf{A} \cdot \mathbf{B} + i(\mathbf{A} \times \mathbf{B}) \cdot \boldsymbol{\sigma}.$$

The first result of interest occurs if $\langle \sigma \rangle_1 = 0$, for which we find

$$\frac{d\sigma}{d\Omega}(1 + \langle \sigma \rangle_2 \cdot \boldsymbol{\sigma}) = |g|^2 + |h|^2 + \mathbf{n} \cdot \boldsymbol{\sigma}\, 2\,\mathrm{Im}\, g^*h.$$

Therefore

$$\frac{d\sigma}{d\Omega} = |g|^2 + |h|^2, \tag{3.45}$$

and

$$\langle \sigma \rangle_2 = \mathbf{n} \frac{2\,\mathrm{Im}\, g^*h}{|g|^2 + |h|^2} = P\mathbf{n}, \tag{3.46}$$

which confirms the results of Eqs. (3.12) and (3.37). The second equality defines P.

The complete evaluation of Eq. (3.44) gives

$$\rho_2 = \frac{1}{2} \frac{d\sigma}{d\Omega} \bigg\{ 1 + P\langle \sigma \rangle_1 \cdot \mathbf{n} + \bigg[P\mathbf{n} + (\langle \sigma \rangle_1 \cdot \mathbf{n})\mathbf{n} \\ - \frac{2\,\mathrm{Re}\, g^*h}{|g|^2 + |h|^2}(\langle \sigma \rangle_1 \times \mathbf{n}) + \frac{|g|^2 - |h|^2}{|g|^2 + |h|^2} \mathbf{n} \times (\langle \sigma \rangle_1 \times \mathbf{n}) \bigg] \cdot \boldsymbol{\sigma} \bigg\}. \tag{3.47}$$

$d\sigma/d\Omega$ is the differential scattering cross section for an unpolarized target. The differential cross section for a polarized target is found by comparing Eqs. (3.47) and (3.42); we find

$$\left.\frac{d\sigma}{d\Omega}\right| = \frac{d\sigma}{d\Omega}(1 + P\langle\sigma\rangle_1 \cdot \mathbf{n}) \qquad (3.48)$$

which thereby concludes a more general proof of Eq. (3.39). The polarization of the scattered beam is

$$\langle\sigma\rangle_2 = \frac{1}{(1 + P\langle\sigma\rangle_1 \cdot \mathbf{n})}\Big\{P\mathbf{n} + (\langle\sigma\rangle_1 \cdot \mathbf{n})\mathbf{n}$$
$$- \frac{2\,\mathrm{Re}\,g^*h}{|g|^2 + |h|^2}(\langle\sigma\rangle_1 \times \mathbf{n}) + \frac{|g|^2 - |h|^2}{|g|^2 + |h|^2}\mathbf{n} \times (\langle\sigma\rangle_1 \times \mathbf{n})\Big\}. \qquad (3.49)$$

Let us consider a target with a polarization $\langle\sigma\rangle_1$. If $\cos\phi = \mathbf{n}\cdot\langle\sigma\rangle$, then Eq. (3.48) becomes

$$\left.\frac{d\sigma}{d\Omega}\right| = \frac{d\sigma}{d\Omega}(1 + |\langle\sigma\rangle_1|\,P\cos\phi), \qquad (3.50)$$

and the scattering asymmetry [cf. Eq. (3.40)] becomes

$$\epsilon = P|\langle\sigma\rangle_1|.$$

The formalism we are using applies equally to scattering of a beam of spin-½ particles by a spin-0 target and to the reverse. In the first case a first scatter of an unpolarized beam will generate a polarization of the scattered beam of

$$\langle\sigma\rangle = P\mathbf{n}.$$

If a second identical (energy and angle) scattering is performed on this scattered beam, a left-right asymmetry will be observed of

$$\epsilon = P|\langle\sigma\rangle| = P^2.$$

Thus a double-scattering experiment will determine P.

We now wish to examine the actual changes in the polarization vector during the scattering: the general case can be done by considering the individual terms in Eq. (3.47). One important property is that the direction of the projection of the initial polarization on the plane of scattering is rotated through an angle β, where

$$\tan\beta = \frac{2\,\mathrm{Re}\,g^*h}{|g|^2 - |h|^2}.$$

The sign of β is positive if a right-handed rotation around $\mathbf{n}$ moves the projection of the initial polarization toward that of the second. In addition there are three special cases:

(I) $\langle\sigma\rangle_1$ lies perpendicular to the plane of scattering; i.e.,

$$\langle\sigma\rangle_1\cdot\mathbf{n} = |\langle\sigma\rangle| \quad\text{and}\quad \langle\sigma\rangle_1\times\mathbf{n} = 0.$$

3.4 Polarization in Spin-0-Spin-½ Scattering

Using Eq. (3.49) we find

$$\langle \sigma \rangle_2 \frac{d\sigma}{d\Omega}\bigg| = [|\langle \sigma \rangle_1|(|g|^2 + |h|^2) + 2 \, \text{Im} \, g^*h] \mathbf{n}$$

or

$$\langle \sigma \rangle_2 \cdot \mathbf{n} = \frac{|\langle \sigma \rangle_1| + P}{1 + P|\langle \sigma \rangle_1|}. \tag{3.51}$$

Thus the polarization perpendicular to the scattering plane is increased or decreased according to the value of P. If $\langle \sigma \rangle_1 = 0$, then $\langle \sigma \rangle_2 = P\mathbf{n}$, as expected.

(II) $\langle \sigma \rangle_1$ is longitudinal to the incident beam; i.e., $= |\langle \sigma \rangle_1| \mathbf{k}_1$. Then the component produced in the direction $\mathbf{k}_2 \times \mathbf{n}$ is $A|\langle \sigma \rangle_1|$, where the parameter A is given by

$$A = (1 - P^2)^{1/2} \sin(\theta - \beta).$$

(III) $\langle \sigma \rangle_1$ is transverse to the incident beam in the direction $\mathbf{k}_1 \times \mathbf{n}$. Then the component produced in the direction $\mathbf{k}_2 \times \mathbf{n}$ is $R|\langle \sigma \rangle_1|$, where the parameter R is given by

$$R = (1 - P^2)^{1/2} \cos(\theta - \beta).$$

We can now enumerate the experiments that determine matrices of the kind M. Of the four real quantities in M (the real and imaginary parts of g and h) it is possible to find three by straightforward experiments.

(1) $\quad d\sigma/d\Omega = |g|^2 + |h|^2$

is determined in a straightforward single-scattering experiment of an unpolarized beam.

(2) $\quad P = \dfrac{2 \, \text{Im} \, g^*h}{|g|^2 + |h|^2}$

is determined by the left–right scattering asymmetry at a polarized target in the case of spin-0 particles on, for example, protons [Eq. (3.50)], or by the normal component of polarization produced in scattering on an unpolarized target (0 on ½) or in the scattering of an unpolarized beam (½ on 0).

(3) β is given by the A or R parameter, and requires the measurement of a transverse polarization produced in scattering from a polarized target (0 on ½) or in scattering of a polarized beam (½ on 0).

The three experiments will give the moduli of g and h and the phase angle between them. Note that this discussion applies at one scattering angle, since g and h are functions of θ. If we are interested in g and h at all angles, then they are best expressed as sums of partial-wave amplitudes [Eqs. (3.20) and (3.21)] and the experimental data analyzed in terms of phase shifts and absorption parameters. This is discussed in Section 3.3.

3.5 Angular Distributions in Spin-0–Spin-$\frac{1}{2}$ Scattering

The two quantities of interest are $d\sigma/d\Omega$ and $P(d\sigma/d\Omega)$. From Eqs. (3.22) and (3.46) we have

$$d\sigma/d\Omega = |g|^2 + |h|^2,$$
$$P(d\sigma/d\Omega) = 2 \operatorname{Im} g^*h, \qquad (3.52)$$

where

$$g = \sum_{l=0} [(l+1)A_{l+} + lA_{l-}] P_l(\cos\theta),$$

$$h = \sum_{l=1} (A_{l+} - A_{l-}) P_l^1(\cos\theta),$$

$$A = \frac{\eta \exp(2i\delta) - 1}{2i},$$

with appropriate subscripts.

The Legendre polynomials are even or odd in powers of $\cos\theta$, and the associated Legendre polynomial $P_l^1(\cos\theta)$ is $\sin\theta$ times odd or even powers of $\cos\theta$ respectively as l is even or odd. It follows that terms in $d\sigma/d\Omega$ that are even in powers of $\cos\theta$ come from amplitudes belonging to partial waves of both even or both odd l; that is, of the same parity. For example in B_0 and B_2 of Eq. (3.31) the terms are $|S_1|^2$, $|P_1|^2$, $|P_3|^2$, $\operatorname{Re} P_3^*P_1$, all of which come from products of amplitudes of states of the same parity. In addition, the highest power of $\cos\theta$ is $2l_{\max}$ when $l_{\max}$ is the highest orbital angular momentum involved.

Conversely the quantity $(P/\sin\theta)(d\sigma/d\Omega)$ has the terms that are even (odd) in powers of $\cos\theta$ formed from the interference of amplitudes that belong to states having the opposite (same) parity. The highest power of $\cos\theta$ is $2l_{\max} - 1$. We can see this by completing the example of $\pi^+ p$ scattering:

$$P(d\sigma/d\Omega) = 2 \operatorname{Im} g^*h$$
$$= (2/k^2) \operatorname{Im}[S_1^* + (2P_3^* + P_1^*)\cos\theta][P_3 - P_1]\sin\theta.$$

Hence

$$(Pk^2/\sin\theta)(d\sigma/d\Omega) = 2 \operatorname{Im}[S_1^*(P_3 - P_1) - 3P_3^*P_1 \cos\theta], \qquad (3.53)$$

which illustrates the statements made above. Using the phase-shift parametrization, we obtain

$$P(d\sigma/d\Omega) = (\sin\theta/k^2)(B_0 + B_1 \cos\theta),$$

3.6 The Ambiguities of Spin-0–Spin-½ Scattering

where
$$B_0 = 2 \sin \delta [\sin \delta_3 \sin(\delta_3 - \delta) - \sin \delta_1 \sin(\delta_1 - \delta)]$$
and
$$B_1 = -6 \sin \delta_3 \sin \delta_1 \sin(\delta_1 - \delta_3).$$

We shall discuss in more detail the information that can be obtained from angular distributions when we consider high-energy pion–nucleon scattering in Section 8.6.

3.6 The Ambiguities of Spin-0–Spin-½ Scattering

We have seen that the determination of the matrix M apart from an overall phase requires three measurements. Since M is a function of scattering angle, these measurements must be done at each angle. The partial-wave analysis permits an analysis of angular distribution into partial-wave amplitudes or into complex phase shifts. The ambiguities that arise in doing analyses of this kind and how they can be resolved are of some importance.

The Minami ambiguity is the first important difficulty. It occurs because a scattering differential cross section is invariant under the change of parity of all the states involved. This means that if we interchange partial-wave amplitudes between the two states of different l (and opposite parity) but of the same j, then there is no change in the angular distribution. The interchanges required are

$$A_{(l-1)+} \rightleftarrows A_{l-}.$$

To see that this is the case we define
$$g' = g + (h \cos \theta / \sin \theta)$$
and
$$h' = (-h/\sin \theta).$$

Then solving for g and h and substituting in Eq. (3.22), we have
$$(d\sigma/d\Omega) = |g'|^2 + |h'|^2 + 2 \cos \theta \, \text{Re} \, g'^* h',$$
and
$$P(d\sigma/d\Omega) = -2 \sin \theta \, \text{Im} \, g'^* h'.$$

It is convenient to put
$$P_l^1(\cos \theta) = \sin \theta \frac{dP_l(\cos \theta)}{d \cos \theta} = \sin \theta \, P_l'(\cos \theta).$$

Then
$$g' = \sum_{l=0} [(l+1)A_{l+} + lA_{l-}]P_l + \cos \theta \sum_{l=1} (A_{l+} - A_{l-})P_l',$$

where we have dropped the arguments of the Legendre polynomials for conciseness. Now

$$(l+1)P_l + \cos\theta\, P_l' = P_{l+1}',$$
$$lP_l - \cos\theta\, P_l' = -P_{l-1}'.$$

Hence

$$g' = A_{0+} + \sum_{l=1}(A_{l+}P_{l+1}' - A_{l-}P_{l-1}')$$
$$= \sum_{l=1}(A_{(l-1)+} - A_{(l+1)-})P_l'.$$

Similarly,

$$h' = \sum_{l=1}(A_{l-} - A_{l+})P_l'.$$

It is evident that under the interchanges

$$A_{(l-1)+} \rightleftarrows A_{l-},$$
$$A_{(l+1)-} \rightleftarrows A_{l+},$$

that

$$g' \rightleftarrows h'.$$

This produces no change in $(d\sigma/d\Omega)$ but changes the sign of P. Thus the Minami ambiguity is resolved by a measurement of the polarization.

A more general form of the Minami ambiguity is the change $A_{(l-1)+}^* \rightleftarrows A_{l-}$, that is, $g'^* \rightleftarrows h'$, which has no effect on either the differential cross section or the polarization. This ambiguity can be resolved only by measuring the R or A parameter at a suitable angle.

The Yang ambiguity is that which arises from the possibility of changing the sign of either g or h thus:

$$g \to -g \quad \text{and} \quad h \to h \qquad \text{or} \qquad g \to g \quad \text{and} \quad h \to -h.$$

Under this change $d\sigma/d\Omega$ remains unchanged, but P changes sign. Either of the more general changes

$$g \to -g^* \quad \text{and} \quad h \to h^* \qquad \text{or} \qquad g \to g^* \quad \text{and} \quad h \to -h^*$$

leaves both $d\sigma/d\Omega$ and P unchanged, and it requires a measurement of R or A to resolve the ambiguity.

These ambiguities in the amplitudes lead to corresponding ambiguities in a phase-shift analysis. For example, in $\pi^+ p$ scattering, which we have treated in Section 3.3, up to 200 MeV there are three amplitudes if we assume $l \leqslant 1$, viz. S_1, P_3, P_1. Any angular distribution observed could, from the Minami ambiguity, have been due to three amplitudes P_1', D_3', S_1' equal to S_1, P_3, P_1, respectively, so that a phase-shift analysis giving values to δ, δ_3, δ_1 must also admit this possible alternative, at least. Notice,

3.6 The Ambiguities of Spin-0–Spin-½ Scattering

however, that allowing this ambiguity means that we have to allow l up to 2. The simplest allowed Yang ambiguity ($g \to g$, $h \to -h$) would permit a new set of phase shifts (δ', δ_3', δ_1') related to a set δ, δ_3, δ_1 by

$$\delta' = \delta,$$
$$\exp 2i\delta_3' = \tfrac{1}{3}(2 \exp 2i\delta_1 + \exp 2i\delta_3),$$
$$\exp 2i\delta_1' = \tfrac{1}{3}(4 \exp 2i\delta_3 - \exp 2i\delta_1),$$

and so on for other ambiguities.

Under the change of sign of all phase shifts $\delta \to -\delta$ the partial wave amplitude $A = [\exp(2i\delta) - 1]/2i$ changes thus: $A \to -A^*$. This causes no change in the differential cross section, but does change the polarization. In the early work on pion–proton scattering this sign ambiguity was resolved by examining small-angle $\pi^+ p$ scattering for interference between the Coulomb scattering and strong-interaction scattering. This was found to be destructive and indicated that the predominant phase shift, in the P_3 state, was positive and fixed the signs of all the other small phase shifts involved (Orear, 1954).

In discussing these ambiguities we have not made use of one restriction. As it stands, $M(\theta)$ [Eq. (3.23b)] and its partial-wave expansion can apply to any spin-0–spin-½ scattering or reaction when there is no change in parity of bosons or fermions between initial and find states (e.g., $\pi^- p \to \pi^- p$ or $K^- p \to \pi^0 \Lambda$). In the case of reactions the ambiguities stand; however, in scattering, the optical theorem requires Im $M(\theta = 0) > 0$; that is, Im $g(\theta = 0) > 0$. In the phase-shift parametrization this is ensured because Im$[(\eta \exp 2i\delta - 1)/2i] > 0$. It follows that in an analysis of scattering, the ambiguities that suggest a negative sign for Im $g(0)$ are not allowed. Thus we cannot apply the following change to already well-behaved solutions:

$$g \to -g, \quad h \to h, \quad \text{or} \quad g \to g^*, \quad h \to -h^*,$$

and we must alter the generalized Minami ambiguity so that the change is $g'^* \rightleftarrows -h'$; i.e., $A^*_{(l-1)+} \rightleftarrows -A_{l-}$, etc. There then remain eight possible solutions. A measurement of P eliminates half of these, and a measurement of R (or A) selects the correct one of the remaining four (see Fig. 3.3). If the analysis is in terms of partial-wave amplitudes, then there could remain an overall phase uncertainty to the limit allowed by the restriction that every amplitude must be on or inside the unitary circle (Section 3.2). This can be solved by measuring interference with Coulomb scattering or by applying the optical theorem. For the case of purely elastic scattering [$\eta = 1$ in Eq. (3.15)], the amplitudes must lie on the unitary circle and no ambiguities remain.

The discussion has neglected experimental errors that may lead to

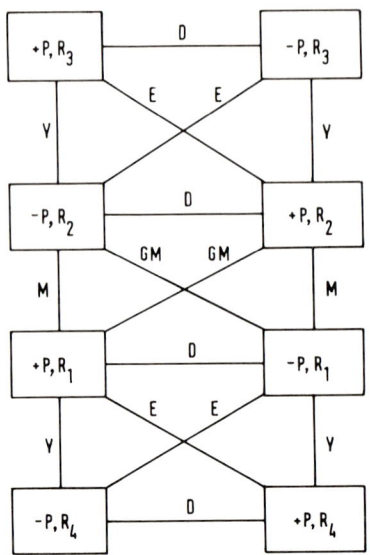

FIG. 3.3. The ambiguities of 0-½ scattering. Each box is a solution of the partial-wave amplitudes, and all give the same $d\sigma/d\Omega$. The labels connecting boxes indicate the changes required to connect the boxes, using the names of the ambiguities:

M (Minami):	$A_{l+} \rightleftarrows A_{(l+1)-}$;
GM (Generalized Minami):	$A_{l+} \rightleftarrows -A^*_{(l+1)-}$;
Y (Yang):	$h \to -h$, $g \to g$;
D (Negative phase shift):	$\delta \to -\delta$ or $A \to -A^*$;
E:	$h \to h^*$, $g \to -g^*$.

$P(d\sigma/d\Omega)$ has the value P or $-P$, and R can have one of the four values as indicated in each box (at one angle). Thus a measurement of $P(d\sigma/d\Omega)$ and R locates the correct solution.

several "correct" solutions. There remain other restrictions that can be applied in this case and when the measurements are not so complete as to include, for example, R or A. First, the energy may suggest that only a few partial waves are involved. Thus below 200 MeV in $\pi^+ p$ scattering it is unlikely that waves with $l \geqslant 2$ are involved. Thus, although the Minami ambiguity would suggest P_1, D_3, S_1 states in place of S_1, P_3, P_1, this is an unlikely alternative. The optical theorem is obviously another restriction if a total cross-section measurement is available. Also, on physical grounds, the amplitudes and phase shifts are expected to vary in monotonic manner with energy so that any analysis will rest upon the results of analyses at lower energy. Finally, the Wigner condition sets a lower limit on the rate of decrease of a phase shift with energy (Section 4.4). We shall discuss the energy dependence of amplitudes in the next chapter.

3.7 The Scattering of Spin-$\frac{1}{2}$ by Spin-$\frac{1}{2}$ Particles

The forces that can exist between two interacting spin-$\frac{1}{2}$ particles are a central, a spin–orbit, and a tensor force. From Section 2.4 we know that the commuting operators are H, J^2, J_z, but not L^2 or S^2. Thus L is not conserved. To see what happens let us consider the states available for various l values (Table 3.1).

TABLE 3.1

ANGULAR-MOMENTUM STATES OF SPIN-$\frac{1}{2}$–SPIN-$\frac{1}{2}$ SYSTEM

State	$^1S_0$	$^3S_1$	$^1P_1$	$^3P_{0,1,2}$	$^1D_2$	$^3D_{1,2,3}$	$^1F_3$	$^3F_{2,3,4}$
l	0	0	1	1	2	2	3	3
s	0	1	0	1	0	1	0	1
j	0	1	1	0, 1, or 2	2	1, 2, or 3	3	2, 3, or 4
Parity	+	+	−	−	+	+	−	−

Consider the states having $j = 2$:
$$^1D_2 \quad ^3D_2 \quad ^3P_2 \quad ^3F_2.$$
Parity: $+$ $+$ $-$ $-$

If l is not conserved but j and parity are conserved, the following transitions among these partial waves can occur:

$$^1D_2 \rightleftharpoons {}^3D_2 \qquad ^3P_2 \rightleftharpoons {}^3F_2.$$

Thus, it is not possible to define phase shifts for the $^1D_2$ and $^3D_2$ or for the $^3P_2$ and $^3F_2$ states. Other states for which transitions can occur are, for example, $^3S_1 \rightleftharpoons {}^3D_1$ (as in D-state mixing into the predominant S state of the deuteron) and $^3D_3 \rightleftharpoons {}^3G_3$, etc. In these cases it is necessary to define what are called eigenphase shifts and a mixing parameter. In the case of nucleon–nucleon scattering, the conservation of isotopic spin (Section 5.4) prevents the transitions between states of opposite symmetry. Thus the $^1D_2 \rightleftharpoons {}^3D_2$ mixing for example, is forbidden and the total spin s becomes a conserved quantity. A full review of nucleon–nucleon scattering has been given by Breit and Haracz (1967).

3.8 The Scattering of Identical Particles

It is well known that a system of identical fermions (bosons) must have a wave function that is antisymmetric (symmetric) under exchange of any two particles in the system. In addition, it is self-evident that any two such identical particles scattering on one another will have an angular distribution with respect to their relative velocity before collision that is symmetric about 90° in the center-of-mass system. Thus if $f_{1,2}(\theta)$ is the amplitude for finding particle 1 at angle θ and particle 2 at $\pi - \theta$ and $f_{2,1}(\theta)$ is the amplitude for finding the particles interchanged, then the amplitude for finding either particle 1 or 2 at angle θ is

$$f_{1,2}(\theta) \pm f_{2,1}(\theta),$$

and hence the differential cross section for finding either particle at θ is given by

$$(d\sigma/d\Omega) = |f_{1,2}(\theta) \pm f_{2,1}(\theta)|^2, \qquad (3.54)$$

where we have plus for bosons (symmetric) and minus for fermions (antisymmetric). $f(\theta)$ contains spin and orbital angular-momentum functions.

Let us consider the symmetry of the partial waves. In Section 5.3 we shall show that even-l states are symmetric and odd-l states are antisymmetric. Thus identical spinless bosons can only have l-even partial waves.

The elastic scattering of α-particles by helium serves as an example. The scattered amplitude is

$$f_{1,2}(\theta) = (1/k) \sum_{l \text{ even}} [4\pi(2l+1)]^{1/2} A_l Y_l(\theta),$$

where $A_l = [\exp(2i\delta_l) - 1]/2i$. Now $f_{2,1}(\theta) = f_{1,2}(\pi - \theta)$; and since $Y_l(\theta) = Y_l(\pi - \theta)$ if l is even, we have

$$d\sigma/d\Omega = (4\pi/k^2) \left| \sum_{l \text{ even}} 2[(2l+1)]^{1/2} A_l Y_l(\theta) \right|^2, \qquad (3.55)$$

which is symmetric about $90°$.

Remember that this is the cross section for seeing either α-particle at angle θ, per unit solid angle. If we discount this factor of 2, we see that there is twice as much scattering in each partial wave as there would be for nonidentical particles, but that only one-half the number of partial waves are present.

The case of identical fermions of spin $\tfrac{1}{2}$ is complicated immediately by the presence of spin. The triplet and singlet spin states are orthogonal and are symmetric and antisymmetric, respectively. For the singlet state, $f_{1,2}(\theta)$ is a sum of functions like $Y_l(\theta)\chi(0,0)$ with l even [$\chi(s, s_z)$ is the total spin function], while for triplet states it is a sum of functions like $Y_l(\theta)\chi(1, s_z)$ with l odd. Both of these are totally antisymmetric, so that for either spin state,

$$f_{1,2}(\theta) - f_{2,1}(\theta) = 2f_{1,2}(\theta)$$

and (for an unpolarized initial state)

$$\frac{d\sigma}{d\Omega} = \left| \begin{array}{c} \text{amplitude due} \\ \text{to } l\text{-even states} \end{array} \right|^2 + \sum_{s_z=-1}^{+1} \left| \begin{array}{c} \text{amplitude due} \\ \text{to } l\text{-odd states} \end{array} \right|.$$
$$\langle \text{singlet} \rangle \qquad\qquad \langle \text{triplet} \rangle$$

There are no cross terms between l even and l odd, and it follows that the whole expression is symmetric about 90°, as required.

The scattering of like particles of higher spin occurs in nuclear physics, e.g., d–d scattering, but it does not occur among elementary particles as a directly observable process.

3.9 The Scattering Matrix (I)

The use of $S = \exp(2i\delta)$ in Section 3.2 can be generalized to include the description of reactions. In addition to elastic scattering,

$$a + b \to a + b,$$

we now consider the binary reactions

$$a + b \to c + d,$$
$$\to \text{etc.}$$

At any energy several final states may be accessible, and one can consider that there is an intermediate system that can lead into any one of these states and is accessible from any one of them. These states can be designated by subscripts $\alpha, \beta, \ldots$, and are called channels. Each channel is described by the type of particle, the relative orbital angular momentum (l), and the total spin of the two particles (channel spin s). Our objective is to express the differential cross section for the reactions in terms of a set of quantities $S_{\alpha\beta,l}$ that are related to the probability that a reaction initiated via channel α with orbital angular momentum l has its final state in channel β. For simplicity we are restricting the analysis to reactions involving particles without spin, and we separate the l value from the other channel labels: then orbital angular momentum is conserved and, if parity is conserved, the reaction will proceed only if the product of the intrinsic parities of all four particles is positive. The case of reactions involving particles with spin is more complicated than is required for our present needs [see, for a general treatment, Blatt and Biedenharn (1952a)].

It is necessary to exercise some care in the construction of wavefunctions. In elastic scattering, the incoming and outgoing waves have the same velocity, reduced mass, and wavenumber. This is not generally true for transitions from one channel to another, and consequently a wave e^{ikr}/r represents an outgoing stream of particles with a flux that depends on k; we shall be in trouble with conservation requirements unless this factor is corrected. Therefore all incoming waves are of the form

$$(1/v^{1/2})(e^{-ikr}/r).$$

where $v = \hbar k/m$ is the channel velocity (nonrelativistic), k is the channel wavenumber, and m is the reduced mass in the channel. Outgoing waves are of the form

$$(1/v^{1/2})(e^{ikr}/r).$$

Subscripts will be added to v, k, and m to indicate the channel. These two waves each represent a flux of one particle per second per steradian. Suppose that the intermediate system is formed via channel α; then the radial part of the asymptotic wavefunction is given by

$$\phi_l = \frac{\exp(-ik_\alpha r)}{v_\alpha^{1/2} r} - \sum_\beta \frac{S_{\alpha\beta,l} \exp(ik_\beta r)}{v_\beta^{1/2} r}. \tag{3.56}$$

where the relative orbital angular momentum is l.

The probability that the incoming flux of one particle per second per steradian induces a reaction into channel β in one second per steradian is $|S_{\alpha\beta,l}|^2$. To convert to cross sections, we need to relate the incoming part of this wavefunction to the incoming part of the plane-wave expansion. We sum terms like Eq. (3.56) by

$$\frac{(\pi v_\alpha)^{1/2}}{k_\alpha} \sum_l (2l+1)^{1/2} i^{l+1} \exp(il\pi/2) Y_l(\cos\theta)$$

and replace the left-hand side by a new wavefunction Ψ, so that

$$\Psi = \frac{(\pi v_\alpha)^{1/2}}{k_\alpha} \sum_l (2l+1)^{1/2} i^{l+1} \left\{ \frac{\exp[-i(k_\alpha r - \tfrac{1}{2}l\pi)]}{v_\alpha^{1/2}} \right.$$
$$\left. - \sum_\beta S_{\alpha\beta,l} \frac{\exp(il\pi)\exp[i(k_\beta r - \tfrac{1}{2}l\pi)]}{v_\beta^{1/2}} \right\} Y_l(\cos\theta). \tag{3.57}$$

A careful comparison of this wavefunction with Eq. (3.9a) shows that it represents an incident plane wave plus a scattered wave, plus outgoing reaction products. The incident plane wave has flux v_α particles per second per unit area; the outgoing wave in channel β, where $\beta \neq \alpha$, is

$$\frac{(\pi v_\alpha)^{1/2}}{k_\alpha} \sum_l (2l+1)^{1/2} i^{l+1} S_{\alpha\beta,l} \frac{\exp[i(k_\beta r - \tfrac{1}{2}l\pi)]}{v_\beta^{1/2} r} Y_l(\cos\theta). \tag{3.58}$$

From our construction of wavefunctions we see that this represents an outgoing flux of particles

$$\frac{v_\alpha \pi}{k_\alpha^2} \left| \sum_l (2l+1)^{1/2} S_{\alpha\beta,l} Y_l(\cos\theta) \right|^2 \quad \text{particles sec}^{-1} \text{ sterad}^{-1}. \tag{3.59}$$

3.9 The Scattering Matrix (I)

The differential cross section is found by dividing this flux by the incident velocity v_α:

$$\left.\frac{d\sigma}{d\Omega}\right|_{\alpha\to\beta} = \frac{\pi}{k_\alpha^2} \left| \sum_l (2l+1)^{1/2} S_{\alpha\beta,l} \, Y_l(\cos\theta) \right|^2. \qquad (3.60)$$

For elastic scattering it is necessary to subtract from Ψ [(Eq. (3.57)] the unaffected incident plane wave [Eq. (3.8)]. The result is that

$$\left.\frac{d\sigma}{d\Omega}\right|_{\alpha\to\alpha} = \frac{\pi}{k_\alpha^2} \left| \sum_l (2l+1)^{1/2} (S_{\alpha\alpha,l} - 1) \, Y_l(\cos\theta) \right|^2, \qquad (3.60\text{a})$$

which is essentially the same as Eq. (3.11).

If there are N channels leading to the compound state, then there are N^2 quantities like $S_{\alpha\beta,l}$ or $2N^2$ real quantities. Taken together, the quantities $S_{\alpha\beta,l}$ make up a matrix S that called the collision or scattering matrix —or simply the S-matrix.

We complete this section by a proof of the unitary property of the S-matrix. An examination of the S-matrix formalism will indicate that if the intermediate system is formed through channel α, where the incoming flux of particles is one per second, then the outgoing flux of particles in channel β is $|S_{\alpha\beta}|^2$ particles per second (we have absorbed the specification l into α and β). Since every intermediate state must disintegrate into one of the available channels, we must have

$$\sum_\beta |S_{\alpha\beta}|^2 = 1. \qquad (3.60\text{b})$$

We now consider the same intermediate state being formed by a simultaneous flux of incoming particles in two channels α and α'; the two incoming waves must be orthonormal:

$$\langle \psi_\alpha | \psi_{\alpha'} \rangle = 0 \qquad (\alpha \neq \alpha').$$

Each incoming wave gives rise to an outgoing wave in each accessible channel, and these two sets of outgoing waves must also be orthonormal. Applying this requirement to the outgoing waves, find we that

$$\sum_\beta S^*_{\alpha\beta} S_{\alpha'\beta} = 0. \qquad \alpha \neq \alpha'.$$

If we combine this with the first property that we proved, we find that

$$\sum_\beta S^*_{\alpha\beta} S_{\alpha'\beta} = \delta_{\alpha\alpha'}. \qquad (3.61)$$

This is the property associated with the elements of a unitary matrix. To show this, we recall that a unitary matrix has (Section 1.2)

$$SS^\dagger = 1 .$$

Writing this multiplication in full, we have

$$\sum_\lambda S_{\kappa\lambda}(S^\dagger)_{\lambda\mu} = \delta_{\mu\kappa} .$$

However, $(S^\dagger)_{\lambda\mu} = S^*_{\mu\lambda}$, and therefore

$$\sum_\lambda S_{\kappa\lambda} S^*_{\mu\lambda} = \delta_{\kappa\mu} ,$$

which is the same as Eq. (3.61). Therefore S is unitary and there are $\tfrac{1}{2}N(N + 1)$ restrictions upon the $2N^2$ real parameters contained in this $N \times N$ matrix.

3.10 Binary Reactions

Equation (3.60) is evidently the first step in the partial-wave analysis of binary reactions

$$a + b \to c + d .$$

As it stands, it applies to spinless nonidentical particles. The extension to the reactions involving spin-0 with spin-$\tfrac{1}{2}$ particles producing like spins and particles (e.g., $\pi^+ + p \to K^+ + \Sigma^+$) is straightforward and yields formulas similar to those of Eqs. (3.20)–(3.22), where k is the initial-state wavenumber and the $A_{l\pm}$ are final-state amplitudes. A matrix M [Eq. (3.23b)] can be used just as for elastic scattering to obtain formulas for the angular distributions of cross section and polarization.

Complications arise in $(0, \tfrac{1}{2})$ binary reactions if there is a change in relative intrinsic parity of the particles involved. An example is the reaction

$$K^- + p \to \pi^0 + \Lambda(1405) .$$
$$J^P: \quad 0^- \quad \tfrac{1}{2}^+ \quad 0^- \quad \tfrac{1}{2}^-$$

Assuming parity is conserved in this strong interaction, it is evident that for a state of given total angular momentum, the orbital angular momentum changes by one unit. Thus in the $J^P = \tfrac{3}{2}^+$ state the initial state has $l = 1$, but the final state has $l = 2$. The partial-wave analysis of this situation is done in the same way as ordinary $(\tfrac{1}{2}, 0)$ scattering, except that the partial-wave amplitudes [Eq. (3.18b)] are replaced by transition amplitudes $T^j_{l+1,l}$ and $T^j_{l-1,l}$, where the final-state orbital angular momentum is $l' = l + 1$, $l - 1$, respectively, l being the initial orbital angular momentum. The initial states having $j = l + \tfrac{1}{2}$ can only change to an l', so that $j = l' - \tfrac{1}{2}$; that is, $l' = l + 1$. Similarly, for $j = l - \tfrac{1}{2}$, l changes to $l' = l - 1$. Thus Eq. (3.19) for a scattered wave becomes one with the $A_{l\pm}$ replaced by $T^j_{l\pm1,l}$.

3.10 Binary Reactions

The final-state total angular-momentum functions must be expressed in terms of the vector sum of the new orbital angular momentum and the spin, so that the scattered wave is given by

$$f_{\pm} = \mp \Big[\sum_{l=0} T^{l+1/2}_{l+1,\bar{l}} \{(l+1) P_{l+1}(\cos\theta)\,\chi(\tfrac{1}{2},\pm\tfrac{1}{2})$$
$$\mp P^1_{l+1}(\cos\theta)\,e^{\pm i\phi}\,\chi(\tfrac{1}{2},\mp\tfrac{1}{2})\}$$
$$+ \sum_{l=1} T^{l-1/2}_{l-1,\bar{l}} \{l\, P_{l-1}(\cos\theta)\,\chi(\tfrac{1}{2},\pm\tfrac{1}{2})$$
$$\pm P^1_{l-1}(\cos\theta)\,e^{\pm i\phi}\,\chi(\tfrac{1}{2},\mp\tfrac{1}{2})\}\Big].$$

The spin-flip and spin-nonflip amplitudes can be extracted from this equation in the usual manner and expressions obtained for angular and polarization distributions as in Section 3.4. However, it may be more convenient to obtain a matrix M as in Eq. (3.23b). In the present case the matrix connects initial and final states whose parities, apart from the intrinsic particle parities, are opposite. It follows that M must now be pseudoscalar under the parity transformation, in contrast to the M for ordinary scattering, which had to be scalar. The vectors available are $\boldsymbol{\sigma}$, $\mathbf{k}_i$, and $\mathbf{k}_f$: from these the scalar that can be constructed is $\boldsymbol{\sigma}\cdot\mathbf{k}_i\times\mathbf{k}_f$ and the pseudoscalars are $\boldsymbol{\sigma}\cdot\mathbf{k}_i$ and $\boldsymbol{\sigma}\cdot\mathbf{k}_f$. Thus ordinary scattering has the scalar

$$M = g - ih\,\boldsymbol{\sigma}\cdot\mathbf{n},$$

where

$$\mathbf{n} = \frac{\mathbf{k}_i \times \mathbf{k}_f}{|\mathbf{k}_i \times \mathbf{k}_f|}.$$

The parity changing reaction requires

$$M = a\,\boldsymbol{\sigma}\cdot\mathbf{k}_i + b\,\boldsymbol{\sigma}\cdot\mathbf{k}_f. \tag{3.62}$$

If we consider the effect of this M on the spin states in our present case in a manner analogous to that used in Section 3.4, we find

$$a = -\sum_{l=1} T^{l-1/2}_{l-1,\bar{l}} P_l{}'(\cos\theta) + \sum_{l=0} T^{l+1/2}_{l+1,\bar{l}} P_l{}'(\cos\theta),$$
$$b = +\sum_{l=1} T^{l-1/2}_{l-1,\bar{l}} P'_{l-1}(\cos\theta) - \sum_{l=0} T^{l+1/2}_{l+1,\bar{l}} P'_{l+1}(\cos\theta).$$

The scattering and polarization angular distributions may be obtained from M using the techniques of the density matrix (Section 3.4). It is worth noting that the scattering angular distribution cannot be more complex than $\cos^{2l_{\max}}\theta$, where $l_{\max}$ is the highest initial orbital angular-momentum state involved, even though there will generally be a final state with orbital angular momentum $l_{\max} + 1$. If there is no $l_{\max} \to l_{\max} + 1$ transition, then the angular distribution is no more complex than $\cos^{2l_{max}-1}\theta$ (this assumes that there is a final state with l_{max} fed from the initial state with $l_{max} - 1$).

3.11 The Scattering Matrix (II)

The scattering matrix can be related to an operator S that acts upon an initial-state vector $|\alpha\rangle$ to produce a state $S|\alpha\rangle$, which is a superposition of all possible final states. The amplitude for finding a particular final state represented by the ket $|\beta\rangle$ is then

$$S_{\alpha\beta} = \langle\beta|S|\alpha\rangle. \qquad (3.63)$$

Considering elastic scattering in orbital angular-momentum state l we see that $S_{\alpha\alpha,l} = \eta_l \exp(2i\delta_l)$. The operator S is expected to be invariant under translations in position and time, and hence momentum and energy will be conserved. It is usual to define the transition matrix T by the relation

$$\langle\beta|S|\alpha\rangle = \delta_{\alpha\beta} + i(2\pi)^4 \delta^4(p_\alpha - p_\beta)\langle\beta|T|\alpha\rangle, \qquad (3.64)$$

where p_α and p_β are the total energy-momentum four-vectors for all the particles in the states α and β, respectively, and $\delta_{\alpha\beta}$ is the Kronecker delta; δ is the Dirac delta function, which applies four times, once for each component of the four vector $p_\alpha - p_\beta$. The matrix element $\langle\beta|T|\alpha\rangle$ is invariant if we normalize all the single-particle states involved in a suitable way. The one chosen here is to normalize to a volume $1/2p_0$, where p_0 is the energy of the particle. To obtain a transition probability it is necessary to take the squared modulus of the second term. This leads to a factor

$$(2\pi)^8 |\delta^4(p_\alpha - p_\beta)|^2,$$

which can be shown to be (Muirhead, 1965)

$$VT(2\pi)^4 \delta^4(p_\alpha - p_\beta), \qquad (3.65)$$

where VT is the space-time volume. Thus the transition probability per unit space-time for the transition $\alpha \to \beta$ is

$$(2\pi)^4 \delta^4(p_\alpha - p_\beta)|\langle\beta|T|\alpha\rangle|^2. \qquad (3.66)$$

Consider now the case of the radioactive decay of state α. The total decay rate is the transition probability summed over all final states. For each particle in the final state there are

$$\int \frac{d\mathbf{p}}{2\pi^3} \frac{1}{2p_0}$$

3.1 The Scattering Matrix (II)

final states, so that if there are r particles in the final state, the sum over final states is a $3r$-fold integral:

$$\omega = 2\pi^4 \frac{1}{(2\pi)^{3r}} \int_1 \int_2 \cdots$$
$$\times \int_r \frac{d\mathbf{p}_1 d\mathbf{p}_2 \cdots d\mathbf{p}_r}{2^r E_1 E_2 E_3 \cdots E_r} \delta^4(p_\alpha - p_\beta) |\langle \beta | T | \alpha \rangle|^2. \quad (3.67)$$

However, we have $2E_\alpha$ initial particles per unit volume, and we must average over initial spin states, $\overline{\sum}_\alpha$, and sum over final spin states, $\sum_\beta$, to give the decay rate of one particle:

$$\omega = \frac{(2\pi)^4}{2E_\alpha} \frac{1}{(2\pi)^{3r}} \int_1 \cdots$$
$$\times \int_r \frac{d\mathbf{p}_1 \cdots d\mathbf{p}_r}{2^r E_1 \cdots E_r} \delta^4(p_\alpha - p_\beta) \overline{\sum}_\alpha \sum_\beta |\langle \beta | T | \alpha \rangle|^2. \quad (3.68)$$

We can now consider the derivation of a cross section for the interaction of two particles a and b in a state α. The cross section $\sigma_{\alpha\beta}$ is the transition probability for one incident particle when there is one target particle per unit area. Putting this another way, it is the transition probability per unit flux of incident particles. The flux is the product of the particle densities ($4E_aE_b$) and the relative velocity v, so that

$$\sigma_{\alpha\beta} = \frac{(2\pi)^4}{4E_aE_b v} \frac{1}{(2\pi)^{3r}} \int_1 \cdots$$
$$\times \int_r \frac{d\mathbf{p}_1 \cdots d\mathbf{p}_r}{2^r E_1 \cdots E_r} \delta^4(p_\alpha - p_\beta) \overline{\sum}_\alpha \sum_\beta |\langle \beta | T | \alpha \rangle|^2. \quad (3.69)$$

The $3r$-fold integral with the δ-function is called the phase space factor. If the matrix element is constant over the interval represented by the integration, it can be taken outside, and the various spectra among final-state particles can be predicted from this factor. For example, the invariant mass distribution of two of the particles in a final state of more than two can be calculated, and deviations from this phase space will indicate structure in the matrix element. Methods for manipulating phase space are discussed in Appendix C. As defined above the phase space is Lorentz invariant. This follows from the fact that $d\mathbf{p}/E$ is an invariant quantity. In fact the phase-space factor can be rewritten in an obviously Lorentz invariant form if we consider instead of a three-dimensional integral over $d\mathbf{p}$, a four-dimensional one over d^4p with a δ-function to keep the particle on the mass shell. Thus

$$\int d^4p\, \delta(p^2 + m^2) = \int d\mathbf{p}\, dE\, \delta(\mathbf{p}^2 - E^2 + m^2) \quad (3.70)$$

$$= \int d\mathbf{p}\, dE\, \delta(E_p^2 - E^2)$$

$$= \int \frac{d\mathbf{p}\, dE}{2E_p} [\delta(E_p - E) + \delta(E_p + E)]$$

$$= \int \frac{d\mathbf{p}}{2E_p} \quad \text{if} \quad E_p > 0. \quad (3.71)$$

So the phase-space integral can be written

$$\int \cdots \int \frac{d\mathbf{p}_1 \cdots d\mathbf{p}_q}{2^q E_1 \cdots E_q} \delta^4(p_\alpha - p_\beta)$$

$$= \int \cdots \int d^4p_1 \cdots d^4p_q\, \delta^4(p_\alpha - p_\beta) \prod_{i=1}^{q} \delta(p_i^2 + m_i^2)\theta(E_i), \quad (3.72)$$

where $\theta(E_i) = 1$ if $E_i > 0$, otherwise 0.

The Lorentz-invariant phase space is easier to manipulate than a noninvariant form that does not have the E factor in the denominator and which is sometimes used.

When partial differential cross sections or transition rates are required, certain of the phase-space integrals are not completed. Formally this may be done by inserting δ-functions into the phase-space integrals to limit one or more variables to a fixed value. Again, further details can be found in Appendix C.

The exact computation of the S or T matrix elements may frequently be impossible, with the nearest approach being through perturbation theory. If the interaction Hamiltonian H_I is known, then first-order time-dependent perturbation theory gives, instead of Eq. (3.68),

$$\omega = 2\pi \sum_{\substack{\text{final} \\ \text{states}}} \delta(E_\alpha - E_\beta) |\langle \beta | H_\text{I} | \alpha \rangle|^2$$

$$= \frac{2\pi}{(2\pi)^{3q}} \int \cdots \int d\mathbf{p}_i \cdots d\mathbf{p}_q\, \delta(E_\alpha - E_\beta) |\langle \beta | H_\text{I} | \alpha \rangle|^2. \quad (3.73)$$

If the matrix element is constant over the region of integration,

$$\omega = 2\pi |\langle \beta | H_\text{I} | \alpha \rangle|^2 (dn/dE_\beta), \quad (3.74)$$

where dn/dE_β is sometimes called the density-of-states factor, which can be expressed as an integral by comparison with Eq. (3.73). We have omitted the sums over final and average over initial spin states, but these should not be forgotten. Equation (3.74) is a statement of Fermi's Golden Rule No. 2 (Segré, 1964). Some elementary applications of this rule are given by Fermi (1951).

3.11 The Scattering Matrix (II)

To complete this section we shall derive from Eq. (3.69) total and differential cross sections for elastic and two-body reactions. First we define some kinematic quantities. A useful and constantly appearing function is

$$\lambda(x, y, z) = (x^2 + y^2 + z^2 - 2xy - 2yz - 2xz)^{1/2}.$$

We anticipate Section 12.2 and use the symbol s for the squared center-of-mass energy. Then for the reaction

$$a + b \to 1 + 2$$

we have

$$s = m_a^2 + m_b^2 + 2m_b E_a,$$

where we have assumed particle a is incident, with total energy E_a, on the stationary target b. It is now convenient to calculate quantities in terms of s. The incident velocity in the laboratory system is

$$\frac{\lambda(s, m_a^2, m_b^2)}{2m_2 E_a}.$$

so that the factor $1/4E_a E_b v$ calculated in the laboratory system is ($E_b = m_b$, etc.)

$$[2\lambda(s, m_a^2, m_b^2)]^{-1}.$$

Therefore

$$\sigma_{\alpha\beta} = \frac{1}{2\lambda(s, m_a^2, m_b^2)} \frac{1}{(2\pi)^2}$$
$$\times \iint \frac{d\mathbf{p}_1 \, d\mathbf{p}_2}{4E_1 E_2} \delta^4(p_1 + p_2 - p_a - p_b) \overline{\sum_\alpha} \sum_\beta |\langle \beta | T | \alpha \rangle|^2. \quad (3.75)$$

The symbols α and β refer to the initial and final states, respectively. All the factors are Lorentz invariant, so we can do the integrals in any reference frame that is convenient. The easiest is the center-of-mass system, for which $\mathbf{p}_1 = -\mathbf{p}_2$. This means that three of four δ-functions can be eliminated immediately, giving

$$\iint d\mathbf{p}_1 \, d\mathbf{p}_2 \, \delta^3(\mathbf{p}_1 + \mathbf{p}_2) = \int d\mathbf{p}_1.$$

We put

$$d\mathbf{p}_1 = p_1^2 \, dp_1 \, d\Omega_1 = p_1 E_1 \, dE_1 \, d\Omega_1.$$

Therefore

$$\sigma_{\alpha\beta} = \frac{1}{2\lambda(s, m_a^2, m_b^2)} \frac{1}{(2\pi)^2}$$
$$\times \int \frac{p_1 \, dE_1 \, d\Omega_1}{4E_2} \delta(E_1 + E_2 - E_a - E_b) \overline{\sum_\alpha} \sum_\beta |\langle \beta | T | \alpha \rangle|^2. \quad (3.76)$$

To do the integral over the δ-function we use

$$\int dx\, \delta(f(x)) = 1\bigg/\left|\frac{\partial f}{\partial x}\right|_{x=x_0},$$

where $f(x_0) = 0$. Here $x = E_1$ and $f(x) = E_1 + E_2 - E_a - E_b$. Hence,

$$\frac{\partial f}{\partial x} = 1 + \frac{\partial E_2}{\partial E_1}.$$

Now conservation of momentum requires, in the center-of-mass system,

$$\mathbf{p}_1^2 = \mathbf{p}_2^2;$$

that is,

$$E_1^2 - m_1^2 = E_2^2 - m_2^2.$$

Therefore,

$$\frac{\partial E_2}{\partial E_1} = \frac{E_1}{E_2},$$

and we find

$$\sigma_{\alpha\beta} = \frac{1}{2\lambda(s, m_a^2, m_b^2)} \frac{1}{(2\pi)^2} \frac{p_1}{4(E_1 + E_2)} \int d\Omega_1 \overline{\sum_a \sum_b} |\langle\beta|T|\alpha\rangle|^2. \quad (3.77)$$

Now

$$p_1 = \frac{1}{2\sqrt{s}} \lambda(s, m_1^2, m_2^2),$$

$$E_1 + E_2 = \sqrt{s}.$$

In addition, if the spin summed squared amplitude is constant, we can do the integral over the solid angle and find

$$\sigma_{\alpha\beta} = \frac{\lambda(s, m_1^2, m_2^2)}{16\pi s \lambda(s, m_a^2, m_b^2)} |T_{\alpha\beta}(s)|. \quad (3.78)$$

Where we have put

$$|T_{\alpha\beta}(s)|^2 = \overline{\sum_\alpha \sum_\beta} |\langle\beta|T|\alpha\rangle|^2.$$

For elastic scattering $\alpha \to \alpha$,

$$\sigma_{\alpha\alpha} = \frac{1}{16\pi s} |T_{\alpha\alpha}(s)|^2. \quad (3.79)$$

From Eq. (3.78) it is evident that the differential cross section is

$$\frac{d\sigma_{\alpha\beta}}{d\Omega} = \frac{1}{64\pi^2 s} \frac{\lambda(s, m_1^2, m_2^2)}{\lambda(s, m_a^2, m_b^2)} |T_{\alpha\beta}(s, \Omega)|^2, \quad (3.80)$$

which becomes, for elastic scattering,

$$\frac{d\sigma_{\alpha\alpha}}{d\Omega} = \frac{1}{64\pi^2 s} |T_{\alpha\beta}(s, \Omega)|^2 . \tag{3.81}$$

If we have a scattering amplitude $f(s, \Omega)$ [see Eq. (3.5)] defined by

$$\frac{d\sigma}{d\Omega} = |f(s, \Omega)|^2 ,$$

then

$$f_{\alpha\beta}(s, \Omega) = \frac{1}{8\pi s^{1/2}} \left[\frac{\lambda(s, m_1^2, m_2^2)}{\lambda(s, m_a^2, m_b^2)} \right]^{1/2} T_{\alpha\beta}(s, \Omega) . \tag{3.82}$$

For elastic scattering the result is

$$f_{\alpha\alpha}(s, \Omega) = \frac{T_{\alpha\alpha}(s, \Omega)}{8\pi s^{1/2}} . \tag{3.83}$$

The optical theorem [Eq. (3.16b)] becomes

$$\operatorname{Im} f_{\alpha\alpha}(s, 0) = \frac{\sigma k}{4\pi} .$$

Substituting $k = \lambda(s, m_a^2, m_b^2)/2s^{1/2}$ (i.e., the momentum in the center-of-mass system) we find

$$\operatorname{Im} T_{\alpha\alpha}(s, 0) = \lambda(s, m_a^2, m_b^2)\sigma = 2ks^{1/2}\sigma , \tag{3.84}$$

where σ is the total cross section.

3.12 Reciprocity

In this section we wish to investigate the relation between a reaction and its reverse. We shall do this through the formalism of the S-matrix, and to proceed it is necessary to anticipate the material of Chapter VI, which deals with the subject of time reversal. However, at this point we shall only state some results. Suppose we have a system represented by the wavefunction $\psi(\mathbf{x}, t)$, which is a solution of the Schrödinger equation.

$$H\psi = i\hbar \frac{\partial \psi}{\partial t} .$$

Then if H, the Hamiltonian, is real, $\psi^*(\mathbf{x}, -t)$ is a solution of the same equation and develops in the direction of reversed time in the same way as does $\psi(\mathbf{x}, t)$ in the direction of normal time. This statement requires some qualification if the system has spin or orbital angular momentum. In these cases it is necessary to reverse the direction of all the angular-momentum vectors in making the time-reversal transformation implied in changing $\psi(\mathbf{x}, t)$ to $\psi^*(\mathbf{x}, -t)$.

We now return to the S-matrix and consider a system without angular momentum having an intermediate system generated via channel α and decaying via channel β. We represent incoming waves by ψ_α. A wave of the phase and amplitude implied by ψ_α gives rise to an outgoing wave in channel β, having the phase and amplitude implied by the wavefunction ψ_β^*. The following equation defines this wavefunction:

$$\psi_\beta^* = -S_{\alpha\beta}\psi_\alpha . \tag{3.85}$$

If we multiply ψ_β^* by $S_{\alpha'\beta}^*$ and sum over β, we have

$$-\sum_\beta S_{\alpha'\beta}^* \psi_\beta^* = \sum_\beta S_{\alpha'\beta}^* S_{\alpha\beta}\psi_\alpha = \delta_{\alpha'\alpha}\psi_\alpha ;$$

hence,

$$\psi_\alpha = -\sum_\beta S_{\alpha\beta}^* \psi_\beta^* .$$

If we take the complex conjugate of this equation, we have

$$\psi_\alpha^* = -\sum_\beta S_{\alpha\beta}\psi_\beta . \tag{3.86}$$

This equation can be interpreted from our statements about time reversal. To a normal observer, ψ_α^* develops like an outgoing wave, while ψ_β develops like an incoming wave; thus Eq. (3.86) expresses a reaction initiated by waves in all channels but with phases and amplitudes such that the intermediate state decays into one channel only. This means that with an incoming wave ψ_β given by Eq. (3.85) in each respective channel, all the outgoing waves in each channel interfere destructively, except in channel α. We can express this interference mathematically. The incoming wave in channel β gives rise to an outgoing wave in channel γ ($\gamma \neq \alpha$) of $-S_{\beta\gamma}\psi_\beta$; summed over all channels, the result is zero, thus:

$$\psi_\gamma^* = -\sum_\beta S_{\beta\gamma}\psi_\beta = 0 .$$

Now the ψ_β that must satisfy this requirement and also satisfy Eq. (3.86) are given by Eq. (3.85); that is,

$$\psi_\beta = -S_{\alpha\beta}^* \psi_\alpha^* . \tag{3.87}$$

Therefore

$$\psi_\gamma^* = \sum_{\beta \neq \alpha} S_{\beta\gamma} S_{\alpha\beta}^* \psi_\alpha^* = 0 , \tag{3.88}$$

or

$$\sum_{\beta \neq \alpha} S_{\alpha\beta}^* S_{\beta\gamma} = 0 .$$

Combining this with Eq. (3.61), we have that

$$\sum_\beta S_{\alpha\beta}^* S_{\beta\gamma} = \delta_{\alpha\gamma} . \tag{3.89}$$

3.12 Reciprocity

If the quantities $S_{\kappa\lambda}$ are the elements of a matrix S, this equation implies

$$S^*S = 1. \tag{3.90}$$

But we have already shown in Eq. (3.61) that

$$SS^\dagger = 1.$$

Transposing Eq. (3.90), we have

$$\tilde{S}S^\dagger = 1;$$

hence

$$\tilde{S} = S,$$

or

$$S_{\alpha\beta} = S_{\beta\alpha}, \tag{3.91}$$

for all subscripts α and β.

Equation (3.91) states that the reaction from channel α to channel β is reversible, and thus the reaction and its reverse proceed with the same S-matrix element. However, we have restricted ourselves to systems without angular momentum. If this is not the case, then the reversed reaction proceeds with the same S-matrix element only if all the angular-momentum vectors ars reversed in direction; that is, $j_z \to -j_z$, $s_z \to -s_z$, and $l_z \to -l_z$. We can represent any channel α in the reversed state by the subscript $-\alpha$; then

$$S_{\alpha\beta} = S_{-\beta-\alpha}. \tag{3.92}$$

This is the law of reciprocity.

If the system is in an external field such as **H**, then reciprocity no longer holds, since quantities like $\boldsymbol{\sigma}\cdot\mathbf{H}$ and $\mathbf{L}\cdot\mathbf{H}$, which are likely to occur in the Hamiltonian, change sign under the time-reversal transformation and prevent the Hamiltonian from being invariant, a necessary condition in the above proof. (We ensured that the Hamiltonian was invariant by stating that it was a real quantity.)

We must consider carefully what is meant by the reversed reaction in the sense of the reciprocity law that we have just introduced. We consider the reaction

$$a + b \to c + d.$$

The channels involved are labeled α for the initial state of particles a and b and β for the final state of particles c and d. The spins and momenta of all particles are specified completely by these labels; thus channel α has momenta and spins $(\mathbf{p}_a, \mathbf{p}_b, \mathbf{s}_a, \mathbf{s}_b)$, and channel β has $(\mathbf{p}_c, \mathbf{p}_d, \mathbf{s}_c, \mathbf{s}_d)$. Let us call the reaction that proceeds from channel α to channel β the "forward reaction" and write it in a form that shows all particle states:

Forward reaction (I): $(\mathbf{p}_a, \mathbf{p}_b, \mathbf{s}_a, \mathbf{s}_b) \to (\mathbf{p}_c, \mathbf{p}_d, \mathbf{s}_c, \mathbf{s}_d)$.

The channels that have labels $-\alpha$ and $-\beta$ are the channels α and β, respectively with all linear and angular momenta reversed. The reaction that proceeds from channel $-\beta$ to channel $-\alpha$ is the time-reversed edition of the direct reaction, and we call this the "time-reversed reaction." In our convention, we have

Time-reversed reaction (II):
$$(-\mathbf{p}_c, -\mathbf{p}_d, -\mathbf{s}_c, -\mathbf{s}_d) \to (-\mathbf{p}_a, -\mathbf{p}_b, -\mathbf{s}_c, -\mathbf{s}_d).$$

The law of reciprocity connects the transition rates for two reactions related in this time-reversed manner.

We also note that since the S-matrix is related to the transition matrix by Eq. (3.64), reciprocity implies

$$|\langle -\alpha | T | -\beta \rangle| = |\langle \beta | T | \alpha \rangle|.$$

3.13 The Principle of Detailed Balance

Let us consider the same forward reaction proceeding from channel α to channel β. If first-order perturbation theory is applicable, then the transition matrix element $T_{\alpha\beta}$ is given by the proportionality

$$T_{\alpha\beta} \propto \langle \beta | H_I | \alpha \rangle,$$

where H_I is the interaction Hamiltonian. Since H_I is a Hermitian operator,

$$T_{\alpha\beta} \propto \langle \beta | H_I | \alpha \rangle = \langle \alpha | H_I | \beta \rangle^*,$$

so that

$$|T_{\alpha\beta}|^2 = |T_{\beta\alpha}|^2.$$

This is not a connection between the forward reaction and its time reverse. This equality connects transition rates for the two reactions:

Reaction I: $(\mathbf{p}_a, \mathbf{p}_b, \mathbf{s}_a, \mathbf{s}_b) \to (\mathbf{p}_c, \mathbf{p}_d, \mathbf{s}_c, \mathbf{s}_d),$

Reaction III: $(\mathbf{p}_c, \mathbf{p}_d, \mathbf{s}_c, \mathbf{s}_d) \to (\mathbf{p}_a, \mathbf{p}_b, \mathbf{s}_a, \mathbf{s}_b).$

We shall call the last reaction the "complementary reaction" in contrast to the first, which is the forward reaction. The equality of their matrix elements implied by Eq. (3.93) is called the "principle of detailed balance."

In Fig. 3.4 we have illustrated the course of events in the forward, time-reversed, and complementary reactions. In this figure the particles, a, b, c, and d are represented by heavy dots and their direction of motion by arrows. The particle spins are indicated by small labeled arrows placed close to the appropriate dots; these spins have been chosen arbitrarily so as to illustrate the relation between the three reactions, and for convenience the spin vectors are parallel to the particle momenta.

3.13 The Principle of Detailed Balance

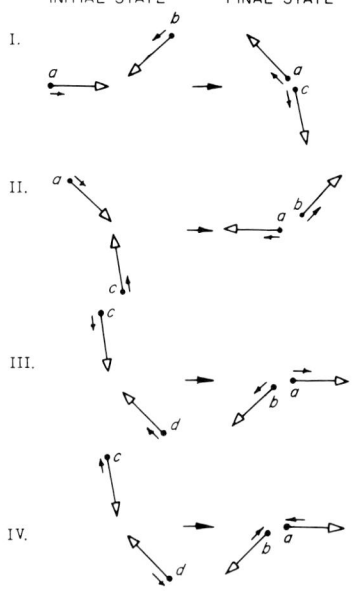

FIG. 3.4. Diagrams illustrating the course of events in the four reactions discussed in the text. The large arrows indicate the momenta and the small arrows the spins of the particles involved. The arrows between initial and final states indicate the reaction direction. I, the forward reaction; II, the time-reversed reaction; III, the complementary reaction; IV, the parity transform of the time-reversed reaction.

The interaction responsible for the reaction will normally be invariant under the parity transformation (parity conservation), and this means that the matrix element for a reaction is the same as that for the reaction that is its mirror image. We can therefore construct Reaction IV, which is the "parity transform of the time-reversed reaction," and by reciprocity and by invariance under the parity transformation we have that the matrix elements for Reactions I, II, and IV are all identical. Reaction IV is shown in Fig. 3.4 and is obtained by taking a mirror image of Reaction II and rotating the image 180° around an axis perpendicular to the mirror: this procedure reverses all linear momenta (vectors) and leaves unchanged all angular and spin momenta (axial vectors):

Reaction IV: $(\mathbf{p}_c, \mathbf{p}_d, -\mathbf{s}_c, -\mathbf{s}_d) \to (\mathbf{p}_a, \mathbf{p}_b, -\mathbf{s}_a, -\mathbf{s}_b)$.

An examination of Reactions III and IV shows that they are the same, apart from spin orientations, and we conclude that in the absence of spin all four matrix elements are identical. Hence detailed balance is correct and follows from reciprocity under those circumstances in which spin is absent and parity is conserved; this is our first conclusion about detailed balancing.

We are not normally in a position to specify precisely the spins of the particles involved in a nuclear reaction, and for unpolarized beams and targets all possible spin orientations will occur with equal probability in the initial state; thus the total reaction probability ω will be proportional to the sum over all initial and final spin states of the squared matrix elements:

$$\omega \propto \sum_{\substack{\text{initial} \\ \text{spins}}} \sum_{\substack{\text{final} \\ \text{spins}}} |T_{\alpha\beta}|^2.$$

We now make such a sum of the Reactions III and IV; since we cover all possible spin orientations, we find that for each term of the sum for one reaction there is a corresponding term in the sum for the other reaction. Hence

$$\sum \sum |T_{\text{III}}|^2 = \sum \sum |T_{\text{IV}}|^2.$$

By reciprocity and parity conservation we must have

$$\sum \sum |T_{\text{I}}|^2 = \sum \sum |T_{\text{IV}}|^2;$$

hence,

$$\sum \sum |T_{\text{III}}|^2 = \sum \sum |T_{\text{I}}|^2. \tag{3.94}$$

This relation between the Reactions I and III is sometimes called the principle of "semidetailed balance." It follows from reciprocity and invariance under the parity transformation.

At this point we must cover some points not previously discussed. Firstly, we examine in Fig. 3.4 the parity transformation that we employed to change Reaction II into Reaction IV in order to compare the former with Reaction III. In the figure it appears that Reaction II is just Reaction III rotated through 180° and that it is unnecessary to make any comparison via Reaction IV; however, this is not correct, and the fault lies in the interpretation of the two-dimensional diagram. If any spin in the forward reaction has a component along a direction perpendicular to the plane of the figures, the derived Reactions II and III are not related by a simple rotation and it is necessary to make the comparison via Reaction IV. This discussion indicates a further conclusion that can be drawn; namely, if all the particle spin vectors lie in the plane of the reaction, then detailed balancing is correct. This follows from the fact that under such circumstances Reactions II and III are related by a simple rotation. The conclusion is independent of the invariance, or otherwise, of the interaction under the parity transformation. In fact, it is impossible to produce a spin state that has the spin vector confined along a given direction or in a given plane; the correct statement is that detailed balance holds if all the polarization vectors lie in the plane of the reaction (Section 2.13). This will normally occur only in the very rare scattering reactions involving weak interactions alone (Chapter XI).

We can now derive the formula used in the application of detailed

3.13 The Principle of Detailed Balance

balance. We work entirely in the center-of-mass coordinates for the two-body reactions

$$a + b \rightleftharpoons c + d.$$

These two reactions are to be compared at the same total center-of-mass energy and in circumstances for which detailed balance is correct. The reaction cross sections are proportional to

$$\frac{\text{density of final states} \times \text{spin multiplicity in final state}}{\text{relative velocity in initial state}}.$$

The constants of proportionality are the same for both reactions under the detailed balancing conditions.

Equation (3.80) shows that the scattering cross section for the forwad reaction is

$$\frac{d\sigma_{\alpha\beta}}{d\Omega} = \frac{1}{64\pi^2 s} \frac{p_\beta}{p_\alpha} \overline{\sum_\alpha} \sum_\beta |\langle \beta | T | \alpha \rangle|^2, \tag{3.95}$$

Where s is the total center-of-mass energy and p_α and p_β are the center-of-mass momenta in the states α and β. The average over the initial spin states $\overline{\sum}_\alpha$ is the straight sum divided by the initial-state spin multiplicity $(2s_a + 1)(2s_b + 1)$, so we have

$$\frac{d\sigma_{\alpha\beta}}{d\Omega} = \frac{1}{64\pi^2 s} \frac{p_\beta}{p_\alpha} \frac{1}{(2s_a + 1)(2s_b + 1)} \sum_\alpha \sum_\beta |\langle \beta | T | \alpha \rangle|^2.$$

Similarly the cross section for the reversed reaction is

$$\frac{d\sigma_{\beta\alpha}}{d\Omega} = \frac{1}{64\pi^2 s} \frac{p_\alpha}{p_\beta} \frac{1}{(2s_c + 1)(2s_d + 1)} \sum_\alpha \sum_\beta |\langle \alpha | T | \beta \rangle|^2.$$

These formulas assume that we do not observe initial or final-state spins, and under this circumstance Eq. (3.93) permits us to relate the two differential cross sections at the same center-of-mass angle: the result is

$$(2s_a + 1)(2s_b + 1) p_\alpha^2 \frac{d\sigma_{\alpha\beta}}{d\Omega} = (2s_c + 1)(2s_d + 1) p_\beta^2 \frac{d\sigma_{\beta\alpha}}{d\Omega}. \tag{3.96}$$

Integration over the solid angle will yield a similar formula for relating total cross sections.

We must exercise care if the particles on one side of the reaction are identical. The requirements of symmetry (Section 5.2) halve the number of states available to identical particles, so that a total cross section leading to that state is halved as the final-state spin multiplicity is reduced from $(2s + 1)^2$ to $\frac{1}{2}(2s + 1)^2$. If we are comparing the differential cross sections, the detector will record particles from two supplementary reaction angles for which the cross sections must be equal; hence the factor of $\frac{1}{2}$ is canceled and the differential form of Eq. (3.85) is correct. On integrating to obtain the total cross sections we must care not to count the identical particles twice, and the result is that the reduced spin multiplicity appears in Eq. (3.85). Readers interested in an immediate application of detailed balance are referred to Section 7.3.

REFERENCES

Blatt, J. M., and Biedenharn, L. C. (1952). *Rev. Mod. Phys.* **24**, 258 (3.9).

Blatt, J. M., and Weisskopf, V. F. (1952). "Theoretical Nuclear Physics." Wiley, New York (3.2).

Breit, G., and Haracz, R. D. (1967). "High Energy Physics" (E. H. S. Burhop, ed.). Academic Press, New York (3.7).

Fermi, E. (1951). "Elementary Particles." Yale Univ. Press, New Haven, Connecticut (3.11).

Mott, N. F., and Massey, H. S. W. (1949). "The Theory of Atomic Collisions," 2nd ed. Oxford Univ. Press, London and New York (3.2).

Muirhead, H. (1965). "The Physics of Elementary Particles." Pergamon Press, New York (3.11).

Orear, J. (1954). *Phys. Rev.* **96**, 1417 (3.6).

Segré, E. (1964). "Nuclei and Particles." Benjamin, New York (3.11).

IV

ENERGY DEPENDENCE IN SCATTERING

4.1 Introduction

In this chapter, we shall consider some of the factors that contribute to energy dependences observed in scattering cross sections. We shall limit ourselves to those that can be considered on general grounds. The effective-range formalism may appear to be an exception since it is discussed with reference to the existence of a potential. However, it can be used without any reference to such a potential and in this way provides a parametrization of low-energy scattering and an insight into the relevant physics.

4.2 Phase-Space Considerations

From our discussions in Section 3.11, it is evident that from the view of the transition-rate formula, apart from the matrix-element variation, there will be a variation of cross sections due to the increased phase space that becomes available as the energy increases. This does not imply continuous increase in cross sections, since if the range of the force existing between two interacting systems is finite, the total cross section is evidently limited by geometrical effects. However, just above reaction thresholds simple considerations will indicate energy variation. Consider, for example, the endothermic reaction

$$\gamma + p \to \pi^0 + p$$

with a threshold at about 150 MeV for the incident γ ray. For the first few tens of million volts above threshold, the total cross-section variation is determined by the final-state properties. The important ones are: (1) the phase-space factor, which is roughly proportional to $\hbar k$, the final-state center-of-mass momentum (see Appendix C); (2) the angular-momentum barrier-penetration factor, which is roughly proportional to $(kR)^{2l}$ where R is the range of the interaction (see Section 4.5). Thus, this reaction is expected to have a total cross section proportional to k^{2l+1}. The fact that the cross section was observed to vary as k^3 helped to fix the dominant orbital angular momentum in the final state just above threshold as $l = 1$ (Section 13.5).

4.3 Phase Shifts at Low Energy

At low energies, there are simple relations that give the behavior of phase shifts in terms of one or more parameters. For example, Goldberger and Watson (1964) give

$$k^{2l+1} \cot \delta_l = a_0 + a_1 E + a_2 E^2 + \cdots, \tag{4.1}$$

where k is the wavenumber and E the energy of the incident particle in the center-of-mass system. A very useful form of this equation is available for low-energy S-wave scattering and is called the effective-range formula [see, for example, Segré (1964)]:

$$k \cot \delta_0 = \frac{1}{a} + \frac{1}{2} r_0 k^2 + O(k^4), \tag{4.2}$$

where a is called the zero-energy scattering length and r_0 the effective range. $O(k^4)$ means that there are other terms, but they are of order k^4 and higher and can be neglected at low energies. To understand the significance of a, it is necessary to consider very-low-energy scattering where $\lambda = 1/k \gg R$, the range of the interaction. The radial part of the wavefunction $\psi(r) = U(r)/r$, where r is the distance between the scattered particles, has the asymptotic form given by

$$U(r) \propto \sin(kr + \delta_0),$$

where δ_0 is the S-wave phase shift. In the absence of any interaction this becomes

$$U(r) \propto \sin kr.$$

Figure 4.1 shows these two functions extrapolated to the region near $r = 0$. In the case of scattering, the wavefunction is zero at a distance $r = -\delta/k$. Looking at the effective-range formula in the limit $k \to 0$ we see that

$$\delta = ak;$$

hence a is the negative of the distance at which the wavefunction becomes zero. This situation is well illustrated by the singlet and triplet neutron–proton scattering at low energy. The approximate form of the wavefunc-

4.3 Phase Shifts at Low Energy

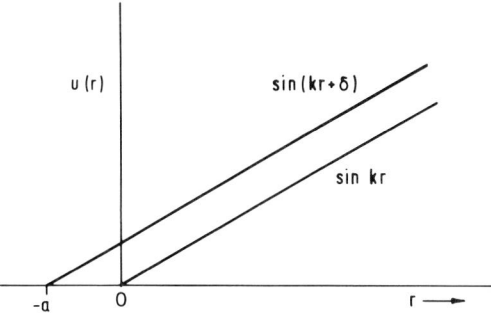

FIG. 4.1. Diagram showing the asymptotic forms of the radial wavefunctions extrapolated to $r = 0$. The phase-shifted wavefunction intercepts the axis at $r = -a$, where a is the scattering length.

tions are shown in Fig. 4.2; in the region near $r = 0$ the wavefunctions are determined by the potential, but the extrapolations of the asymptotic wavefunctions intersect the abscissa to give the zero-energy scattering lengths a_s and a_t. These asymptotic wavefunctions are deduced from the experimentally observed values of these parameters, which are [see the review by Engelke et al. (1963)]

$$a_t = -5.400 \pm 0.011 \quad \text{Fm},$$
$$a_s = +23.677 \pm 0.029 \quad \text{Fm}$$

(Fm = fermi = 10^{-13} cm). For completeness we give the effective ranges:

$$r_{0t} = 1.732 \pm 0.014 \quad \text{Fm},$$
$$r_{0s} = 2.46 \pm 0.12 \quad \text{Fm}.$$

It is remarkable that these four parameters describe low-energy n–p scattering up to incident energies of several million volts. Early attempts to derive the form of the n–p potential from a knowledge of the deuteron properties and the measured cross sections were not successful because all

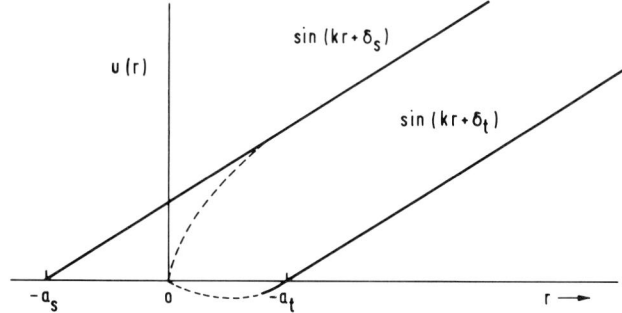

FIG. 4.2. Diagram showing the singlet and triplet neutron–proton wavefunctions in the region of $r = 0$. The actual wavefunction in the region of interaction will be of the form indicated by the broken lines.

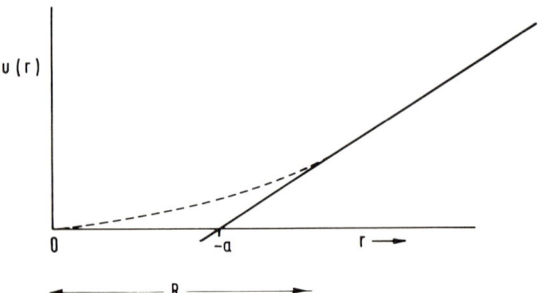

FIG. 4.3. Diagram showing the wavefunction for a square repulsive potential of range R.

likely potentials could be adjusted in their parameters, depth and range, to give the observed scattering lengths and effective ranges. We note that in fact the effective range is not related directly to the actual range of the potential, although it normally has about the same value.

The negative triplet scattering length is typical of a situation in which there is a bound state. The form of the wavefunction dictates that the asymptotic wavefunction has a zero at a distance greater than the range so that the magnitude of the scattering length is also greater than the range. This is illustrated for the n–p triplet state in Fig. 4.2. The other situation in which there is a negative scattering length is when there is a repulsive potential. The wavefunction falls off exponentially inside the potential, and it follows that matching with the asymptotic wavefunction will give a negative scattering length with a magnitude less than the range. Figure 4.3 shows the wavefunction as it would be for a square repulsive potential of range R.

The total cross section for S-wave scattering is

$$\sigma = \frac{4\pi}{k^2} \sin^2 \delta_0$$

$$= \frac{4\pi}{k^2} \frac{1}{1 + \cot^2 \delta_0}. \qquad (4.3)$$

In the limit $k \to 0$ this becomes, using Eq. (4.2)

$$\sigma = 4\pi a^2.$$

This applies to spin-independent scattering. Neutron–proton scattering takes place in the two spin states with multiplicities that give, in the low-energy limit,

$$\sigma = 4\pi(\tfrac{3}{4}|a_\text{t}|^2 + \tfrac{1}{4}|a_\text{s}|^2). \qquad (4.4)$$

4.3 Phase Shifts at Low Energy

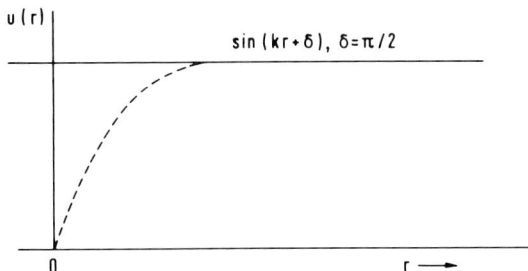

FIG. 4.4. Diagram showing the radial wavefunction near $r = 0$ for the case of a resonance, when $\delta = 90°$ and $\alpha = \infty$.

At energies away from zero we can define an S-wave scattering length α by

$$k \cot \delta_0 = \frac{1}{\alpha}, \quad (4.5)$$

in which case

$$\left| \frac{e^{2i\delta_0} - 1}{2ik} \right|^2 = \frac{\alpha^2}{1 + k^2\alpha^2},$$

and hence

$$\sigma = \frac{4\pi\alpha^2}{1 + k^2\alpha^2}. \quad (4.6)$$

As the energy increases, α will change from the zero-energy scattering length. If δ increases from zero α will increase, and if there is a resonance ($\delta = 90°$, see next section) it will become infinite, as illustrated in Fig. 4.4. For $\delta > 90°$ the scattering length becomes negative, decreasing in magnitude from $|-\infty|$ as δ increases.

An important example of S-wave scattering involves K^- and $\bar{K}^0$ mesons on nucleons, at low energies. This illustrates the complication caused by absorption, in this case into channels such as

$$K^- + p \to \Sigma^0 + \pi^0,$$
$$\bar{K}^0 + p \to \Sigma^+ + \pi^0,$$

and so on. There is an additional complication caused by the presence of two isotopic-spin channels in the K^-p system (see Section 8.7). This we can avoid by indicating how S-wave $\bar{K}^0p$ scattering would be treated. This is entirely isotopic spin 1, although we should note that this scattering cannot be observed without complication (see Sections 7.5 and 8.7). We have already included the possibility of absorption in scattering by making the phase shift complex, so that the elastic scattering amplitude can be expressed in terms of two real parameters η and δ:

$$A = \frac{\eta \exp(2i\delta) - 1}{2ik}.$$

The scattering length defined by Eq. (4.5) can likewise be made complex to give a complex phase shift in this equation. Extracting the real parameters we find

$$2\delta = \tan^{-1}\frac{2\,\text{Re}(k\alpha)}{1 - k^2|\alpha|^2}, \tag{4.7}$$

$$\eta = \left|\frac{1 + ik\alpha}{1 - ik\alpha}\right|, \tag{4.8}$$

and the elastic scattering amplitude is

$$A = \frac{\eta e^{2i\delta} - 1}{2ik} = \frac{\alpha}{1 - ik\alpha}. \tag{4.8a}$$

We can put

$$\alpha = a + ib,$$

where a and b are real. The total elastic scattering cross section is then [using the S-wave term from Eq. (3.12)]

$$\sigma_s = 4\pi|A|^2 = \frac{4\pi(a^2 + b^2)}{(1 + kb)^2 + k^2 a^2}, \tag{4.9}$$

and the reaction cross section [Eq. (3.13)]

$$\sigma_r = \frac{\pi}{k^2}(1 - \eta^2) = \frac{4\pi}{k}\frac{b}{(1 + kb)^2 + k^2 a^2}. \tag{4.10}$$

As $k \to 0$, σ_s approaches a constant, $4\pi(a^2 + b^2)$, but σ_r increases as $1/k$, or more familiarly, as $1/v$, the typical reaction-cross-section variation in circumstances of strong absorption, where v is the incident-particle velocity.

The effective-range formula can be applied to scattering in orbital momentum states other than S-wave if $k \cot \delta_0$ is replaced by $k^{2l+1} \cot \delta_l$. The Chew–Low formula for pion–nucleon scattering in the $TJ^P = \frac{3}{2}, \frac{3}{2}^+$ state is an example (Section 8.4).

4.4 The Wigner Condition

By considering the behavior of a wave packet undergoing scattering it is possible to put an upper limit on the magnitude of the rate decrease of phase shift with k (Wigner, 1955; Goldberger and Watson, 1964). To do

4.4 The Wigner Condition

this we shall consider spherical waves: to construct a wave packet the incoming and outgoing waves must be superpositions of infinite waves of different k. To avoid complication we can consider two infinite waves of wave number k, $k + \Delta k$, and observe the velocity of one point at which the waves interfere constructively. This "group velocity" is the velocity of a wave packet and is distinct from the "phase velocity" of the infinite waves of fixed k. The two incoming waves k and $k + \Delta k$, combine thus:

$$\psi(r, t) = \exp(-i[kr + \omega t]) + \exp(-i[(k + \Delta k)r + (\omega + \Delta \omega)t])$$
$$= \exp(-i[kr + \omega t])\{1 + \exp(-i[r \Delta k + t \Delta \omega])\}.$$

This represents a modulated incoming spherical wave. The maximum in the modulation travels so that

$$r \Delta k + t \Delta \omega = 0. \tag{4.11}$$

Hence $r = -t(\Delta \omega/\Delta k)$ and the group velocity is

$$\lim_{\Delta k \to 0} \frac{\Delta \omega}{\Delta k} = \frac{\partial \omega}{\partial k}.$$

The maximum would be at $r = 0$ at a time $t = 0$, and if the scattering center has a potential of range R the maximum will reach the potential at time $t = -R/(\partial \omega/\partial k)$.

Each of the two incoming infinite waves gives rise to outgoing scattered waves, which combine thus:

$$\psi(r, t) = \exp i(kr - \omega t + 2\delta)$$
$$+ \exp i([k + \Delta k]r - [\omega + \Delta \omega]t + 2[\delta + \Delta \delta])$$
$$= \exp i(kr - \omega t + 2\delta)[1 + \exp i(r \Delta k - t \Delta \omega + 2 \Delta \delta)].$$

The maximum travels so that

$$r \Delta k - t \Delta \omega + 2 \Delta \delta = 0, \tag{4.12}$$

that is,

$$r = \frac{t \Delta \omega}{\Delta k} - \frac{2 \Delta \delta}{\Delta k}.$$

We can think of Eqs. (4.11) and (4.12) as representing the equations of motion of the incoming and outgoing wave packets, so by causality the outgoing wave packet certainly cannot be at $r = R$ before the incoming wave packet has reached $r = R$; hence at $t = -R/(\Delta \omega/\Delta k)$ the outgoing wave packet must have $r \leqslant R$. That is,

$$R \geqslant -R - \frac{2 \Delta \delta}{\Delta k};$$

hence
$$d\delta/dk \gtrsim -R. \quad (4.13)$$

Thus the rate of change of δ with k is lower limited. Consider, for example, the case of $\pi^+ p$ scattering where the range of interaction is about 10^{-13} cm.

$$d\delta/dk > -10^{-13} \text{ cm} = -10^{-13}\hbar/(1.97 \times 10^{-11}) \text{ radians/MeV}/c,$$

or

$$\frac{1}{\hbar}\frac{d\delta}{dk} \gtrsim -0.3 \text{ degrees/MeV}/c.$$

Straddling the $\frac{3}{2}$, $\frac{3}{2}$ resonance by a 120 → 260 MeV incident energy range we find the $\hbar k$ changes about 100 MeV/c so that $\Delta\delta \gtrsim -30°$. In fact, over this range the phase shift must change by about $\pm 90°$, so that δ must be increasing (+) and not decreasing with energy through this resonance. An example of a decreasing phase shift passing through $\delta = 90°$ occurs in low-energy neutron–proton scattering.

4.5 Resonance and the Breit–Wigner Formula

Figure 4.5 shows the variation with energy of the total cross section of π^+ mesons on protons. At 190 MeV the cross section reaches $8\pi/k^2$, the value expected if the elastic scattering is dominated by one state of total angular momentum $j = \frac{3}{2}$ [Eq. (3.28a): $\sigma = 4\pi\lambda^2(j + \frac{1}{2})$] with a phase shift passing through 90° and no inelastic channels. The shape of the cross-section curve above and below this energy indicates that δ passes through 90°, where the scattering amplitude is a maximum, with $d\delta/dk$ positive; in this situation the system is said to be resonant, and in this par-

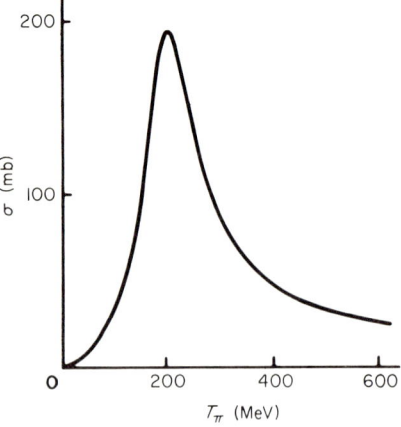

FIG. 4.5. The total cross section for the π^+-meson on protons as a function of the incident-pion kinetic energy.

4.5 Resonance and the Breit-Wigner Formula

ticular case the resonant state is one having $j = \frac{3}{2}$. Similar resonances of various widths occur in elementary-particle physics, and the view taken is that they are members of the elementry-particle spectrum and that each width reflects the lifetime of the particle. If Γ (MeV) is the full width at half height of a resonance, then the uncertainty principle gives a mean life τ of

$$\tau \simeq (6.6 \times 10^{-22})/\Gamma \quad \text{sec}. \qquad (4.14)$$

The $j = \frac{3}{2}$, $\pi^+ p$ resonance has $\Gamma \simeq 120$ MeV, so that $\tau \simeq 5.5 \times 10^{-24}$ sec.

One formula for the energy variation of the scattering amplitude is that due to Breit and Wigner (see Segré, 1964). Suppose we take the zero-effective-range formula:

$$\cot \delta_l = 1/ak^{2l+1} + \cdots = 1/F(k).$$

If $\delta = \pi/2$, $F(k) = \infty$ and we can expand the variation of $1/F(k)$ about this point ($1/F = 0$) by a Taylor series:

$$1/F = 0 + (E_0 - E)(d/dE)1/F_{E_0}.$$

If we define

$$(d/dE)(1/F)_{E_0} = 2/\Gamma,$$

then

$$\cot \delta_l = (E_0 - E)(2/\Gamma) \equiv \epsilon.$$

Now the scattering amplitude is given by

$$Ak = \frac{\exp 2i\delta_l - 1}{2i} = \frac{1}{\cot \delta_l - i}.$$

Therefore,

$$Ak = \frac{\Gamma/2}{(E_0 - E) - i\Gamma/2}. \qquad (4.15)$$

The total elastic cross section in the case when the scattering takes place only in the resonant state is then given by

$$\sigma_s = \pi \lambda^2 \frac{(2j+1)}{(2s_1+1)(2s_2+1)} \frac{\Gamma^2}{(E_0 - E)^2 + \Gamma^2/4}, \qquad (4.16)$$

where j is the spin of the resonant state and s_1 and s_2 are spins of the incident and target particles. The factor containing the spins comes from the statistical weight of the resonant state and an average over initial spin states: for 0-$\frac{1}{2}$ scattering the factor reduces to $j + \frac{1}{2}$, and the expression for σ_s to that of Eq. (3.27). The elastic cross section is a maximum, $4\pi\lambda^2(2j+1)/(2s_1+1)(2s_2+1)$, at $E = E_0$, and it has half this value at energies E_1, E_2 such that $E_1 - E_2 = \Gamma$, which is therefore the full width at half height.

The energies E_0 and E refer to the total energy of the resonant system; in $\pi^+ p$ scattering, for example, this is the total center-of-mass energy, and at the $j = \frac{3}{2}$ resonance has the value 1236 MeV. If the resonant state has several channels through which it can decay, one of these is same as the incident channel and gives elastic scattering, while the remainder will contribute to particle reactions. If the partial widths are Γ_s, Γ_r, respectively, we have

$$\Gamma = \Gamma_s + \Gamma_r,$$

and putting g for the spin factor we have

$$\sigma_s = \pi \lambdabar^2 g \frac{\Gamma_s^2}{(E_0 - E)^2 + \Gamma^2/4}$$

$$\sigma_r = \pi \lambdabar^2 g \frac{\Gamma_s \Gamma_r}{(E_0 - E)^2 + \Gamma^2/4}.$$

If we put $x = \Gamma_s/\Gamma$, called the elasticity, then the scattering amplitude is given by

$$Ak = \frac{x}{\epsilon - i} = \frac{(\epsilon + i)x}{\epsilon^2 + 1}$$

and we have

$$\sigma_s = \frac{4\pi \lambdabar^2 g x^2}{\epsilon^2 + 1}, \qquad (4.17)$$

$$\sigma_r = \frac{4\pi \lambdabar^2 g x(1 - x)}{\epsilon^2 + 1}, \qquad (4.18)$$

and the total cross section is

$$\sigma = \frac{4\pi \lambdabar^2 g x}{\epsilon^2 + 1}, \qquad (4.19)$$

where $\epsilon = 2(E_0 - E)/\Gamma$.

The parameter x has physical significance which follows from the meaning of the partial widths. Γ_s is the width the resonance would have if the elastic channel were the only one open and is a measure of the probability of decay into that channel per unit time. Thus, if other channels are open, the total width Γ is greater than Γ_s and the ratio $x = \Gamma_s/\Gamma$ is the probability that, once formed, the resonance decays by the elastic channel. The occurrence of the factors x^2, $x(1 - x)$, x in Eqs. (4.17)–(4.19) can be understood. For elastic scattering the resonance is approached and left by the elastic channel, each action involving a probability x, thus giving x^2. The reaction cross section involves approach by the elastic channel (factor x) and exit by any other (factor $1 - x$). The total involves approach by the elastic (factor x) and exit by all (factor 1).

4.5 Resonance and the Breit-Wigner Formula

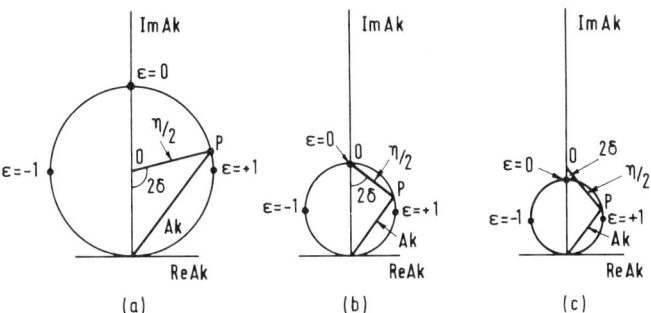

FIG. 4.6. Argand diagrams of the scattering amplitude Ak (point P) due to a single resonance. The diagrams show the connection between the parameters x and ϵ and the parameters δ and η. The circle has diameter x and as the energy increases (ϵ decreasing) is traversed counterclockwise. The point O has coordinates $(0, i/2)$. (a) $x = 1$; (b) $x = \frac{1}{2}$; (c) $x < \frac{1}{2}$.

Consider now the energy dependence of Ak. In the case of elastic scattering only ($x = 1$), Ak plotted on an Argand diagram lies on a circle, radius $\frac{1}{2}$ (Fig. 4.6a). As ϵ decreases from $+\infty$ to $-\infty$, Ak increases, changing most rapidly as ϵ goes from $+1$ to -1, thereafter decreasing slowly. At the resonant energy, Ak is purely imaginary and of unit magnitude.

When $x = 1$ we have that the absorption parameter $\eta = 1$. Since η must always be less than or equal to one to conserve matter, the circle traced out by Ak as energy varies defines the limits on this amplitude. Ak must lie on this "unitary" circle for purely elastic scattering and inside the circle ($x < 1$, $\eta < 1$) if there is absorption. In this case if x is independent of energy, Ak will trace out a circle of radius $x/2$, center $ix/2$, in the same manner as it does when $x = 1$ (Figs. 4.6b and c). Figure 4.7 shows the energy variation of the real and imaginary parts of Ak for elastic scattering as functions of energy, passing through a resonance with with $x = 1$.

The parametrization of a resonance in terms of a phase shift δ and an absorption parameter η is not so useful, as can be seen from Fig. 4.6. For $x = 1$ and $\eta = 1$, δ passes through $90°$ at $\epsilon = 0$. If $\frac{1}{2} < x < 1$, then δ still passes through $90°$ at resonance, but for $x < \frac{1}{2}$ it is decreasing through $0°$ at resonance. It is also obvious that η varies in a rapid manner as the resonance is traversed and that a given η can correspond to two values of x, namely $x = \frac{1}{2}(1 \pm \eta)$ at resonance.

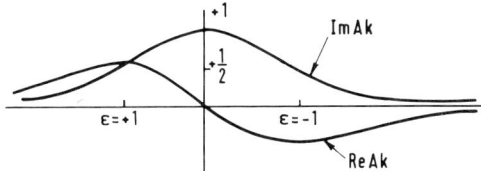

FIG. 4.7. The energy dependence of the real and imaginary parts of the one-channel Breit-Wigner resonant amplitude.

In discussing the energy variation we have neglected the possibility that the parameters x and Γ vary with energy. This, in fact, is bound to occur, since as E increases the phase space available to final-state particles increases and it becomes easier to penetrate angular-momentum and Coulomb barriers. Consider a resonance with $x = 1$ so that the final state is the same as the initial state; the phase space is proportional to final-state momentum and hence is proportional to k (nonrelativistic). An angular-momentum barrier l will introduce a factor

$$(kR)^{2l}/D_l \qquad (4.20)$$

(Blatt and Weisskopf, 1952), where R is a radius of interaction and D_l are approximately

$$D_0 = 1,$$
$$D_1 = 1 + (kr)^2,$$
$$D_2 = 9 + 3(kR)^2 + (kR)^4, \quad \text{etc.}$$

So approximately,

$$\Gamma = \gamma[(kR)^{2l+1}/D_l]$$

and γ will contain the dynamic factors relevant to the internal structure of the resonant state. Such variation in Γ can cause a resonance to have an asymmetric shape, with the high-energy half-height point farther from the resonance maximum than the half-height point on the low-energy side. However, the values of x do not seem to vary strongly, presumably since all the partial widths are increasing, if not exactly *pro rata*, approximately so in the resonance region. It follows that in the resonance region Ak will lie on a curve in the Argand diagram that is very close to the upper half of a circle.

A resonance may occur in the presence of a nonresonant background of the same quantum numbers: this situation is indicated in Fig. 4.8. OX represents the slowly varying background scattering amplitude. XY represents the amplitude due to a resonance. As the resonance energy is approached, Y moves around a circle to the point Y_0 at which the scattering amplitude is OY_0. Notice that, at the resonance, the phase shift δ is not $\pi/2$ and may never reach $\pi/2$ if the resonant circle does not enclose C.

In addition, the absorption parameter will vary in a manner quite different from the case of a resonance without background. Note that Y will not trace out a circle if the nonresonant background amplitude is varying or if the elasticity of the resonance is changing. It is important to note that there must be a relative phase between the background (phase δ_B) and the resonant amplitude (phase δ_R) so that the resonant phase reaches $90°$ at an energy (Y_0 in Fig. 4.8) that is different from the point at which the cross section is a maximum (Y_M). That such a phase difference must exist

4.5 Resonance and the Breit-Wigner Formula

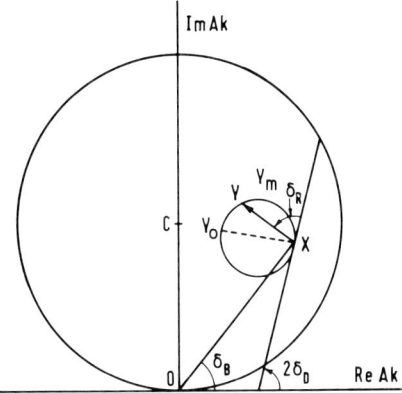

FIG. 4.8. Amplitude diagram showing the addition of resonant and non-resonant amplitudes.

follows from unitarity and can be seen from a simple example: consider a two-body scattering in one channel in which there is a background amplitude $A_B = \exp(i\delta_B)\sin\delta_B/k$ varying slowly with energy. On the amplitude diagram, this is represented by a point moving slowly counterclockwise around the unitary circle. Suppose that the energy changes so that the system crosses a narrow resonance with amplitude

$$A_R = \exp(i\delta_R)\sin\delta_R/k,$$

where δ_R varies rapidly from 0 to 180°. The total amplitude a cannot be the sum of these, as A_R would make an excursion outside the unitary circle, which is forbidden. In fact, we must have

$$A = A_B + A_R \exp(2i\delta_D). \tag{4.21}$$

In this simple case, the only possible value for δ_D is δ_B, which follows from the geometry of the unitary circle. The same is true if the background is entirely elastic and the resonance is inelastic (i.e., $x < 1$). If both are inelastic the situation with respect to δ_D is not so clear. Dalitz (1963) discusses the problem by using the K-matrix formula and Michael (1966) by requiring that each channel amplitude satisfy unitary requirements. The S_{11} amplitude in the pion-nucleon scattering phase-shift solution of Bareyre et al. (1965) is a good example of such behavior.

Jackson (1964) has given relativistic formulas for the line shape of resonances produced under different conditions. In discussing the Breit-Wigner formula above we assumed that it was observed in a two-body collision either in the elastic channel or in an inelastic channel. This situation is that of the "formation" of a resonance. At the "production" of a resonance it is observed produced along with one or more other particles. Thus the formation of the $\Delta(1236)$ is observed in pion-proton elastic scattering at 200 MeV,

$$\pi^+ + p \to \Delta^{++} \to \pi^+ + p,$$

while the production of the Δ is observed in inelastic K^+p scattering at momenta about 1.0 GeV/c:

$$K^+ + p \to K^0 + \Delta^{++}$$
$$\downarrow$$
$$\pi^+ + p. \qquad (4.22)$$

The formation cross section given by Jackson is

$$\sigma(E) = 4\pi\lambda^2 \frac{(2j+1)}{(2s_1+1)(2s_2+1)} \frac{E_0^2 \, \Gamma_s^2(E)}{(E^2 - E_0^2)^2 + E_0^2 \, \Gamma^2(E)}. \qquad (4.23)$$

This corresponds to Eq. (4.16). We have emphasized the energy variations of Γ by writing $\Gamma(E)$, which will be discussed shortly. To observe a resonance in production it is necessary to observe the distribution of the values of the invariant mass of the particles believed to be resonating [π^+p in Eq. (4.22)]. In the case of a resonance observed decaying into two particles the distribution will deviate from that expected by phase-space arguments (Appendix C) by a factor

$$\phi(E) = C \frac{E}{p} \frac{\Gamma_s(E)}{(E^2 - E_0^2)^2 + E_0^2 \, \Gamma^2(E)}, \qquad (4.24)$$

where p is the three-momentum of each decay product in the rest frame of the resonance. C is a factor that depends upon the details of the production mechanism and is normalized when fitting to observed data.

The energy variation of $\Gamma(E)$ and of $\Gamma_s(E)$ can lead to a shifting of the maximum in the resonant peak away from E_0. The variation $\Gamma_s \propto (kR)^{2l+1}$ shifts the maximum to an energy below E_0. This can be seen by considering the behavior of Eqs. (4.23) and (4.24) near resonance when $E^2 - E_0^2 \ll E_0^2 \, \Gamma^2$. Then Eq. (4.23) is varying as λ^2, which is decreasing with increasing E, and Eq. (4.24) is varying roughly as $1/\Gamma$, which is also decreasing. Thus, if there is a maximum, it must occur for $E < E_0$ in both cases. Jackson (1964) gives theoretical and empirical formulas for the variation of $\Gamma_s(E)$ for various common combinations of spin and parity of resonance and decay products.

4.6 Unitarity

The conservation of matter leads to the S-matrix being unitary (Section 3.9), and this has the effect of placing restrictions upon the amplitudes in scattering and reactions when more than one channel is involved. We shall not derive the equations that express these restrictions but refer readers to Källen (1964) or Gasiorowicz (1966).

4.6 Unitarity

However, we shall briefly consider the permitted variation of reaction amplitudes, which is influenced by unitarity. For simplicity we consider a two-channel situation, elastic scattering and one reaction channel. The final-state amplitudes (A for scattering and T for the reaction) are given by (cf. Eqs. (3.60) and (3.60a)]

$$Ak = (S_{\alpha\alpha} - 1)/2i = (\eta e^{2i\delta} - 1)/2i,$$
$$Tk = S_{\alpha\beta}/2i = \eta' e^{2i\delta'}/2i,$$

where k is the incident-channel (α) wave number, and $\alpha\alpha$ refers to elastic scattering and $\alpha\beta$ to the reaction. Equation (3.60b) tells us that

$$\eta^2 + \eta'^2 = 1. \qquad (4.25)$$

Thus the reaction amplitudes Tk plotted on an Argand diagram is restricted to lie on or inside a circle of radius $\frac{1}{2}$, center at the origin. If Tk lies on this circle, then $\eta' = 1$ and it follows that $\eta = 0$ and Ak is purely imaginary with value $+i/2$. This corresponds to complete absorption and the necessary presence of diffraction scattering. These considerations can be applied to more channels, and it is clear that each reaction amplitude becomes more restricted as unity has to be spread among more η's [Eq. (4.25)].

Now let us consider a resonance: the amplitudes for scattering and reaction in the two-channel case have energy variations given by [Eqs. (4.17)–(4.19)]

$$Ak = \frac{x}{\epsilon - i},$$

$$Tk = \frac{[x(1-x)]^{1/2}}{\epsilon - i}.$$

As Ak traces out a circle of radius $x/2$, center $ix/2$, then Tk traces out a circle of radius $[x(1-x)]^{1/2}/2$, center $i[x(1-x)]^{1/2}/2$. For more than two channels with fractional widths x_α, x_β, where α is the entry channel, the circles for the inelastic channels obviously have radius $(x_\alpha x_\beta)^{1/2}$, center $i(x_\alpha x_\beta)^{1/2}/2$, etc. In the two-channel case the greatest reaction amplitude at any energy occurs when $x = \frac{1}{2}$, and in this case when $\epsilon = 0$ (at resonance) the reaction amplitude reaches the limit allowed by the circle discussed in the previous paragraph and the elastic amplitude becomes entirely imaginary.

REFERENCES

Bareyre, P., Brickman, C., Stirling, A. V., and Villet, G. (1965). *Phys. Lett.* **18**, 342 (4.5).
Blatt, J. M., and Weisskopf, V. F. (1952). "Theoretical Nuclear Physics." Wiley, New York (4.5).

Dalitz, R. H. (1963). *Ann. Rev. Nucl. Sci.* **13**, 339 (4.4).
Engelke, C. E., Benenson, R. E., Melkonian, E., and Lebowitz, J. M. (1963). *Phys. Rev.* **129**, 324 (4.3).
Gasiorowicz, S. (1966). "Elementary Particle Physics." Wiley, New York (4.6).
Goldberger, M. L., and Watson, K. M. (1964). "Collision Theory." Wiley, New York (4.3, 4.4).
Jackson, J. D. (1964). *Nuovo Cimento* **34**, 1644 (4.5).
Källen, G. (1964). "Elementary Particle Physics." Addison-Wesley, Reading, Massachusetts (4.6).
Michael, C. (1966). *Phys. Lett.* **21**, 93 (4.5).
Segré, E. (1964). "Nuclei and Particles." Benjamin, New York (4.3, 4.5).
Wigner, E. P. (1955). *Phys. Rev.* **98**, 145 (4.4).

V

SYMMETRY, ISOTOPIC SPIN, AND HYPERCHARGE

5.1 Introduction

In Chapter II, we discussed at length the implications of the invariance of nature under rotations in ordinary space; that is, conservation of angular momentum. There is another important invariance that strong interactions possess, namely, invariance under rotations in isotopic-spin space or "charge independence" as it is called. We shall introduce this concept by considering the symmetry of two-nucleon states and show that it can be extended to other particles. The chapter concludes by introducing strangeness and hypercharge.

5.2 Symmetry and Antisymmetry

The theory of atomic structure and of the periodic table invokes the exclusion principle due to Pauli, which states that no physical state may be occupied by more than one electron; the description of the state includes all the quantum numbers, including spin. The property expressed by this principle can be shown to be related to the properties of the electron–positron field. Pauli was able to show that physical inconsistencies would occur unless the field operators satisfied various commutation or anticommutation laws. In the case of particles with a spin that is an odd number of units of $\hbar/2$, the rules lead immediately to the exclusion principle; such particles include electrons and baryons, and they are given the generic name Fermi–Dirac particles or fermions since they must obey the Fermi–Dirac statistics. In the case of particles having a spin that is zero or an even number of units of $\hbar/2$, the commutation rules do not lead to an exclusion principle and the particles obey Bose-Einstein statistics; they are called bosons.

It is well known that in wave mechanics the exclusion principle implies that the wavefunction for an assembly of like fermions must be antisymmetric under exchange of any two of the fermions and that the wavefunction for an assembly of like bosons must be symmetric under exchange of any two of the bosons. The fact that wavefunctions must have one or the other property follows from the exact indentity of the particles involved. In the bra-ket notation we have to put it in the following way. Let us consider two particles, 1 and 2, in a state given by the state vector $|1, 2\rangle$. This implies that the particle with label 1 is in a state given by the position occupied by 1 within the ket, and the particle with label 2 in another state specified by that position in the ket. The state $|2, 1\rangle$ has the particles interchanged. We wish to consider the amplitude for finding one particle in a position that, along with its momentum and spin, is specified by the label a and another particle with a momentum and spin that is labeled by b. This amplitude for finding particle 1 in a and 2 in b is

$$\langle a, b | 1, 2 \rangle .$$

The amplitude for finding 2 in a and 1 in b will be

$$\langle a, b | 2, 1 \rangle .$$

Since the particles are identical, we must have

$$|\langle a, b | 2, 1 \rangle|^2 = |\langle a, b | 1, 2 \rangle|^2 ,$$

which leads to

$$\langle a, b | 2, 1 \rangle = e^{i\alpha} \langle a, b | 1, 2 \rangle .$$

Thus, the effect of exchanging the particles is to multiply the amplitude by $e^{i\alpha}$. If we exchange twice, we must return to the original amplitude so that $e^{2i\alpha} = 1$ and $e^{i\alpha} = \pm 1$. The amplitude for finding either particle in a and the other in b is the sum of these two amplitudes, namely,

$$\langle a, b | 1, 2 \rangle \pm \langle a, b | 2, 1 \rangle . \tag{5.1}$$

If the two states occupied by the particle are the same—i.e., $|1, 2\rangle = |2, 1\rangle$—and the sign is negative, then the amplitude is zero and the situation is impossible. This is just the exclusion-principle case, so we conclude that the sign is minus for fermions and plus for bosons.

In writing amplitudes for situations involving identical particles it is necessary to have a total amplitude that, in the case of fermions, changes sign under the exchange of any two identical particles or, in the case of two bosons, does not change sign. Equation (5.1) satisfies these requirements for two particles. State vectors for the systems must be odd or even (antisymmetric or symmetric) under the exchange of identical fermions or bosons. Thus, a state of two fermions could be represented by

$$|\phi\rangle = |1, 2\rangle - |2, 1\rangle. \qquad (5.2)$$

5.3 Two-Nucleon State Vectors

We can now construct the state vectors for a system of two identical fermions. For simplicity, we can consider two protons, which are fermions with spin $\hbar/2$; they must therefore have an antisymmetric total state vector. We have previously used orbital and spin state vectors in conjunction to describe completely the state of a particle; for example, Eq. (2.59). We can construct a similar state vector $|\phi\rangle$ to describe the two protons. This total state vector is the product of the orbital state vector $|\psi\rangle$ and the spin state vector $|\chi\rangle$:

$$|\phi\rangle = |\psi\rangle|\chi\rangle.$$

The $|\phi\rangle$ must be antisymmetric under the interchange of the two protons; thus if $|\psi\rangle$ is symmetric, then $|\chi\rangle$ must be antisymmetric and vice versa. Let us construct the physically possible spin state vectors:

$$|1, 1\rangle = |\tfrac{1}{2}, \tfrac{1}{2}\rangle_a |\tfrac{1}{2}, \tfrac{1}{2}\rangle_b, \qquad (5.3)$$

$$|1, 0\rangle = \sqrt{\tfrac{1}{2}} \{|\tfrac{1}{2}, \tfrac{1}{2}\rangle_a |\tfrac{1}{2}, -\tfrac{1}{2}\rangle_b + |\tfrac{1}{2}, -\tfrac{1}{2}\rangle_a |\tfrac{1}{2}, \tfrac{1}{2}\rangle_b\}, \qquad (5.4)$$

$$|1, -1\rangle = |\tfrac{1}{2}, -\tfrac{1}{2}\rangle_a |\tfrac{1}{2}, -\tfrac{1}{2}\rangle_b, \qquad (5.5)$$

$$|0, 0\rangle = \sqrt{\tfrac{1}{2}} \{|\tfrac{1}{2}, \tfrac{1}{2}\rangle_a |\tfrac{1}{2}, -\tfrac{1}{2}\rangle_b - |\tfrac{1}{2}, -\tfrac{1}{2}\rangle_a |\tfrac{1}{2}, \tfrac{1}{2}\rangle_b\}. \qquad (5.6)$$

The notation on the right is such that the state-vector kets contain the quantum numbers associated with the operators S^2 and S_z for the particle indicated by the letter in the subscript. The notation on the left will be clarified shortly. These four vectors satisfy the symmetry requirements; vectors (5.3)–(5.5) are symmetric and (5.6) is antisymmetric. If we apply our knowledge of angular momentum, we discover that the quantum numbers of the total spin operator S^2 are 1, 1, 1, and 0, respectively, and that the quantum numbers of S_z are 1, 0, -1, and 0, respectively. This explains the figures in the kets on the left of these equations: $|s, s_z\rangle$. The state vectors (5.3)–(5.5) are the three vectors describing the three "triplet" spin states, which have $s = 1$; vector (5.6) is the state vector describing the "singlet" spin state, which has $s = 0$. Returning to the business of symmetry, we observe that in order to maintain overall antisymmetry if the protons are in the singlet spin state, they must be in a symmetric orbital state, whereas if they are in a triplet spin state, their orbital state must be antisymmetric.

We can examine now the symmetry of the orbital states. Any such state will be a superposition of states of different orbital angular momentum

l. The symmetry will be that of the corresponding spherical harmonic function:

$$Y_l^{l_z}(\theta, \phi) \propto \exp(il_z\phi) P_l^{|l_z|}(\cos\theta).$$

The transformation that interchanges the particles is that which reverses the direction of the vector that joins them. Under this change the polar coordinates of the vector change $\theta \to \pi - \theta$ and $\phi \to \phi - \pi$. Then we have

$$Y_l^{|l_z|}(\theta, \phi) \to Y_l^{|l_z|}(\pi - \theta, \phi - \pi) = (-1)^l Y_l^{|l_z|}(\theta, \phi).$$

Thus the states with l even are symmetric and those with l odd are antisymmetric.

The final effect on the two-proton system can now be stated. If the two protons are in the singlet spin state, then their orbital state can only be one that contains states with l even—that is, in spectroscopic notation, the ^1S, ^1D, ^1G, etc., states; the superscript 1 indicates the total spin degeneracy, which is one in these singlet spin states. On the other hand, if the two protons are in the triplet spin state, only states with l odd are allowed; these states are ^3P, ^3F, ..., etc. Again, the superscript indicates the spin degeneracy, triplet in this case. Thus any two-proton state can only be a linear superposition of the series of states that commences

$$^1S, {}^3P, {}^1D, {}^3F, {}^1G, {}^3H, \ldots, \text{ etc.}$$

The case of two neutrons is exactly the same; the case of a neutron and proton is different in that the particles are different and there are no symmetry requirements; thus, the n-p system can be a linear superposition of the series of states

$$^1S, {}^3S, {}^1P, {}^3P, {}^1D, {}^3D, {}^1F, {}^3F, {}^1G, {}^3G, \ldots, \text{ etc.}$$

The vector addition of the total spin angular momentum with the orbital angular momentum to find the total angular momentum and the eigenstates present can be done by the means described in Chapter II. Thus the ^3P state can have $j = 2, 1, 0$, and this state is therefore a linear superposition of the three corresponding eigenvectors. The ^1P state can have $j = 1$ only.

The case of more than two identical fermions becomes more complicated but is still governed by the total antisymmetry requirement under the exchange of any two particles.

5.4 Isotopic Spin

We are now in a position to consider a concept that simplifies further the construction of nucleon–nucleon state vectors. An examination of the

5.4 Isotopic Spin

energy levels of mirror nuclei, e.g., ^{13}C and ^{13}N, indicates similarities that could be explained by assuming that the force between two neutrons is identical to that between two protons if the Coulomb force is neglected; this is the postulate of charge symmetry. An extension of this, the postulate of charge independence, states that the force between any two nucleons depends only on their state of spin and relative orbital angular momentum. Thus the force between two protons in the ^1S state is expected to be identical, apart from Coulomb forces, to that between a neutron and proton or to that between two neutrons in the same state. The first equality is found to be true from an analysis of low-energy p–p and n–p scattering measurements. The consequences of charge independence include all the consequences of charge symmetry and others not given by that postulate.

A neutron and a proton are supposed to be two manifestations of the same particle, the nucleon. The nondegeneracy that makes them different particles is assumed to disappear if we switch off the Coulomb field. Since the system is a doublet, the ordinary spin properties suggest that the two states of the doublet correspond to the two different orientations of an "isotopic" spin vector of value $\frac{1}{2}$. The space in which this orientation is described is not physically realizable but is called "isotopic-spin space." In direct analogy with ordinary spin, we construct three operators T_1, T_2, T_3 corresponding to S_x, S_y, S_z; also operators $\mathbf{T}$ and T^2 corresponding $\mathbf{S}$ and S^2. A pure spin state is usually specified by the quantum numbers associated with the operators S^2 and S_z, and this convention is followed for isotopic-spin space so that a pure isotopic-spin state is specified by the quantum numbers t, t_3 associated with the operators T^2 and T_3. Thus a nucleon has $t_3 = +\frac{1}{2}$ if it is a proton and $t_3 = -\frac{1}{2}$ if it is a neutron. The charge is an eigenvalue of the operator $e(T + \frac{1}{2})$. In all this we no longer have the $\hbar$ that occurs in angular-momentum eigen equations.

The operators T^2, $\mathbf{T}$, T_1, T_2, and T_3 satisfy the same commutation relations as do the corresponding angular-momentum operators; thus the analogy between spin and isotopic spin is very close, and the isotopic-spin operators can be manipulated in the same way as spin operators. For example, two isotopic-spin vectors can be added in the same way as two angular-momentum vectors. To illustrate this and to continue the matter of the last sections, we consider the case of two nucleons. We can write down combinations of the individual isotopic-spin state vectors to give the total isotopic-spin state vector. These total state vectors must be symmetric or antisymmetric and must be eigenfunctions of the total isotopic-spin operators; by complete analogy with Eqs. (5.3)–(5.6), we have that

$$|1, 1\rangle = |\tfrac{1}{2}, \tfrac{1}{2}\rangle_a |\tfrac{1}{2}, \tfrac{1}{2}\rangle_b , \tag{5.7}$$

$$|1, 0\rangle = \sqrt{\tfrac{1}{2}} \{ |\tfrac{1}{2}, \tfrac{1}{2}\rangle_a |\tfrac{1}{2}, -\tfrac{1}{2}\rangle_b + |\tfrac{1}{2}, -\tfrac{1}{2}\rangle_a |\tfrac{1}{2}, \tfrac{1}{2}\rangle_b \} , \tag{5.8}$$

$$|1, -1\rangle = |\tfrac{1}{2}, -\tfrac{1}{2}\rangle_a |\tfrac{1}{2}, -\tfrac{1}{2}\rangle_b, \tag{5.9}$$

$$|0, 0\rangle = \sqrt{\tfrac{1}{2}} \{|\tfrac{1}{2}, \tfrac{1}{2}\rangle_a |\tfrac{1}{2}, -\tfrac{1}{2}\rangle_b - |\tfrac{1}{2}, -\tfrac{1}{2}\rangle_a |\tfrac{1}{2}, \tfrac{1}{2}\rangle_b\}. \tag{5.10}$$

The notation on the right is such that the quantities in the kets give the quantum number associated with the operators T^2 and T_3 for the particle labeled by the letter in the subscript. The notation on the left is such that the numbers in the kets are the quantum numbers of the total isotopic-spin operators. The kets (5.7)–(5.9) with $t = 1$ are symmetric and belong to the triplet isotopic-spin state; that of (5.10) with $t = 0$ is antisymmetric and belongs to the singlet isotopic-spin state.

We must relate this description of two nucleons to physical situations. Two protons have individual isotopic-spin state vectors $|\tfrac{1}{2}, \tfrac{1}{2}\rangle_a$ and $|\tfrac{1}{2}, \tfrac{1}{2}\rangle_b$; thus their total isotopic-spin vector is $|1, 1\rangle$. Similarly, two neutrons have vector $|1, -1\rangle$. A proton and a neutron do not make a state that is an eigenstate of T^2 and T_3. Actually such a pair of nucleons has an isotopic spin state that is a superposition of the two eigenstates $|1, 0\rangle, |0, 0\rangle$. Thus

$$|\tfrac{1}{2}, \tfrac{1}{2}\rangle_a |\tfrac{1}{2}, -\tfrac{1}{2}\rangle_b = \sqrt{\tfrac{1}{2}} \{|1, 0\rangle + |0, 0\rangle\}.$$

This follows from solving Eqs. (5.8) and (5.10) and is equivalent to finding the total angular-momentum states present in the vector addition of two spins.

If we now construct a total state vector to describe space, the spin, and the isotopic spin of two nucleons, we have

$$|\phi\rangle = |\psi\rangle |\chi\rangle |\tau\rangle.$$

Since the nucleons are fermions, this complete state vector must be antisymmetric under exchange of the two nucleons. We have already shown how this is done in the case of two protons or two neutrons; the vector $|\psi\rangle |\chi\rangle$ is antisymmetric. When we add the isotopic-spin vector, the symmetry must not change and we are therefore restricted to symmetric isotopic-spin vectors. Thus for the proton pair the vector is given by Eq. (5.7), and for the neutron pair the vector is that of Eq. (5.9). If the system is a neutron and proton with antisymmetric space-spin vectors, the vector $|\tau\rangle$ must be symmetric and it can only be that given by Eq. (5.8). We now notice that the antisymmetric space-spin vectors for two nucleons go with isotopic-spin vectors having total isotopic spin of one. Finally, if the neutron–proton system is in a symmetric space-spin state, then the isotopic-spin vector must be antisymmetric. This requirement is fulfilled by the vector of Eq. (5.10), which has zero total isotopic spin.

So far we have added nothing to our description of the forces between nucleons, as these are already completely specified by the spin and orbital momentum states and do not require any further parameters. However,

5.4 Isotopic Spin

we notice that charge independence implies that for a given orbital momentum, the force between two nucleons depends on the total isotopic spin, not upon their charge states. In addition we now have a uniformly antisymmetric description of two nucleon states and a formalism that will permit a simple description of some postulated nucleon–nucleon forces and an important extension of charge independence to other particles.

We will now show that the postulate of charge independence implies that the interaction Hamiltonian is invariant under rotations in isotopic-spin space. Charge independence requires that the interaction between two nucleons depends on their total isotopic-spin state but not upon their individual charge states if we neglect Coulomb forces. Hence the interaction Hamiltonian contains terms depending on the total isotopic spin, that is, operators such as T^2 or $\mathbf{T}_a \cdot \mathbf{T}_b$. (The operator $\mathbf{T}_a \cdot \mathbf{T}_b$ has eigenvalues $-\frac{1}{4}$ and $+\frac{3}{4}$ for the triplet and singlet isotopic-spin two-nucleon state, respectively.) These operators commute with the operators that are the generators of rotations in isotopic-spin space, namely, $\mathbf{T}$ (T_1, T_2, and T_3). By complete analogy with the situation in angular momentum (Section 2.4) we can state that the eigenvalues of T^2 and T_3 will be conserved quantities in transitions caused by the Hamiltonian. Thus for the two-nucleon system, charge independence requires the conservation of total isotopic spin and of its third component. The conservation of the eigenvalue of T_3 is a natural outcome of the conservation of electric charge, since we associated electric charge with this third component.

We can indicate the use of charge independence in nucleon–nucleon scattering by deriving an inequality that must hold if nucleon–nucleon forces are charge independent. We consider n–p and p–p scattering: the S-matrix commutes with the isotopic-spin operator and so can be divided into two parts S_0 and S_1, which are responsible for transitions in the isotopic-spin states 0 and 1 respectively. The initial isotopic spin state for n–p scattering is the vector $\sqrt{\frac{1}{2}} \{|1, 0\rangle + |0, 0\rangle\}$; the final-state isotopic-spin state for scattering at angle θ can be taken as the same; the state for scattering at an angle $(\pi - \theta)$ corresponds to interchanging the final-state nucleons and so has the state vector $\sqrt{\frac{1}{2}} \{|1, 0\rangle - |0, 0\rangle\}$. The p–p isotopic-spin state vectors are $|1, 1\rangle$.

Therefore,

$$\left(\frac{d\sigma_{np}}{d\Omega}\right)_\theta \propto \tfrac{1}{4}|[\langle 1, 0| + \langle 0, 0|](S_1 + S_0)[|1, 0\rangle + |0, 0\rangle]|^2,$$

$$\left(\frac{d\sigma_{np}}{d\Omega}\right)_{\pi-\theta} \propto \tfrac{1}{4}|[\langle 1, 0| - \langle 0, 0|](S_1 + S_0)[|1, 0\rangle + |0, 0\rangle]|^2,$$

$$\left(\frac{d\sigma_{pp}}{d\Omega}\right)_\theta \propto |\langle 1, 1|S_1|1, 1\rangle|^2.$$

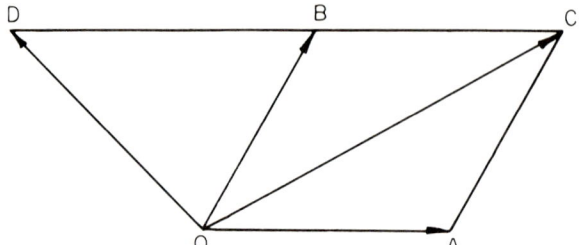

FIG. 5.1. An Argand diagram of the four matrix elements involved in nucleon-nucleon scattering.

We write the matrix elements thus:

$$\langle 1, t_3 | S_1 | 1, t_3 \rangle = M_1 ,$$
$$\langle 0, 0 | S_0 | 0, 0 \rangle = M_0 .$$

Then

$$\left(\frac{d\sigma_{np}}{d\Omega}\right)_\theta^{1/2} \propto \tfrac{1}{2} | M_1 + M_0 | ,$$

$$\left(\frac{d\sigma_{np}}{d\Omega}\right)_{\pi-\theta}^{1/2} \propto \tfrac{1}{2} | M_1 - M_0 | ,$$

$$\left(\frac{d\sigma_{pp}}{d\Omega}\right)_\theta^{1/2} \propto | M_1 | .$$

The matrix elements are complex quantities and so we can represent them as vectors on an Argand diagram. In Fig. 5.1 we have

$$OA = M_1, \quad OB = M_0, \quad OC = M_1 + M_0, \quad OD = M_0 - M_1.$$

We can apply the triangular inequality to $\triangle ODC$,

$$| M_1 - M_0 | + | M_1 + M_0 | \geqslant 2M_1$$

or

$$2\left(\frac{d\sigma_{np}}{d\Omega}\right)_{\pi-\theta}^{1/2} + 2\left(\frac{d\sigma_{np}}{d\Omega}\right)_\theta^{1/2} \geqslant 2\left(\frac{d\sigma_{pp}}{d\Omega}\right)_\theta^{1/2}.$$

At 90° we have $\theta = \pi - \theta$ and hence we expect that

$$4\frac{d\sigma_{np}}{d\Omega}(90°) \geqslant \frac{d\sigma_{pp}}{d\Omega}(90°).$$

This relation should hold between cross sections at the same energy.

We must note at this point that it appears that the two-nucleon interaction exists as a result of meson exchange processes, in particular, pion exchange. The charge independence of the nucleon–nucleon interaction is

5.4 Isotopic Spin

a consequence of the charge independence of the interaction of nucleons with pions and with other mesons. We discuss the evidence in the case of the pion–nucleon interaction in Section 8.4, but at this point we wish to widen the use of isotopic spin to other particles.

Three pions have been observed, two charged π^+, π^-, of equal mass (139.6 MeV), and a neutral π^0 (135.0 Mev). It is assumed that in the absence of the electromagnetic interactions these three particles would become degenerate and indistinguishable and are, therefore, identified as the three states of an isotopic-spin triplet. Thus, the pion has isotopic-spin 1 and the charge is the eigenvalue of the operator eT_3, where T_3 is the third component of the isotopic-spin operator (e is the magnitude of the electronic charge). This should be compared with the charge operator for the nucleon doublet

$$Q = e(T_3 + \tfrac{1}{2}).$$

We are now in a position to construct state vectors for a state containing two pions as an example of the two-boson system. We can start with two neutral pions. These are identical and spinless, so that the state vector must contain information about the orbital angular momentum state and the radial separation. The latter part has no effect on the symmetry, which is therefore given by $(-1)^l$. This must be $+1$, so l must be even; that is, only $S, D, G, \ldots$, etc. states are allowed. The isotopic-spin state of two neutral pions will be given by treating the addition of two states having isotopic spin 1 in the same way as the vector addition of two angular momenta both having $j = 1$. We have that $|t_a, t_{a3}, t_b, t_{b3}\rangle$ is the state containing two particles of isotopic spins t_a and t_b with third components t_{a3}, t_{b3} and $|t, t_3\rangle$ is the state of total isotopic spin t, t_3. Then for two π^0-mesons we have $t_a = t_b = 1$, $t_{a3} = t_{b3} = 0$, and

$$|\pi^0 \pi^0\rangle = |1, 0, 1, 0\rangle = \sqrt{\tfrac{2}{3}} \,|2, 0\rangle + \sqrt{\tfrac{1}{3}} \,|0, 0\rangle.$$

Thus the two-π^0 system is a mixture of total isotopic spin 0 and 2.

The $\pi^+ \pi^-$ system is more complicated. The two particles are identical in the absence of the electromagnetic field and must have an overall state vector that is symmetric. Thus, the orbital state and isotopic-spin state are either both even or both odd. The total isotopic-spin states are

$$|2, 0\rangle = \sqrt{\tfrac{1}{6}} \,|\pi^+\pi^-\rangle + \sqrt{\tfrac{2}{3}} \,|\pi^0\pi^0\rangle + \sqrt{\tfrac{1}{6}} \,|\pi^-\pi^+\rangle, \quad (5.11)$$

$$|1, 0\rangle = \sqrt{\tfrac{1}{2}} \,|\pi^+\pi^-\rangle - \sqrt{\tfrac{1}{2}} \,|\pi^-\pi^+\rangle, \quad (5.12)$$

$$|0, 0\rangle = \sqrt{\tfrac{1}{3}} \,|\pi^+\pi^-\rangle - \sqrt{\tfrac{1}{3}} \,|\pi^0\pi^0\rangle + \sqrt{\tfrac{1}{3}} \,|\pi^-\pi^+\rangle. \quad (5.13)$$

These are, respectively, symmetric, antisymmetric, and symmetric under exchanges of the two particles. Thus the even orbital states $S, D, G, \ldots$ have $t = 0$ or 2, and the odd states, $P, F, H, \ldots$ have $t = 1$.

136   V. Symmetry, Isotopic Spin, and Hypercharge

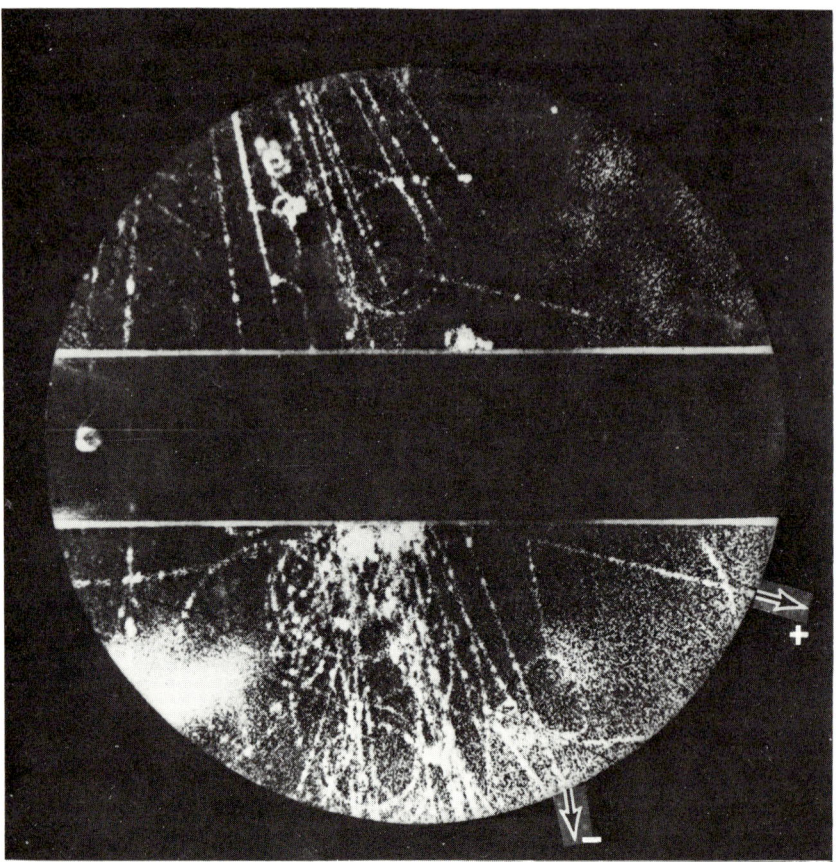

FIG. 5.2. A cloud-chamber photograph of the production and decay of a neutral strange particle. It was produced in one of the nuclear interactions that have occurred in the metal plate placed across the chamber and has decayed below the plate into two charged particles. [From Rochester, G. D., and Butler, C. C. (1947). *Nature* **160**, p. 855 (Fig. 1).]

5.5 Strangeness and Hypercharge

The strange particles were first observed in cloud-chamber studies of cosmic radiation at high altitudes. In Fig. 5.2, we reproduce a photograph of an event in which one is observed to decay. The general nature of such photographs indicates that the correct interpretation of the V event is that a neutral particle was produced in the high-energy interaction in the metal plate and after traveling several centimeters decayed into two oppositely charged particles. The rate at which such events were observed indicated that the production cross section was of the order of 10^{-28} cm^2, which is the

5.5 Strangeness and Hypercharge

order of magnitude associated with strong interactions. Such interactions take place in time of about 10^{-23} sec, i.e., the proton radius divided by the velocity of light. At first, it was presumed that such particles were produced by reactions such as

$$\pi^- + p \to \Lambda^0 + \pi^0, \tag{5.14}$$

where Λ^0 was the symbol given to the neutral particle. The distance moved by the particle before decay indicated that its half-life was of the order of 10^{-10} sec. One of the decay channels observed was

$$\Lambda^0 \to \pi^- + p. \tag{5.15}$$

If reaction (5.14) is the fast (10^{-23} sec) production process, then the following virtual reactions are also fast:

$$\Lambda^0 \to \pi^- + p + \pi^0,$$

followed by Yukawa absorption of the π^0 by the proton

$$\pi^- + p + \pi^0 \to \pi^- + p.$$

Therefore if reaction (5.14) is fast, then reaction (5.15) must also be fast, and the fact that the decay takes 10^{-10} sec instead of an expected 10^{-23} sec is a considerable anomaly. It is this that gave these particles the qualification "strange."

We shall not discuss the careful experimental work that established the existence of the various strange particles. Interested readers are referred to the articles by Rochester and Butler (1953) and by Dalitz (1957), which review the discovery of the "stable" strange particles. By stable we mean long-lived ($\sim 10^{-10}$ sec) on the strong-interaction time scale ($\sim 10^{-23}$ sec). Table 5.1 lists these particles and gives their more important properties.

At this point, we define some further generic names for particle groups. Mesons originally meant particles intermediate in mass between the electron and the proton, but has come to mean all particles, other than electrons and neutrinos, that are not baryons. Apart from the muon these are bosons, and some are heavier than the proton. Baryons are all those fermions that have a mass equal to or greater than that of the proton. The hyperons are the baryons that are heavier than the nucleons and are strange particles. Antibaryons include the antiproton and antineutron, which have been observed, and the antihyperons, some of which have been observed.

The law of conservation of baryons has never been observed to be broken; it states that the number of baryons less the number of antibaryons is a conserved quantity; thus we attribute to every baryon a baryon number 1, and to every antibaryon a baryon number -1, and require the total baryon number to be conserved. We notice the close similarity to the lepton conservation law (see Section 11.2).

TABLE 5.1

THE "STABLE" STRANGE PARTICLES

	Mass (MeV)	Mean life (sec)	Isotopic spin t_3	Isotopic spin t	Strangeness	Spin
Mesons						
K^+	493.8	1.235×10^{-8}	$+\tfrac{1}{2}$		$+1$	
K^0	497.7	—	$-\tfrac{1}{2}$	$\tfrac{1}{2}$	$+1$	0
$\bar{K}^0$	497.7	—	$+\tfrac{1}{2}$		-1	
K^-	493.8	1.235×10^{-8}	$-\tfrac{1}{2}$		-1	
Hyperons						
Λ^0	1115.6	2.51×10^{-10}	0	0	-1	
Σ^+	1189.4	0.80×10^{-10}	$+1$			
Σ^0	1192.5	$<1 \times 10^{-14}$	0	1	-1	$\tfrac{1}{2}$
Σ^-	1197.3	1.49×10^{-10}	-1			
Ξ^0	1314.7	3.0×10^{-10}	$+\tfrac{1}{2}$	$\tfrac{1}{2}$	-2	
Ξ^-	1321.2	1.66×10^{-10}	$-\tfrac{1}{2}$			
Ω^-	1673.0	$\sim 1.3 \times 10^{-10}$	0	0	-3	$\tfrac{3}{2}$

The prolonged decay behavior of the strange particles suggests that some mechanism is inhibiting decay. Early in the history of these particles, it was suggested that they possess a very large intrinsic angular momentum and that the decay was retarded by the angular-momentum barrier. The alternative, suggested by Pais (1952), is that strange particles possess some internal degree of freedom specified by a quantum number, which we call strangeness, and that various selection rules based upon the conservation or nonconservation of strangeness are operating in the production and decay. The retarded decay suggests that this process takes place through an interaction that does not conserve strangeness; this interaction has approximately the same strength as the weak interaction responsible for β-decay.

On the other hand, the strong interaction responsible for production is supposed to conserve strangeness, and it follows that the strong production of one strange particle is impossible and that two of opposite strangeness must be produced in the same interaction This hypothesis of associated production was confirmed by cloud-chamber work at the Brookhaven Cosmotron (Fowler *et al.*, 1954, 1955).

Taking the strangeness assignments given in Table 5.1, we can give examples of reactions that conserve strangeness, baryon number, and electric charge; thus

$$\begin{aligned}
\pi^- + p &\rightarrow K^+ + \Sigma^-, \\
\pi^- + p &\rightarrow K^0 + \Sigma^0 \text{ (or } \Lambda^0), \\
\gamma + p &\rightarrow K^+ + \Sigma^0 \text{ (or } \Lambda^0), \\
\gamma + p &\rightarrow K^0 + \Sigma^+.
\end{aligned} \quad (5.16)$$

5.5 Strangeness and Hypercharge

These reactions have been observed. Reactions such as

$$\pi^- + p \to K^- + \Sigma^+,$$
$$n + n \to \Lambda^0 + \Lambda^0, \quad (5.17)$$

do not conserve strangeness; they have not been observed.

Let us now examine these strangeness assignments in detail. Such assignments were, of course, preceded by the observation of reactions such as (5.16) and by the concept of associated production. We proceed by considering the isotopic-spin assignments to the mesons and baryons. The electric charge Q in units of e, the electronic charge, is given by

$$Q = t_3 + \tfrac{1}{2} y, \quad (5.18)$$

where y is some integer number; for example, it is 0 for pions and 1 for nucleons. Now let us consider the reaction

$$\pi^- + p \to K^+ + \Sigma^-,$$

which is observed. This satisfies the laws of conservation of charge and baryon number. The isotopic-spin quantum numbers of the intial state are $t = \tfrac{1}{2}$ or $\tfrac{3}{2}$, $t_3 = -\tfrac{1}{2}$. If we assume charge independence, then the final state must also have $t = \tfrac{1}{2}$ or $\tfrac{3}{2}$, which implies that either the K or the Σ particle belongs to a multiplet of half odd-integer isotopic spin. The Σ hyperon has three charge states and to it we can assign $t = 1$, $y = 0$, so that Σ^- has $t_3 = -1$; from this it follows that the K^+ must have $t_3 = +\tfrac{1}{2}$ and could belong to a doublet, $t = \tfrac{1}{2}$, $y = 1$. Compared to the pion, this is an unusual assignment for a meson, as we might expect that the K^+, K^0, and K^- belong to a triplet; however, it appears that there are two types of neutral kaon with opposite strangeness, K^0 and $\bar{K}^0$, so that K^+, K^0 is one doublet and $\bar{K}^0$, K^- is a second doublet. The first has $y = 1$; the second has $y = -1$. If the two types of neutral meson had not been observed, it would still be impossible to interchange the doublet and triplet assignments between K and Σ. For example, suppose the Σ was a doublet (Σ^-, Σ^0). We would expect that the Σ^- was the baryon and Σ^+ its antibaryon, in which case the conservation of baryons would forbid the reaction

$$\pi^+ + p \to K^+ + \Sigma^+.$$

This reaction is observed; in addition, the Σ^- and Σ^+ have unequal masses, which precludes their being related by charge conjugation. Thus the proposed isotopic-spin assignments at least lead to a scheme consistent with observation. By analyzing the observed strange-particle reactions in this manner Gell-Mann and Pais (1955) and Nishijima (1954) were able to assign values of t, t_3, and y to all particles such that all the observed associated production reactions satisfied the conservation of Q and of t_3. All the reactions of (5.16) satisfy these conservation laws, while the unobserved reactions such as those of (5.17) do not; the reader can verify these facts using the data presented in Table 5.1. For example,

$$\pi^- + p \to K^- + \Sigma^+ ;$$

t_3 values:

$$(-1) + (+\tfrac{1}{2}) \neq (-\tfrac{1}{2}) + (+1).$$

This reaction conserves electric charge but not t_3 and is therefore forbidden. This illustrates the essence of the Gell-Mann–Nishijima scheme: the conservation of t_3 has a significance that is different from the conservation of charge Q. In pion physics conservation of t_3 meant the same as the conservation of Q; in the physics of strange particles this is no longer the case.

We can now examine the constant y of Eq. (5.18); it has the values

$$\begin{aligned}
y &= 1 &&\text{for} && K^+, K^0, n, p, \overline{\Xi^-}, \overline{\Xi^0}, \\
y &= 0 &&\text{for} && \Sigma, \overline{\Sigma}, \Lambda, \overline{\Lambda}, \pi, \\
y &= -1 &&\text{for} && K^-, \overline{K}^0, \bar{n}, \bar{p}, \Xi^-, \Xi^0.
\end{aligned} \tag{5.19}$$

The Gell-Mann scheme puts

$$y = B + S. \tag{5.20}$$

The number y is called the hypercharge where B is the baryon number and S is the strangeness of the particle involved. From (5.19) it follows that the strangeness quantum number assignments are

$$\begin{aligned}
S &= -2 &&\text{for} && \Xi^-, \Xi^0, \\
S &= -1 &&\text{for} && \Sigma, \Lambda, \overline{K}^0, K^-, \\
S &= 0 &&\text{for} && n, \bar{n}, p, \bar{p}, \pi, \\
S &= +1 &&\text{for} && \overline{\Sigma}, \overline{\Lambda}, K^0, K^+, \\
S &= +2 &&\text{for} && \overline{\Xi^-}, \overline{\Xi^0}.
\end{aligned} \tag{5.21}$$

Equation (5.18) becomes

$$Q = t_3 + \tfrac{1}{2}(B + S) \tag{5.22}$$

and we observe that the conservation of total charge, third component of total isotopic spin, and total baryon number also implies the conservation of total strangeness and hypercharge. This law of conservation of strangeness is simple and direct to apply and agrees exactly with the observations of associated production and other strange-particle strong interactions. All the reactions of (5.16) conserve strangeness.

5.5 Strangeness and Hypercharge

The decays of strange particles manifestly do not obey the conservation of strangeness; for example,

$$K^+ \to \pi^+ + \pi^0,$$
$$\Sigma^- \to \pi^- + n.$$

The decay of strange particles will be discussed in Chapter XI.

Before closing this section, we shall examine one of the interesting consequences of strangeness conservation. It concerns the scattering and absorption reactions of the charged kaons by hydrogen and by neutrons; there are many more reaction channels open in the K^- meson reactions than are open in the K^+ meson reactions. The following list of reactions in hydrogen illustrates this point:

$$\left. \begin{array}{l} K^+ + p \to K^+ + p \\ K^- + p \to K^- + p \end{array} \right\} \text{elastic scattering,}$$

$$\to \bar{K}^0 + n \quad \text{charge-exchange scattering,}$$

$$\left. \begin{array}{l} \to \Sigma^+ + \pi^- \\ \to \Sigma^0 + \pi^0 \\ \to \Sigma^- + \pi^+ \\ \to \Lambda^0 + \pi^0 \\ \to \Lambda^0 + \pi^0 + \pi^0 \\ \to \Lambda^0 + \pi^+ + \pi^- \end{array} \right\} \text{strangeness-exchange reactions.} \quad (5.23)$$

All these channels are open for incident kaon kinetic energies greater than 7 MeV, and all the strangeness-exchange channels are open for K^- mesons absorbed from Bohr-like orbits around protons. We have not included other reaction channels that open at higher incident energies (e.g., extra pion production) or the double strangeness-exchange reactions such as

$$K^- + p \to K^0 + \Xi^0.$$

A similar list of reactions can be written for the case of a neutron target. The striking difference between K^+ and K^- is maintained. We see that strangeness conservation allows the K^- to produce hyperons by absorption in nucleons, whereas the K^+ cannot do this. This basic asymmetry in behavior has the result that the mean free paths of the K^+ and K^- meson are very different; for example, in photographic emulsions,

K^+: ~95 cm in the energy range 30 → 120 MeV (Lannutti *et al.*, 1956),

K^-: ~30 cm in the energy range 16 → 160 MeV (Webb *et al.*, 1958).

The mean free path for K^- mesons corresponds to an interaction cross section equal to the geometrical area of the nuclei. This behavior is in contrast to that of medium-energy π^+ and π^- mesons, which both have mean free paths of approximately 30 cm. The difference in kaon mean free paths becomes less at greater energies because of the number of inelastic channels that open for both charges.

We have found that t_3 is conserved in the strong interactions involving strange particles. In addition, we expect these interactions to be charge independent, that is, to conserve total isotopic spin. A large number of equalities and inequalities connecting strange-particle reaction cross sections can be derived, assuming charge independence. For example, consider the reactions:

$$\text{(a)} \quad \pi^+ + p \to K^+ + \Sigma^+, \tag{5.24}$$

$$\text{(b)} \quad \pi^- + p \to K^+ + \Sigma^-, \tag{5.25}$$

$$\text{(c)} \quad \pi^- + p \to K^0 + \Sigma^0. \tag{5.26}$$

It is easy to confirm by the use of the methods used in Sections 5.4 and 7.7 that the following inequality holds if the interaction is charge-independent

$$\left(\frac{d\sigma_a}{d\Omega}\right)^{1/2} + \left(\frac{d\sigma_b}{d\Omega}\right)^{1/2} - \left(\frac{2d\sigma_c}{d\Omega}\right)^{1/2} \geqslant 0,$$

where the differential cross sections apply to the same center-of-mass energy and angle. Crawford *et al.* (1959) report some measurements of reactions (5.25) and (5.26), which they have combined with earlier measurements on reaction (5.24); the results satisfy the inequality. It is now believed that the strong interactions are charge independent.

5.6 Conclusion

In this chapter, we have introduced, by way of symmetry, the concept of isotopic spin. The strong interactions appear to conserve isotopic spin; that is, the interaction is invariant under rotations in isotopic-spin space. In addition, we have introduced the concept of strangeness and its alternative, hypercharge. The latter will be related more closely to isotopic spin when we consider the classification of elementary particles by unitary symmetry.

REFERENCES

Crawford, F. S., Douglass, R. L., Good, M. L., Kalbfisch, G. R., Stevenson, M. L., and Ticho, H. K. (1959). *Phys. Rev. Lett.* **3**, 394 (5.5).
Dalitz, R. H. (1957). *Rep. Progr. Phys.* **20**, 163 (5.5).

References

Fowler, W. B., Shutt, R. P., Thorndike, A. M., and Whittemore, W. L. (1954). *Phys. Rev.* **93,** 861 (5.5).
Fowler, W. B., Shutt, R. P., Thorndike, A. M., and Whittemore, W. L. (1955). *Phys. Rev.* **98,** 121 (5.5).
Gell-Mann, M., and Pais, A. (1955). *Proc. Glasgow Conf. Nucl. Meson Phys.* (1954), p. 342, Pergamon Press, New York (5.5).
Lannutti, J. E., Chupp, W. W., Goldhaber, G., Goldhaber, S., Helmy, E., Iloff, E. L. Pevsner, A., and Ritson, D., (1956). *Phys. Rev.* **101,** 1617 (5.5).
Nishijima, K. (1954). *Progr. Theor. Phys.* **12,** 107 (5.5).
Pais, A. (1952). *Phys. Rev.* **86,** 513 (5.5).
Rochester, G. D., and Butler, C. C. (1947). *Nature (London)* **160,** 855 (5.5).
Rochester, G. D., and Butler, C. C. (1953). *Rep. Progr. Phys.* **16,** 364 (5.5).
Webb, F. H., Iloff, E. L., Featherstone, F. H., Chupp, W. W., Goldhaber, G., and Goldhaber, S. (1958). *Nuovo Cimento* [10], **8,** 899 (5.5).

VI

PARITY, TIME REVERSAL, CHARGE CONJUGATION, AND *G*-PARITY

6.1 Introduction

In this chapter, we consider three discrete transformations, namely, the parity transformation, time reversal, and charge conjugation. The elementary-particle interactions were originally assumed to have symmetry under these transformations; but since 1956, when parity nonconservation was discovered, these symmetries have been under close scrutiny. We shall discuss the consequences of the three symmetries and of violations, where they exist.

6.2 Parity

If this section were our first introduction to parity, it would contain the matter of Section 1.7. To avoid repetition, readers may find it helpful to read that section before proceeding.

First, we have to prove that the parity of a one-component function is the eigenvalue of the unitary operator P that describes the effects of a parity coordinate transformation. Consider the function $f(\mathbf{x})$; let us apply the coordinate transformation that changes the sign of the space coordinates of each space–time point. The function of the new coordinates, $\mathbf{x}'$, that has the same numerical value at a point as $f(\mathbf{x})$ at that point is $f'(\mathbf{x}')$, where

$$f'(\mathbf{x}') = f(\mathbf{x}),$$

and

$$\mathbf{x}' = -\mathbf{x}.$$

$f'(\mathbf{x}')$ defines a function $f'(\mathbf{x}) = Pf(\mathbf{x})$. Now the point with coordinates

6.2 Parity

x in the transformed system had coordinates $-x$ in the untransformed system; therefore

$$Pf(x) = f'(x) = f(-x),$$

but

$$f(-x) = \pm f(x),$$

for functions having even $(+)$ or odd $(-)$ parity. Therefore,

$$Pf(x) = \pm f(x),$$

and $f(x)$ is an eigenfunction of P with eigenvalues $+1$ or -1 for even or odd parity.

We shall now stop talking about the coordinate transformation in the context of parity and devote our attention to the active parity transformation; the results of any analysis are independent of these two views of the transformation. The active transformation inverts the physical system so that what was at point x is now at the point $-x$, etc.; this procedure reverses the direction of all vectors (linear momentum, etc.) but leaves the direction of all axial vectors (e.g., angular momentum, spin, etc.) unchanged. Thus if the physical system is described by a state vector that is an eigenstate of momentum and total angular momentum with eigenvalues p, j, and j_z, then the parity-transformed state is an eigenstate with eigenvalues $-p, j, j_z$; i.e.,

$$P|p, j, j_z\rangle = \eta|-p, j, j_z\rangle.$$

η is a constant such that $\eta^2 = 1$, as can be seen by operating twice with P. Therefore $\eta = \pm 1$. Now it is evident that $|p, j, j_z\rangle$ is not an eigenstate of P unless $p = 0$, so we must conclude that a particle state is not a parity eigenstate unless the particle is at rest. However, in the center-of-mass system the particle has zero momentum and the intrinsic parity is η. Consider now a one-particle state that is described as a superposition of spherical waves; then it can be shown (Roman, 1964) that a spherical wave of orbital angular momentum l has a parity $\eta(-1)^l$. The case of two particles of intrinsic parities η_1 and η_2 is similar: if the state is expanded into spherical waves, the lth wave has parity $\eta_1\eta_2(-1)^l$, where l will be given by the relative orbital angular momentum of the particles. If there is a third particle present of intrinsic parity η_3 with orbital angular momentum L with respect to the center of mass of the first two, the parity of the whole system is $\eta_1\eta_2\eta_3(-1)^{l+L}$. The more general case of many particles is discussed by Roman (1964).

In Section 1.7, we suggested that the Schrödinger Hamiltonian is invariant under the passive parity transformation; it is also invariant under the active transformation, for we expect the total energy of the system to be unchanged under the inversion. It follows that H and P commute. H

contains the energy of interaction between particles in the system, which induces particle reactions, and it follows that the parity of the complete system is conserved whatever reaction it undergoes. Therefore the statement that the (Hamiltonian of the) system is invariant under the parity transformation is identical in meaning to the statement that the parity of the system is conserved.

In discussing the party of a system, the intrinsic parity of the vacuum is taken to be positive. Spinless particles of even or odd parity are often referred to as scalar or pseudoscalar particles. Since a vector changes sign under the parity transformation, particles of spin 1 and odd parity are called vector particles, whereas those of even parity are called axial-vector particles. Bosons of greater spin are also known. The intrinsic parity of a fermion cannot be defined relative to the vacuum: this is not serious, since the conservation of baryon number means that an odd number of baryons interacting will give a final state containing the same odd number (plus a number of baryon–antibaryon pairs), and all that needs to be known is the relative parity between baryons and between baryon and antibaryon. It is normal to express all baryon parities relative to the proton, although this can involve ambiguity (see below); the relative parity of a baryon and its antibaryon is negative (Bjorken and Drell, 1965). Thus a proton-antiproton system in a state of relative orbital angular momentum l has parity $-(-1)^l$. Contrast this with a boson–antiboson system, which has parity $(-1)^l$. The pairs $\pi^+\pi^-$ and $K^0\bar{K}^0$ are examples of such a system.

There will always remain certain ambiguities in the definition of parity due to the existence of superselection rules (Roman, 1964). The conservation laws of baryon number B, charge Q, and hypercharge y are supposed to be due to the invariance of systems under various gauge transformations (Section 10.4). Strong interactions possess this invariance, and no strong transition can take place between two systems that differ in any of their quantum numbers, B, Q, and y. Suppose $|\phi_1\rangle$ and $|\phi_2\rangle$ are state vectors representing two such systems. The matrix element of any observable operator O taken between these states is zero:

$$\langle \phi_2 | O | \phi_1 \rangle = 0 .$$

The result is that it is impossible to define a relative phase between the two states and the relative parity becomes impossible to define. Thus the relative parity of π^- and π^0 or that of neutron and proton cannot, in principle, be found. However, the reaction

$$\pi^- + p \to \pi^0 + n$$

is due to strong interactions, which conserve B, Q, and y. It is known that the orbital angular-momentum state is the same initially and finally, so that the intrinsic parity of the π^-p system is the same as that of the $\pi^0 n$ system. Thus either the p–n relative parity and the π^-–π^0 relative parity are both even or they are both odd. By convention the parities within an isotopic-spin multiplet are defined to be the same. This is consistent with experimental observation, and so it is usual to assign even relative parity to neutron and proton and odd parity to all three charge states of the π-meson (since the π^0 is known to have odd parity; see Section 7.3).

Consider now the reaction

$$K^- + p \to \Lambda + \pi^0.$$

Given that the orbital angular-momentum states are known, this reaction will determine the product of the K and Λ parities relative to the proton (or nucleon) parity. The result is that the $K\Lambda N$ parity is odd. Since the classification of particles into unitary multiplets was recognized, it has been conventional to assume the same parity within such a supermultiplet. The Λ appears to belong to the same supermultiplet as the nucleon, so the ΛN relative parity is even and the K-meson will have odd parity. This is then consistent with the fact that the K- and π-mesons also appear to belong to the same supermultiplet. The reaction

$$K^- + p \to \Xi^- + K^+$$
$$y:\ -1\ \ +1\ \ -1\ \ +1$$

can be used to find the Ξ^-p relative parity, although these two particles have hypercharges differing by two. This result depends only upon assuming the same parity for the boson and antiboson, K^- and K^+.

The weak interaction does not conserve hypercharge, but cannot be used to make a parity connection between systems of differing hypercharge because it does not conserve parity either, as we shall discuss in Section 6.4.

6.3 Parity Conservation

Parity conservation places restraints upon nuclear or particle reactions that must be maintained in any theoretical investigation of the reactions; this permits us to make tests of the validity of proposed interactions or reactions. There are several ways of making these tests, which we shall now discuss.

The first method of testing requires that the initial- and final-state wavefunctions have the same parity. In Section 7.3, we shall deduce the intrinsic parity of the negative π-meson by considering the total wavefunction parities before and after π^- absorption in deuterium.

We can introduce now a second method, which is of great use in making simple tests on reactions involving polarization and which uses the invariance under the active parity transformation. Consider a simple nuclear or elementary-particle reaction. If the system is invariant under P, then any reaction and its parity transform will occur with equal probability and therefore with equal cross section. Let us see how to apply this. The parity transformation reverses the space coordinates of a point, and this is identical to reversing one coordinate and rotating through 180° around the axis of that coordinate; since the conservation of angular momentum is almost beyond doubt, the system is invariant under the final rotation and parity invariance reduces to invariance under reversal of one coordinate. A simple pictorial way of making such a reversal is to take a mirror image of the system to be transformed. There is no difficulty in taking the mirror image of vectors. Axial vectors such as magnetic field or angular momentum are more difficult, and the easiest way is to consider the corresponding vector; e.g., a loop of electric current for the examples mentioned. The result is that the axial-vector components parallel to the mirror surface reverse their direction in going from object to image, while axial-vector components perpendicular to the mirror are unchanged. A 180° rotation of the image in the plane of the mirror turns an axial-vector image into its object, thus confirming that the parity transformation leaves axial vectors unchanged. Let us apply this to meson–proton scattering to show that the differential cross section is independent of any proton polarization in the direction of the incident π-meson. We refer to Fig. 6.1, which has a mirror set in the plane containing the x and z axes. In the foreground a stationary target proton awaits the arrival of a π-meson with momentum $\mathbf{p}_\pi$ parallel to the z axis; the meson scatters from the proton with final momentum $\mathbf{p}_\pi'$ at an angle θ to $\mathbf{p}_\pi$ and in the yz plane. The mirror image of this system is seen in the mirror on the far side. The object has the proton spin $\boldsymbol{\sigma}_p$ pointing along the $-z$ direction, while its image points along the $+z$ direction. A simple 180° rotation around $\mathbf{p}_\pi$ will transform the image into the same form as the object, apart from the proton spin. If the interaction responsible for scattering is invariant under the parity transformation, these two reactions will have equal probability. Thus the differential cross section at a given scattering angle is independent of any polarization of the target protons in the direction of the incident π-mesons. This fact was used in the partial-wave analysis of positive meson scattering made in Section 3.3 and in other places in this book. The differential cross section is, in fact, independent of any proton polarization in the scattering plane. However, it does depend upon a polarization of the target protons in a direction perpendicular to the scattering plane. Taking a mirror image of a scatter generates an exactly identical system (identical after a simple rotation), so there is nothing to tell us how its cross section is related to that for an opposite proton polari-

6.3 Parity Conservation

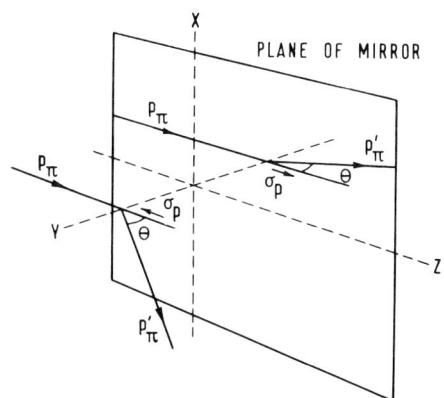

FIG. 6.1. A diagram of a meson-nucleon scattering and its mirror image to illustrate the parity transformation.

zation, and there can be a difference in cross sections; it is this difference that we have called the left-right scattering asymmetry in Chapter III. By drawing similar diagrams of hypothetical reactions we can show that a parity-conserving interaction can polarize reaction products only in a direction perpendicular to the plane of the reaction; this applies to two-body products from unpolarized initial states. These results are in complete accord with those derived in Chapter III.

The third method of examining the effects of parity invariance concerns the structure of the Hamiltonian, which must commute with the parity operator in the case of parity conservation, that is

$$PHP^{-1} = H.$$

This places constraints upon the form of the Hamiltonian, which must be a scalar operator. This can often be verified by inspection. For example, the following operators often occur in Hamiltonians:

$\mathbf{L} \cdot \boldsymbol{\sigma}$ (spin-orbit interaction),

$\boldsymbol{\sigma}_1 \cdot \boldsymbol{\sigma}_2$ (spin-spin interaction),

$[3(\boldsymbol{\sigma}_1 \cdot \mathbf{r})(\boldsymbol{\sigma}_2 \cdot \mathbf{r})/r^2] - \boldsymbol{\sigma}_1 \cdot \boldsymbol{\sigma}_2$ (tensor force).

$\mathbf{L}$ and $\boldsymbol{\sigma}$ are axial vectors and $\mathbf{r}$ is a polar vector, so that all these terms are scalars. There could also be terms depending upon the linear momentum of the particles involved; for example, in the case of π-meson–nucleon scattering there may be more specific terms such as

$$\mathbf{p}_\pi \cdot \mathbf{p}_\pi',$$
$$(\mathbf{p}_\pi \times \mathbf{p}_\pi') \cdot \boldsymbol{\sigma}_p.$$

All these examples are scalars. For the meson–nucleon scattering the two vectors $\mathbf{p}_\pi$, $\mathbf{p}_\pi'$ and the axial vector $\boldsymbol{\sigma}_p$ can be combined to form only two independent scalars, namely the two just given. Thus the only dependence on $\boldsymbol{\sigma}_p$ is via the scalar $(\mathbf{p}_\pi \times \mathbf{p}_\pi') \cdot \boldsymbol{\sigma}_p$, which is zero unless the proton polarization has a component perpendicular to the scattering plane. It follows that the scattering cross section cannot depend upon the presence of any initial-state polarization in the scattering plane. Thus we can apply parity conservation by requiring that the Hamiltonian, the transition rate, and the cross section depend upon scalar quantities alone. Let us now consider the restrictions placed upon matrices of the kind M of Eq. (2.86), which connect the spin-density matrices of the initial and final states of some reaction. Consider scattering or a reaction involving one spin-$\tfrac{1}{2}$ particle in both initial and final states. If parity is conserved the S matrix element for a reaction and its parity transform are equal:

$$\langle \mathbf{p}_f, \mathbf{s}_f | S | \mathbf{p}_i, \mathbf{s}_i \rangle = \langle -\mathbf{p}_f, \mathbf{s}_f | S | -\mathbf{p}_i, \mathbf{s}_i \rangle \tag{6.1}$$

where $\mathbf{p}_i$ and $\mathbf{p}_f$ are the initial- and final-state momenta, $\mathbf{s}_i$ and $\mathbf{s}_f$ are the initial and final spins. Now the matrix M is a function of $\mathbf{p}_i$ and $\mathbf{p}_f$, and connects the spinors representing the spin states $\mathbf{s}_f$ and $\mathbf{s}_i$ which remain unchanged under a parity transformation (we imply an active transformation). Thus we can write the left-hand side of Eq. (6.1) as

$$\langle \mathbf{p}_f, \mathbf{s}_f | S | \mathbf{p}_i, \mathbf{s}_i \rangle = u_{s_f}^\dagger M(\mathbf{p}_f, \mathbf{p}_i, \boldsymbol{\sigma}) u_{s_i},$$

and the right-hand side

$$\langle -\mathbf{p}_f, \mathbf{s}_f | S | -\mathbf{p}_i, \mathbf{s}_i \rangle = (Pu_{s_f}) M(-\mathbf{p}_f, -\mathbf{p}_i, \boldsymbol{\sigma})(Pu_{s_i})$$
$$= u_{s_f}^\dagger M(-\mathbf{p}_f, -\mathbf{p}_i, \boldsymbol{\sigma}) u_{s_i}.$$

Hence

$$M(\mathbf{p}_f, \mathbf{p}_i, \boldsymbol{\sigma}) = M(-\mathbf{p}_f, -\mathbf{p}_i, \boldsymbol{\sigma}). \tag{6.2}$$

In Section 3.3 we used

$$M = g - ih\boldsymbol{\sigma} \cdot \mathbf{n}, \tag{6.3}$$

where

$$\mathbf{n} = \frac{\mathbf{p}_i \times \mathbf{p}_f}{|\mathbf{p}_i \times \mathbf{p}_f|}.$$

6.4 Parity Nonconservation

This M satisfies Eq. (6.2) and so conserves parity. Eq. (6.3) applies where there is no change in the intrinsic parity of the particles. If there is a change, then

$$M(\mathbf{p}_f, \mathbf{p}_i, \boldsymbol{\sigma}) = -M(-\mathbf{p}_f, -\mathbf{p}_i, \boldsymbol{\sigma}) ; \tag{6.4}$$

see, for example, Section 3.10. We have deduced these properties of M assuming only one spin-$\frac{1}{2}$ particle present: this is an unnecessary restriction, for Eq. (6.2) also holds in more complicated situations such as nucleon–nucleon scattering. In this case the operator $\boldsymbol{\sigma}$ will appear twice, once as $\boldsymbol{\sigma}_1$, acting on the spinor for one nucleon, again as $\boldsymbol{\sigma}_2$ acting on the spinor for the second nucleon,

There is an interesting consequence of parity conservation that we can mention at this point. First, we notice that a reaction proceeding from a state of no defined parity (e.g., a plane wave) gives final states of different parity (e.g., partial waves) that do interfere. Sometimes, nuclear and particle reactions can be analyzed in terms of intermediate or compound states of defined parity and total angular momentum. If only one such state is involved, then the outgoing waves must have the same parity; that is, only all-even- or all-odd-l spherical harmonics are present in the final state amplitude. This has the result that the reaction differential cross-section is symmetrical about 90° in the center-of-mass system. Reversing the argument, we can say that reactions not having this symmetry must have at least two states (partial waves) of opposite parity operative.

However, we cannot say anything about reactions having this symmetry except that if states of opposite parity are operative, they are incoherent; for example, proton–proton scattering.

6.4 Parity Nonconservation

Let us briefly consider the situation that led to the discovery of the nonconservation of parity. In 1956 there appeared to be two kinds of K-meson that were, within experimental accuracy, identical in mass, charge, and lifetime; their difference was in their decay:

$$\theta^+ \to \pi^+ + \pi^0 ,$$
$$\tau^+ \to \pi^+ + \pi^+ + \pi^- .$$

Dalitz (1957) suggested that these two decays were not alternative channels for one particle and hence the θ and τ were different particles: his argument goes as follows (we assume that the π-meson is pseudoscalar). If the spin of the θ is s_θ in units of $\hbar$, then the relative orbital angular momentum of the two π-mesons must be s_θ and the final-state parity is $(-1)^{s_\theta}$; therefore, if parity is conserved, the parity of the θ-meson is $P_\theta = (-1)^{s_\theta}$.

Consider now the τ decay; in the final state we divide the motion into two parts, that of the two π^+ mesons about their center of mass (angular momentum l) and that of the two-π center-of-mass relative to the π^- (angular momentum L). We have

and
$$|l - L| \leqslant s_\tau \leqslant l + L$$
$$P_\tau = (-1)^3(-1)^l(-1)^L.$$

The identity of the two π^+ mesons requires their relative wavefunction to be symmetric, which means l is even; hence

$$P_\tau = -(-1)^L.$$

Let us suppose the θ and τ are identical and work through various spin assignments until we can obtain the same parity:

$s_\theta = s_\tau = 0$ gives $P_\theta = +1$ but $P_\tau = -1$ $(l = L)$;

$s_\theta = s_\tau = 1$ gives $P_\theta = -1$ and $P_\tau = \mp 1$,

as L is even (negative sign) or odd (positive sign). L is even ($P_\tau = -1$) in the following combinations: $l = 2, L = 2; l = 4, L = 4$; etc. L is odd ($P_\tau = +1$) in the following combinations: $l = 0, L = 1; l = 2, L = 1$ or 3; $l = 4, L = 3$, or 5; etc. Thus if $\theta \equiv \tau$, the lowest assignments are $s_\theta = s_\tau = 1$, $P_\theta = P_\tau = -1$, and $l = L = 2$. These orbital motions in the τ decay mode would give the π^- an anisotropic distribution with respect to the direction of relative momentum of the two π^+, which is not observed, and would inhibit the decay mode very strongly on account of angular-momentum barrier. In addition, this assignment is not favored if we try to include θ^0, the neutral counterpart of the θ^+ in the scheme. It is observed to be able to decay into two π^0-mesons and it can do this only if it has even spin and even parity. Higher τ and θ spins can be matched, but similar contrary arguments apply and it is therefore difficult to reconcile the apparent identity of the θ and τ mesons with the conservation of angular momentum and parity (Dalitz, 1957).

These difficulties led Lee and Yang (1956) to the suggestion that parity is not conserved in certain interactions. They examined the field of β-decay and found that no experiments had ever been done that indicated that parity is conserved in the β-interaction. They suggested experiments that would test their suggestion, and these were performed immediately. It was found that parity is not conserved either in β-decay (Wu et al., 1957) or in the decay of the π- and μ-mesons (Garwin et al., 1957). It appears that parity is not conserved in reactions in which four fermions interact at a single vertex and that such vertices with virtual fermions are involved in the τ and θ decays (Section 11.14).

Let us now examine the experiment that first showed parity is not conserved. It is evident that nonconservation makes it possible for an

6.4 Parity Nonconservation

elementary-particle reaction to depend upon a pseudoscalar combination of observables. We apply this to the intensity of β-particles emitted from a radioactive specimen in which the nuclei have been polarized; this polarization is described by the expectation value of **J**, namely $\langle \mathbf{J} \rangle$, which is an axial vector. Let the vector **p** be the momentum of the outgoing electrons. The quantity $\langle \mathbf{J} \rangle \cdot \mathbf{p}$ is pseudoscalar, and if parity is not conserved the intensity of the electrons can be a function of this quantity, among others. This leads to an electron distribution proportional to $1 + \alpha \cos \theta$, where

$$\cos \theta = \frac{\langle \mathbf{J} \rangle \cdot \mathbf{p}}{|\langle \mathbf{J} \rangle||\mathbf{p}|},$$

that is, the cosine of the angle between the polarization and the emitted electron. The electons are therefore emitted preferentially along the polarization vector ($\alpha > 0$) or against ($\alpha < 0$). Such asymmetry was found in the decay of a polarized specimen of ^{60}Co (Wu et al., 1957) and was unequivocal proof of the nonconservation of parity.

We find it convenient at this point to illustrate some of the effects of parity nonconservation by considering the decay of the Λ^0 hyperon:

$$\Lambda^0 \rightarrow \pi^- + p + 37 \text{ MeV}.$$

We do this because it is easy to deal with the meson–nucleon system and we can avoid obscuring the essential properties in the field-theoretical language used in β-decay. We treat the decay in the rest system of the Λ^0 and assume that the intrinsic parity of the Λ^0 is even (or odd) with respect to that of the proton. If parity is conserved, the meson–nucleon system will have to be in a P (or S) state of relative orbital motion. If parity is not conserved, both states S and P can occurs, with respective amplitudes S and P. We wish to construct the amplitude for finding a proton at a polar angle $0°$ and of spin $s_z' = \pm \frac{1}{2}$ for the two possible orientations of the Λ spin, $s_z = \pm \frac{1}{2}$. The final S-wave state will be represented by the state vector $S|0, 0, \frac{1}{2}, \pm\frac{1}{2}\rangle$ (labeling: $|l, l_z, s, s_z\rangle$) while the final P-wave will be

$$P\{\langle 1, 0, \tfrac{1}{2}, \pm\tfrac{1}{2} | \tfrac{1}{2}, \pm\tfrac{1}{2}\rangle | 1, 0, \tfrac{1}{2}, \pm\tfrac{1}{2}\rangle$$
$$+ \langle 1, \pm 1, \tfrac{1}{2}, \mp\tfrac{1}{2} | \tfrac{1}{2}, \pm\tfrac{1}{2}\rangle | 1, \pm 1, \tfrac{1}{2}, \mp\tfrac{1}{2}\rangle\}$$
$$= P\{\mp \sqrt{\tfrac{1}{3}} | 1, 0, \tfrac{1}{2}, \pm\tfrac{1}{2}\rangle \pm \sqrt{\tfrac{2}{3}} | 1, \pm 1, \tfrac{1}{2}, \mp\tfrac{1}{2}\rangle\}.$$

The final state $|\psi_\text{f}\rangle$ is the superposition of these two states so that the amplitude for finding a proton at angle θ, ϕ with spin state $|s', s_z'\rangle$ is

$$\langle \theta, \phi, s', s_z' | \psi_\text{f}\rangle = S\langle \theta, \phi | 0, 0\rangle \langle s', s_z' | \tfrac{1}{2}, \pm\tfrac{1}{2}\rangle$$
$$+ P\{\mp \sqrt{\tfrac{1}{3}} \langle \theta, \phi | 1, 0\rangle \langle s', s_z' | \tfrac{1}{2}, \pm\tfrac{1}{2}\rangle$$
$$\pm \sqrt{\tfrac{2}{3}} \langle \theta, \phi | 1, \pm 1\rangle \langle s', s_z' | \tfrac{1}{2}, \mp\tfrac{1}{2}\rangle\}.$$

At $\theta = 0$, $\phi = 0$ this amplitude is given by

$$\langle 0, 0, s', s_z' | \phi_f \rangle = (1/\sqrt{4\pi})(S \mp P)\langle s', s_z' | \tfrac{1}{2}, \pm\tfrac{1}{2}\rangle. \quad (6.5)$$

Since $\langle s', s_z' | \tfrac{1}{2}, \pm\tfrac{1}{2}\rangle = \delta_{s'1/2}\delta_{s_z \pm 1/2}$, we find that a Λ that has $s_z = +\tfrac{1}{2}$ gives, at this angle, a proton having spin, $s_z' = +\tfrac{1}{2}$ with an amplitude $(S - P)/\sqrt{4\pi}$; similarly a Λ having spin $s_z = -\tfrac{1}{2}$ gives a proton having spin $s_z' = -\tfrac{1}{2}$ with an amplitude $(S + P)/\sqrt{4\pi}$. If the Λ is in a pure spin state, that is, fully polarized with the spin pointing in the direction θ', ϕ', the amplitude for finding $s_z = +\tfrac{1}{2}$ is $\cos(\theta'/2)\exp(-i\phi'/2)$ and the amplitude for finding $s_z = -\tfrac{1}{2}$ is $\sin(\theta'/2)\exp(i\phi'/2)$. Thus this state has an amplitude for emitting a proton along the z axis of

$$(1/\sqrt{4\pi})\{\cos(\theta'/2)\exp(-i\phi'/2)(S - P) \\ + \sin(\theta'/2)\exp(i\phi'/2)(S + P)\}, \quad (6.6)$$

which gives

$$I(\theta') = (1/4\pi)\{|S|^2 + |P|^2 - 2\,\mathrm{Re}\,S^*P\cos\theta'\}, \\ \propto 1 + \alpha \cos\theta'. \quad (6.7)$$

Thus there is an up–down decay asymmetry with respect to the direction of polarization given by the parameter α which depends upon Re S^*P. Thus, as expected, there is such an asymmetry only if there are both S and P waves in the final state, a situation that will occur if parity is not conserved in the decay. This decay asymmetry was first observed by Crawford and co-workers (1958). If the Λ are only partially polarized, then the apparent decay asymmetry is reduced from α by a factor that is the magnitude of the Λ polarization.

Equation (6.6) indicates that if the initial spin state is represented by a column vector, this can be transformed into the column vector representing the final proton spin by the matrix M given by

$$M = \frac{1}{\sqrt{4\pi}} \begin{bmatrix} S - P & 0 \\ 0 & S + P \end{bmatrix} \\ = \frac{1}{\sqrt{4\pi}}(S - P\sigma_z)$$

or more generally

$$M = \frac{1}{\sqrt{4\pi}}(S - P\mathbf{k}\cdot\boldsymbol{\sigma}), \quad (6.8)$$

where $\mathbf{k}$ is a unit vector along the direction of the proton. We notice that $M(\mathbf{k})$ is not equal to $M(-\mathbf{k})$, as expected for a parity-nonconserving transition. We can use this matrix and the formalism of the density matrix to

6.4 Parity Nonconservation

calculate the proton polarization $\langle\sigma\rangle_p$, given the initial Λ polarization $\langle\sigma\rangle_\Lambda$. The respective density matrices ρ_p and ρ_Λ are given by [Eq. (2.90)]

$$\rho_\Lambda = \tfrac{1}{2}\{1 + \langle\sigma\rangle_\Lambda \cdot \sigma\},$$
$$\rho_p = \tfrac{1}{2} I(\mathbf{k})\{1 + \langle\sigma\rangle_p \cdot \sigma\},$$

and

$$\rho_p = M\rho_\Lambda M^\dagger.$$

That is,

$$\tfrac{1}{2} I(\mathbf{k})\{1 + \langle\sigma\rangle_p \cdot \sigma\} = (1/8\pi)\{|S|^2 + |P|^2 + \langle\sigma\rangle_\Lambda \cdot \mathbf{k}(-2\,\text{Re}\,SP^*) \\
+ \sigma \cdot [\mathbf{k}(-2\,\text{Re}\,SP^*) + \mathbf{k}(\mathbf{k}\cdot\langle\sigma\rangle_\Lambda) \\
+ \langle\sigma\rangle_\Lambda \times \mathbf{k}\, 2\,\text{Im}\,SP^* \\
+ (|S|^2 + |P|^2)\mathbf{k} \times (\langle\sigma\rangle_\Lambda \times \mathbf{k})]\}. \quad (6.9)$$

We put

$$\alpha = \frac{-2\,\text{Re}\,SP^*}{|S|^2 + |P|^2},$$
$$\beta = \frac{2\,\text{Im}\,SP^*}{|S|^2 + |P|^2},$$
$$\gamma = \frac{|S|^2 - |P|^2}{|S|^2 + |P|^2},$$

where $\alpha^2 + \beta^2 + \gamma^2 = 1$. Then we have

$$I(\mathbf{k}) = (1/4\pi)(|S|^2 + |P|^2)\{1 + \alpha\langle\sigma\rangle_\Lambda \cdot \mathbf{k}\}, \quad (6.10)$$

$$\langle\sigma\rangle_p = (1 + \alpha\langle\sigma\rangle_\Lambda \cdot \mathbf{k})^{-1}\{(\alpha + \mathbf{k}\cdot\langle\sigma\rangle_\Lambda)\mathbf{k} \\
+ \beta\langle\sigma\rangle_\Lambda \times \mathbf{k} + \gamma\mathbf{k} \times (\langle\sigma\rangle_\Lambda \times \mathbf{k})\}. \quad (6.11)$$

The three parameters connect various components of the initial Λ polarization with those of the final proton polarization. This formula is very similar to the one that occurs in $0-\tfrac{1}{2}$ scattering (Section 3.4), where initial and final polarization are likewise connected. The most noteworthy feature is that if the initial polarization is zero, i.e., $\langle\sigma\rangle_\Lambda = 0$, we have

$$\langle\sigma\rangle_p = \alpha\mathbf{k}.$$

This means that the outgoing proton is longitudinally polarized. This can occur only if parity is not conserved, since the quantity $\sigma_p \cdot \mathbf{k}$ is a pseudoscalar. The existence of the polarization is useful, since its measurement permits a determination of α independently of the incident Λ polarization, which is not the case in a measurement of the up–down decay asymmetry. The weighted mean value of α is 0.646 ± 0.016 (Particle Data Group, 1969), and $\gamma = 0.76$. β is possibly slightly negative, as expected (see Section 6.6).

We notice that the asymmetry and the proton polarization require that both S and P be nonzero and thus depend upon interference between decay channels of opposite parity. If the initial state and final state are of the same or completely opposite parity, no such effects can occur. These statements apply generally to parity-nonconserving interactions.

The discovery of parity nonconservation in the decay of particles cast doubt upon the parity invariance of the strong interactions, such as those responsible for nuclear forces. This invariance has now been tested to a considerable degree of accuracy (for example: Wilkinson, 1958; Jones et al., 1958; Crawford et al., 1959; Haas et al., 1959; Bock and Schopper, 1965), and we now have considerable confidence in its validity in strong interactions. In addition, the upper limit of about 10^{-7} in the intensity of opposite parity admixture in atomic energy levels indicates that parity-nonconserving amplitudes are less than 10^{-3} of parity conserving amplitudes in the electromagnetic interaction.

6.5 Time Reversal

The laws of classical mechanics and of electromagnetism are invariant under the time-reversal transformation; that is, a system can develop backwards in time in the same way as it normally does forwards in time. There is no reason, however, why this should also be true for elementary-particle physics. We shall briefly investigate the time-reversal transformation and examine the requirements of invariance or noninvariance (Wick, 1958; Kemmer et al., 1959; Sachs, 1953).

We can start by considering the time-dependent Schrödinger equation and the wavefunction for the system whose Hamiltonian is H:

$$i\hbar \frac{\partial \psi(\mathbf{x}, t)}{\partial t} = H\psi(\mathbf{x}, t) \qquad (6.12)$$

At time $t = 0$, the system will be described by the wavefunction $\psi(\mathbf{x}, t = 0) = \psi(0)$, and develops so that by the time τ, it is described by the function $\psi(\mathbf{x}, t = \tau) = \psi(\tau)$. It we reverse the sense of time and label by ψ' these states in their time-reversed form, then the system will be invariant under time reversal if the state described by $\psi'(\tau)$ develops during a time interval τ into the state $\psi'(0)$. Let us define the function

$$\phi(t') = \psi(\tau - t') . \qquad (6.13)$$

t' is now the new time coordinate (reversal of previous t) and $\phi(t')$ has the same boundary conditions as does $\psi(t)$, as can be seen by putting $t' = 0$ and τ into Eq. (6.13). Let us transform Eq. (6.12) to the new time coordinate system:

6.5 Time Reversal

$$i\hbar \frac{\partial \phi(\tau - t')}{\partial(\tau - t')} = H\phi(\tau - t'),$$

that is,

$$-i\hbar \frac{\partial \phi(t')}{\partial t'} = H\phi(t') \qquad (6.14)$$

This means that $\phi(0) \big(= \psi(\tau)\big)$ will develop into $\phi(\tau) \big(= \psi(0)\big)$ if $\phi(t')$ satisfies Eq. (6.14). This is not Schrödinger's equation. However, if we take the complex conjugate of Eq. (6.14), we have

$$i\hbar \frac{\partial \phi^*(t')}{\partial t'} = H^* \phi^*(t').$$

If $H = H^*$, then this is Schrödinger's equation and in this circumstance $\phi^*(t')$ is a wavefunction that will describe the system that develops backwards in time in the same way as $\psi(t)$ develops forwards. Since $\phi^*(t') = \psi^*(\tau - t')$, we can define the time-reversed wavefunction ψ' by

$$\psi'(\mathbf{x}, t') = \psi^*(\mathbf{x}, \tau - t'). \qquad (6.15)$$

Complex conjugation of the wavefunction does not alter the state described by the wavefunction; thus when $t = 0$, $\psi(0)$ describes the system. The time-reversed system at $t' = \tau$ is described by $\psi'(\tau) = \psi^*(\tau - t') = \psi^*(0)$, which has the same normalization as the unreversed state. Our conclusion is that a system is invariant under time reversal if its Hamiltonian is real.

If the system contains angular momentum or spin, then the above discussion must be extended. We introduce the time-reversal operator T, which has the effect of producing a new wavefunction and a new Hamiltonian; thus, under time reversal,

$$\begin{aligned} \psi &\to \psi' = T\psi, \\ H &\to H' = THT^{-1}. \end{aligned} \qquad (6.16)$$

We require that ψ' satisfy the Schrödinger equation with $t \to -t$ and Hamiltonian H' as does ψ with Hamiltonian H. This can be done if T is the product of two operators U and K, where U is a unitary operator and K the operator that changes a quantity into its complex conjugate. Substituting these in the equation that ψ' must satisfy, namely

$$i\hbar \frac{\partial \psi'}{\partial(-t)} = H'\psi', \qquad (6.17)$$

gives

$$-i\hbar \frac{\partial U\psi^*}{\partial t} = UH^*U^{-1}U\psi^*.$$

Taking the complex conjugate and factoring out the time-independent operator U gives the correct unreversed Schrödinger equation, Eq. (6.12), so that the factorization of T into U and K gives a consistent scheme. If the system is invariant under time reversal, then Eq. (6.17) must have the same Hamiltonian as the unreversed Schrödinger equation, that is,

$$H' = THT^{-1} = UH^*U^{-1}. \qquad (6.18)$$

T is not a unitary operator, since the presence of K causes the complex conjugation of the coefficients in the expansion of a wavefunction in terms of orthonormal functions, a thing linear operators cannot do. T is said to be antiunitary. In the normal description of elementary-particle reactions or decays the initial and final states are not eigenvalues of T for the simple reason that this operator interchanges incoming and outgoing states.

Under time reversal, operators transform

$$O \to O' = TOT^{-1}. \qquad (6.19)$$

Some operators, such as position and energy, do not change; but others, such as momentum, orbital spin, and total angular momentum, will change sign. In the case of spin $\frac{1}{2}$ we expect the operator $\boldsymbol{\sigma}$ to transform to $-\boldsymbol{\sigma}$.

In the usual representation of the Pauli matrices the operator $U = \sigma_y$ ($U^{-1} = \sigma_y$) has the required properties, namely,

$$T\boldsymbol{\sigma} T^{-1} = \sigma_y \boldsymbol{\sigma}^* \sigma_y = -\boldsymbol{\sigma}. \qquad (6.20)$$

As expected, the eigenstates of σ_z change under time reversal, as can be confirmed by using the Pauli representation:

$$|\tfrac{1}{2}, \pm\tfrac{1}{2}\rangle \to T|\tfrac{1}{2}, \pm\tfrac{1}{2}\rangle = \sigma_y|\tfrac{1}{2}, \pm\tfrac{1}{2}\rangle = \pm i|\tfrac{1}{2}, \mp\tfrac{1}{2}\rangle$$

or

$$T|s, s_z\rangle = (i)^{2s_z}|s, -s_z\rangle. \qquad (6.21)$$

An analogous relation holds for orbital angular-momentum eigenstates:

$$T Y_l^{l_z}(\theta, \phi) = Y_l^{l_z*}(\theta, \phi) = (i)^{2l_z} Y_l^{-l_z}(\theta, \phi). \qquad (6.22)$$

If we take the case of a total angular momentum $\mathbf{j} = \mathbf{l} + \mathbf{s}$ say, then if $|j, j_z\rangle$ is the total angular-momentum eigenstate, we have

$$|j, j_z\rangle = \sum_{s_z, l_z} \langle l, l_z, s, s_z | j, j_z\rangle |l, l_z\rangle |s, s_z\rangle,$$

where $|s, s_z\rangle$ are the eigenstates of an angular momentum s. Applying time reversal,

$$T|j, j_z\rangle = \sum_{s_z, l_z} \langle l, l_z, s, s_z | j, j_z\rangle |l, -l_z\rangle |s, -s_z\rangle (i)^{2(l_z+s_z)}$$

$$= (i)^{2j_z} \sum_{s_z, l_z} (-1)^{l+s-j} \langle l, -l_z, s, -s_z | j, -j_z\rangle |l, -l_z\rangle |s, -s_z\rangle,$$

6.5 Time Reversal

which follows from the asymmetry properties of the Clebsch–Gordan coefficients. Hence

$$T|j, j_z\rangle = (i)^{2j_z}(-1)^{l+s-j}|j, -j_z\rangle. \tag{6.23}$$

The phase relation is not consistent with the spin-$\frac{1}{2}$ or orbital angular-momentum cases. It can be made so by redefining the Clebsch–Gordan coefficients so as to absorb the factor $(-1)^{l+s-j}$, but this is not normally done.

Equations (6.21)–(6.23) show that all states are eigenstates of the operator T^2 with eigenvalues $+1$ if the total angular momentum is an integer and -1 if it is a half odd-integer. The conservation of this eigenvalue corresponds to the conservation of statistics.

We can now examine the effects of time reversal on the matrix elements of operators. To do this we shall consider a set of base vectors $|i\rangle$, $i = 1, 2\cdots$, which are normalized

$$\langle i|j\rangle = \delta_{ij}. \tag{6.24}$$

We represent the time-reversal state by $|Ti\rangle$, etc., and because the normalization will not be affected, we have

$$\langle Ti|Tj\rangle = \delta_{ij}. \tag{6.25}$$

Consider the states $|\phi_1\rangle$ and $|\phi_2\rangle$ expanded in terms of the base states, for example:

$$|\phi_1\rangle = \sum_i \langle i|\phi_1\rangle |i\rangle$$

If we apply time reversal, then

$$|T\phi_1\rangle = \sum_i \langle i|\phi_1\rangle^* |Ti\rangle.$$

Also

$$\langle \phi_2| = \sum_j \langle \phi_2|j\rangle \langle j|,$$

and

$$\langle T\phi_2| = \sum_j \langle \phi_2|j\rangle^* \langle Tj|.$$

Then

$$\langle T\phi_2|T\phi_1\rangle = \sum_i \sum_j \langle \phi_2|j\rangle^* \langle Tj|Ti\rangle \langle i|\phi_1\rangle^*$$
$$= \sum_i \langle \phi_2|i\rangle^* \langle i|\phi_1\rangle^*$$
$$= \langle \phi_2|\phi_1\rangle^*, \tag{6.26}$$

where we have used the orthonormal condition, Eq. (6.25), and Eq. (1.7). Thus we have that

$$\langle T\phi_2|T\phi_1\rangle = \langle \phi_1|\phi_2\rangle. \tag{6.27}$$

Consider now an operator O: it has the matrix element $\langle \phi_2 | O | \phi_1 \rangle$ between the states $\langle \phi_2 |$ and $| \phi_1 \rangle$. From Eq. (6.26) we can write

$$\langle T\phi_2 | TO | \phi_1 \rangle = \langle \phi_2 | O | \phi_1 \rangle^* . \qquad (6.27a)$$

Now the left-hand side can be written

$$\langle T\phi_2 | TOT^{-1}T | \phi_1 \rangle$$

or

$$\langle T\phi_2 | O' | T\phi_1 \rangle . \qquad (6.28)$$

If O is invariant under time reversal, $O' = TOT^{-1} = O$ and we have

$$\langle T\phi_2 | O | T\phi_1 \rangle = \langle \phi_2 | O | \phi_1 \rangle^* . \qquad (6.29)$$

6.6 The Consequences of Time-Reversal Invariance

In Section 3.12 we applied time-reversal invariance (TRI) to prove the law of reciprocity, but there are other consequences of this symmetry that we must now discuss.

We have noted that TRI requires the Hamiltonian to satisfy

$$H \to H' = THT^{-1} = H,$$

which means that the Hamiltonian has real coefficients and contains no operators that are odd under time reversal. Examples of odd operators are $\boldsymbol{\sigma}$, $\mathbf{p}$, $\mathbf{r} \cdot \boldsymbol{\sigma}$ and $(\mathbf{p}_1 \times \mathbf{p}_2) \cdot \boldsymbol{\sigma}$, whereas, for example, $\mathbf{r}$, $\mathbf{L} \cdot \boldsymbol{\sigma}$ and $\mathbf{p}_1 \cdot \mathbf{p}_2$ are even and thus allowed. This does not mean, in fact, that the observables of the odd operators will be zero in all elementary-particle reactions. For example, the observed final-state nucleon polarization perpendicular to the plane of elastic pion–nucleon scattering is the observable of the odd operator $(\mathbf{p}_1 \times \mathbf{p}_2) \cdot \mathbf{s}_n$. This happens because the invariance of H does not lead to the absence of time odd observables except in first-order perturbation theory (Born approximation). This is a bad approximation in strong interactions, so that odd observables are not excluded. However, in weak interactions the Born approximation is good, and thus odd observables are not expected, except as a consequence of final-state interactions. The decay $K \to \pi + \mu + \nu$ defines a plane containing the two momenta $\mathbf{p}_\pi$ and $\mathbf{p}_\mu$. Any μ-meson polarization with a component perpendicular to this plane represents a nonzero observable of the form $(\mathbf{p}_\pi \times \mathbf{p}_\mu) \cdot \mathbf{s}_\mu$ and would imply a breakdown of time-reversal invariance. The only final-state interaction in the $K_{\mu 3}$ decay is electromagnetic in the case of $K_2 \to \pi^- \mu^+ \nu$ and gives a very small effect. This is not the case in the decay $\Lambda \to p\pi^-$. If the Λ is polarized, then an observable value for the time odd operator $(\mathbf{s}_\Lambda \times \mathbf{p}_\pi) \cdot \mathbf{s}_p$ is to be expected as a consequence of the final-state interaction between the pion and proton. This interaction is well known and gives rise to a predictable

6.6 The Consequences of Time-Reversal Invariance

effect. In Eq. (6.11) the polarization of the proton parallel to the direction $\mathbf{s}_\Lambda \times \mathbf{p}_\pi$ is given by the parameter β. A real Hamiltonian gives a real S and P in the Born approximation and consequently a zero β. However, the amplitudes S and P are shifted in phase by approximately $e^{i\delta}$, where δ is the appropriate π-meson–nucleon scattering phase shift, so that S and P are no longer real. We shall prove this assertion shortly. This decay is not a good place to look for a small breakdown of TRI because the polarization expected would be swamped by the polarization due to the final-state interaction.

We wish now to look at the consequences we can find in TRI for elementary-particle interactions. To do this we can consider, as an example, matrices of the kind M, which we have met in connection with pion–nucleon scattering (Section 3.4) and Λ decay (Section 6.4). In general, M is a function of the initial- and final-state momenta $\mathbf{p}_i$, $\mathbf{p}_f$, and the initial and final spins $\mathbf{s}_i$, $\mathbf{s}_f$. The S-matrix element between the initial and final states is invariant under time reversal, thus:

$$\langle \mathbf{p}_f, \mathbf{s}_f | S | \mathbf{p}_i, \mathbf{s}_i \rangle = \langle -\mathbf{p}_i, -\mathbf{s}_i | S | -\mathbf{p}_f, -\mathbf{s}_f \rangle. \tag{6.30}$$

If $\mathbf{s}_i = \mathbf{s}_f = \frac{1}{2}$, we represent the initial and final spin states by the appropriate spinors u_{s_i}, u_{s_f}, and M is a 2×2 matrix that can be written as a linear sum of the four Pauli matrices; here these matrices appear in a purely formal role and are not immediately connected with angular momentum. In addition, we write $M(\mathbf{p}_i, \mathbf{p}_f, \boldsymbol{\sigma})$ to indicate the dependence on $\mathbf{p}_i$ and $\mathbf{p}_f$. The left-hand side of Eq. (6.30) can be written in terms of M, u_{s_i}, and u_{s_f}, thus:

$$\langle \mathbf{p}_f, \mathbf{s}_f | S | \mathbf{p}_i, \mathbf{s}_i \rangle = u_{s_f}^\dagger M(\mathbf{p}_i, \mathbf{p}_f, \boldsymbol{\sigma}) u_{s_i}. \tag{6.31}$$

We can now construct the matrix element of S in the time-reversed situation by taking the appropriate spinors and substituting $\mathbf{p}_i = -\mathbf{p}_f$ and $\mathbf{p}_f = -\mathbf{p}_i$ in M. Thus

$$\langle -\mathbf{p}_i, -\mathbf{s}_i | S | -\mathbf{p}_f, -\mathbf{s}_f \rangle = (T u_{s_i})^\dagger M(-\mathbf{p}_f, -\mathbf{p}_i, \boldsymbol{\sigma})(T u_{s_f})$$
$$= (\sigma_y u_{s_i}^*)^\dagger M(-\mathbf{p}_f, -\mathbf{p}_i, \boldsymbol{\sigma})(\sigma_y u_{s_f}^*).$$

By the rules of matrix algebra this complex number will be the complex conjugate of the Hermitian conjugate of the matrices taken in the reverse order; that is,

$$\langle -\mathbf{p}_i, -\mathbf{s}_i | S | -\mathbf{p}_f, -\mathbf{s}_f \rangle = [(\sigma_y u_{s_f}^*)^\dagger M^\dagger(-\mathbf{p}_f, -\mathbf{p}_i, \boldsymbol{\sigma})(\sigma_y u_{s_i}^*)]^*$$
$$= [u_{s_f}^{*\dagger} \sigma_y M^\dagger(-\mathbf{p}_f, -\mathbf{p}_i, \boldsymbol{\sigma}) \sigma_y u_{s_i}^*]^*.$$

Since $\sigma^\dagger = \sigma$ and $\sigma_y \boldsymbol{\sigma} \sigma_y = -\boldsymbol{\sigma}$, we have

$$\langle -\mathbf{p}_i, -\mathbf{s}_i | S | -\mathbf{p}_f, -\mathbf{s}_f \rangle = [(u_{s_f}^*)^\dagger M^*(-\mathbf{p}_f, -\mathbf{p}_i, -\boldsymbol{\sigma}) u_{s_i}^*]^*$$
$$= u_{s_f}^\dagger M(-\mathbf{p}_f, -\mathbf{p}_i, -\boldsymbol{\sigma}) u_{s_i}. \tag{6.32}$$

Thus, comparing Eqs. (6.31) and (6.32) we find time-reversal invariance requires

$$M(-\mathbf{p}_f, -\mathbf{p}_i, -\boldsymbol{\sigma}) = M(\mathbf{p}_i, \mathbf{p}_f, \boldsymbol{\sigma}). \quad (6.33)$$

In Section 3.3 we used

$$M = g - ih\boldsymbol{\sigma} \cdot \mathbf{n},$$

where $\mathbf{n} = (\mathbf{p}_i \times \mathbf{p}_f)/(|\mathbf{p}_i \times \mathbf{p}_f|)$. This M satisfies Eq. (6.33). Notice that although TRI requires the Hamiltonian to have real coefficients, no such restriction applies to M in this case [f and g are generally complex amplitudes, Eqs. (3.20) and (3.21)]. For Λ decay

$$M = (1/\sqrt{4\pi})(S - P\mathbf{k} \cdot \boldsymbol{\sigma}),$$

where $\mathbf{k} = \mathbf{p}_f/|\mathbf{p}_f|$. Under time reversal $\mathbf{p}_f \to -\mathbf{p}_i$, where $|\mathbf{p}_i| = |\mathbf{p}_f|$ in this case, so that this M also satisfies Eq. (6.33). The fact that in both these cases it is possible to have a final state with a nonzero observable to a time odd operator [$(\mathbf{s}_\Lambda \times \mathbf{k}) \cdot \mathbf{s}_p$ in Eq. (6.11) and $\boldsymbol{\sigma} \cdot (\mathbf{p}_i \times \mathbf{p}_f)$ in Eq. (3.47)] confirms the statements made at the beginning of this section. The restrictions we have found for the form of the matrix M can be extended to other more complicated cases; for example, nucleon–nucleon scattering. In this case M is a function of the operators $\boldsymbol{\sigma}_1$ and $\boldsymbol{\sigma}_2$, which act on the spinor representing the spins of the two nucleons. TRI requires that M does not change under $\boldsymbol{\sigma}_1 \to -\boldsymbol{\sigma}_1$, $\boldsymbol{\sigma}_2 \to -\boldsymbol{\sigma}_2$, $\mathbf{p}_i \to -\mathbf{p}_f$, and $\mathbf{p}_f \to -\mathbf{p}_i$.

Another useful consequence of TRI is the Watson theorem (Gell-Mann and Watson, 1954; Hamilton, 1959), which permits approximate statements to be made about certain weak transitions in the presence of strong transitions. Consider the transitions

$$\Lambda \to \pi^- + p,$$
$$\pi^- + p \to \pi^- + p,$$
$$\pi^- + p \to \Lambda.$$

The elastic $\pi^- p$ scattering can take place through many channels of differing angular momentum and parity, but for the present purpose we are interested in those that can communicate with the Λ, namely $S_{1/2}$ and $P_{1/2}$. We can construct an S-matrix that has rows and columns labeled by the three channels of interest, namely Λ, $\pi^- p$ in an $S_{1/2}$ state, and $\pi^- p$ in a $P_{1/2}$ state. In the presence of strong interactions alone the Λ remains a Λ and the $\pi^- p$ scattering conserves parity, so that the S-matrix is

$$S_0 = \begin{array}{c} \begin{array}{ccc} \Lambda & S_{1/2} & P_{1/2} \end{array} \\ \begin{bmatrix} 1 & 0 & 0 \\ 0 & e^{2i\delta} & 0 \\ 0 & 0 & e^{2i\delta_1} \end{bmatrix} \end{array},$$

6.6 The Consequences of Time-Reversal Invariance

where we have suppressed isotopic spin indices in the pion–nucleon phase shifts. When we add weak interactions, the S-matrix will have off-diagonal elements and we can now show that these take a particularly simple form. In circumstances of this kind the addition of the weaker transition changes the diagonal matrix S_0 into $S = S_0 + i\varepsilon$ where we expect the elements of the matrix ε to be small. The S-matrices S_0 and S have the following properties [Eq. (3.61)]:

$$S_0 S_0^\dagger = 1 \quad \text{and} \quad S S^\dagger = 1 .$$

Substituting, we have

$$(S_0 + i\epsilon)(S_0^\dagger - i\epsilon^\dagger) = 1 ,$$

or

$$S_0 S_0^\dagger + i\epsilon S_0 - i S_0 \epsilon^\dagger + \epsilon \epsilon^\dagger = 1 .$$

Now $\epsilon \epsilon^\dagger$ is of second-order smallness, so we find that

$$\epsilon S_0^\dagger = S_0 \epsilon^\dagger .$$

This is an equality between matrices and we examine an element of the left-hand side

$$(\epsilon S_0^\dagger)_{mn} = (\epsilon)_{mn}(S_0^\dagger)_{nn} = (\epsilon)_{mn}(S^*)_{nn} .$$

These relations follow from the diagonal nature of S_0. The on-diagonal elements of S_0 are of the form $\exp 2i\delta$, where δ is a scattering phase shift; hence

$$(\epsilon S_0^\dagger)_{mn} = (\epsilon)_{mn} \exp(-2i\delta_n) .$$

Similarly,

$$(S_0 \epsilon^\dagger)_{mn} = (S_0)_{mm}(\epsilon^\dagger)_{mn} = \exp(2i\delta_m)(\epsilon^\dagger)_{mn} .$$

Now $(\epsilon^\dagger)_{mn} = (\epsilon^*)_{nm}$ and we have that

$$(\epsilon)_{mn} \exp(-2i\delta_n) = (\epsilon^*)_{nm} \exp(2i\delta_m) .$$

In the absence of angular momentum the S-matrix is symmetric if TRI holds (Section 3.12). Our system contains angular momentum, but nonetheless the S-matrix can be made symmetric by the correct choice of the representation of the incoming and outgoing waves (Hamilton, 1959). Assuming that this choice has been made, we have $(\epsilon)_{mn} = (\epsilon)_{nm}$, and multiplying the last equation above by $\exp[i(\delta_n - \delta_m)]$, we have that

$$(\epsilon)_{mn} \exp[-i(\delta_n + \delta_m)] = (\epsilon^*)_{mn} \exp[i(\delta_n + \delta_m)] .$$

We see immediately that this is an equality between real quantities: therefore

$$(\epsilon)_{mn} = \rho_{mn} \exp[i(\delta_n + \delta_m)] ,$$

where ρ_{mn} is a real quantity.

For our example the S-matrix is now given by

$$S = \begin{bmatrix} 1 & i\rho_{12}e^{i\delta} & i\rho_{13}e^{i\delta_1} \\ i\rho_{12}e^{i\delta} & e^{2i\delta} & i\rho_{23}e^{i(\delta+\delta_1)} \\ i\rho_{13}e^{i\delta_1} & i\rho_{23}e^{i(\delta+\delta_1)} & e^{2i\delta_1} \end{bmatrix}, \quad (6.34)$$

and it follows that the amplitudes S and P of

$$M = (1/\sqrt{4\pi})(S - P\boldsymbol{\sigma}\cdot\mathbf{k})$$

can be written

$$S = se^{i\delta}, \quad P = pe^{i\delta_1}$$

where s and p are real quantities. The parameter β, which will give a nonzero result for the time-odd operator in Eq. (6.11), can be seen to depend upon the phase shifts δ and δ_1, thus manifesting the effect of a final-state interaction. These phase shifts are known from the analysis of pion–nucleon scattering and this permits, in theory, the final-state contribution to β to be separated from the contribution due to any breakdown of time-reversal invariance.

The Watson theorem has been used to relate pion photoproduction amplitudes to pion–nucleon scattering amplitudes (Watson et al., 1956).

In recent years there have been many tests of time-reversal invariance stimulated by the discovery of CP nonconservation in neutral K-meson decay (C = charge conjugation; see Section 6.7). If the CPT theorem is valid (Section 6.9), then this CP-violation indicates a violation of TRI, although probably very small. For example, Bodansky et al. (1966) have tested detailed balancing in nuclear reactions, and Handler et al. (1967) have examined polarization correlations in proton–proton scattering without detecting any violation. These experiments look for symmetry breaking in strong interactions, but they cannot put upper limits on symmetry-breaking amplitudes of less than about 10^{-3} of the invariant amplitudes. Tests for violations due to the electromagnetic interaction of hadrons have been made, for example, by Atac et al. (1968) in nuclear transitions, by Glasser et al. (1966) in the decay $\Sigma^\circ \to \Lambda e^+e^-$, and by Chen et al. (1968) in inelastic electron–proton scattering, again without detecting any significant effect. In weak interactions the phase of the axial-vector to the vector transition (Section 11.5) should be π: Burgy et al. (1958) find $180 \pm 8°$. Young et al. (1967) have not found any μ-polarization perpendicular to the $K^\circ_{\mu3}$ decay plane. For the decay $\Lambda \to \pi^-p$, Overseth and Roth (1967) measured β and this gave $\delta_S - \delta_P = 9 \pm 5.5°$, where the known pion–nucleon scattering phase shifts predict $6.5 \pm 1.5°$. These tests are hardly stringent, but do indicate that any breaking of time-reversal invariance is small.

The upper limit on the electric dipole moment of the neutron is also relevant to tests of time-reversal invariance. However, its significance will be discussed in Section 6.10 after we have considered charge conjugation and the *CPT* theorem.

6.7 Charge Conjugation

The operation of charge conjugation transforms particles into antiparticles and vice versa; thus electrons transform into positrons, positrons into electrons, and so on. Perhaps a more correct name for this operation is particle–antiparticle conjugation, as there is not always a change in electric charge (e.g., neutron $\rightleftarrows$ antineutron); however, present usage favors the name charge conjugation, and we define this to be the transformation that changes the sign of all the internal quantum numbers: charge, strangeness (hypercharge), baryon number, lepton number, etc. Thus, this operation is not a Lorentz transformation as are the parity transformation and time reversal; charge conjugation changes the nature of the particles involved, but leaves unchanged their linear and angular momenta, and position. In the Schrödinger wave mechanics, the wavefunction does not necessarily specify the nature of the particles involved; however, in order to satisfy a continuity equation for charge and current the wavefunction must be complex, and it is found that the particle represented by the complex conjugate wavefunction has opposite charge and current densities. Thus it appears that under charge conjugation

$$|\psi\rangle \to |\psi'\rangle = C|\psi\rangle,$$
$$\langle \mathbf{x}|\psi\rangle \to \langle \mathbf{x}|\psi'\rangle = \langle \mathbf{x}|C|\psi\rangle = \langle \mathbf{x}|\psi\rangle^*. \quad (6.35)$$

In the case of charged scalar or pseudoscalar particles the state vector $|\psi\rangle$ can be represented as the sum of two vectors

$$|\psi\rangle = |\psi_1\rangle + i|\psi_2\rangle \quad (6.36)$$

with the properties

$$C|\psi_1\rangle = +|\psi_1\rangle, \quad (6.37)$$
$$C|\psi_2\rangle = -|\psi_2\rangle. \quad (6.38)$$

Then

$$\langle \mathbf{x}|C|\psi\rangle = \langle \mathbf{x}|\psi\rangle^*,$$

as required to satisfy the continuity equation. Particles that have zero charge, strangeness, and baryon number transform into themselves (self-conjugate) under charge conjugation and are eigenstates of C with eigenvalues (charge parity, C) $+1$ or -1. The photon is self-conjugate, and since all the components of the vector potential of the electromagnetic field must change sign under charge conjugation, it follows that the state vector representing one photon must be odd ($C = -1$) under charge conjugation. Two photons have $C = +1$; and so on.

If we postulate the invariance of a physical system under charge conjugation, it follows that charge parity is a conserved quantity. This leads to selection rules in the decay of particles and permits the charge partity of self-conjugate particles to be established. The interactions responsible for the decay of the π°- and η-mesons do not violate C invariance in a quantitatively large way, so that the observed decay of these particles into two photons means that both have $C = +1$ and that their decay into three photons will be a measure of a C-violation (see below).

Apart from self-conjugate particles there are also combinations of particles that are self-conjugate. For example, $\pi^+\pi^-$, K^+K^-, $p\bar{p}$, $K^0\bar{K}^0\pi^0$, etc. The particle–antiparticle combinations are particularly interesting, and we shall now discuss their charge parity. We consider first a pair of non-self-conjugate bosons so that they are distinguishable. The state vector describing the system then has to specify the relative orbital angular momentum (quantum number l) and the total spin (s). The operation of C has the same effect as interchanging the particle coordinates. Under this exchange, the orbital part contributes $(-1)^l$ and the total spin contributes $(-1)^s$. This last fact follows from the properties of the Clebsch–Gordan coefficients for the vector addition of two integer spins to make a total spin s. Thus the symmetry under coordinate exchange and the charge parity are $(-1)^{l+s}$. The case of two self-conjugate bosons (e.g., $\pi^0\pi^0$, $\rho^0\rho^0$) is similar. Since the particles are indentical, the overall state vector describing total spin, orbital state, and isotopic spin state must be symmetric under exchange of the particle coordinates and hence C is even; since $(-1)^{l+s}$ is always even in these cases, we can again put $C = (-1)^{l+s}$. The baryon-antibaryon case also follows in a similar way. Under exchange of the particle coordinates the factors contributing are $(-1)^l$ from the orbital part, (-1) from the odd relative parity, and $(-1)^{s+1}$ from the total spin. (This last is a consequence of the properties of the coefficients for the vector addition of two equal half odd-integer spins to make spin s.) Collecting the factors, we find that the charge parity is $(-1)^{l+s}$.

Consider a simple example: the charge parity for positronium is $(-1)^{l+s}$. This is $+1$ for the S-state singlet positronium, and consequently this state can only decay into an even number of photons. C is -1 for the S-state triplet positronium, which can therefore only decay into an odd number of photons. Thus two-photon decay is from the singlet state and the three-photon decay from the triplet state of S-state positronium if the electromagnetic interaction is charge-conjugation invariant.

Charge-conjugation invariance breaks down in weak interactions, as we shall find in Section 6.9, and until 1964 was thought not to be violated by the strong and electromagnetic interactions. However, the discovery of *CP*-violation (Chapter XIV) also stimulated an examination of the status of *C*-invariance. Baltay *et al.* (1965) examined the products of antiproton annihilation for charge asymmetries in the pion and kaon spectra: they concluded that in strong interactions *C*-violating amplitudes were less than 1% for pions and 3% for kaons of the *C*-invariant amplitudes. The status of *C*-invariance in the electromagnetic interactions of hadrons has been tested by Duclos *et al.* (1965), who put an upper limit of 5×10^{-6} on the branching ratio $\pi^0 \to 3\gamma$. This result does not set a very low limit on the *C*-violating electromagnetic amplitude (2×10^{-2}), and a potentially more useful place to search is among the decay channels of the η-meson. Several tests that have been made are not consistent. However, the statistically heaviest result is on the spectra in $\eta \to \pi^+\pi^-\pi^0$ and $\eta \to \pi^+\pi^-\gamma$ and indicates no violation, although this is not a stringent test. The situation in η-decays has been reviewed by Bazin *et al.* (1968). Liu and Roberts (1966) have measured the branching ratio of ^1S positronium decay into three photons to be $5 \pm 3 \times 10^{-4}$. This indicates that the violation of *C*-invariance, if any, in the purely electromagnetic interaction, is not large.

6.8 *G*-Parity

The eigenvalue of *C* is a useful quantum number, but it is restricted to systems that have $Q = B = y = 0$. It is possible to extend it to systems that have $Q \neq 0$ but $B = y = 0$; for example, all the nonstrange mesons. The extension is made by combining invariance under charge conjugation with charge independence. To illustrate this, consider the isotopic-spin triplet of π-mesons π^+, π^0, π^-. Charge conjugation has the effect of transforming $\pi^+ \to \pi^-$, $\pi^0 \to \pi^0$, and $\pi^- \to \pi^+$, so that only the π^0 can be an eigenstate of *C*. However, we assume that the phases of the three eigenvectors representing these particles are such that

$$C|\pi^+\rangle = -|\pi^-\rangle, \qquad C|\pi^-\rangle = -|\pi^+\rangle. \qquad (6.39)$$

An alternative way of transforming a nonstrange boson into its antiparticle is to rotate through 180° around the second axis in isotopic-spin space. This is done by operating on the state vectors with the operator $\exp(iT_2\pi)$. The transformation properties of an eigenstate of isotopic spin $|t, t_3\rangle$ under this operator are the same as those of an angular-momentum state $|j, j_z\rangle$ under the operator $\exp(iJ_y\pi)$. With the phases chosen in Chapter II we have

$$\exp(iJ_y\pi)|j, j_z\rangle = (-1)^{j-j_z}|j, -j_z\rangle.$$

Similarly we have

$$\exp(iT_2\pi)|t, t_3\rangle = (-1)^{t-t_3}|t, -t_3\rangle. \qquad (6.40)$$

Now $\pi^\pm$, π^0 have eigenvectors $|1, \pm 1\rangle$, $|1, 0\rangle$, respectively, so that

$$\exp(iT_2\pi)|\pi^\pm\rangle = +|\pi^\mp\rangle, \qquad \exp(iT_2\pi)|\pi^0\rangle = -|\pi^0\rangle.$$

If we define

$$G = C\exp(iT_2\pi), \tag{6.41}$$

it is evident that

$$G|\pi^\pm\rangle = -|\pi^\pm\rangle, \qquad G|\pi^0\rangle = -|\pi^0\rangle.$$

The eigenvalue of G is called the G-parity, which is therefore odd for π-mesons. If we put G as the G-parity, then from Eqs. (6.40) and (6.41) we have for the neutral member of a multiplet

$$G = (-1)^l C. \tag{6.42}$$

By suitably choosing the relative phases of the charged states [as in Eq. (6.39)] all members of a multiplet will have the same G-parity. For any boson-antiboson or any fermion-antifermion systems having $Q = B = y = 0$,

$$C = (-1)^{l+s}, \tag{6.43}$$

so that for those that have $B = y = 0$ but any value of Q, we have

$$G = (-1)^{l+s+t}, \tag{6.44}$$

where t is the total isotopic spin, not that of the individual particles. For example, the $^1S_0$ state of protonium has $G = +1$ when $t = 0$ and $G = -1$ for $t = 1$. A system containing n particles all of which have well-defined G-parity has an overall G-parity that is the product of the individual G-parities. Thus an odd number of π-mesons has odd G-parity and an even number has even G-parity; a system containing a neutron and an antiproton ($t = 1$) in a $^1S_0$ state ($G = -1$) plus a π^+-meson must have overall $G = +1$.

G-parity is a conserved quantum number whenever charge parity and isotopic spin are conserved. This is the case when only strong interactions are involved, so that in these circumstances, for example, the following are forbidden: inelastic scattering of π-meson by π-meson producing an extra π-meson,

$$\pi^+ + \pi^- \to \pi^+ + \pi^- + \pi^0,$$

proton-antiproton annihilation from the $t = 0$, $^1S_0$ state into an odd number of π-mesons, etc. However, isotopic spin is not conserved in electromagnetic interactions (charge parity appears to be conserved), and it follows that G-parity is also not conserved. As an example we can quote

6.8 G-Parity

the case of the nonstrange η-meson: There are no charged η-mesons, so $t = 0$, and from the observed decay

$$\eta \to \gamma + \gamma,$$

we deduce $C = +1$ and hence $G = +1$. It follows that the decay

$$\eta \to \pi^+ + \pi^- + \pi^0$$

does not conserve G, and the conclusion is that this transition is electromagnetic in origin. (The weak interaction cannot be responsible because the 3π decay mode would not compete with the 2γ mode.)

It is possible to derive selection rules governing the changes in isotopic spin and G-parity in electromagnetic transitions. To do this we must consider the isotopic-spin changes that occur in photon emission or absorption, $A \rightleftharpoons B + \gamma$. The charge on a particle, in units of the electronic charge, is given by

$$Q = y/2 + t_3.$$

Under rotations in isotopic spin space this quantity transforms like a scalar and the third component of a vector. Since it is basically charge that interacts with the electromagnetic field, we expect photon emission (or absorption) will cause the following changes in isotopic spin of the source (or sink)

$$\Delta \mathbf{t} = 0 \text{ or } 1, \qquad \Delta t_3 = 0.$$

Since a photon has $C = -1$, it acts on emission or absorption like the emission or absorption of a particle with $T^G = 0^-$ or 1^+. Thus the process $A \to B + \gamma$ has either

$$\text{Change in } G, \qquad \Delta \mathbf{t} = 0$$

or

$$\text{No change in } G, \qquad \Delta \mathbf{t} = 1. \tag{6.45}$$

If A and B are both neutral, then we can define their charge parities, which must be opposite. From Eq. (6.42) it follows that the second selection rule then becomes

$$\text{No change in } G, \qquad \Delta t = \pm 1. \tag{6.46}$$

Now consider a change $A \to B$ that is electromagnetic but does not involve the emission of any real photons. We assume the transition occurs by the emission followed by the absorption of a virtual photon (higher-order transitions will be increasingly forbidden). If A and B are charged, then the selection rules that apply are those obtained by making all possible double combinations from Eq. (6.45). Thus we have

$$\text{Change in } G, \quad \Delta t = 1$$
$$\text{No change in } G, \quad \Delta t = 0 \text{ or } 2 \,. \tag{6.47}$$

One apparent possibility has been excluded. It appears that applying no G change, $\Delta t = 1$ twice can give $\Delta t = 0, 1,$ or 2. However, $\Delta t = 1$ requires the vector addition of two vectors (one for each photon vertex), each having $t = 1, t_3 = 0$ to give another vector also having $t = 1, t_3 = 0$. The Clebsch–Gordan coefficient for this addition is zero, so the virtual photon cannot act to cause that transition and $\Delta t = 1$ is disallowed. If A and B are neutral, the double application of Eq. (6.46) gives the selection rules

$$\text{Change in } G, \quad \Delta t = \pm 1$$
$$\text{No change in } G, \quad \Delta t = 0 \text{ or } \pm 2 \tag{6.47a}$$

These results are summarized in Table 6.1.

TABLE 6.1

ISOTOPIC SPIN SELECTION RULES IN TRANSITIONS INVOLVING THE ELECTROMAGNETIC INTERACTION

	Change in G	
Transition	Yes	No.
$A^\pm \rightleftharpoons B^\pm + \gamma$	$\Delta t = 0$	$\Delta t = 1$
$A^0 \rightleftharpoons B^0 + \gamma$	$\Delta t = 0$	$\Delta t = \pm 1$
$A^\pm \to B^\pm$	$\Delta t = 1$	$\Delta t = 0$ or 2
$A^0 \to B^0$	$\Delta t = \pm 1$	$\Delta t = 0$ or ± 2

The use of G-parity does not give any more information than does the use of charge-conjugation invariance and charge independence together. However, it does obtain the same results more simply and quickly.

6.9 The *CPT* Theorem

There exists a very important theorem that relates the properties of particle fields under proper Lorentz transformations with their properties under the improper transformations and charge conjugation (Lüders, 1954; Pauli, 1956). The theorem states that the interaction Lagrangian (or Hamiltonian) for locally interacting fields, which is Hermitian and invariant under proper Lorentz transformations, will commute with any of the six operators that can be formed by the product of T, C, and P taken in any order. This theorem is true so long as certain arbitrary phases are chosen

6.9 The CPT Theorem

correctly, and this can always be done. This theorem has some very important consequences, which we shall now discuss.

We have seen that the weak interactions are not invariant under the parity transformation. Since the interaction must be invariant under the transformation CPT (since H commutes with CPT), the interaction must also be noninvariant under at least one of the two transformations, T and C. When the postulate of parity nonconservation was experimentally demonstrated, it became obvious from the theory of β-decay that the responsible weak interaction was not invariant under C (Lee and Yang, 1956); no conclusions could be drawn as to the invariance, or otherwise, of the interaction under T. A question that arises immediately is why, if C invariance breaks down, are the mass and lifetime of any elementary particle and its antiparticle identical? For example, the lifetimes of the free μ^+ and μ^- appear to be identical, although the decay interaction is not invariant under C or P (see Section 11.8). That there quantities will be invariant is a consequence of the CPT theorem, as we shall now show.

(1) To show that particle and antiparticle have the same mass we consider the transformation CPT applied to a free particle. This generates an antiparticle with the same total energy and momentum, and therefore of the same rest mass. This rather trivial proof fails to be precise about the meaning of the mass of a particle; the correct proof has been given by Lüders and Zumino (1957), and it contains a proof of the equality of lifetimes. This happens because the rest mass is truly a complex quantity having the observed rest mass as the real part and $\hbar$ multiplied by the reciprocal of the lifetime as the complex part. However, we give another proof of the lifetimes equality due to Lee *et al.* (1956).

(2) To show the equality of lifetimes for particle and antiparticle we consider particle A and its decay products B, which are represented by eigenstates $|A\rangle$ and $|B\rangle$ of the Hamiltonian H_S. The complete Hamiltonian is, in fact, $H_S + H_\omega$ where H_ω is due to the interaction responsible for the transition from $A \to B$; H_ω is negligible compared to H_S, and its presence does not affect the eigenfunctions of H_S. We assume H_S is invariant under T, C, and P separately; H_ω must be invariant under CPT or permutations thereof. If the initial and final states are contained in a normalization volume, we find that the eigenstates $|A\rangle$ and $|B\rangle$ are stationary states, in which case

$$T|A\rangle = |A'\rangle \quad \text{and} \quad T|B\rangle = |B'\rangle, \tag{6.48}$$

where $|A'\rangle$ and $|B'\rangle$ represent the stationary states $|A\rangle$ and $|B\rangle$ with spins reversed; i.e., with opposite eigenvalues of the operator J_z. From Eq. (6.27a) we can write the matrix element of H_ω:

$$\langle B|H_\omega|A\rangle = \langle TB|TH_\omega|A\rangle^*$$
$$= \langle TB|TH_\omega T^{-1}T|A\rangle^*$$
$$= \langle B'|TH_\omega T^{-1}|A'\rangle^*. \tag{6.49}$$

Now the operator PCT commutes with H_ω; therefore
$$PCTH_\omega T^{-1}C^{-1}P^{-1} = H_\omega,$$
or
$$TH_\omega T^{-1} = C^{-1}P^{-1}H_\omega PC.$$

Substituting this equality into Eq. (6.49) after taking the complex conjugate gives
$$\langle B|H_\omega|A\rangle^* = \langle B'|C^{-1}P^{-1}H_\omega PC|A'\rangle$$
$$= \langle CB'|P^{-1}H_\omega P|CA'\rangle. \tag{6.50}$$

The states $|CB'\rangle$ and $|CA'\rangle$ are the charge conjugates of the states $|B'\rangle$ and $|A'\rangle$. In general, H_ω is not invariant under P and we can divide it into two parts: one parity-conserving, the other parity-changing,
$$H_\omega = H_1 + H_2,$$
where
$$PH_1 P^{-1} = H_1,$$
$$PH_2 P^{-1} = -H_2.$$

H_1 only has nonzero matrix elements between states of the same parity, while H_2 only has nonzero matrix element between states of opposite parity. Equation (6.50) can now be written

$$|\langle B|H_1|A\rangle + \langle B|H_2|A\rangle|^2 = |\langle CB'|H_1|CA'\rangle - \langle CB'|H_2|CA'\rangle|^2. \tag{6.51}$$

The left-hand side is the partial transition rate for the decay $A \to B$, while the right-hand side is the partial rate for the corresponding antiparticle decay $\bar{A} \to \bar{B}$. The transition rate for the decay is found by summing over all states of J_z and all accessible eigenstates of B and finally integrating over all directions. In this final integration the interference terms between states of opposite parity go to zero and the effect of the opposite sign between the matrix elements on left and right disappears. Thus the equality of Eq. (6.51) leads to the equality of lifetimes of particle and antiparticle; we have not required H_ω to be invariant under C.

There is a third theorem of importance concerning the effects of charge-conjugation invariance.

(3) To show that the parity conserving and parity nonconserving final states cannot interfere if H_ω is invariant under charge conjugation, we use the notation employed in the previous proof but must restrict ourselves to

6.9 The CPT Theorem

decay without spin in initial or final states. In this case Eqs. (6.48) reduce to

$$T|A\rangle = |A\rangle, \qquad T|B\rangle = |B\rangle.$$

We have that CPT commutes with H_ω; hence

$$TH_\omega T^{-1} = P^{-1}C^{-1}H_\omega CP = P^{-1}H_\omega P, \qquad (6.52)$$

since we are investigating the effects of invariance under C. Substituting Eq. (6.52) into Eq. (6.49) modified for these spinless conditions gives

$$\langle B|H_\omega|A\rangle^* = \langle B|P^{-1}H_\omega P|A\rangle. \qquad (6.53)$$

Let us divide $|B\rangle$ into two parts: $|B_+\rangle$, which has the same parity as $|A\rangle$, and $|B_-\rangle$, which has opposite parity. Eq. (6.53) becomes

$$\langle B_+|H_1|A\rangle^* + \langle B_-|H_2|A\rangle^* = \langle B_+|H_1|A\rangle - \langle B_-|H_2|A\rangle. \qquad (6.54)$$

It follows (in the case of C-invariance) that the amplitude for the decay $A \to B_+$ is entirely real and that for the decay $A \to B_-$ is entirely imaginary. This is the condition that excludes interference between the states of opposite parity, B_+ and B_-. This is the required result. It does not hold if there is any interaction between the decay products. The more general proof, including spin, has been given by Coester (1957).

We can conclude from the foregoing theorem that, in the absence of final state interactions, it is impossible to observe the interference effects due to the nonconservation of parity if the decay interaction is invariant under charge conjugation. In the case of the β-decay asymmetry observed by Wu and collaborators (1957), the final-state interaction, which is between the charge of the residual nucleus and the charge of the β particle, is too small to account for the observed asymmetry, and it becomes evident that the β-decay interaction is not invariant under charge conjugation. In the case of the Λ^0 decay, there is a considerable interaction in the final state, but again it is insufficient to accout for the observed asymmetry and we must conclude that the decay interaction is not invariant under charge conjugation. The case of K^+-meson decay is different: if we assume that the θ^+ and τ^+ have spin zero, the apparent nonconservation of parity is detected by the appearance of decay states having opposite parity; this is not an interference effect, and we can draw no conclusions about the invariance of the decay interaction under C.

Fitch (1967) reviewed the experimental comparisons of lifetimes, masses, and magnetic moments for particle and antiparticle that have been made as a test of CPT invariance. Apart from one test none of these can place low limits on CPT-violation. The exception is the upper limit deduced for the K^0-$\bar{K}^0$ mass difference. From the measured K_L-K_S mass difference it is possible to infer that the K^0 and $\bar{K}^0$ have the same mass to within one part in 10^{14}. This means that the ratio of CPT nonconserving amplitude to conserving amplitude is less than approximately 10^{-14} in strong interactions, 10^{-12} in electromagnetic, and 10^{-9} in weak interactions.

6.10 Conclusion

The study of weak interactions suggested that although C and P are not separately conserved the eigenvalues of the operator CP are conserved. This conclusion about the weak interactions may have been proved incorrect by the discovery of the CP-nonconserving decay

$$K_2 \to \pi^+ + \pi^-$$

(see Section 14.6). However, this decay amplitude is about 1/500 of that for the CP-conserving decay

$$K_1 \to \pi^+ + \pi^-,$$

so that the CP-nonconserving effect is small. The origin of CP-nonconservation has not yet been identified, and various suggestions have been made. As far as this chapter is concerned it is worth noting that due to Lee (1965). He shows that the definition of the operators T, C, and P are interaction dependent and that there are therefore three types of each, one for each of the three important elementary-particle interactions. The "nonequality" of the different types of one transformation implies symmetry breaking.

We complete this chapter with a brief discussion of the neutron electric dipole moment. If time-reversal invariance holds, all particles must have zero electric dipole moment. If the symmetry is broken, then a nonzero moment is possible as long as parity invariance is violated. The search for a nonzero moment is at present most easily done on the neutron and experiments (Miller *et al.*, 1967; Shull and Nathans 1967) have set an upper limit on μ/e of about 5×10^{-22} cm. This is difficult to interpret, but we can argue as follows: if there were no symmetry restriction, we would expect a moment (divided by e) of about 10^{-14} cm (the neutron Compton wavelength). If the weak interactions violated T, or if the electromagnetic interaction caused maximum C violation, we would expect this times the dimensionless weak-interaction coupling constant $GM_p^2 = 10^{-5}$, that is, 10^{-19} cm (Feinberg, 1966). The observed upper limit suggests, therefore, that the weak-interaction violation of TRI and the electromagnetic violation of C are small or zero. However, quantitative predictions of the electric dipole moment are so model dependent that it is obviously impossible to draw strong conclusions about any symmetry breaking from the upper limit.

Our treatment of these symmetry operations has not been extensive, and we refer readers to various reviews and their bibliographies (Lee and Wu, 1965, 1966; Nilsson, 1967; Kemmer et al., 1959).

REFERENCES

Atac, M., Chrisman, B., Debruinner, P., and Frauenfelder, H. (1968). *Phys. Rev. Lett.* **20**, 691 (6.6).
Baltay, C. et al. (1965). *Phys. Rev. Lett.* **15**, 591 (6.7).
Bazin, M. J., Goshaw, A. T., Zacher, A. R., and Sun, C. R. (1968). *Phys. Rev. Lett.* **20**, 895 (6.7).
Bock, R., and Schopper, H. F. (1965). *Phys. Lett.* **16**, 284 (6.4).
Bodansky, D., Braithwaite, W. J., Shreve, D. C., Storm, D. W., and Weitkamp, W. G. (1966). *Phys. Rev. Lett.* **17**, 589 (6.6).
Bjorken, J. D., and Drell, S. D. (1965). "Relativistic Quantum Fields." McGraw-Hill, New York (6.2).
Burgy, M. T., Krohn, V. E., Novey, T. B., Ringo, G. R., and Telegdi, V. L. (1958). *Phys. Rev. Lett.* **1**, 324 (6.6).
Chen, J. R. et al. (1968). *Phys. Rev. Lett.* **21**, 1279 (6.6).
Coester, F. (1957). *Phys. Rev.* **107**, 299 (6.9).
Crawford, F. S. et al. (1958). *Proc. Int. Conf. High Energy Phys., CERN, 1958*, p. 323. CERN, Geneva (6.4).
Crawford, F. S., Cresti, M., Good, M. L., Solmitz, F. T., and Stevenson, M. L. (1959). *Phys. Rev. Lett.* **2**, 11 (6.4).
Dalitz, R. H. (1957). *Rep. Progr. Phys.* **20**, 163 (6.4).
Duclos, J., Freytag, D., Schlüpman, K., Soergel, V., Heintze, J., and Rieseberg, H. (1965). *Phys. Lett.* **19**, 253 (6.7).
Feinberg, G. (1966). *Phys. Rev.* **140**, B1402 (6.10).
Fitch, V. L. (1967). *Proc. Int. Conf. High Energy Phys., 13th, Berkeley, 1966*, p. 63. Univ. of California Press, Berkeley, California (6.9).
Garwin, R. L., Lederman, L. M., and Weinrich, M. (1957). *Phys. Rev.* **105**, 1415 (6.4).
Gell-Mann, M., and Watson, K. M. (1954). *Annu. Rev. Nucl. Sci.* **4**, 219 (6.6).
Glasser, R. G., Kehoe, B., Engelmann, P., Schneider, H., Alff, C., and Kirsch, L. (1966). *Proc. Int. Conf. Weak Interactions, Argonne, 1965*, p. 13. U.S. Dept. of Commerce, Washington, D.C. (6.6).
Haas, R., Leipuner, L. B., and Adair, R. K. (1959). *Phys. Rev.* **116**, 1221 (6.4).
Hamilton, J. (1959). "The Theory of Elementary Particles," p. 358. Oxford Univ. Press, London and New York (6.6).
Handler, R., Wright, S. C., Pondrom, L., Limon, P., Olsen, S., and Kloeppel, P. (1967). *Phys. Rev. Lett.* **19**, 933 (6.6).
Jones, D. P., Murphy, P. G., and O'Neill, P. L. (1958). *Proc. Phys. Soc. London* **72**, 429 (6.4).
Kemmer, N., Polkinghorne, J. C., and Pursey, D. L. (1959). *Rep. Progr. Phys.* **22**, 368 (6.5, 6.10).
Lee, T. D., (1965). *Proc. Int. Conf. Elementary Particles, Oxford, 1965*, p. 225. Rutherford High Energy Lab., England (6.10).
Lee, T. D., and Wu, C. S. (1965). *Annu. Rev. Nucl. Sci.* **15**, 381 (6.10).
Lee, T. D., and Wu, C. S. (1966). *Annu. Rev. Nucl. Sci.* **16**, 471 (6.10).
Lee, T. D., and Yang, C. N. (1956). *Phys. Rev.* **104**, 254 (6.4, 6.9).

Lee, T. D., Oehme, R., and Yang, C. N. (1956). *Phys. Rev.* **106**, 340 (6.9).
Liu, D. C., and Roberts, W. K. (1966). *Phys. Rev. Lett.* **16**, 67 (6.7).
Lüders, G. (1954). *Kgl. Dan. Vidensk. Selsk. Mat Fys. Medd.* **28**, No. 5 (6.9).
Lüders, G., and Zumino, B. (1957). *Phys. Rev.* **106**, 385 (6.9).
Miller, P. D., Dress, W. B., Baird, J. K., and Ramsey, N. F. (1967). *Phys. Rev. Lett.* **19**, 381 (6.10).
Nilsson, J. (1967). *Proc. CERN School of Phys. 1967*, **2**. (6.10).
Overseth, O., and Roth, R. (1967). *Phys. Rev. Lett.* **19**, 391 (6.6).
Particle Data Group (1969). *Rev. Mod. Phys.* **41**, 109 (6.4).
Pauli, W. (1956). "Niels Bohr and the Development of Physics," p. 30. Pergamon Press, New York (6.9).
Roman, P. (1964). "Theory of Elementary Particles," 2nd ed. North-Holland Publ., Amsterdam (6.2).
Sachs, R. G. (1953). "Nuclear Theory." Addison-Wesley, Reading, Massachusetts (6.5).
Shull, C. G., and Nathans, R. (1967). *Phys. Rev. Lett.* **19**, 384 (6.10).
Watson, K. M., Keck, J. C., Tollestrup, A. V., and Walker, R. L. (1956). *Phys. Rev.* **101**, 1159 (6.6).
Wick, G. C. (1958). *Annu. Rev. Nucl. Sci.* **8**, 41 (6.5).
Wilkinson, D. H. (1958). *Phys. Rev.* **109**, 1603, 1610, 1614 (6.4).
Wu, C. S., Ambler, E., Hayward, R. W., Hoppes, D. D., and Hudson, R. P. (1957). *Phys. Rev.* **105**, 1413 (6.4, 6.9).
Young, K. K., Longo, M. J., and Helland, J. A. (1967). *Phys. Rev. Lett.* **18**, 806 (6.6).

VII

THE BOSONS

7.1 Introduction

So far in this book, we have emphasized general concepts applicable to elementary-particle physics and in doing this, have not hesitated to introduce experimental facts such as the existence of the π-mesons (or pions) to illustrate, for example, isotopic spin and G-parity. In addition, we have introduced the strange particles. In this chapter, we wish to take a longer and more careful look at the family of strongly interacting bosons.

7.2 The Pions

The nucleons are partially described by the Dirac equation, but have anomalous magnetic moments indicating that this theory is inadequate and that nucleons have closely associated meson fields. The meson field and the particle that must exist in the quantized field were predicted by Yukawa (1935) to explain the strong forces that exist between nucleons. From the properties of these forces Yukawa predicted that the meson would have a mass of about 200 m_e. The π-meson discovered in 1947 (Lattes *et al.*, 1947) is certainly the Yukawa meson. We shall now enumerate the more immediately observable properties of the π-mesons. The discovery of π-mesons was made in cosmic radiation; their properties were evaluated in detail by the use of accelerators able to produce mesons artificially.

1. *Electric charge.* Three kinds of π-meson exist, namely, positive, negative and uncharged, which we indicate by π^+, π^-, and π^0, respectively. The triplet nature of the charge properties of the meson field permit an immediate isotopic spin assignment, which is discussed in Section 5.4

2. *Mass.* The charged π-mesons have a mass of 139.58 MeV, the uncharged π-meson 134.98 MeV. We shall not discuss the determination

of the masses, but refer readers to the references given by the Particle Data Group (1969).

3. *Lifetime.* The direct measurement of the mean lifetime of the π^+ at rest in scintillating material gives $2.604 \pm 0.007 \times 10^{-8}$ sec (Particle Data Group, 1969); nuclear capture prevents a similar measurement of the π^- lifetime, but measurements made in flight show the π^+ and π^- lifetimes to be the same within error, as expected from the *CPT* theorem. The π^0 lifetime is too short to be measured accurately by a direct method and must be determined using the Primakoff effect (e.g., Bellettini *et al.*, 1965). The presently accepted value is $0.89 \pm 0.18 \times 10^{-16}$ sec.

4. *Decay products.* The π^+ is observed to decay into a μ-meson and a neutrino:

$$\pi^+ \to \mu^+ + \nu .$$

The μ-meson is a spin-$\frac{1}{2}$ fermion of mass 105.66 MeV (Particle Data Group, 1969), which we shall discuss later. It appears to play no part in the π-meson–nucleon interaction. Other decay modes are

$$\pi^+ \to e^+ + \nu ,$$
$$\pi^+ \to \mu^+ + \nu + \gamma ,$$
$$\pi^+ \to e^+ + \nu + \gamma ,$$
$$\pi^+ \to \pi^0 + e^+ + \nu .$$

These are all rare decay modes, although the first and last are particularly important in the theory of weak interaction and will be discussed in Chapter XI. The π^0 is observed to decay

$$\pi^0 \to \gamma + \gamma ,$$
$$\pi^0 \to \gamma + e^+ + e^- , \quad \text{branching ratio } 1.17\% ,$$

and

$$\pi^0 \to e^+ + e^- + e^+ + e^- , \quad \text{branching ratio } 3.2 \times 10^{-5}$$

(Samios, 1961; Samios *et al.*, 1962). Other decays that are energetically possible but have not been observed are

$$\pi^0 \to \mu^+ + e^- ,$$
$$\pi^0 \to \mu^- + e^+ ,$$
$$\pi^0 \to e^+ + e^- , \quad \text{etc.}$$

Decay into more than two photons has not been observed; the two-photon decay permits us to conclude that the spin of the π^0-meson is zero or an even integer—see Section 7.3.

5. *Spin.* In fact, the π-meson appears to have no spin; this is the result of an analysis discussed in Section 7.3.

7.3 The Spin and Parity of the Pions

6. *Parity.* A study of the absorption of π^--mesons in deuterium indicates that this particle carries odd intrinsic parity (Section 7.3). As we expect all the π-mesons to have the same parity, this is normally assumed.

7. *Interaction properties.* The π-mesons interact strongly with nucleons, and the properties of the interaction are of considerable interest; some aspects will be discussed in the next chapter.

7.3 The Spin and Parity of the Pions

The spin of the positive π-meson is deduced from a study of a π-meson production reaction and its inverse:

$$p + p \rightarrow \pi^+ + d, \qquad (7.1)$$

$$\pi^+ + d \rightarrow p + p. \qquad (7.2)$$

These reactions are observed for unpolarized beams and without distinguishing different final spin states. We can therefore apply detailed balance arguments in comparing cross sections measured at the same center-of-mass total energy.

Let s_π, s_p, and s_d be the spins (in units of $\hbar$) of the π-meson, proton, and deuteron, respectively. Let σ_1 and σ_2 be the cross sections for the reactions (7.1) and (7.2), respectively. At the same center-of-mass energy the cross sections are related by the appropriate form of Eq. (3.96), namely,

$$\frac{d\sigma_1}{d\Omega} = \left(\frac{p_\pi}{p_p}\right)^2 \frac{(2s_\pi + 1)(2s_d + 1)}{(2s_p + 1)^2} \frac{d\sigma_2}{d\Omega}, \qquad (7.3)$$

where p_π is the π^+ (or d) momentum and p_p is the proton momentum, in the center-of-mass system. Now $s_d = 1$ and $s_p = \frac{1}{2}$, so that Eq. (7.3) becomes

$$\frac{d\sigma_1}{d\Omega} = \frac{3}{4}\left(\frac{p_\pi}{p_p}\right)^2 (2s_\pi + 1)\frac{d\sigma_2}{d\Omega}.$$

We find the relation for the total cross sections by integrating, taking care not to count each proton twice, as described in Section 3.13. We find

$$\sigma_1 = \tfrac{3}{2}(p_\pi/p_p)^2(2s_\pi + 1)\sigma_2. \qquad (7.4)$$

We use this equation to compare the observed cross sections and hence determine the value of s_π. It is sufficient for our immediate purpose to use the following data:

(1) The total cross section for reaction (7.1) at an incident proton energy of 340 MeV is 0.18 ± 0.06 mb (Cartwright *et al.*, 1953). This incident energy corresponds to π^+ mesons produced with an energy of 22.3 MeV in the center-of-mass system.

(2) The total cross section for reaction (7.2) at an incident π-meson

energy of 29 MeV is 3.1 ± 0.3 mb (Durbin *et al.*, 1951). The incident π-meson 29-MeV energy in the laboratory system corresponds to 25-MeV π-mesons in the center-of-mass system. Since s_π is an integer, we do not require great precision for its determination, and we can use nonrelativistic approximations. The center-of-mass energies are sufficiently equal for the reactions to be compared.

Evaluating the center-of-mass momenta gives

$$p_\pi/p_p = 0.201 ,$$

and substituting the data in Eq. (7.4), we find that

$$(0.18 \pm 0.06) = (3.1 \pm 0.3)0.0606(2s_\pi + 1) .$$

The nearest integer for s_π is obviously zero, and we conclude that the π^+ meson has zero spin. The analysis would be invalid if, by chance, the π^+ did have spin and the meson beams used were polarized; however, it is very unlikely that the π-meson beams produced by cyclotrons are sufficiently polarized (Clark *et al.*, 1951) to give an accidental spin-0 result, and our conclusion is almost certainly correct.

Since the π^+ and π^- are charge conjugates, we deduce that the π^- spin is also zero. Assuming that the charged pions and the neutral pion belong to an isotopic-spin triplet, it follows that the π^0 also has spin zero, since the rotations in isospin space, which convert the pions from one charge state to another, have nothing to do with the ordinary space properties of the pions. Further justification comes from the fact that there are no inconsistencies in the subsequent elementary-particle physics based upon this spin zero result.

The charged pion parity can be determined by a study of the absorption of slow π^--mesons in liquid hydrogen and liquid deuterium. The steps that lead to the nuclear absorption of low-energy negative π-mesons incident upon the liquid target are

(1) Slowing down of mesons by ordinary stopping-power mechanism, which finally leaves the incident pion with a velocity comparable to the electron orbital velocities; the time taken is about 10^{-10} sec. Nuclear interaction in flight can take place but the probability is small compared with stopping if the meson range is small.

(2) Capture into an excited Bohr orbit around a proton (or deuteron).

(3) Deexcitation of this mesonic atom by inelastic collisions with neighboring molecules to a level having a principal quantum number of about 7 takes about 10^{-12} sec. At this point, the presence of atomic electric fields causes Stark mixing of different l-values (Day *et al.*, 1960). Absorption from S-states is strongly favored because they have large wavefunction densities at the nucleus and the consequence is that S-wave nuclear absorp-

7.3 The Spin and Parity of the Pions

tion takes place from levels of high principal quantum number in preference to deexcitation. This has been confirmed by observing that the number of "at rest" π^- decays in liquid hydrogen is less than expected if deexcitation occurred before absorption (Fields et al., 1960).

An understanding of this mechanism is important in establishing from which state absorption occurs. In the absence of Stark mixing, processes such as absorption from a P state may occur in preference to deexcitation to an S state. If such uncertainties existed, it would be difficult to be sure of the parity of the wavefunction of the preabsorption pion state and of its angular momentum.

In hydrogen the absorption reactions that occur are as follows (Panofsky et al., 1951):

$$\pi^- + p \to \pi^0 + n, \qquad (7.5)$$

$$\pi^- + p \to \gamma + n. \qquad (7.6)$$

The ratio of the probability that reaction (7.5) takes place to the probability that reaction (7.6) takes place is 1.50 ± 0.11 (Cassels et al., 1957). This is the Panofsky ratio. In deuterium the energetically possible reactions are

$$\pi^- + d \to \pi^0 + 2n, \qquad (7.7)$$

$$\pi^- + d \to \gamma + 2n, \qquad (7.8)$$

$$\pi^- + d \to 2n. \qquad (7.9)$$

Reaction (7.7) is unobserved, and the probability ratio for reactions (7.9) to (7.8) is 2.36 ± 0.36 (Kuehner et al., 1958; Panofsky et al., 1951).

The hydrogen absorption reactions lead immediately to the conclusion that the π^--meson is a boson. From that point we proceed directly to the deuterium absorption, as it gives us immediately more information. We assume that the proton and neutron have even relative parity and that the π^- spin is zero. We know absorption takes place from the S-state of orbital motion, so we have that the total angular momentum (j) of the initial state is just the deuteron spin, which is one. Therefore $j = 1$ and the initial state is $^3S_1$, in spectroscopic notation. The deuteron has an even-parity wave function (it is a superposition of the even states $^3S_1$ and $^3D_1$); the orbital S state is even, so the parity of the initial state is that of the meson. We now examine the final-state of reaction (7.9); from Section 5.3 we know that the only permitted states for two neutrons are, in spectroscopic notation,

$$^1S_0, {}^3P_0, {}^3P_1, {}^3P_2, {}^1D_2, {}^3F_2, {}^3F_3, \ldots, \text{etc.}$$

The only state with $j = 1$ is $^3P_1$, which has odd parity; hence the fact that reaction (7.9) occurs indicates that the π^--meson has odd parity.

An odd-parity assignment to the π^0 is weakly supported by the nonoccurrence of reaction (7.7). In this case the energy available to the two neutrons is very small, and it is very unlikely that they can be in any state of relative motion other than $^1S_0$ (cf. the argument of Section 3.2). From conservation of angular momentum it follows that the meson must be in an orbital angular momentum state $l = 1$, i.e., a P state, relative to the center of mass of the two neutrons; the angular-momentum barrier strongly inhibits the reaction going into this state because of the low π^0 kinetic energy. However, this final state is forbidden by conservation of parity if the π^0 and the π^- have the same parity. Taken together, we see that we cannot definitely assign odd parity to the π^0, since the nonoccurrence of reaction (7.7) could be due to the angular-momentum barrier. Nevertheless, it is usual to assume that all three types of π-meson have odd intrinsic parity.

Note that we have assumed that the proton and neutron have even relative parity. This has been discussed in Section 6.2.

We now return to the absorption in hydrogen and find that for reaction (7.5) the initial and final states are $^2S_{1/2}$ with odd parity. For reaction (7.6) this implies that the outgoing photon is due to an electric dipole transition in which the nucleon spin has flipped. The surprising point about these reactions is that (7.5) is only 50% more probable than (7.6) when the former is expected to be about six times as frequent as the latter. This expectation comes from the knowledge that the electromagnetic interaction has about 1/137 the strength of the meson–nucleon interaction and the weight of the density of states factors. The observed ratio suggests that (7.5) is inhibited in some way. In fact, the meson–nucleon interaction is much weaker when the relative orbital angular momentum is zero compared with its value when the relative motion is a P state. Thus we have anticipated a result that is brought out more strongly by the meson scattering experiments.

In fact, an independent and absolute determination of the parity of the neutral pion can be made by considering the significance of its decay into two photons and the related decay into two electron–positron pairs. The first process is governed by various symmetry and conservation laws with results that were first discussed by Yang (1950). We follow his analysis in a modified way. We proceed by setting up a Cartesian coordinate system in the center of mass, having its z axis along the direction of motion of one of the outgoing photons. We are then interested in the properties of the initial state (the π^0-meson) and of the final state (two photons) under three transformations. The first is the parity transformation, the second is rotation around the z axis, and the third is rotation through 180° around the x axis. The effect of each transformation on the state vectors is represented by an operator, and in some cases the vectors are eigenvectors of these oper-

7.3 The Spin and Parity of the Pions

ators. We expect that these transformations do not change the physical consequences and, as a result of our arguments of Section 1.6, that the operators commute with the Hamiltonian operator for the complete system; it follows that the initial and final states must have the same eigenvalues. Thus we shall investigate the eigenvalues for both initial and final states to discover the values of spin and parity that the π^0-meson can have if it decays into two photons. Actually, the selection rules we derive can be applied to any decay into two photons; for example, to positron annihilation.

We first examine the properties of the transformations.

(1) *Parity transformation.* This transformation P has the effect of reversing the direction of motion of all particles while leaving all angular momenta unchanged.

(2) *Rotation around the z axis.* A vector is only an eigenvector of this rotation operator R_z if it is a pure state having a discrete value of the z component of angular momentum, j_z. For a rotation through an angle α the eigenvalue is $\exp i j_z \alpha$.

(3) *Rotation through 180° around the x axis.* For two photons moving along the z axis this rotation R_x has the effect of reversing the direction of movement and the intrinsic angular momentum of each photon. In the case of a single-particle state this rotation has the same effect as it would have on the spherical harmonic having the same angular-momentum eigenvalues.

We shall consider the final state of two photons. From our knowledge of the electromagnetic field we know that a photon carries intrinsic angular momentum (spin) $\pm \hbar$ directed along its direction of motion. We call the photon having $+\hbar$ right-hand polarized, while that having $-\hbar$ is called left-hand polarized. In our case the two photons are moving in opposite directions along the z axis and we describe the two photons by two arrows, the first applying to the photon moving in the $+z$ direction, the second applying to that moving in the $-z$ direction; the arrow points up ($\uparrow$) if the z component of the photon spin is $+\hbar$ and points down ($\downarrow$) if it is $-\hbar$. Thus $\uparrow\downarrow$ indicates that both photons are right-hand, and so on. There are four such state vectors, and we are interested in those or their combination that are eigenstates of P and R_z; they are

$$|\uparrow\uparrow\rangle, \quad |\downarrow\downarrow\rangle, \quad |\uparrow\downarrow + \downarrow\uparrow\rangle, \quad \text{and} \quad |\uparrow\downarrow - \downarrow\uparrow\rangle,$$

The properties of our transformation operators are such that:

(1) P exchanges the arrows;
(2) R_z does nothing to the arrows but multiplies the vectors by $\exp i j_z \alpha$;
(3) R_x exchanges the arrows and turns each arrow upside down.

We can now construct a table of eigenvalues of these states (Table 7.1).

TABLE 7.1

EIGENVALUES OF TWO-PHOTON FINAL STATES

Spin function	Transformation		
	1 P	2 R_z	3 R_x
$\lvert\uparrow\uparrow\rangle$	+1	$e^{2i\alpha}$	
$\lvert\downarrow\downarrow\rangle$	+1	$e^{-2i\alpha}$	
$\lvert\uparrow\downarrow + \downarrow\uparrow\rangle$	+1	+1	+1
$\lvert\uparrow\downarrow - \downarrow\uparrow\rangle$	−1	+1	+1

We can now do the same for the initial state, which is the neutral π-meson at rest. If this particle has spin j and z component of spin j_z, then its properties under rotation are the same as those of the spherical harmonic having $l = j$ and $l_z = j_z$ [Eq. (2.57)]. Then R_z has eigenvalues $\exp ij_z\alpha$. If $j_z = 0$, then R_x has eigenvalues $(-1)^j$. We can now construct a second table, Table 7.2, which sets out the eigenvalues for initial π^0 states of different spin and parity. This initial state can have any value of j_z from $-j$ to $+j$ and will be coherent sum of all these states if in a pure spin state and an incoherent sum if unpolarized; therefore the column of R_z eigenvalues includes $2j + 1$ possibilities.

TABLE 7.2

EIGENVALUES OF POSSIBLE π^0 STATES

Spin j	Transformation		
	1 P	2 R_z	3 R_x
0	$\begin{cases} +1 \\ -1 \end{cases}$	+1 +1	+1 +1
1	$\begin{cases} +1 \\ -1 \end{cases}$	$e^{i\alpha}, 1, e^{-i\alpha}$	$\begin{cases} -1 \\ -1 \end{cases}$
2, 4, 6, ...	$\begin{cases} +1 \\ -1 \end{cases}$	$e^{ij\alpha}, \ldots, 1, \ldots, e^{-ij\alpha}$	$\begin{cases} +1 \\ +1 \end{cases}$
3, 5, 7, ...	$\begin{cases} +1 \\ -1 \end{cases}$	$e^{ij\alpha}, \ldots, 1, \ldots, e^{-ij\alpha}$	$\begin{cases} -1 \\ -1 \end{cases}$

We now compare the rows of the two Tables 7.1 and 7.2, to construct a final Table 7.3, which indicates the two photon states that are accessible to a decaying neutral meson having the spin and parity indicated. We shall

7.3 The Spin and Parity of the Pions

TABLE 7.3

Two Photon States for Decay of Various Types of π^0-Meson

	Spin j			
Parity	0	1	2, 4, ...	3, 5, ...
Even	$\lvert\uparrow\downarrow + \downarrow\uparrow\rangle$	Forbidden	$\lvert\uparrow\uparrow\rangle, \lvert\downarrow\downarrow\rangle$, and $\lvert\uparrow\downarrow + \downarrow\uparrow\rangle$	$\lvert\uparrow\uparrow\rangle$ and $\lvert\downarrow\downarrow\rangle$
Odd	$\lvert\uparrow\downarrow - \downarrow\uparrow\rangle$	Forbidden	$\lvert\uparrow\downarrow - \downarrow\uparrow\rangle$	Forbidden

describe this comparison in two cases. Consider the meson having spin 1 and of either intrinsic parity; only the state having $R_z = 1$ can decay into two photons, and since this corresponds to $j_z = 0$, we may also examine the eigenvalues of R_x which is -1 for the meson. There is no two-photon state having this eigenvalue, and we conclude that a spin-1 state cannot decay into two photons. Our second example is that of a meson having even spin of 2 or greater and even parity. If $j_z = 0$ ($R_z = 1$), then the particle can decay into $\lvert\uparrow\downarrow + \downarrow\uparrow\rangle$; if $j_z = +2$ or -2 ($R_z = e^{2i\alpha}, e^{-2i\alpha}$), it decays into $\lvert\uparrow\uparrow\rangle$ or $\lvert\downarrow\downarrow\rangle$, respectively, for which the R_x eigenvalue has no influence.

As it is almost certain that the π^0 has zero spin, we find that a pseudoscalar π^0 will decay into the two photons with the state vector

$$\lvert\uparrow\downarrow - \downarrow\uparrow\rangle.$$

For convenience we can rewrite this state vector in a way that indicates the polarizations of the two photons. Using the relation between polarization, direction of motion, and spin z component we find that this wavefunction corresponds to

$$\sqrt{\tfrac{1}{2}}\,(\lvert RR\rangle - \lvert LL\rangle). \tag{7.10}$$

In this and the subsequent equation, the first letter in the ket indicates right or left circular polarization for the photon moving along the z axis, the second letter indicates the polarization of the opposite photon.

Similarly, the scalar meson decays into the state

$$\sqrt{\tfrac{1}{2}}\,(\lvert RR\rangle + \lvert LL\rangle). \tag{7.11}$$

We now wish to examine the effect of making polarization measurements on these two outgoing photons. If we were to make a measurement of the polarization of one of the outgoing photons, then we would find left-hand or right-hand polarization with equal probability, independent of the parity of the decaying meson. This follows from the fact that R and L have amplitudes with the same modulus throughout both the photon state vectors (7.10) and (7.11). The next step would be to measure the circular polarization state of both outgoing photons, and we can predict the results of such measurements by finding the probability of obtaining a particular result. Thus the probability of finding both photons left-hand polarized is the square of the matrix element between the bra $\langle LL|$ and the ket of the state under observation; for scalar or pseudoscalar mesons this is

$$|\langle LL|\sqrt{\tfrac{1}{2}}(|RR\rangle \pm |LL\rangle)|^2 = \tfrac{1}{2}.$$

The probability is the same for finding both photons right circularly polarized, but the probability for finding the photons oppositely polarized is zero. These results are all independent of the state vectors for the two photons, and we conclude that it is impossible to determine the parity of the π^0-meson by measuring the circular polarization of its decay photons.

However, it is possible to deduce the parity by measuring the state of plane polarization of both photons. We proceed by rewriting the two state-vectors (7.10) and (7.11). To do this we recall that a right (or left) circularly polarized photon can also be described by a sum of two state-vectors with correct relative phase, where we use $|X\rangle$ to represent a photon plane polarized with its electric vector in xz plane and $|Y\rangle$ to represent a photon polarized with its electric vector in the yz plane. Thus, using Eqs. (13.18) and (13.20),

$$|R\rangle = -\sqrt{\tfrac{1}{2}}(|X\rangle + i|Y\rangle), \qquad (7.12)$$

$$|L\rangle = \sqrt{\tfrac{1}{2}}(|X\rangle - i|Y\rangle). \qquad (7.13)$$

We substitute (7.12) and (7.13) into the photon state vectors (7.10) and (7.11) to find that the two photon state vector is

$$\pi^0 \text{ pseudoscalar:} \quad -i\sqrt{\tfrac{1}{2}}(|YX\rangle + |XY\rangle),$$

$$\pi^0 \text{ scalar:} \quad \sqrt{\tfrac{1}{2}}(|XX\rangle + |YY\rangle).$$

An examination of these state vectors indicates immediately that if the π^0-meson is pseudoscalar, the two photons are always oppositely plane polarized. If the meson is scalar, the two photons have the same plane polarization. This determination of photon polarization has not been done directly, as there is no way known at present of measuring the plane polarzation of high-energy photons with high efficiency. However, the decay of π^0-meson into two electron–positron pairs does reflect this polarization correlation in a correlation between the planes of the pairs. If the angle between these planes is ϕ, the correlation is

$$P_p(\phi) = 1 - a\cos 2\phi,$$

$$P_s(\phi) = 1 + a\cos 2\phi.$$

a is a positive quantity that is a function of the energies and angles of the pairs. The subscript s or p indicates the correlation for even- or odd-parity meson respectively. The observed correlation (Plano et al., 1959) is unambiguously of the kind expected for an odd-parity π^0. The remaining uncertainties are due to any uncertainties there may be in the use of quantum electrodynamics applied to the internal conversion in π^0-decay. It is believed that the theory is reliable.

7.4 The K-Mesons

We have already indicated in Section 5.5 that there are four K-mesons (or kaons):

$$y = 1, \quad \text{isospin doublet:} \quad K^+, K^0$$
$$y = -1, \quad \text{isospin doublet:} \quad \bar{K}^0, K^-$$

The $y = -1$ members are antiparticles to the $y = 1$ members. The kaons require accelerators in the energy range of 1 GeV and greater for their production, and their properties were studied systematically as these became available in the 1950s.

1. *Mass.* The neutral kaons have a mass 497.7 ± 0.3 MeV, determined by kinematic measurements on the decay $K^0 \to \pi^+ + \pi^-$ observed in bubble chambers. The charged kaons have a mass 493.78 ± 0.17 MeV, determined by kinematic measurements on the decay $K^+ \to \pi^+\pi^+\pi^-$ observed in photographic emulsions [for references see Particle Data Group, 1969].

2. *Lifetime.* The charged kaons have mean life of 1.235×10^{-8} sec, determined electronically. The neutral kaons produced strongly are mixtures of two states that have exponential decays. Consequently neutral decays show two lifetimes, due to a short-lived kaon K_S with a mean life 0.862×10^{-10} sec and a long-lived kaon K_L with a mean life 5.4×10^{-8} sec. We shall discuss this phenomenon in Chapter XIV and Section 7.5.

3. *Decay Products.* The kaons have a large variety of decay channels, which are described in Section 11.10. This wealth of decay channels has been of great use in probing the strangeness-nonconserving parts of the weak interaction, as will become clear in Chapter XI.

4. *Spin.* No direct test by detailed balance is possible in the case of kaons, because of the impossibility of arranging the required projectiles and targets. All methods therefore rely upon conservation of angular momentum applied to decays. Since symmetry requires a $2\pi^0$ system to have an even orbital angular momentum, the decay $K_S \to 2\pi^0$ indicates the spin of the K is even. The decay $K^+ \to \pi^+ + \gamma$ is forbidden if the K^+ spin is zero, but

would be expected to be strongly competitive with other modes if the spin were not zero: this mode is not observed. The decay $K^+ \to \mu^+ + \nu$ gives muons with the same polarization as muons from $\pi^+ \to \mu^+ + \nu$, which indicates spin 0 for the K^+. Evidence from the decay mode $K^+ \to \pi^+ + \pi^- + \pi^+$ is also in favor of spin 0 (Section 7.9). The conclusion is that the spin of the kaons is 0, and this is entirely consistent with all data.

5. *Parity.* The nonconservation of parity in weak interactions prevents a direct determination of the kaon parity. Conservation of hypercharge in strong interactions means that it is only possible to determine the relative parity for the three particles K, Λ, N. It is now conventional to assign the same parity to Λ and N, so this relative parity will be called the kaon parity. Consider the reaction

$$K^- + {}^4\text{He} \to {}^4_\Lambda\text{H} + \pi^0,$$

where the hyperfragment ${}^4_\Lambda\text{H}$ is a proton, two neutrons, and a Λ hyperon bound together. This object and this reaction have been observed. If ${}^4_\Lambda\text{H}$ has spin 0, then the reaction can occur only if π^0 and K^- have the same parity (assuming both have spin 0). Thus an odd-parity assignment to the kaons depends on the ${}^4_\Lambda\text{H}$ spin (j): this hyperfragment decays

$$ {}^4_\Lambda\text{H} \to \pi^- + {}^4\text{He}$$

with a branching fraction of about 0.67. This is as expected if $j = 0$, whereas if $j = 1$ this fraction is expected to be much smaller. If there were an excited state of the ${}^4_\Lambda\text{H}$ that could decay by gamma emission to a $j = 0$ ground state 0, this could then be the state produced on K^- absorption, and the original argument leading to odd parity for the kaons would become invalid. However, arguments based upon reasonable assumptions about the Λ–N interaction indicate that an excited state cannot exist.

The recent success of unitary symmetry has strengthened the odd-parity (and spin-0) assignment to the kaons. They are believed to be partners with the pions in an octet (Section 9.4), and all particles within a multiplet must have the same spin and parity. We shall from now on assume the kaons are pseudoscalar.

7.5 The Neutral *K*-Meson System

In this section we wish to introduce the details of the neutral K-meson system, which has some remarkable properties. We have already observed that there exist two neutral K-mesons, which are related by charge conjugation and differ in their hypercharge (i.e., strangeness) assignment. Strong interactions conserve hypercharge, so the neutral kaon produced by a strong

7.5 The Neutral K-Meson System

interaction is determined. However, kaons are involved in the weak interactions, which need not conserve strangeness. Thus, in addition to causing the decay of kaons, the weak interaction can cause the transition of a neutral kaon to its antiparticle:

$$K^0 \rightleftharpoons \bar{K}^0.$$

Let us consider the decay properties: to do this we shall anticipate some of the material of Chapter XI. We consider the two operators C and P responsible for charge conjugation and the parity transformation, respectively. The weak interactions do not conserve the eigenvalues of P or C separately, but do conserve the eigenvalue of the combined operator CP—with one qualification that there exists a small CP-nonconserving interaction, which we shall ignore until Chapter XIV, since its effect is small compared with the effect of the normal weak interaction. It follows that if we consider the nonleptonic decays of the neutral kaons, we may be able to classify them according to their CP eigenvalue. The decay final-states of interest are

(a) $\pi^0 + \pi^0$
(b) $\pi^+ + \pi^-$
(c) $\pi^0 + \pi^0 + \pi^0$
(d) $\pi^+ + \pi^- + \pi^0$

From Section 6.7 we know that the eigenvalues of C are $(-1)^{l+s}$ and of P are $(-1)^l$ for a two-boson system. Thus the two-pion states (a) and (b) have $CP = (-1)^s$, which is $+1$ since $s = 0$. From the discussion in Section 6.4 we know that a state of three pions, two of which are identical, cannot have simultaneously spin 0 and even parity; therefore it follows that $P = -1$ for state (c) and, since $C = +1$ for the π^0, that $CP = -1$. The situation in state (d) is slightly more complicated; for the $\pi^+\pi^-$ part $CP = +1$. The π^0 brings in other factors, $C = +1$, and $P = -(-1)^L$ for its intrinsic parity and orbital angular momentum L with respect to the $\pi^+\pi^-$ center of mass. Since the kaon is spin 0, this L must be matched by another L in the $\pi^+\pi^-$ system, but this does not change the overall CP. Thus if $L = 0$, or is even, $CP = -1$; and if L is odd, $CP = +1$. L is probably zero, since $L = 1$ introduces a large angular-momentum barrier which would depress the decay rate in a manner not observed, and there is no evidence for $L \neq 0$ in $K^+ \to \pi^+\pi^-\pi^+$ decay. Thus $CP = -1$ is the most likely value for (d).

The strong interaction produces K^0 and $\bar{K}^0$ mesons, which we represent by base states $|K^0\rangle$ and $|\bar{K}^0\rangle$, where we can choose the phase so that

$$C|K^0\rangle = -|\bar{K}^0\rangle, \qquad C|\bar{K}^0\rangle = -|K^0\rangle.$$

The weak interactions, by conserving CP, will emphasize base states that are eigenvalues of CP. We call these $|K_1\rangle$ and $|K_2\rangle$ with eigenvalues $+1$ and -1, respectively:

$$CP|K_1\rangle = |K_1\rangle, \qquad CP|K_2\rangle = -|K_2\rangle.$$

Since $P|K^0\rangle = -|K^0\rangle$ and $P|\bar{K}^0\rangle = -|\bar{K}^0\rangle$, the linear combinations of $|K^0\rangle$ and $|\bar{K}^0\rangle$ that are eigenstates of CP are

$$|K_1\rangle = \sqrt{\tfrac{1}{2}}\{|K^0\rangle + |\bar{K}^0\rangle\}, \quad |K_2\rangle = \sqrt{\tfrac{1}{2}}\{|K^0\rangle - |\bar{K}^0\rangle\}. \qquad (7.14)$$

Inverting these equations we have

$$|K^0\rangle = \sqrt{\tfrac{1}{2}}\{|K_2\rangle + |K_1\rangle\}, \quad |\bar{K}^0\rangle = \sqrt{\tfrac{1}{2}}\{|K_1\rangle - |K_2\rangle\}. \qquad (7.15)$$

The $\sqrt{\tfrac{1}{2}}$ factors are put in to preserve the orthonormality of all the states involved. What do Eqs. (7.14) and (7.15) mean? Equation (7.15) shows that a measurement that shows up the CP eigenvalue of a neutral kaon of well-defined strangeness will find $CP = +1$ will probability of 50 % and $CP = -1$ with probability of 50 %. Nonleptonic decay can do just this. Similarly, Eq. (7.14) shows that a measurement of the hypercharge of a K_1 (or of a K_2) meson will find $+1$ with probability 50 % and -1 with equal probability.

Conservation of CP tells us that it is the K_1 fraction of a neutral kaon that can decay into two pions ($CP = +1$), whereas the K_2 part can only decay into three pions ($CP = -1$, in these circumstances). (This does not include the leptonic decay channels, which are open to both but contribute only a fraction of the decay probability and do not effect these arguments.) There is no reason why the decay rates into these $CP = +1$ and -1 channels should be the same, and in fact they are different by a factor of about 600; the consequence is that the K_1 fraction decays rapidly with a lifetime of about 9×10^{-11} sec, while the K_2 fraction decays with a lifetime of about 6×10^{-8} sec. That neutral kaons would show this particle-mixture behavior was predicted by Gell-Mann and Pais (1955).

Let us follow a neutral kaon beam from its source. Let us suppose it was produced so that all kaons had hypercharge $+1$ (K^0). Within the next few periods of 10^{-10} sec the K_1 fraction will decay into the two-pion final state (the leptonic decay fractions are too small to be important for K_1), leaving the beam finally as entirely K_2 mesons, decaying slowly into 3π and other channels with a very much longer lifetime. That the K-mesons behave this way was first demonstrated by Brown et al. (1960). Using a xenon bubble chamber that could detect both the $\pi^+\pi^-$ and the $2\pi^0$ decay modes of the K_1, they showed that 0.53 ± 0.05 of all K^0 produced decayed by these modes, in agreement with the expected 0.5. The remaining K^0 known to have been produced (by association with Λ) left the chamber before a significant number had decayed.

7.5 The Neutral K-Meson System

Let us return to our neutral K-meson beam and discuss the properties of the pure K_2 beam that remains after the K_1 part has been lost. It this beam is passed through an absorber, strangeness-conserving reactions can take place; for example,

$$\bar{K}^0 + p \to \pi^0 + \Sigma^+,$$
$$\bar{K}^0 + n \to \pi^0 + \Sigma^0,$$
$$K^0 + p \to K^+ + n, \quad \text{etc.}$$

These reactions attenuate the beam, and in considering their effect on the beam it is necessary to consider its K^0 and $\bar{K}^0$ qualities. Equation (7.14) tells us that the pure K_2 beam can appear to be 50% K^0 and 50% $\bar{K}^0$. These fractions have different strangeness and, as in the case of K^+ and K^- (Section 5.5), can have absorption cross sections different by a factor of as much as 3 at low energies. The result is the $\bar{K}^0$ fraction is attenuated more rapidly than the K^0 fraction and the beam emerges from the absorber more K^0 than $\bar{K}^0$. Subsequently, the beam is free to decay, but as a result of the change in K^0 to $\bar{K}^0$ ratio, it now contains some K_1 fraction, although the beam was pure K_2 before passing through the absorber. This K_1 fraction rapidly decays and leads to the reappearance of the 2π decay modes immediately after the absorber. Pais and Piccioni (1955) first suggested the existence of this process of regeneration, and its experimental existence (see Section 14.2) confirms the particle-mixture hypothesis.

That K_2 particles can enter an absorber and emerge as K_1 particles may at first sight seem very unexpected behavior. It can only happen if there is a difference in absorption of the K^0 and $\bar{K}^0$ fractions. There is a classical analogy that will be more familiar to many readers. Consider a beam of plane-polarized light that traverses two filters. The first absorbs only circularly (for example, right-hand) polarized light, and to describe the behavior of the beam it is necessary to represent the plane polarization by the sum of two circular polarizations of correct amplitude and relative phase; one of these fractions (right-hand) is absorbed, and the light emerges predominantly circularly polarized (left-hand). The second filter absorbs only plane-polarized light, and to describe the attenuation it is necessary to represent the circular polarization by the sum of two mutually perpendicular plane polarizations of correct amplitude and phase; one of these components is absorbed, and the emerging beam is once more predominantly plane-polarized. These changes in the nature of the beam are analogous to those that occur in the neutral K-meson beam, and we notice that in both cases the necessary representation depends upon the treatment that the beam undergoes; the intensities of light beams become probabilities for single neutral K-mesons.

We shall discuss the behavior of single neutral K-mesons further in Chapter XIV.

An interesting and important application of the particle-mixture formalism is to systems consisting of a $K^0\bar{K}^0$ pair. Since the total system has zero baryon number and hypercharge, it can be an eigenstate of G and C; we are interested in the latter. We use the notation that makes the vector $|K^0\bar{K}^0\rangle$ represent a state in which particle 1 is K^0 and particle 2 is $\bar{K}^0$, and so on. The state that has $C = +1$ is

$$|+1\rangle = \sqrt{\tfrac{1}{2}}\{|K^0\bar{K}^0\rangle + |\bar{K}^0 K^0\rangle\}, \tag{7.16}$$

and that with $C = -1$ is

$$|-1\rangle = \sqrt{\tfrac{1}{2}}\{|K^0\bar{K}^0\rangle - |\bar{K}^0 K^0\rangle\}. \tag{7.17}$$

Substituting from Eq. (7.15) we have

$$|\pm 1\rangle = \tfrac{1}{2}\sqrt{\tfrac{1}{2}}\{(|K_1 K_1\rangle - |K_2 K_2\rangle + |K_2 K_1\rangle - |K_1 K_2\rangle) \\ \pm (|K_1 K_1\rangle - |K_2 K_2\rangle - |K_2 K_1\rangle + |K_1 K_2\rangle)\}.$$

Therefore

$$|+1\rangle = \sqrt{\tfrac{1}{2}}\{|K_1 K_1\rangle - |K_2 K_2\rangle\}, \tag{7.18}$$

$$|-1\rangle = \sqrt{\tfrac{1}{2}}\{|K_2 K_1\rangle - |K_1 K_2\rangle\}. \tag{7.19}$$

Consider a state having $C = +1$, which decays into $K^0 + \bar{K}^0$. Equation (7.18) tells us that if one of these K^0 decays as a $K_1(K_2)$, the other must also decay as a $K_1(K_2)$. In the case of $C = -1$, if one K^0 decays as a K_1, the other must decay as a K_2. This situation can now be turned around to provide a method of finding the eigenvalue of C of a state that decays into two neutral kaons. Unambiguous observation of two K_1 decays fixes $C = +1$ and, in contrast, the observation that two K_1 decays never occur together fixes $C = -1$.

The observation of K_1 and K_2 decays is particularly appropriate to bubble chambers. The K_1 lifetime is such that if the neutral kaon is produced in an interaction within the chamber liquid, it will normally decay within a few centimeters of the point of production. The $2\pi^0$ mode is not normally detected in a liquid-hydrogen chamber, but the $\pi^+\pi^-$ mode has very clear characteristics—a V event pointing back to the source. The presence of two such $\pi^+\pi^-$ decays belonging to the same event would indicate two neutral kaons, and if strangeness conservation implied a $K^0\bar{K}^0$ pair the charge-conjugation eigenvalue for this $K\bar{K}$ system would be $+1$. The K_2 mode is noticeable by its absence, since its mean life implies that most will leave a normal-sized bubble chamber before decaying. Thus the observation of only one K_1 decay when a $K^0\bar{K}^0$ pair is known to be present is strong indication of $C = -1$, once correction has been made for the possibility that $K_1 \to 2\pi^0$ decays are being missed.

This situation arises in the decay of the ϕ-meson (Particle Data Group, 1969), which is readily produced in K^- interactions; for example,

$$K^- + p \to \phi + \Lambda.$$

The data indicate that ϕ decays with a branching fraction of about 36% into the state $K^0\bar{K}^0$. These particles are always observed as $K_1 K_2$, and this fixes $C = -1$ for the ϕ, since it decays by C-conserving strong interactions.

7.6 Meson Resonances

Many bosons have been discovered, mainly in bubble chambers, and they are normally revealed by plotting the invariant mass of combinations of two or more mesons, when several are produced in a final state. Consider, for example, pion–nucleon interactions at high energy where there are two pions in the final state, such as

$$\pi^- + p \to \pi^+ + \pi^- + n. \tag{7.20}$$

The invariant mass M of the two pions is given by

$$M^2 = (E_1 + E_2)^2 - (\mathbf{p}_1 + \mathbf{p}_2)^2, \tag{7.21}$$

where subscripts 1 and 2 refer to π^+ and π^-. The distribution of values of M normally shows structure as in Fig. 7.1. There is a general background with two peaks superimposed, which can be fitted with relativistic Breit–Wigner shapes (Section 4.5) with resonance masses 765 and 1264 MeV and widths (Γ) 125 and 151 MeV, respectively. These are the ρ^0- and f-mesons.

A plot of totally neutral three-pion invariant-mass distributions such as occur, for example, in

$$\pi^- + p \to \pi^+ + \pi^- + \pi^0 + \pi^- + p \tag{7.22}$$

frequently show the existence of two resonances, the ω and η mesons, with masses 783 and 549 MeV, widths 12 MeV and <10 keV, respectively (Fig. 7.2). These are several examples among many.

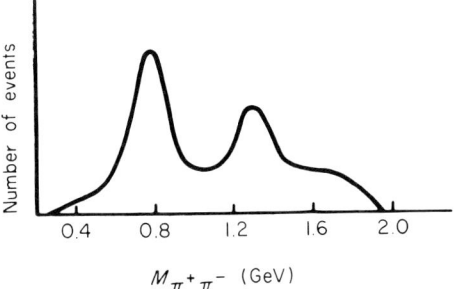

FIG. 7.1. Typical distribution of the invariant mass of two pions produced in $\pi^- p \to \pi^+ \pi^- n$.

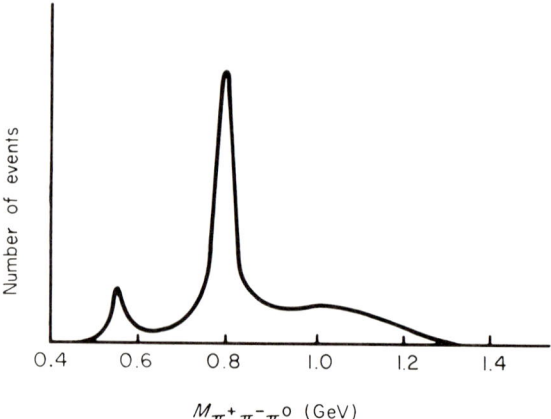

FIG. 7.2. Typical distribution of the invariant mass of neutral three-pion systems produced in $\pi^+ p \to \pi^+\pi^-\pi^0\pi^+ p$.

In general, these resonances have more than one decay mode. A knowledge of what decays occur and their branching fractions will contribute to a determination of the properties of the parent resonances.

We now wish to examine the methods of obtaining information on the quantum numbers of such resonances. The first thing is to decide, if possible, the interaction responsible for the decay: widths greater than 1 MeV indicate lifetimes of less than 6×10^{-22} sec, and this is the realm of strong interactions. Smaller widths cannot normally be resolved in bubble chambers, so it may be difficult to make an immediate decision. Mean lives corresponding to visible distances moved before decay mean weak-interaction decays.

Let us consider the strong-interaction decays. All quantum numbers are conserved, so that those for the decaying particle can be determined if the quantum numbers of the final state can be found. There are some decays that are due to the electromagnetic interaction, which does not conserve isotopic spin or G parity, so it is necessary to consider the changes that can occur; this is sometimes a less certain procedure for determining the initial quantum numbers. If weak interactions are responsible for the decay, then it is possible to rely only on the conservation of angular momentum, charge, and baryon and lepton numbers (Section 11.2).

There is one interesting point that is worth mentioning. A resonance R may be observed in two ways: by "production,"

$$A + B \to R + C$$

[Eq. (7.21), for example]; or by "formation,"

$$A + B \to R \to A + B$$
$$\searrow$$
$$\text{other channels}$$

7.6 Meson Resonances

as, for example, the $\Delta(1236)$ is formed and observed in pion–nucleon scattering. Until recently, all boson resonances have been observed by production. However, there appear to be some bosons heavier than two protons, so it should become possible to observe these in formation with antiprotons scattering on protons:

$$\bar{p} + p \to R \to \text{various channels}.$$

In addition, the ρ^0 meson has been observed in formation by means of colliding-beam experiments:

$$e^+ + e^- \to \rho^0 \to \pi^+ + \pi^-$$

(Auslander et al., 1967; Augustin et al., 1968). This is a particulary interesting way of observing the ρ, as here it is free of any other strongly interacting particles at production and decay. This is not normally the case, and the observation of ρ-meson properties is often spoiled by interference effects.

It is worth mentioning the method of observing resonances by a missing-mass technique (Maglic and Costa, 1965). Consider the reaction

$$\pi^- + p \to p + x^-,$$
$$\;\;1\quad\;\;2\quad\;\;3\quad 4$$

where x^- may consist of several particles. If the incident momentum is known and a measurement made of the momentum and angle of the recoil proton, then the mass of x is determined. To see this, consider the four-vector relation for energy and momentum conservation

$$p_1 + p_2 = p_3 + p_4,$$

where $p_4 = p_a + p_b + p_c + \cdots$, $a, b, c, \ldots$ being the particles in x. Now the mass of x (m_4) is given by

$$m_4^2 = (p_a + p_b + p_c + \cdots)^2$$
$$= (p_1 + p_2 - p_3)^2.$$

Evaluating this in the laboratory where we assume particle 2 is the target and at rest, we have

$$m_4^2 = m_1^2 + m_2^2 + m_3^2 + 2E_1 m_2 - 2E_3 m_2 - 2E_1 E_3 + 2\mathbf{p}_1 \cdot \mathbf{p}_3. \quad (7.23)$$

All the quantities on the right are known. This technique has been used to explore the $y = 0$, $t > 0$ meson mass spectrum up to 2400 MeV by identifying and measuring the energy of protons recoiling from a liquid-hydrogen target bombarded by high-energy pions [for references to this work, consult Particle Data Group (1969)].

7.7 Meson Resonances Decaying into Two Mesons

Let us first consider resonances decaying into two pions. In Table 7.4 we have tabulated the quantum numbers of the two-π systems with some of the lowest values of l, the orbital angular momentum. P and C are both

TABLE 7.4

THE QUANTUM NUMBERS OF TWO-PION STATES

	l	P	C	t	G
$\pi^+\pi^-$	0, 2, 4, ...	+1	+1	0 or 2	+1
	1, 3, 5, ...	−1	−1	1	+1
$\pi^0\pi^0$	0, 2, 4, ...	+1	+1	0 or 2	+1
$\pi^+\pi^0, \pi^-\pi^0$	0, 2, 4, ...	+1	—	2	+1
	1, 3, 5, ...	−1	—	1	+1
$\pi^+\pi^+, \pi^-\pi^-$	0, 2, 4, ...	+1	—	2	+1

$(-1)^l$. G must be $+1$, as G is -1 for a single pion. The values of t, the total isotopic spin, can only be 0, 1, or 2, as the isospin of the pion is 1; the total isospin for a given l is further restricted by the need to satisfy Bose statistics. Conservation of angular momentum requires that the l of a two-pion state will be the same as the spin of the parent state.

Let us now apply this to the ρ^0- and f-mesons (Fig. 7.1). These two particles have widths that indicate a strong-interaction decay, so their quantum numbers are those of the two-pion final state. Of course, l will be the spin of the parent. Table 7.4 shows that we must determine l and t to fix all the quantum numbers. The charged counterparts of the ρ^0 have been observed; for example,

$$\pi^- + p \to \rho^- + p$$
$$\searrow \pi^- + \pi^0$$
$$\pi^+ + p \to \rho^+ + p$$
$$\searrow \pi^+ + \pi^0$$

[see Particle Data Group (1969) for references]. No doubly charged ρ has been observed, nor has the following decay been observed:

$$\rho^0 \to \pi^0 + \pi^0 .$$

These facts are consistent only with $t = 1$ and an odd l. The f-meson is different. No charged f-mesons have been observed, but the decay

$$f \to \pi^0 + \pi^0$$

7.7 Meson Resonances Decaying into Two Mesons

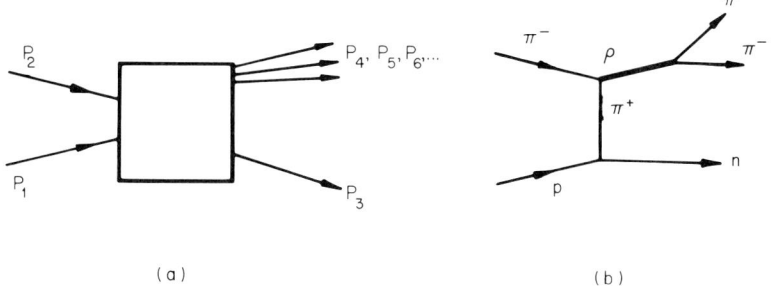

FIG. 7.3. (a) Diagram of four-momenta in the production of a resonance. (b) The Feynman diagram for one-pion exchange in ρ production.

has been observed (Sodickson et al., 1964); thus $t = 0$ and l is even. We are now left with the problem of finding l of the two pions in the final state.

This problem requires some consideration of the mechanism by which the ρ and f are produced. Let us start by looking at the kinematics of a simple production process. In Fig. 7.3 we have particle 1 incident on particle 2, producing particles 3, 4, 5, ... in the final state. The p_i, $i = 1, 2, \ldots$ represent the four-momenta of the particles involved. Suppose we are involved with a resonance production process, so that we find that particles 4, 5, 6, ... have an invariant mass m:

$$(p_4 + p_5 + p_6 + \cdots)^2 = m^2. \tag{7.24}$$

For example: 1, a target proton; 2, an incident π; 3, a recoil nucleon; and 4 and 5, pions from the decay of a ρ. As we shall show in Section 12.2, if we consider this as a two-body reaction, the kinematics are completely determined by two invariants:

$$s = (p_1 + p_2)^2 = (p_3 + p_4 + p_5 + \cdots)^2, \tag{7.25}$$

$$t = (p_1 - p_3)^2 = (p_2 - p_4 - p_5 - p_6 - \cdots)^2, \tag{7.26}$$

where s is the square of the total center-of-mass energy and t is the square of a momentum transfer (do not confuse s and t with ordinary and isotopic spin). In the physical region t is negative. If we apply this to ρ or f production by π-mesons on protons, so that p_1 is the target proton four-momentum and p_3 is that of the recoiling nucleon, then small $|t|$ corresponds to small angles of production of the ρ- or f-meson and to small recoil energies for the nucleon. As the production angle increases, the recoil energy increases and t becomes more negative. It is found that the differential cross section is sharply peaked at small values of t and falls off with t roughly as $\exp(+9t)$ if t is in GeV^2 (Morrison, 1966; also see Fig. 7.4). There are various interpretations of this phenomenon and, although none is on a sound quantitative basis, it seems that the idea of peripheral

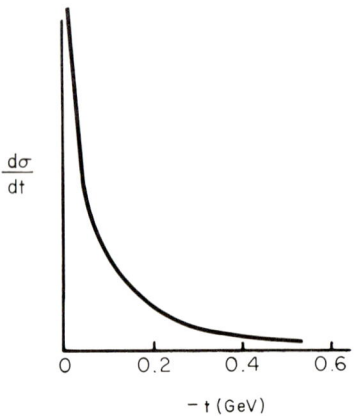

FIG. 7.4. Typical plot of $d\sigma/dt$ as a function of t for the reaction $\pi^- p \to \rho^0 n$.

production is qualitatively correct. The idea is that there are various reactions for which the most likely mechanism is the interaction of the incoming particle with a virtual particle in the field of the target proton. This will happen more readily, the more nearly real the particle is. This would occur at $t = +m^2$, where m is this particle's mass. However, as defined above, t must be less than zero and thus the reaction favors small t. One particle that must be present in the proton field is the pion, so that the process could be illustrated by a Feynman diagram of the kind shown in Fig. 7.3b, where we have shown an incident pion interacting with a pion provided by the target and producing a ρ. This is the one-pion exchange (OPE) process. Simple field theory predicts that this process would be proportional to $t/(t - m_\pi^2)^2$, which is strongly peaked at low t. Chew and Low (1959) first discussed these mechanisms and showed that the reaction will depend upon the cross section for interaction of the incident particle with the virtual particle. Thus a resonance in π-π scattering would be expected to show up as predominant production of π-π pairs with an invariant mass equal to the resonance energy, at low t values. This is in fact just what happens in the case of the ρ and f. Pion exchange is not the only possible peripheral process. For example,

$$\pi + \pi \to \omega$$

peripherally is not allowed because G-parity is not conserved; thus the observed peripheral ω production must take place by exchange of another particle, for example, the ρ:

$$\pi + \rho \to \omega.$$

Returning to the ρ- and f-mesons and using the peripheral model, we can transform into the rest frame of the ρ- or f-meson (Fig. 7.5) and consider

7.7 Meson Resonances Decaying into Two Mesons

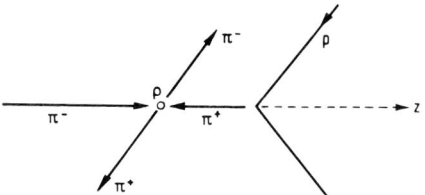

FIG. 7.5. Three-vectors of particles in the ρ rest frame for the reaction $\pi^- p \to \rho^0 n$.

the production mechanism being such that the incident pion and a virtual pion collide head-on to produce the particle at rest. If we put our axis of quantization along the incident-pion direction, then for the $\pi\pi$ collision $l_z = 0$ and the meson (spin j) must be produced in a state with $j_z = 0$ and its two-body decay must have angular distribution $|Y_j^0(\cos\theta)|^2$, where θ is the angle between outgoing π^- and the z axis. This is an oversimplified picture because it ignores other production mechanisms and unrelated mechanisms not producing a π-π resonance, which interfere and make a full interpretation difficult. However, there appears to be sufficient truth in the picture to permit definite conclusions about l. In Figs. 7.6 (a) and (b) we show typical f- and ρ-decay angular distributions in the forms described above. Only events in which the momentum transfer $|t|$ is less than a certain amount are used. This ensures that only events are accepted for which the

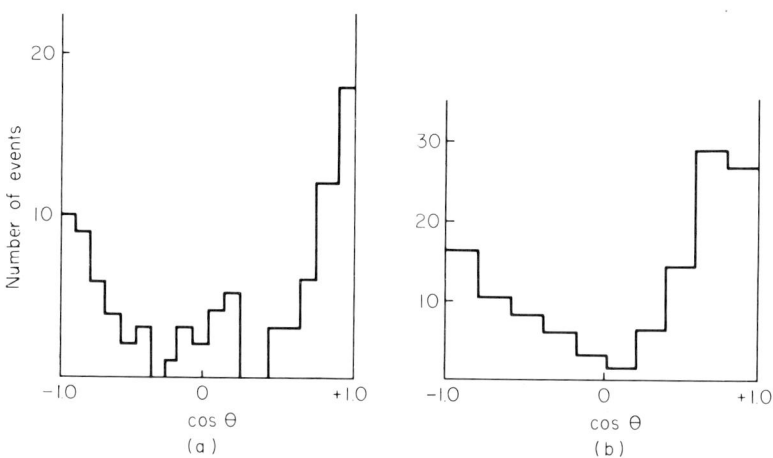

FIG. 7.6. (a) Angular distribution of decay of $f \to \pi^+\pi^-$ in $\pi^- p \to fn$ at 3 GeV/c (Lee et al., 1964); (b) angular distribution of decay $\rho^0 \to \pi^+\pi^-$ in $\pi^- p \to \rho^0 n$ at 1.25 GeV/c (Pickup et al., 1961). In both cases θ is the angle between incident π^- and decay π^- in the f or ρ rest frame, and in both cases events have been plotted only if the production momentum transfer was less than a certain amount. For a pure isolated resonance decay the distributions should be symmetric; the asymmetry indicates that background and other production processes are interfering.

peripheral model is expected to be a good description, and hence we can expect the decay angular distribution to give some information on the spin. That for the ρ is clearly consistent with the lowest permitted value, $l = 1$, while the f is consistent with $l = 2$ but not $l = 0$. The data cannot rule out $l = 3$ for the ρ or $l = 4$ for the f, but these are unlikely and all subsequent data are in agreement with the first values.

We are therefore in a position to give the quantum numbers for the ρ and f. These are conventionally written $T^G J^P$. The ρ has $1^+ 1^-$ and the f has $0^+ 2^+$.

We discuss the peripheral model of resonance production more fully in Chapter XII.

There are several other resonances that have two-pion decay channels; for details of these we refer readers to compilations of data (Particle Data Group, 1969). However, there are two places where weak decays into 2π have interesting consequence. The weak decay

$$K^\pm \to \pi^\pm + \pi^0$$

conserves angular momentum, so that the final state must have $l = 0$. Referring to Table 7.1 we see that this means $t = 2$. The initial state has $t = \frac{1}{2}$, so that this transition has $\Delta t = \frac{3}{2}$ at least. This situation will be discussed in Chapter XI. The decay of the long-lived neutral K-meson into the channels

$$K_L \to \pi^+ + \pi^-, \qquad K_L \to \pi^0 + \pi^0$$

(Christenson et al., 1964; Cronin et al., 1967; Gaillard et al., 1967) gives final states having $CP = +1$. Before these decays were observed, it was believed that the long-lived neutral K-meson had $CP = -1$ and that CP was conserved in weak decays. Observation of these decays has revealed the presence of a small CP-nonconserving interaction. This situation will be discussed in Chapter XIV.

In addition to resonances leading to states containing two pions, there can also be resonance decay into a kaon plus a pion. There are no restrictions due to Bose statistics, so that the allowed quantum numbers are less restricted than in the two-pion case. The total angular momentum of a $K\pi$ system is equal to l, the orbital angular momentum, and the parity is $(-1)^l$. The total isotopic spin can be $\frac{1}{2}$ or $\frac{3}{2}$. The hypercharge is nonzero, so that C and G have no relevance.

In Fig. 7.7 we show the invariant mass distribution for $\bar{K}^0 \pi^-$ combinations in the reaction

$$K^- + p \to \bar{K}^0 + \pi^- + p \qquad (7.27)$$

(Alston et al., 1961). The peak at 890 MeV is a prominent feature and is frequently found in reactions in which K-π systems are produced. This

7.7 Meson Resonances Decaying into Two Mesons

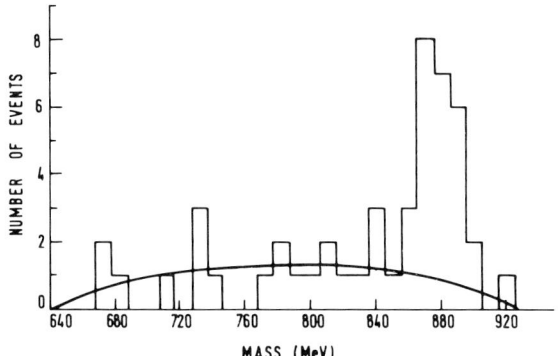

FIG. 7.7. The mass distribution of $\bar{K}^0\pi^-$ systems produced in $K^-p \to \bar{K}^0\pi^-p$ at 1.15 GeV/c. This distribution was the first evidence obtained for the existence of the K^*. (Alston et al., 1961.)

peak is now attributed to a resonance K^*, mass 892 MeV, width 50 MeV. Four charge–hypercharge combinations have been observed as with the K-meson (assuming hypercharge conservation at production). They are

$$y = +1 \quad K^{*+} \text{ and } K^{*0},$$
$$y = -1 \quad \bar{K}^{*0} \text{ and } K^{*-}.$$

The nonobservation of other charge states suggests that these are members of two isospin doublets. This has been confirmed by comparing the rates for the two reactions

$$K^- + p \to K^{*-} + p$$
$$\searrow K^- + \pi^0$$
or
$$\searrow \bar{K}^0 + \pi^-.$$

Since the decay is due to strong interactions, isotopic spin is conserved. From Eq. (5.18) the K^{*-} has $t_3 = -\tfrac{1}{2}$. If $t = \tfrac{1}{2}$, then the $K\pi$ combination with these quantum numbers is

$$|\tfrac{1}{2}, -\tfrac{1}{2}\rangle = \sqrt{\tfrac{1}{3}} \, |\pi^0 K^-\rangle - \sqrt{\tfrac{2}{3}} \, |\pi^- \bar{K}^0\rangle.$$

whereas if $t = \tfrac{3}{2}$,

$$|\tfrac{3}{2}, -\tfrac{1}{2}\rangle = \sqrt{\tfrac{2}{3}} \, |\pi^0 K^-\rangle + \sqrt{\tfrac{1}{3}} \, |\pi^- \bar{K}^0\rangle.$$

Thus $t = \tfrac{1}{2}$ predicts a branching ratio for $(K^{*-} \to K^-\pi^0)/(K^{*-} \to \bar{K}^0\pi^-)$ of $\tfrac{1}{2}$, whereas $t = \tfrac{3}{2}$ predicts 2. The value observed by Alston and co-workers

(1961) was 0.75 ± 0.35, which supports the $t = \frac{1}{2}$ assignment. Later data also supports this assignment. Studies of the reaction

$$K^+ + p \to K^+ + \pi^- + \pi^+ + p$$

by Chinowsky et al. (1962) showed that the reaction proceeds predominantly through the channel

$$K^+ + p \to K^{*0} + \Delta^{++}$$

(Δ is the $t = \frac{3}{2}$, $j = \frac{3}{2}$ πN resonance at 1236 MeV, Section 8.5) and that it was markedly peripheral. This allowed a study of the K^*-decay angular distribution as in the case of the ρ-meson, with the conclusion that $j = 1$. Thus the K^* is a strange meson with $TJ^P = \frac{1}{2}1^-$. There is a strange counterpart to the f with a mass of 1400 MeV that also decays into $K\pi$, but we refer readers to the literature for details (Hague et al., 1965).

TABLE 7.5

THE QUANTUM NUMBERS OF $K\bar{K}$ STATES

	l	P	C	t	G
$K^+K^-, \bar{K}^0K^0$	0, 2, 4, ...	+1	+1	0	+1
				1	−1
	1, 3, 5, ...	−1	−1	0	−1
				1	+1
$K^+\bar{K}^0, K^0K^-$	0, 2, 4, ...	+1	—	1	−1
	1, 3, 5, ...	−1	—	1	+1

Other interesting systems are those nonstrange resonances that decay into a pair of kaons. In Table 7.5 we show the possible quantum numbers of the $K\bar{K}$ system. In Section 7.5 we mentioned the ϕ meson observed, for example, in the reaction

$$K^- + p \to \Lambda + \phi$$
$$\searrow K^+ + K^-$$
or
$$\searrow K_1 + K_2.$$

The mass is 1019 MeV and the width about 4.0 MeV. Although this is a narrow resonance compared with the ρ, for example, the Q value in the center-of-mass system is so small that the decay is inhibited by the small phase space available and we still conclude that the decay is by strong interactions. In Section 7.5 we indicated that observation of the K_1K_2 mode fixes $C = -1$. No charged ϕ has been observed, which suggests isospin 0.

7.8 Meson Resonances Decaying into Three Mesons (I)

This would explain the absence of a $\pi\pi$ decay mode, as there is no $t = 0$, $C = -1$ state of the neutral two-pion system. Since $C = (-1)^{l+s}$ and s is zero, we have l odd (see Table 7.5). Angular correlation studies favored $l = 1$. In addition, the angular-momentum barrier is significant in its effect on the branching ratio

$$\frac{\phi \to K^+K^-}{\phi \to K^0 \bar{K}^0} \simeq 1.2$$

because of the different Q values (31 or 23 MeV), so that the higher j, the higher this ratio is expected to be. The observed ratio, 1.2, is consistent with the value 1.7 expected for $l = 1$ but not with the value ~ 2.9 for $l = 3$ (Schlein et al., 1963). Thus the ϕ has $T^G J^P = 0^- 1^-$.

This discussion has not exhausted the possibilities. Heavier resonances exist that have two-meson decay channels such as $\rho\pi$ or $\eta\pi$. The discussion of the quantum numbers is a straightforward continuation of what we have done in this section.

7.8 Meson Resonances Decaying into Three Mesons (I)

We shall first discuss decay into three pions. The final state is specified by the charges on the pions and by the momenta of the three particles. The former will be a result of the isospin of the final state, and the distribution in momenta will be a consequence of the dynamics and of the parity and total angular momentum of the final state. There is a limitation implicit in the requirement of Bose statistics, which affect the distributions, and we shall investigate all these effects to show how information on the spin and parity of the parent can be obtained. This problem has been discussed in detail by Zemach (1964), and we shall follow his method. We have a boson X decaying

$$X \to \pi_1 + \pi_2 + \pi_3.$$

In the center-of-mass system

$$\mathbf{p}_1 + \mathbf{p}_2 + \mathbf{p}_3 = 0,$$

and

$$\epsilon_1 + \epsilon_2 + \epsilon_3 = m_X$$

with obvious notation. One independent axial vector $\mathbf{q}$ can be constructed from the three momenta:

$$\mathbf{q} = \mathbf{p}_1 \times \mathbf{p}_2 = \mathbf{p}_2 \times \mathbf{p}_3 = \mathbf{p}_3 \times \mathbf{p}_1. \quad (7.28)$$

The distribution in momenta is best shown on a Dalitz plot (Appendix B.4). We use the arrangement of axes that places all events within a region inscribed within an equilateral triangle. The region boundary is between

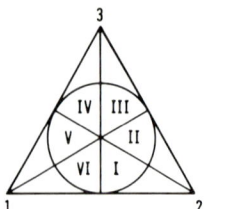 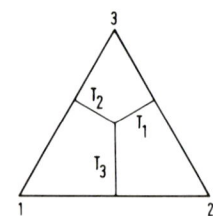

FIG. 7.8. The division of the allowed region of a three-pion Dalitz plot into sextants I–VI, and a diagram to show how the kinetic energies determine the position of an event on the plot.

the inscribed circle for the nonrelativistic case and an equilateral triangle for the extreme relativistic case. We are interested in dividing the allowed region into sextants (Fig. 7.8). The density of events is given by the square of an amplitude M, which in general is a sum of amplitudes, each being a product of three terms:

$$M = \sum M_T M_J M_F . \tag{7.29}$$

M_T contains the isotopic-spin effects, M_J the spin–parity effects, and M_F is a form factor for other energy-dependent effects.

The first problem to be considered is how the spin–parity of the final state affects the distribution of momentum among the three final particles. If the initial state has spin j, then the final state will also have the same total angular momentum. Since we are considering three-pion states (all spin zero), the final-state momentum configuration will transform under rotations in the same way as does the initial state, that is, as $Y_j(\cos\theta)$. Apart from parity the problem thus reduces to constructing objects out of $\mathbf{p}_1$, $\mathbf{p}_2$, $\mathbf{p}_3$, and $\mathbf{q}$ that transform like $Y_j(\cos\theta)$. We have that for any $\mathbf{p}$ or $\mathbf{q}$,

$$p_z \propto p Y_1^0(\cos\theta) ,$$
$$p_x + ip_y \propto p Y_1^1(\cos\theta) ,$$
$$p_x - ip_y \propto p Y_1^{-1}(\cos\theta) ,$$

so that these vectors as they stand transform like a state having $j = 1$. Obviously any combination of scalars or pseudoscalars ($\mathbf{p}_1 \cdot \mathbf{p}_2$, $\mathbf{p}_1 \cdot \mathbf{q}$, etc.) transforms like a state having $j = 0$. An object such as $\mathbf{p}_1\mathbf{p}_2$ (a dyadic) is evidently like the direct product of two $j = 1$ states and, as we know from angular momentum, will transform as a linear superposition of states with $j = 0, 1,$ and 2. This dyadic is in fact a second-rank tensor, and it is possible to subtract from it parts such that the resultant tensor $T^{(2)}$ is symmetric in any two indices and traceless (that is, contraction over any two indices gives zero): the result is said to be irreducible. Thus a second-rank irreducible tensor has $\alpha\beta$ component:

$$T^{(2)}_{\alpha\beta} = \tfrac{1}{2}[(\mathbf{p}_1)_\alpha(\mathbf{p}_2)_\beta + (\mathbf{p}_2)_\alpha(\mathbf{p}_1)_\beta] - \tfrac{1}{3}\delta_{\alpha\beta}\,\mathbf{p}_1 \cdot \mathbf{p}_2 , \tag{7.30}$$

7.8 Meson Resonances Decaying into Three Mesons (I)

where α and β range over x, y, z. This has five independent components and transforms like a system having $j = 2$. The reduction of a reducible tensor corresponds to decomposing it into a Clebsch–Gordan series (Chapter IX) or, in the familiar language of angular momentum, to constructing eigenstates of J^2 and J_z from the eigenstates of two angular momenta. For full details of this tensor method we refer readers to Hamermesh (1962). We can now write down those irreducible tensors for the rank of 0, 1, and 2 when constructed from one or two independent momenta.

$$T^{(0)} = 1, \quad \mathbf{p}_1 \cdot \mathbf{p}_2, \quad \text{or} \quad \mathbf{p}_1^2, \quad \text{etc.};$$
$$T_\alpha^{(1)} = (\mathbf{p}_1)_\alpha \quad \text{or} \quad (\mathbf{p}_2)_\alpha, \quad \text{etc.};$$
$$T^{(2)} \quad \text{as in Eq. (7.30).}$$

In the case we are discussing there are three pseudoscalar particles and momenta $\mathbf{p}_1, \mathbf{p}_2, \mathbf{p}_3$, and $\mathbf{q}$. We only wish to retain the essential parts of the tensor and remove the scalar parts depending upon momenta into M_F. Thus $T^{(0)} = 1$ is the only zero-rank tensor of interest. The first-rank tensors we represent by $T^{(1)}(1), T^{(1)}(2), T^{(1)}(3), T^{(1)}(q)$ ($= \mathbf{p}_1, \mathbf{p}_2, \mathbf{p}_3, \mathbf{q}$) and the second-rank by $T^{(2)}(11), T^{(2)}(12), \ldots, T^{(2)}(1q)$ with obvious notation. The parity of these tensors follows from the odd parity of the vectors $\mathbf{p}_1, \mathbf{p}_2$, and $\mathbf{p}_3$ and the even parity of the axial vector $\mathbf{q}$. Thus the J^P properties of the tensors, including $(-1)^3$ for the three pseudoscalar pions, are

$$\begin{aligned} T^{(0)}: &\quad 0^-, \\ T^{(1)}(1), T^{(1)}(2), T^{(1)}(3): &\quad 1^+, \\ T^1(q): &\quad 1^-, \\ T^{(2)}(1q), T^{(2)}(2q), T^{(2)}(3q): &\quad 2^+, \\ T^{(2)}(11), T^{(2)}(12), \text{etc.}: &\quad 2^-. \end{aligned} \quad (7.31)$$

We shall use these tensors as the M_J in constructing final-state amplitudes with required transformation properties. We notice that there is no 0^+ state of three pions (or of any three pseudoscalar bosons), as we have already discovered in discussing the τ–θ puzzle (Section 6.4). From the fact that, for any $\mathbf{p}$, $\mathbf{p} \cdot \mathbf{q} = 0$, it follows that the second-rank tensors containing $\mathbf{q}$ are of the form [cf. Eq. (7.30)]

$$T^{(2)}(pq) = \tfrac{1}{2}(\mathbf{pq} + \mathbf{qp}). \quad (7.32)$$

In addition, $T^{(2)}(\mathbf{qq})$ is a sum of the other second-rank tensors not containing $\mathbf{q}$ so does not have to be listed among the second-rank tensors. There are only three independent 2^- tensors because of Eq. (7.28). A suitable choice of three is $T^{(2)}(11), T^{(2)}(22),$ and $T^{(2)}(33)$.

The tensors contain some energy dependence, since they are functions of the momenta; however, these factors will not normally contain all the dependence, and we must allow for this by constructing appropriate M_F. We shall require factors that are symmetric and antisymmetric under the exchange of pairs of particles. We denote by f_s a function that is entirely symmetric in the variables ϵ_1, ϵ_2, and ϵ_3 but is otherwise left undefined so that its appearance in two places does not imply the same function. A function that is entirely antisymmetric in the energy variables can only be of the form

$$(\epsilon_1 - \epsilon_2)(\epsilon_2 - \epsilon_3)(\epsilon_3 - \epsilon_1) f_s . \tag{7.33}$$

Similarly we denote by f_3 a function of the energies that is symmetric in the exchange of ϵ_1 and ϵ_2. Two other similar functions are obtained by cyclically permuting the energies, and in any given amplitude the appearance of f_1, f_2, and f_3 means three functions related in this way. The appearance of f_1, f_2, and f_3 in another matrix element does not imply the same functions as in the first, only the same symmetry properties. Evidently the function $(\epsilon_2 - \epsilon_3) f_1$ has opposite symmetry properties to f_1, etc. Under an exchange $2 \leftrightarrow 3$, $f_1 \leftrightarrow f_1$ by definition, but $f_3 \leftrightarrow f_2$, and so on.

Let us now consider the isotopic-spin factor. It is possible to avoid the complications of the vector coupling of three isotopic spins by the following method (Zemach, 1964). We consider three Hermitian operators ϕ_1, ϕ_2, and ϕ_3 that have the property that the operators

$$\begin{aligned} \phi_+ &= \sqrt{\tfrac{1}{2}} \, (\phi_1 + i\phi_2) , \\ \phi_- &\equiv \phi_+^\dagger = \sqrt{\tfrac{1}{2}} \, (\phi_1 - i\phi_2) , \\ \phi_0 &\equiv \phi_3 \end{aligned} \tag{7.34}$$

can be interpreted to create, in the sense of field theory, particles that carry charge plus, minus, and zero, respectively (see Section 10.7). Thus, if $|0\rangle$ is the vacuum ket, we have

$$\phi_+ |0\rangle = -|\pi^+\rangle, \qquad \phi_- |0\rangle = |\pi^-\rangle, \qquad \phi_0 |0\rangle = |\pi^0\rangle . \tag{7.35}$$

The minus sign must be chosen to make the phase consistent with that used in Chapter II.

The operator ϕ, having the three components (ϕ_1, ϕ_2, ϕ_3), is evidently a vector in isotopic-spin space. Let us consider three such operators **a**, **b**, **c** with components subscripted 1, 2, and 3 and combinations thereof subscripted $+$, $-$, and 0 related as in Eqs. (7.34). Vector **a** creates the first pion, **b** the second, and so on. A scalar combination of two would be

$$\mathbf{a} \cdot \mathbf{b} = a_0 b_0 + a_+ b_- + a_- b_+ . \tag{7.36}$$

Notice that the components have been defined so that the scalar operator

7.8 Meson Resonances Decaying into Three Mesons (I)

a · b creates $\pi^0\pi^0$, $\pi^+\pi^-$, or $\pi^-\pi^+$ but no charged combinations. Allowing for the phase of the positive pion states, Eq. (7.36) is essentially the same as Eq. (5.13), which gives that superposition of two isovectors needed to make an isoscalar.

The only independent scalar that can be made from the three vectors is

$$(\mathbf{a} \times \mathbf{b}) \cdot \mathbf{c} = i \begin{vmatrix} a_+ & b_+ & c_+ \\ a_- & b_- & c_- \\ a_0 & b_0 & c_0 \end{vmatrix}. \qquad (7.37)$$

Again every term is "neutral." We can now examine some possibilities for the total isotopic spin of our three-pion state. Let us consider the final state having $t = 0$; it must be created by a scalar operator, and the only one available is $(\mathbf{a} \times \mathbf{b}) \cdot \mathbf{c}$, which contains no term in which all three pions do not have different charge. Thus $\pi^+\pi^-\pi^0$ is the only $t = 0$ state possible; $3\pi^0$ cannot have $t = 0$. A small digression is of interest at this point. We can check that $3\pi^0$ cannot have $t = 0$ simply, since this happens to be an easy case of vector addition. Any one pion of the three has $t = 1$, $t_3 = 0$, so the remaining two isospins must be vectorially added to have $t = 1$, $t_3 = 0$. However, the Clebsch–Gordan coefficient for adding two $t = 1$ states both having $t_3 = 0$ to give a state having $t = 1$, $t_3 = 0$ is zero [(Eq. (2.53b)], the required result. Thus we see that the simple method of Zemach (1964) is giving the correct result.

Returning to the $t = 0$ matrix element we note that the scalar $M_T = (\mathbf{a} \times \mathbf{b}) \cdot \mathbf{c}$ is antisymmetric under the exchange of any two pions, so must be associated with an antisymmetric spin–parity contribution $M_J M_F$. The M_T gives the particle charges and $|M_J M_F|^2$ gives the plot density which, in this case, is symmetric under particle exchange. Exchange corresponds to passing between two points that are equidistant but on opposite sides of one median of the triangle. Since this symmetry exists for all pairs, it exists for all three medians and the plot has sextant symmetry when $t = 0$.

For the case of $t = 1$, we have to create vectors from three isotopic-spin vectors. In adding the three isotopic vectors we can form $t = 0$, or 1, or 2 with two of the vectors before adding in the third to make a total $t = 1$. We can use the tensor notation to wite down combinations of two vectors that are zeroth- (scalar), first- (vector), and second-rank tensors. They are

$$T^{(0)}(\mathbf{ab}) = \mathbf{a} \cdot \mathbf{b},$$
$$T^{(1)}(\mathbf{ab}) = \mathbf{a} \times \mathbf{b}, \qquad (7.38)$$
$$T^{(2)}(\mathbf{ab}) = \mathbf{ab} + \mathbf{ba} - \tfrac{2}{3} \mathbf{I}(\mathbf{a} \cdot \mathbf{b}).$$

The last equation is Eq. (7.30) in dyadic notation with a factor of $\tfrac{1}{2}$

removed. [The reader will have no difficulty in passing through the next stage if he writes $T^{(2)}$ in the original form of Eq. (7.30).] Now we add in another vector **c** in such a way as to make three final vectors. They are

$$(\mathbf{a} \cdot \mathbf{b})\mathbf{c} \tag{7.39a}$$

$$(\mathbf{a} \times \mathbf{b}) \times \mathbf{c} = (\mathbf{a} \cdot \mathbf{c})\mathbf{b} - \mathbf{a}(\mathbf{b} \cdot \mathbf{c}), \tag{7.39b}$$

$$\mathbf{a}(\mathbf{b} \cdot \mathbf{c}) + \mathbf{b}(\mathbf{a} \cdot \mathbf{c}) - \tfrac{2}{3}\mathbf{c}(\mathbf{a} \cdot \mathbf{b}). \tag{7.39c}$$

The last vector comes from contracting on two indices the tensor product $\mathbf{c}T^{(2)}(\mathbf{ab})$. We remember that the vectors **a**, **b**, **c** go with particles 1, 2, 3, respectively, and it follows that under the exchange $1 \leftrightarrow 2$, vectors (7.39) are even, odd, even, respectively. Thus, a general $t = 1$ matrix element may be formed by multiplying vector (7.39a) by a function even in 1 and 2, adding (7.39b) multiplied by an odd function and (7.39c) by another even function, and finally adding in terms formed by cyclically permuting the indices. If this is done in detail, a very considerable simplification is found in that the whole reduces to the form

$$M = A\mathbf{a}(\mathbf{b} \cdot \mathbf{c}) + B\mathbf{b}(\mathbf{a} \cdot \mathbf{c}) + C\mathbf{c}(\mathbf{a} \cdot \mathbf{b}), \tag{7.40}$$

where A, B, and C are general functions of momenta and energy that are symmetric under exchanges $2 \leftrightarrow 3$, $1 \leftrightarrow 3$, and $1 \leftrightarrow 2$, respectively and have the appropriate tensor form to satisfy the spin–parity requirements.

To use Eq. (7.40) consider the decay into a $t = 1$ state of three pions. For $t_3 = +1$ we have two possibilities: $\pi^+\pi^+\pi^-$, and $\pi^0\pi^0\pi^+$. We put the labels 1 and 2 with the like pions, so that the first corresponds to $a = a_+$, $b = b_+$, and $c = c_-$, which is to be found only in the first and second terms of Eq. (7.40). Thus

$$M(\pi^+\pi^+\pi^-) = A + B. \tag{7.41}$$

Similarly, $\pi^0\pi^0\pi^+$ corresponds to a_0, b_0, c_+ which is to be found only in the last term; thus

$$M(\pi^0\pi^0\pi^+) = C. \tag{7.42}$$

The ratio of the squares of these amplitudes gives the branching ratio into these modes at the appropriate point on the Dalitz plot. The ratio of suitable sums over the plot will give the overall branching ratio. The $t_3 = -1$ cases are the same as the $t_3 = +1$ cases and are given by Eqs. (7.41) and (7.42) with the interchange $+ \leftrightarrow -$. The $t = 1$, $t_3 = 0$ cases are

$$M(\pi^0\pi^+\pi^-) = A, \tag{7.43}$$

$$M(\pi^0\pi^0\pi^0) = A + B + C. \tag{7.44}$$

The $t = 2$ and 3 cases are discussed by Zemach (1964).

7.8 Meson Resonances Decaying into Three Mesons (I)

We now have to collect all these results together to obtain distributions for various spin, parity, and isotopic assignments to three-pion states. The appropriate $M_J M_F$ are given in Table 7.6. All of these entries have the correct symmetry under two-particle exchange and have the tensor properties required by the J^P.

TABLE 7.6

Decay Amplitudes to Three-Pion, $t = 0$ States

J^P	$M_J M_F$
0^-	$(\epsilon_1 - \epsilon_2)(\epsilon_2 - \epsilon_3)(\epsilon_3 - \epsilon_1) f_s$
1^+	$\mathbf{p}_1(\epsilon_2 - \epsilon_3) f_1 + \mathbf{p}_2(\epsilon_3 - \epsilon_1) f_2 + \mathbf{p}_3(\epsilon_1 - \epsilon_2) f_3$
1^-	$\mathbf{q} f_s$
2^+	$f_1 T^{(2)}(1q) + f_2 T^{(2)}(2q) + f_3 T^{(2)}(3q)$
2^-	$(\epsilon_2 - \epsilon_3) f_1 T^{(2)}(11) + (\epsilon_3 - \epsilon_1) f_2 T^{(2)}(22) + (\epsilon_1 - \epsilon_2) f_3 T^{(2)}(33)$

TABLE 7.7

Decay Amplitudes to Three-Pion, $t = 1$ States

J^P	A	B	C
0^-	f_1	f_2	f_3
1^+	$f_1 \mathbf{p}_1$	$f_2 \mathbf{p}_2$	$f_3 \mathbf{p}_3$
1^-	$(\epsilon_2 - \epsilon_3) f_1 \mathbf{q}$	$(\epsilon_3 - \epsilon_1) f_2 \mathbf{q}$	$(\epsilon_1 - \epsilon_2) f_3 \mathbf{q}$
2^+	$(\epsilon_2 - \epsilon_3) f_1 T^{(2)}(1q)$	$(\epsilon_3 - \epsilon_1) f_2 T^{(2)}(2q)$	$(\epsilon_1 - \epsilon_2) f_3 T^{(2)}(3q)$
2^-	$f_1 T^{(2)}(23)$	$f_2 T^{(2)}(13)$	$f_3 T^{(2)}(12)$

To form expressions for the expected density of events on the Dalitz plot we must relate the amplitudes to the required intensities. The intensity for the $j = 0$ case is the squared modulus of the scalar matrix element. For the $j = 1$ cases the intensity is the squared modulus of the amplitude vector. For the $j = 2$ cases the amplitude is a second-rank tensor M_{ij} and the intensity is the quantity $\sum_{ij} M_{ij} M_{ij}$. Using Eqs. (7.40)–(7.44) and Table 7.7 the intensity for any given J^P for $t = 1$ states may be written down. The intensity for $t = 0$ is given by the appropriate entry from Table 7.6. There is nothing to be gained by giving these intensities, but we shall discuss the general form of the results because, independent of the details of the intensity distributions, there are features that are unique to various spin–parity–isotopic-spin combinations. One of these we have mentioned, namely, the sextant symmetry that exists for the $t = 0$ three-pion states (this symmetry also exists for $t = 3$). Another feature that is common to the J^P states 1^-, 2^+, 3^-, 4^+, etc. is that their amplitudes contain $\mathbf{q}(=\mathbf{p}_1 \times \mathbf{p}_2$, etc.). Now on the boundary of the kinematically allowed region $\mathbf{p}_1$, $\mathbf{p}_2$, and $\mathbf{p}_3$ are

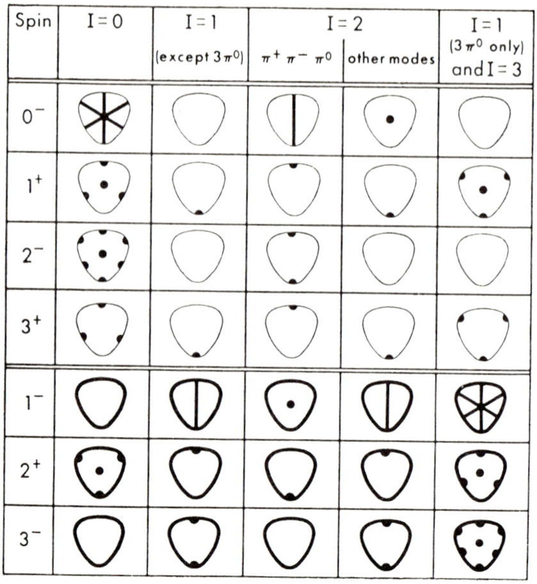

FIG. 7.9. Regions of the Dalitz plot where the intensity must vanish are shown in black. The $t = 1$ (except $3\pi^0$) cases are oriented so that the charges of particles 1, 2, 3 are as in Eqs. (7.72)–(7.76) with the plotting convention shown in Fig. 7.8. Where dots and heavy lines occur together, the vanishing is of higher order. This figure is taken from the article by Zemach, 1964. Note that he uses the symbol I for the isotopic spin.

collinear, so **q** is zero and the intensity of events must be zero for these J^P states at the boundary. The $J^P = 0^-$ and $t = 0$ final state must vanish when any two pions have the same energy—that is, along all medians of the Dalitz triangle. These results are summarized in Fig. 7.9, which is taken from Zemach's (1964) article. It includes the results for spins and isotopic spins that we have not discussed. Stevenson *et al.* (1962) have published typical distributions for $t = 0$, $J^P = 0^-$, 1^+, and 1^- cases. We show their results in Fig. 7.10. There is, evidently, a marked difference between these three J^P-values.

We have oriented the discussion in this section to the three-pion cases. Evidently, symmetry requirements are relaxed if the particles do not belong to the same isotopic-spin multiplet. Discussion of these cases has been given, for example, by Zemach (1964) and by Berman and Jacob (1965). In addition, if the decay is a two-step process, the final state-distributions will be altered. For example, there will be ρ-meson bands in the Dalitz plot for the decay of the A_2 meson

$$A_2 \to \rho^{\pm} + \pi^{\mp}$$
$$\searrow$$
$$\pi^{\pm} + \pi^0.$$

7.8 Meson Resonances Decaying into Three Mesons (I)

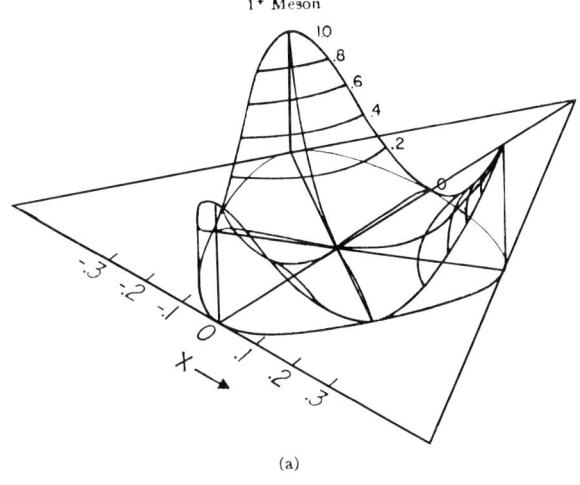

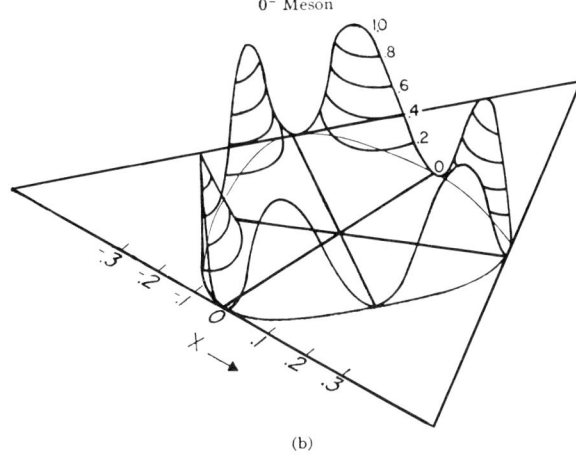

FIG. 7.10. Relative intensities on the Dalitz plot for the $t = 0$, 3π state in the three cases $J^P = 1^+, 0^-, 1^-$. (Stevenson et al., 1962.)

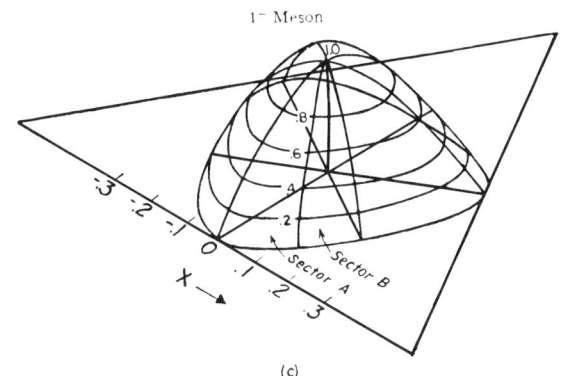

7.9 Meson Resonances Decaying into Three Mesons (II)

Let us now apply what we have learned in the previous section to various mesons decaying into three pions.

The ω-meson was discovered in antiproton annihilation at rest (Maglic et al., 1961). The particular reaction studied was

$$\bar{p} + p \rightarrow \pi^+ + \pi^+ + \pi^- + \pi^- + \pi^0.$$

The mass distribution of neutral three-pion states showed a peak at 783 MeV, which was interpreted as a neutral unstable particle decaying into $\pi^+\pi^-\pi^0$. No corresponding peak was found in charged combinations, so that it is very likely that the particle has isospin zero. In Fig. 7.11 we show the Dalitz plot of a large sample of ω-mesons obtained by Alff et al. (1962). This has sextant symmetry, and the intensity, allowing for background, disappears at the boundary (cf. Fig. 7.9). The distribution is consistent with final-state assignments $t = 0, J^P = 1^-$. Since isotopic spin is conserved, we assign $T^G J^P = 0^- 1^-$ to the ω-meson. The ω is narrow (12.2 ± 1.3 MeV) compared to the ρ-meson ($\sim$120 MeV) because of the smaller phase space available to the final state and to the effect of the centrifugal barrier. This has the result that it is very likely that an electromagnetic decay can compete and suggests that the neutral decay mode, which has a branching ratio of about 10 %, is $\omega \rightarrow \pi^0 \gamma$. The $\pi^+\pi^-$ mode, which is an electromagnetic transition, has been observed (Flatté et al., 1966).

The next case is not so simple or straightforward as that of the ω. In the reaction at 1.23 GeV/c,

$$\pi^+ + d \rightarrow p + p + \pi^+ + \pi^- + \pi^0, \tag{7.45}$$

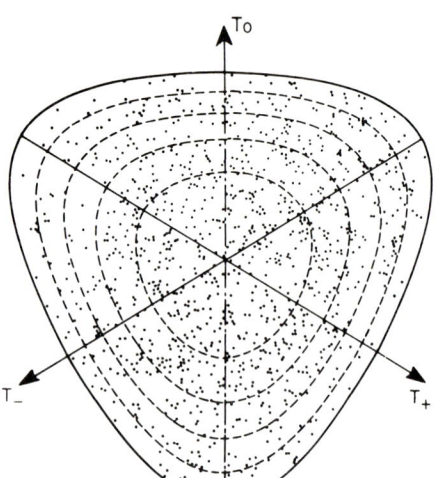

FIG. 7.11. The Dalitz plot for a sample of the decay $\omega \rightarrow \pi^+\pi^-\pi^0$. (Alff et al., 1962.)

7.9 Meson Resonances Decaying into Three Mesons (II)

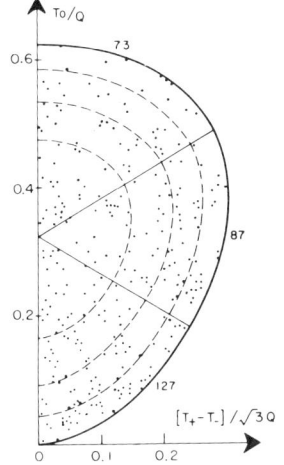

FIG. 7.12. Dalitz plot for 287 examples of the $\pi^+\pi^-\pi^0$ decay of the η-meson. (Compiled by Alff *et al.*, 1962.)

the mass distribution of the $\pi^+\pi^-\pi^0$ system showed a peak at 549 MeV, which is now attributed to a meson called η (Pevsner *et al.*, 1961).

An unsuccessful search has been made for charged η-mesons: since the η^0 production cross section is large (~ 1 mb), it is produced by a strong interaction so that charge independence should hold. Given the cross section for

$$\pi^+ + n \to p + \eta^0$$

from reaction (7.45) it is possible to calculate the cross section for reactions such as

$$\pi^+ + n \to n + \eta^+ ,$$

assuming an isospin other than zero for the η. The results indicate that the charged η should have been observed if it existed. Thus this particle has isospin 0.

In Fig. 7.12 we show the Dalitz plot for a sample of η-mesons decaying to $\pi^+\pi^-\pi^0$ obtained by Alff *et al.* (1962). The plot has been folded, assuming that charge-conjugation invariance holds. If it did not, the plot could indicate a different energy spectrum for the π^+ and the π^-. The decay mode has been examined for such an effect, but none has been observed with certainty (see Section 6.7) and the effect can be neglected here. The numbers of events in the three sextants are 73, 87, 127, so there is no sextant symmetry and thus the three pions are not in a pure isospin-0 state. Since the initial state has $t = 0$, this indicates a decay interaction violating charge independence. The obvious candidate is the electromagnetic interaction. This is supported by the existence of other decays, some obviously electromagnetic:

$$\eta \to 2\gamma\,,$$
$$\to \pi^0 \gamma \gamma\,,$$
$$\to 3\pi^0\,,$$
$$\to \pi^+ \pi^- \gamma\,.$$

Since these compete, all decays, whether or not they have a γ-ray in the final state, occur by the electromagnetic interaction and we expect a lifetime comparable to the π^0 ($\sim 10^{-16}$ sec). The 2γ-decay mode indicates that $C = +1$ for the η (Section 6.7) and, since $t = 0$, that $G = +1$. In addition, it shows that the η spin cannot be 1 (Section 7.3). Returning to the decay

$$\eta \to \pi^+ + \pi^- + \pi^0\,,$$

we note that the final state must have $G = -1$, so that the decay transition changes G. The electromagnetic interation will do this and in lowest order occurs by the emission followed by the absorption of a virtual photon. From the selection rules given in Table 6.1, we find that only $\Delta t = 1$ gives a change in G, hence the final state can only be one having $t = 1$. Comparing Figs. 7.9 and 7.12, we see that the two lowest J^P possibilities for the final state are 0^- and 2^-. Spins other than 0 would give more marked centrifugal barrier effects than are observed, so the conclusion is that the spin is zero. The electromagnetic interaction conserves angular momentum and parity, and it follows that the η quantum number assignments are $T^G J^P = 0^+ 0^-$.

The last case of three-pion decay we shall discuss is

$$K^+ \to \pi^+ + \pi^- + \pi^+\,.$$

The decay is by weak interactions, and an analysis of this and other K^+ decay modes lead to the discovery of parity nonconservation in weak interactions (Section 6.4). Figure 7.13 shows the Dalitz plot for a sample of this decay (Orear *et al.*, 1956). It has been folded into one semicircle, since two of the pions are identical. Since the density of points is very close to, if not completely, uniform, the only possibilities available from Fig. 7.9 are, assuming a pure isospin final state,

$$TJ^P \quad 10^-,\quad 30^-,\quad 12^-,\quad 22^-,\quad 32^-,\quad \text{and higher } j\,.$$

The spins other than 0 would introduce centrifugal barrier effects that would cause considerable deviation from uniformity, and we conclude that the final state has $J^P = 0^-$ and t probably 1. (This follows from the initial state having $t = \frac{1}{2}$ and the known importance of the $\Delta t = \frac{1}{2}$ weak transition; see Section 11.12.) The only conclusion that can be passed back to the parent K is that $j = 0$.

7.10 Conclusion

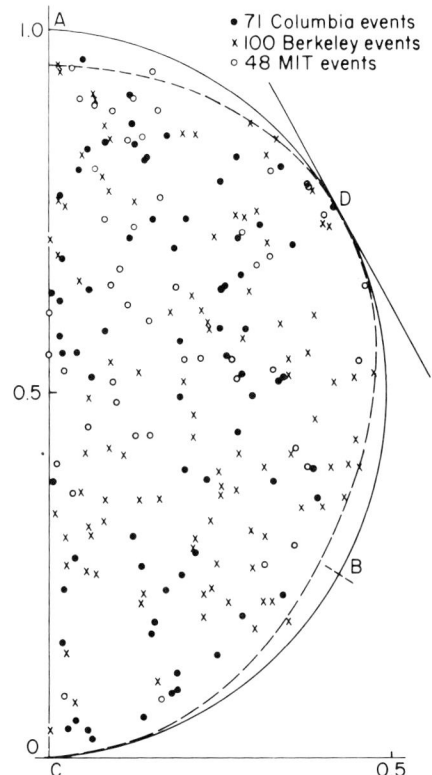

Fig. 7.13. The Dalitz plot for 219 decays $K^+ \to \pi^+\pi^+\pi^-$. The boundary of the plot is the broken line. The solid line is the inscribed circle to the triangle and is the boundary for a nonrelativistic treatment of the kinematics. (Orear et al., 1956.)

7.10 Conclusion

We have discussed the properties of the π- and K-mesons and shown how the simpler properties of some of the short-lived mesons can be investigated. In anticipation of the material of Chapter X, we wish to note some facts about the mesons we have mentioned. The three π-mesons, the four K-mesons, and the η-meson are all pseudoscalar. Of the remaining particles introduced, the ρ-mesons, the K^*-mesons, and the ω-meson are all vector particles. The fact to note about these two groups of particles is that they have the same isotopic-spin structure, namely, one singlet, two doublets, and a triplet. The singlet ϕ-meson is also vector and is matched by a ninth pseudoscalar meson, the X^0, which we have not mentioned. This pattern of nine is repeated for mesons having $J^P = 2^+$, of which the f-meson is one of the singlets. For a full listing of references on the discovery of these particles and their properties we refer readers to the data compilation of the Particle Data Group (1969).

REFERENCES

Alff, C. et al. (1962). *Phys. Rev. Lett.* **9**, 325 (7.9).
Alston, M. H. et al. (1961). *Phys. Rev. Lett.* **6**, 300 (7.7).
Augustin, J. E. et al. (1968). *Phys. Rev. Lett.* **20**, 126 (7.6).
Auslander, V. L., Budker, G. I., Pestov, Ju. N., Sidorov, V. A., Skrinsky, A. N., and Khabakhpashev, A. G. (1967). *Phys. Lett.* B **25**, 433 (7.6).
Bellettini, G., Bemporad, C., Braccini, P. L., and Foà, L. (1965). *Nuovo Cimento A* **40**, 1139 (7.2).
Berman, S. M., and Jacob, M. (1965). *Phys. Rev.* B **139**, 1023 (7.8).
Brown, J. et al. (1960). Reported by Glaser, D. A., *Proc. Conf. High Energy Nuclear Physics, Kiev, 1959.* p. 242. U.S.S.R. Academy of Science (7.5).
Cassels, J. M., Fidecaro, G., Wetherell, A. M., and Wormald, J. R. (1957). *Proc. Phys. Soc. London, Sect. A* **70**, 405 (7.3).
Cartwright, W. F., Richman, C., Whitehead, M. N., and Wilcox, H. A. (1953). *Phys. Rev.* **91**, 677 (7.3).
Chew, G. F., and Low, F. E. (1959). *Phys. Rev.* **113**, 1640 (7.7).
Chinowsky, W., Goldhaber, G., Goldhaber, S., Lee, W., and O'Halloran, T. (1962). *Phys. Rev. Lett.* **9**, 930 (7.7).
Christenson, J. H., Cronin, J. W., Fitch, V. L., and Turlay, R. (1964). *Phys. Rev. Lett.* **13**, 138 (7.7).
Clark, D. L., Roberts, A., and Wilson, R. (1951). *Phys. Rev.* **83**, 649 (7.3).
Cronin, J. W., Kunz, P. F., Risk, W. S., and Wheeler, P. C. (1967). *Phys. Rev. Lett.* **18**, 25 (7.7).
Day, T. B., Snow, G. A., and Sucher, J. (1960). *Phys. Rev.* **188**, 864 (7.3).
Durbin, R., Loar, H., and Steinberger, J. (1951). *Phys. Rev.* **84**, 581 (7.3).
Fields, T. H., Yodh, G. B., Derrick, M., and Fetkovich, J. G. (1960). *Phys. Rev. Lett.* **5**, 69 (7.3).
Flatté, S. M. et al. (1966). *Phys. Rev.* **145**, 1050 (7.9).
Gaillard, J. M. et al. (1967). *Phys. Rev. Lett.* **18**, 20 (7.7).
Gell-Mann, M., and Pais, A. (1955). *Phys. Rev.* **97**, 1387 (7.5).
Hague, N. et al. (1965). *Phys. Lett.* **14**, 338 (7.7).
Hamermesh, M. (1962). "Group Theory." Addison-Wesley, Reading, Massachusetts (7.8).
Kuehner, J. A., Merrison, A. W., and Tornabene, S. (1958). *Proc. Phys. Soc. London* **73**, 551 (7.3).
Lattes, C. M. G., Occhialini, G. P. S., and Powell, C. F. (1947). *Nature London* **160**, 453 (7.2).
Lee, Y. Y., Roe, B. P., Sinclair, D., and Van der Velde, J. C. (1964). *Phys. Rev. Lett.* **12**, 342 (7.7).
Maglic, B. C., and Costa, G. (1965). *Phys. Lett.* **18**, 185 (7.6).
Maglic, B. C., Alvarez, L. W., Rosenfeld, A. H., and Stevenson, M. L. (1961). *Phys. Rev. Lett.* **7**, 178 (7.9).
Morrison, D. R. O. (1966). "Review of Inelastic Two-Body Reactions." CERN TC Physics, 66-20 (7.7).
Orear, J., Harris, G., and Taylor, S. (1956). *Phys. Rev.* **102**, 1676 (7.9).
Pais, A., and Piccioni, O. (1955). *Phys. Rev.* **100**, 1487 (7.5).
Panofsky, W. K. H., Aamodt, R. L., and Hadley, J. (1951). *Phys. Rev.* **81**, 565 (7.3).
Particle Data Group (1969). *Rev. Mod. Phys.* **41**, 109 (7.2, 7.4, 7.5, 7.6, 7.7, 7.10).
Pevsner, A. et al. (1961). *Phys. Rev. Lett.* **7**, 421 (7.9).

REFERENCES

Pickup, E., Robinson, D. K., and Salant, E. O. (1961). *Phys. Rev. Lett.* **7**, 192 (7.7).
Plano, R., Prodell, A., Samios, N. P., Schwartz, M., and Steinberger, J. (1959). *Phys. Rev. Lett.* **3**, 525 (7.3).
Samios, N. P. (1961). *Phys. Rev.* **121**, 275 (7.2).
Samios, N. P., Plano, R., Prodell, A., Schwartz, M., and Steinberger, J. (1962). *Phys. Rev.* **126**, 1844 (7.2).
Schlein, P., Slater, W. E., Smith, L. T., Stork, D. H., and Ticho, H. K. (1963). *Phys. Rev. Lett.* **10**, 368 (7.7).
Sodickson, L., Wahlig, M., Manelli, I., Frisch, D., and Fackler, O. (1964). *Phys. Rev. Lett.* **12**, 485 (7.7).
Stevenson, M. L., Alvarez, L. W., Maglic, B. C., and Rosenfeld, A. H. (1962). *Phys. Rev.* **125**, 687 (7.8).
Yang, C. N. (1950). *Phys. Rev.* **77**, 242 (7.3).
Yukawa, H. (1935). *Proc. Phys. Math. Soc. Jap.* **17**, 48 (7.2).
Zemach, C. (1964). *Phys. Rev. B* **133**, 1201 (7.8).

VIII

THE BARYONS

8.1 Introduction

The baryons are all those particles that have half-odd spin and a mass equal to or greater than the proton mass. The observation of the stable members is, in most cases, a fairly straightforward matter of having the correct production reaction. However, there are many baryon resonances that can be called baryons on their own right and that are observed in formation experiments. It follows that the scattering of kaons and pions by nucleons is very relevent to baryons and is, therefore, discussed in this chapter.

8.2 The Stable Baryons

In Chapter V, we introduced the strange particles, among which are the strange baryons (hyperons). Table 5.1 gives those that are stable against strong decays and that we therefore call stable. In this section, we wish to consider the nucleons and these hyperons (apart from the Ω^-, which is a special case and will be discussed in Section 8.11). We have the following isotopic-spin multiplets

$$t = \tfrac{1}{2}: \quad p, n,$$
$$t = 0: \quad \Lambda,$$
$$t = 1: \quad \Sigma^+, \Sigma^0, \Sigma^-,$$
$$t = \tfrac{1}{2}: \quad \Xi^-, \Xi^0.$$

We notice that the isotopic-spin and hypercharge structure of these eight particles is the same as that of the eight pseudoscalar and eight vector mesons. This fact will be discussed in Chapter IX.

We now discuss the determination of the spin and parity of these

8.2 The Stable Baryons

particles. Work that has been done on the magnetic moments of the neutron and proton indicates that there are only two energy levels for these particles when isolated in a magnetic field. This fixes the spin as $\frac{1}{2}$. This is entirely consistent with the hyperfine structure of atomic hydrogen and deuterium and with the energy levels and spins of nuclear states. The parity of the nucleon is assumed to be even, which is a matter of convention, as we have discussed in Section 6.2.

The spin of the Λ was originally determined by two methods. One of these methods is due to Adair (1955). Consider, for example, the reaction

$$\pi^- + p \to K^0 + \Lambda$$
$$\searrow$$
$$\pi^- + p.$$

We take the z axis along the direction of the incident π and select a sample of events with a production angle for the K^0 close to 0 or 180°, Angular-momentum conservation ensures that the intensity comes mainly from those waves with $l_z = 0$, i.e., no spin flip. It follows that the Λ has the same z spin component as the proton and will be, for an unpolarized target, in a state that is an incoherent sum of the states with $j_z = \pm \frac{1}{2}$. If the Λ spin, j, is $\frac{1}{2}$ and parity is conserved in the decay, both of these magnetic substates will give a uniform decay distribution for the Λ in its own rest frame. However if the Λ-spin is $\frac{3}{2}$ or greater then this is no longer the case. Consider $j = \frac{3}{2}$; then the two possible states for our sample of Λ particles are $|\frac{3}{2}, \pm\frac{1}{2}\rangle$. The final state of $\pi^- p$ from Λ decay will have $l = 1$ or $l = 0$, depending upon the parity. and the decomposition will be, for $l = 1$,

$$|\tfrac{3}{2}, \pm\tfrac{1}{2}\rangle = \sqrt{\tfrac{1}{3}}\,|1, \pm 1, \tfrac{1}{2}, \mp\tfrac{1}{2}\rangle + \sqrt{\tfrac{2}{3}}\,|1, 0, \tfrac{1}{2}, \pm\tfrac{1}{2}\rangle,$$

where the kets on the right are $|l, l_z, s, s_z\rangle$ and s now refers to the final proton spin. Thus the decay intensity will be given by

$$\tfrac{1}{3}|Y_1^{\pm 1}(\theta, \phi)|^2 + \tfrac{2}{3}|Y_1^0(\theta, \phi)|^2 = (1/8\pi)(1 + 3\cos^2\theta).$$

The same distribution is obtained for the opposite parity ($l = 2$) case, by the Minami ambiguity. If $j = \frac{5}{2}$ the distribution is proportional to $(5\cos^4\theta - 2\cos^2\theta + 1)$. Thus the decay distribution of this sample of Λ-hyperons gives a measure of j, the spin. This method was used in the case of the Λ and gave a uniform distribution, thus indicating that the spin was $\frac{1}{2}$. The fact of parity nonconservation makes the decay of the Λ asymmetric if it is polarized. The Adair method does not give a polarized sample if the target proton is unpolarized (for $j > \frac{1}{2}$ it gives an aligned sample, which must decay symmetrically). Thus the indication of spin $\frac{1}{2}$ by the Adair method was not altered by the discovery of parity nonconservation. However, this did indicate another method of finding the spin. The decay asymmetry for a sample of polarized Λ was analyzed in Section 6.6, assuming the Λ spin was $\frac{1}{2}$. This assumption is not necessary to give an asymmetry, and Lee and Yang (1958) have shown how the actual decay distribution can be used to find the spin without making any assumptions other than conservation of angular momentum. The method depends upon the use of a certain test function of $\cos\theta$ whose average over the distribution must satisfy certain inequalities for a given spin assumption. For example: let $N(\cos\theta)\,\Delta(\cos\theta)$ be the number of decays observed having the π-meson in the interval $\Delta(\cos\theta)$ and let ξ be the average value of $\cos\theta$ for the sample, i.e., $\xi = \{\sum \cos\theta\, N(\cos\theta)/\sum N(\cos\theta)\}$, where $\sum$ indicates summation over all intervals $\Delta(\cos\theta)$. Lee and Yang show that ξ must satisfy the following inequalities:

$$-1/(2j+1) \leqslant \xi \leqslant 1/(2j+1),$$

where j is the hyperon spin in units of $\hbar$. If we assume the decay distribution is linear in $\cos\theta$, that is,

$$N(\cos\theta) \propto (1 + a\cos\theta),$$

the inequalities become

$$-(1/6j) \leqslant \xi \leqslant (1/6j).$$

Crawford *et al.* (1959) were the first to apply this technique. Their sample contained 614 charged decays of the Λ-hyperon. The data satisfied the spin $\frac{1}{2}$, failed the spin $\frac{3}{2}$ by 3.2 standard deviations and failed spin $\frac{5}{2}$ by 4.9 standard deviations. The assignment of spin $\frac{1}{2}$ was strong indicated and is supported by all subsequent data. The spins of the Σ and Ξ hyperons can be found in a similar manner to that used for the Λ. All the evidence is consistent with a spin $\frac{1}{2}$ assignment to all these particles.

In Section 6.2, we discussed the ambiguities that can occur in assigning hyperon parities. In Section 7.4, we showed that the relative $K\Lambda N$ parity was odd and indicated that it is usual, from this, to assign even parity to the Λ-hyperon and odd parity to the K-meson. In Section 8.9 we shall show that a study of the baryon resonance $\Lambda(1520)$ shows that the $K\Sigma N$ relative parity is the same as the $K\Lambda N$ relative parity. Consequently we assign even parity to the Σ particles. Notice we are assuming the same spin–parity within isotopic-spin multiplets. As we indicated in Section 6.2 the ΞN relative parity can, in principle, be determined. This has not been done. The application of unitary symmetry to the classification of particles suggests that the stable baryons belong to a supermultiplet and should all have the same spin and parity, thus supporting our assignment of even parity to the N, Λ, and Σ and indicating that the Ξ parity is also even.

There is one interesting proposal for determining the relative $\Xi^- p$ parity

8.2 The Stable Baryons

(Treiman, 1959). It involves the capture of slow Ξ^- by protons in the reaction

$$\Xi^- + p \to \Lambda + \Lambda.$$

The arguments that suggest π^- mesons are absorbed from S-states (Section 7.3) can be applied here, in which case the initial state is $^1S_0$ or $^3S_1$ with parity even or odd as the Ξp parity is even or odd. The identity of the two lambdas restricts them to final states $^1S_0$, $^3P_{0,1,2}$, $^1D_2$, $^3F_{2,3,4}$, etc, with parity $+$, $-$, $+$, $-$, etc, Thus, the permitted transition are

$^1S_0 \to {}^1S_0$	for Ξ parity $+$,
$^1S_0 \to {}^3P_0$	for Ξ parity $-$,
$^3S_1 \to {}^3P_1$	for Ξ parity $-$,
$^3S_1$	no final state for Ξ parity $+$.

The correlation between the Λ-decay momenta is different for singlet and triplet spin states in the final state. We arrange axes so that the z direction is along the direction of flight of one Λ, and if $\mathbf{p}_1$ is the momentum of the decay proton of this Λ in its rest frame and $\mathbf{p}_2$ is the momentum of the decay proton of the other Λ in its rest frame, then we define θ as the angle between $\mathbf{p}_1$ and $\mathbf{p}_2$. (The two rest frames are not the same, but this angle is defined as if they were.) The decay angular distribution of a fully polarized Λ is

$$dN/d\Omega = (1/4\pi)(1 + \alpha \cos \phi),$$

where ϕ is the angle between the Λ polarization and the outgoing proton [Eq. (6.10)]. Then the distribution of the double Λ decay events as a function of $\cos \theta$ is

$$dN/d\Omega = (1/4\pi)(1 - \alpha^2 \cos \theta)$$

for the $^1S_0$ final state, and

$$dN/d\Omega = (1/4\pi)[1 + (\alpha^2/3) \cos \theta]$$

for the $^3P_0$ and $^3P_1$ final states. These distributions correspond to even and odd Ξp parity, respectively, and thus a measurement of the experimental distribution would determine this parity.

The Σ^0 decays thus:

$$\Sigma^0 \to \Lambda + \gamma,$$

and, in principle, a study of the photon and Λ polarizations from polarized Σ^0 baryons determines the $\Sigma^0 \Lambda$ relative parity. This, however, is impossible with presently available techniques, but the rarer decay mode

$$\Sigma^0 \to \Lambda + e^+ + e^-$$

can be used in its place. The smaller the invariant mass of the e^+e^- pair, the closer this mode is in its dynamics to the real photon decay. The plane of the e^+e^- pair reflects the polarization that the photon would have. In addition the predicted distribution in invariant mass of the e^+e^- pairs is markedly different for the two possibilities of even and odd relative parity. Courant et al. (1963) have measured this distribution and find that it is consistent only with even relative parity.

We have used the decay properties of these particles to determine their spin. Apart from the $\Sigma^0 \to \Lambda\gamma$, all the decay modes observed occur by weak interactions, and we defer until Chapter XI a full discussion of these modes and their significance.

We will not discuss the determination of some of the other properties of these baryons, such as mass, lifetime, or magnetic moment, because that involves detailed experimental techniques and would not add greatly to our discussion of the physics of these particles.

8.3 Baryon Resonances

In the last section we briefly reviewed the eight stable baryons. All these particles interact strongly with the mesons and manifest in these interactions excited states, often called resonances. There are several families of these resonances, and their quantum numbers frequently determine how they are to be observed. As in the case of mesons, there are two methods of observing these resonances (see also Section 4.5): by formation, as for example in

$$\pi^+ + p \to \Delta^{++} \to \pi^+ + p;$$

and by production, as in

$$K^+ + p \to K^0 + \Delta^{++}$$
$$\searrow$$
$$\pi^+ + p.$$

With conventional targets (p, or n in deuterium) and bombarding particles (n, p, γ, $\pi^\pm$, $K^\pm$, K_2) the resonances that are accessible by formation are limited. In Table 8.1 we list the channels leading to resonances with quantum numbers accessible by formation experiments and the notation used to indicate resonances with these quantum numbers. Thus $\Lambda(1520)$ is the $y = 0$, $t = 0$ resonance, which has $\Sigma\pi$ as one of its important decay channels. Energy conservation also limits the resonances, even though the quantum numbers are correct. Thus no $y = 2$ or 0 resonance can be formed if its mass is less than 1432 MeV. The two well-known cases are the $\Sigma(1385)$ and $\Lambda(1405)$. These must be observed in production. There do not appear to be any $y = 1$ resonances with an energy less than the πN-

8.3 Baryon Resonances

TABLE 8.1

BARYON RESONANCE TYPES ACCESSIBLE BY FORMATION[a]

Channel	y	t	Notation	Channel threshold (MeV)
KN	+2	0 1	Z	1432
or πN γN	+1	$\frac{1}{2}$ $\frac{3}{2}$	N Δ	or 1078 938
$\bar{K}N$	0	0 1	Λ Σ	1432

[a] Provided mass is greater than channel threshold.

channel mass 1078 Mev. If the incident channel is weakly coupled to a resonance, it may be difficult to observe it in a formation experiment.

In Table 8.2, we give the quantum numbers and notation for the resonance types that are accessible by production alone. The Ω is not a family but one particle, the only one known so far with $y = -2$.

TABLE 8.2

KNOWN BARYON RESONANCES NOT ACCESSIBLE BY FORMATION

y	t	Notation	Production example
-1	$\frac{1}{2}$	Ξ	$\bar{K}N \to K\Xi$
-2	0	Ω	$\bar{K}N \to KK\Omega$

The resonances we have listed as being observable by formation can, of course, also be observed in production experiments.

A resonance will be observed in formation experiments, given that the quantum numbers are correct, at an energy of the incident particle such that the total center-of-mass energy is equal to the mass. The resonance will show up in several ways. Measurements of total cross section will show a peak at the resonant energy, normally superimposed on a smoothly varying nonresonant cross section. From Eq. (4.19) we see that the resonant contribution will be

$$\sigma = 4\pi \lambda^2 gx/(\epsilon^2 + 1),$$

with a height at resonance of $4\pi\lambda^2 gx$. Obviously this will show up so long as this quantity can be distinguished from other resonant or nonresonant contributions. Resonances are difficult to see in total cross sections if they are weakly coupled to the incident channel (small x) or if there are several resonances in the same region. At high energies both these conditions occur, aggravated by the fact that the resonances tend to become broader and therefore overlap more, even at the same density. The information about resonances that can be obtained from total cross-section measurements is limited. If the resonant peak is clear of other peaks, then the variation with incident momentum gives the mass and width. If the height can be measured, then this gives a value of gx. In the case of pions and kaons incident on protons or neutrons, $g = j + \frac{1}{2}$, where j is the resonance spin.

Resonances also affect the angular distributions of elastic and inelastic scattering reactions. Thus the angular distribution in π^+ elastic scattering

$$\pi^+ + p \to \pi^+ + p$$

at 300 MeV/c is almost entirely determined by the $\Delta(1236)$ resonance, and the reaction

$$K^- + p \to \Sigma^+ + \pi^-$$

at 390 MeV/c is strongly influenced by the $\Lambda(1520)$. These are clear examples; at higher energies the resonances become broader, overlap, and become more inelastic, so that simple two-body channels become weaker and more difficult to study. The information that can be obtained from angular distributions is considerable: once a phase-shift analysis has been done, the resonant channels can be completely specified as to energy, width, spin, parity, and elasticity. The phase-shift analysis will require experimental information on angular distributions of scattering intensity and polarization over a wide range of energies in order to give complete and unambiguous results. Often, less-complete information is available, or at high energies the phase-shift analysis becomes too complicated for definite results. We have hinted at these problems in Chapter III, and we shall meet them again in this chapter.

We shall discuss the πN and KN channels at several energies to illustrate these methods. Sections 8.4–8.6 are devoted to πN scattering, which essentially explores the $y = 1$, $t = \frac{1}{2}$ and $\frac{3}{2}$ resonances (as long as they are sufficiently strongly coupled to the πN system). Sections 8.8 and 8.9 are devoted to KN and $\bar{K}N$ scattering; these processes explore the $y = +2$ and $y = 0$ states, respectively, that have $t = 0$ or 1, and are strongly coupled to these meson–nucleon systems.

Resonances observed in production must be examined by different techniques. Consider for example, the reaction

$$K^+ + p \to K^0 + \pi^+ + p. \tag{8.1}$$

A measurement of the invariant mass of the $\pi^+ p$ system in a sample of events (obtained in a bubble chamber) will show a broad spectrum of masses spread over the kinematically allowed region with a superimposed peak

8.3 Baryon Resonances

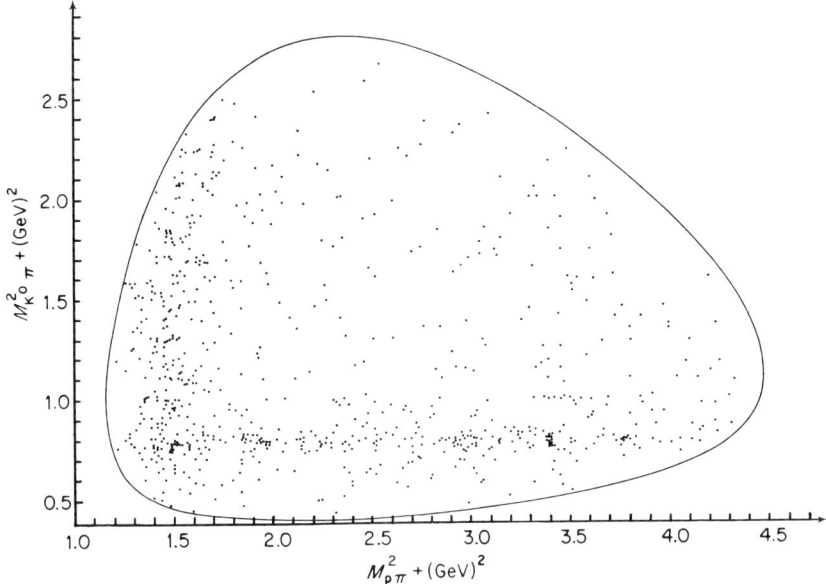

FIG. 8.1. Dalitz plot for the three-body final state of the reaction $K^+ + p \to K^0 + \pi^+ + p$. (From Ferro-Luzzi et al., *Il Nuovo Cimento* **36**, 1011, 1965.)

having a mass of about 1236 MeV and width of about 120 MeV. This peak shows the presence of the Δ^{++} (1236), so that a large fraction of the reaction must be proceeding through the channel

$$K^+ + p \to K^0 + \Delta^{++}$$
$$\searrow \pi^+ + p.$$

A Dalitz plot of the events of the reaction (8.1) will give even more information, since it displays events by placing them with coordinates that are the masses of two of the pairs, $M(\pi p)$ and $M(K\pi)$ say. Figure 8.1 shows such a plot; there is a broad band of a high density of events due to the Δ^{++} and a second band at right angles due to the $K^{*+}(892)$ (Section 7.7). This shows that the reaction

$$K^+ + p \to K^{*+} + p$$
$$\searrow K^0 + \pi^+$$

is also an important channel in reaction (8.1). If there were a KN resonance, it might be expected to show up along a straight band diagonally across the plot. Thus the distribution in mass of the events will give information on the masses and widths of the resonances involved.

The events in which the Δ^{++} is produced can be isolated, with some background, by taking those events in a correctly placed band across the Dalitz plot with a width that straddles the central Δ-mass. These events can now be studied as a two-step process with the Δ decaying separately from the production process. In the case we are discussing, this may not be entirely true because the mean life of the Δ is such that it does not always decay clear of the production-interaction region; however, many other baryon resonances do live long enough to be considered almost free. The production angular distribution and, in particular, the decay distribution will now give information on the spin and parity of the Δ.

These methods are not limited to three-body reactions or to resonances decaying into two particles. The data obtained in bubble chambers are often sufficiently complete to allow an investigation of the mass distributions of various combinations of two or more particles arising from reactions that produce three or more particles. Of course the number of combinations increases rapidly, and real resonances may become submerged because of the number of ways a particular combination may be chosen.

The missing-mass method (Section 7.6) of searching for resonances has also been used to investigate baryon resonances. For example, in

$$\gamma + p \to K^- + Z, \qquad p + p \to p + B$$

the momentum spectrum of the K^--mesons or of the protons at a fixed angle may be used to investigate the mass spectrum of the system Z or B, respectively.

8.4 Isotopic Spin of the Pion–Nucleon System

We have indicated in the last section, particularly in Table 8.1, that a study of the interactions of the πN system permits an exploration of the baryon resonances having $y = 1$ and $t = \frac{1}{2}$ or $\frac{3}{2}$. It is now appropriate to start looking at this system in detail. In this section we look at the isotopic-spin possibilities and how charge independence affects pion–nucleon scattering.

In Section 5.4 we assigned $t = 1$ to the π-meson, and in Section 5.3 we assigned $t = \frac{1}{2}$ to the nucleons. Thus t_3 is $+1, 0, -1$ for π^+, π^0, π^-, respectively, and $+\frac{1}{2}, -\frac{1}{2}$ for proton and neutron. We can apply what we know about the vector addition of angular momentum to the system consisting of one pion and one nucleon. This system can have total isotopic spin $t = \frac{3}{2}$ or $\frac{1}{2}$ and will in general consist of a superposition of the two states with coefficients that are the Clebsch–Gordan coefficients. These are given for this case by Table 2.1; the resulting state vectors for the pion–nucleon system are given in Table 8.3. The equations can be inverted to give the pure isospin states in terms of pion–nucleon states (Table 8.4).

8.4 Isotopic Spin of the Pion-Nucleon System

TABLE 8.3

ISOTOPIC-SPIN STATE VECTORS FOR THE PION-NUCLEON SYSTEM

Particle states	$\|t_a, t_{a3}, t_b, t_{b3}\rangle$	$\|t, t_3\rangle$
$\|\pi^+ p\rangle$	$= \|1, +1, \tfrac{1}{2}, +\tfrac{1}{2}\rangle =$	$\|\tfrac{3}{2}, +\tfrac{3}{2}\rangle$
$\|\pi^0 p\rangle$	$= \|1, 0, \tfrac{1}{2}, +\tfrac{1}{2}\rangle = \sqrt{\tfrac{2}{3}} \|\tfrac{3}{2}, +\tfrac{1}{2}\rangle - \sqrt{\tfrac{1}{3}} \|\tfrac{1}{2}, +\tfrac{1}{2}\rangle$	
$\|\pi^- p\rangle$	$= \|1, -1, \tfrac{1}{2}, +\tfrac{1}{2}\rangle = \sqrt{\tfrac{1}{3}} \|\tfrac{3}{2}, -\tfrac{1}{2}\rangle - \sqrt{\tfrac{2}{3}} \|\tfrac{1}{2}, -\tfrac{1}{2}\rangle$	
$\|\pi^+ n\rangle$	$= \|1, +1, \tfrac{1}{2}, -\tfrac{1}{2}\rangle = \sqrt{\tfrac{1}{3}} \|\tfrac{3}{2}, +\tfrac{1}{2}\rangle + \sqrt{\tfrac{2}{3}} \|\tfrac{1}{2}, +\tfrac{1}{2}\rangle$	
$\|\pi^0 n\rangle$	$= \|1, 0, \tfrac{1}{2}, -\tfrac{1}{2}\rangle = \sqrt{\tfrac{2}{3}} \|\tfrac{3}{2}, -\tfrac{1}{2}\rangle + \sqrt{\tfrac{1}{3}} \|\tfrac{1}{2}, -\tfrac{1}{2}\rangle$	
$\|\pi^- n\rangle$	$= \|1, -1, \tfrac{1}{2}, -\tfrac{1}{2}\rangle =$	$\|\tfrac{3}{2}, -\tfrac{3}{2}\rangle$

TABLE 8.4

ISOTOPIC-SPIN STATE VECTORS FOR THE PION-NUCLEON SYSTEM

Total isotopic state $\|t, t_3\rangle$	Particle states
$\|\tfrac{3}{2}, +\tfrac{3}{2}\rangle =$	$\|\pi^+ p\rangle$
$\|\tfrac{3}{2}, +\tfrac{1}{2}\rangle =$	$\sqrt{\tfrac{1}{3}} \|\pi^+ n\rangle + \sqrt{\tfrac{2}{3}} \|\pi^0 p\rangle$
$\|\tfrac{3}{2}, -\tfrac{1}{2}\rangle =$	$\sqrt{\tfrac{2}{3}} \|\pi^0 n\rangle + \sqrt{\tfrac{1}{3}} \|\pi^- p\rangle$
$\|\tfrac{3}{2}, -\tfrac{3}{2}\rangle =$	$\|\pi^- n\rangle$
$\|\tfrac{1}{2}, +\tfrac{1}{2}\rangle =$	$\sqrt{\tfrac{2}{3}} \|\pi^+ n\rangle - \sqrt{\tfrac{1}{3}} \|\pi^0 p\rangle$
$\|\tfrac{1}{2}, -\tfrac{1}{2}\rangle =$	$\sqrt{\tfrac{1}{3}} \|\pi^0 n\rangle - \sqrt{\tfrac{2}{3}} \|\pi^- p\rangle$

Since nucleon–nucleon forces appear to be charge independent, we expect the pion–nucleon interaction to have the same symmetry. All the experimental facts about this system are consistent with making this assumption, and we shall now consider how to derive some of the consequences of charge independence. As usual, this is done by requiring that isotopic spin be a conserved quantity. The conservation of the eigenvalue of T_3 is automatically satisfied by charge conservation; thus we only need add the requirement that the eigenvalue of T^2 be conserved. This implies that the pion–nucleon interaction depends upon the total isotopic spin of the system and not upon its charge state. Let us apply this to pion–nucleon scattering. The properties of pions and nucleons permit us to make direct experimental observations on only the following scattering reactions:

$$\pi^+ + p \to \pi^+ + p \quad \text{elastic scattering,} \tag{8.2}$$

$$\pi^- + p \to \pi^- + p \quad \text{elastic scattering,} \tag{8.3}$$

$$\pi^- + p \to \pi^0 + n \quad \text{charge-exchange scattering,} \tag{8.4a}$$

$$\pi^- + p \to \gamma + n \quad \text{radiative scattering.} \tag{8.4b}$$

The first experimental results obtained by Fermi and his co-workers

(Anderson et al., 1953) indicated that at the same angle the differential cross section for the reactions (8.2), (8.3), and (8.4) were approximately in the ratio 9:1:2. This result applies to measurements made at the same center-of-mass energy for all reactions and can be immediately interpreted as a manifestation of charge independence. We shall examine the analysis that leads to this result; we commence by assuming that charge independence is correct and that isotopic spin is a conserved quantity. It follows that the interaction Hamiltonian for the system H consists of two parts, $H_3 + H_1$. The first part, H_3, describes the interaction between meson and nucleon in the total isotopic-spin state $t = \frac{3}{2}$ and causes transitions between initial and final states having $t = \frac{3}{2}$; the second part, H_1, does the same for the $t = \frac{1}{2}$ states.

The cross section for any reaction (σ) is proportional to the square of the matrix element between initial and final states:

$$\sigma \propto |\langle \beta | H | \alpha \rangle|^2 .$$

Applying this to our three reactions and expanding the initial and final states into total isotopic-spin eigenstates, we have

$$\sigma_+ \propto |\langle \pi^+ p | H | \pi^+ p \rangle|^2 = |\langle \tfrac{3}{2}, \tfrac{3}{2} | H_3 | \tfrac{3}{2}, \tfrac{3}{2} \rangle|^2 , \tag{8.5}$$

$$\sigma_- \propto |\langle \pi^- p | H | \pi^- p \rangle|^2 = |\tfrac{1}{3} \langle \tfrac{3}{2}, -\tfrac{1}{2} | H_3 | \tfrac{3}{2}, -\tfrac{1}{2} \rangle + \tfrac{2}{3} \langle \tfrac{1}{2}, -\tfrac{1}{2} | H_1 | \tfrac{1}{2}, -\tfrac{1}{2} \rangle|^2 , \tag{8.6}$$

$$\sigma_0 \propto |\langle \pi^0 p | H | \pi^- p \rangle|^2 = 2 |\tfrac{1}{3} \langle \tfrac{3}{2}, -\tfrac{1}{2} | H_3 | \tfrac{3}{2}, -\tfrac{1}{2} \rangle - \tfrac{1}{3} \langle \tfrac{1}{2}, -\tfrac{1}{2} | H_1 | \tfrac{1}{2}, -\tfrac{1}{2} \rangle|^2 . \tag{8.7}$$

The subscripts $+$, $-$, and 0 refer to reactions (8.2), (8.3), and (8.4), respectively. If all other things, such as energy and angle of scattering, are equal, the constants of proportionality are the same and we have

$$\sigma_+ : \sigma_- : \sigma_0 = |M_3|^2 : |\tfrac{1}{3} M_3 + \tfrac{2}{3} M_1|^2 : 2 |\tfrac{1}{3} M_3 - \tfrac{1}{3} M_1|^2 \tag{8.8}$$

where we have written

$$M_3 = \langle \tfrac{3}{2}, t_3 | H_3 | \tfrac{3}{2}, t_3 \rangle ,$$
$$M_1 = \langle \tfrac{1}{2}, t_3 | H_1 | \tfrac{1}{2}, t_3 \rangle .$$

Charge independence means that M_1 and M_3 are independent of the value of t_3. If M_1 is small compared with M_3, we see immediately that

$$\sigma_+ : \sigma_- : \sigma_0 = 9 : 1 : 2 . \tag{8.9}$$

That this result was observed at energies around 150 MeV indicates that charge independence was at least approximately valid and that the meson–nucleon interaction was particularly strong in the state having total isotopic spin $t = \tfrac{3}{2}$. A more complete analysis shows that M_1 is nonzero but does

8.4 Isotopic Spin of the Pion–Nucleon System

not change this conclusion about the strength of the $t = \tfrac{3}{2}$ interaction at this energy.

Let us now look at how the total cross sections for π^+ and π^- in hydrogen depend upon isotopic spin. From Eq. (3.16b) we have a relation between total cross section σ and the imaginary part of the forward elastic-scattering amplitude

$$\sigma = (4\pi/k)\,\mathrm{Im}\,f(0)\,.$$

We can redefine the matrix elements M_1 and M_3 to be scattering amplitudes; then for π^+

$$\sigma_+ = (4\pi/k)\,\mathrm{Im}\,M_3\,. \qquad (8.10)$$

Similarly for π^- we shall have

$$\sigma_- = (4\pi/k)\,\mathrm{Im}(\tfrac{1}{3}M_3 + \tfrac{2}{3}M_1)\,. \qquad (8.11)$$

Now we can separate the contributions to the total cross section due to the two isotopic parts if we define the pion isotopic-spin cross sections

$$\sigma_1 = (4\pi/k)\,\mathrm{Im}\,M_1\,, \qquad \sigma_3 = (4\pi/k)\,\mathrm{Im}\,M_3\,.$$

Then we have

$$\sigma_3 = \sigma_+\,, \qquad (8.12)$$

$$\sigma_1 = \tfrac{1}{2}(3\sigma_- - \sigma_+)\,. \qquad (8.13)$$

This is a useful procedure, since it allows the effects of $t = \tfrac{1}{2}$ structure, which occurs only in the π^-p system, to be isolated from that due to $t = \tfrac{3}{2}$, which is also present. In Fig. 8.2 we show the two cross sections as a function of energy.

In the next section, we shall look at a full phase-shift analysis of pion–nucleon scattering, applying charge independence.

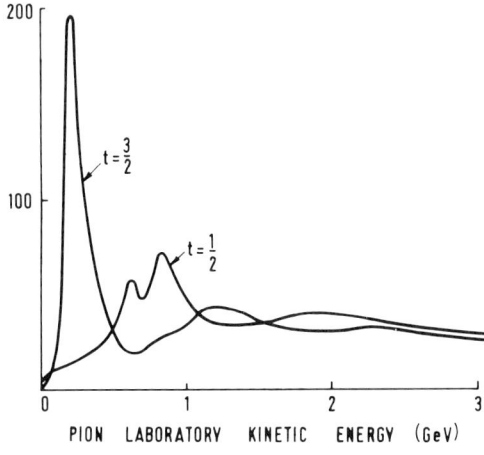

Fig. 8.2. The $t = \tfrac{1}{2}$ and $t = \tfrac{3}{2}$ pion–nucleon total cross-sections as a function of pion kinetic energy.

8.5 Low-Energy Pion–Nucleon Scattering

In Section 3.3, we derived an expression for the differential scattering cross section in terms of partial-wave scattering amplitudes [Eqs. (3.20)–(3.22)]:

$$d\sigma/d\Omega = |g(\theta)|^2 + |h(\theta)|^2,$$

where

$$g(\theta) = \sum_{l=0} \big((l+1)A_{l+} + lA_{l-}\big) P_l(\cos\theta),$$

$$h(\theta) = \sum_{l=1} (A_{l+} - A_{l-}) P_l^1(\cos\theta),$$

and each A is given by a phase shift δ and an absorption parameter η:

$$A = (\eta\, e^{2i\delta} - 1)/2ik.$$

The subscripts refer to the orbital angular momentum l, and the plus and minus are for $j = l + \tfrac{1}{2}$ and $j = l - \tfrac{1}{2}$.

This formula can be applied directly to $\pi^+ p$ scattering, as was done in Section 3.3, because this system is pure $t = \tfrac{3}{2}$. In the case of $\pi^- p$ scattering we have a mixture of $t = \tfrac{3}{2}$ and $t = \tfrac{1}{2}$. The $t = \tfrac{3}{2}$ amplitudes will be the same as in the $\pi^+ p$ scattering. To distinguish amplitudes we add an extra subscript, $2t$, in front of $l+$, or on its own in g and h; thus

$$d\sigma_+/d\Omega = |g_3(\theta)|^2 + |h_3(\theta)|^2, \tag{8.14}$$

where

$$g_3(\theta) = \sum_{l=0} \big((l+1)A_{3l+} + lA_{3l-}\big) P_l(\cos\theta),$$

and so on. From Eq. (8.8) we have that the π^- elastic-scattering amplitude is one-third $t = \tfrac{3}{2}$ and two-thirds $t = \tfrac{1}{2}$: hence

$$d\sigma_-/d\Omega = \tfrac{1}{9}|g_3(\theta) + 2g_1(\theta)|^2 + \tfrac{1}{9}|h_3(\theta) + 2h_1(\theta)|^2, \tag{8.15}$$

and similarly the charge-exchange amplitude is $\sqrt{\tfrac{2}{9}}$ times the difference of the two isospin amplitudes, which gives

$$d\sigma_0/d\Omega = \tfrac{2}{9}|g_3(\theta) - g_1(\theta)|^2 + \tfrac{2}{9}|h_3(\theta) - h_1(\theta)|^2, \tag{8.16}$$

where

$$g_1(\theta) = \sum_{l=0} \big((l+1)A_{1l+} + lA_{1l-}\big) P_l(\cos\theta),$$

and so on. Thus the three observable scattering differential cross sections should be parametrizable in terms of an absorption parameter and a phase shift for each partial wave.

Consider now low-energy scattering; that is, up to ~ 200 MeV incident pion energy. By the angular-momentum-barrier argument only partial

8.5 Low-Energy Pion-Nucleon Scattering

waves having $l = 0$ or 1 should be significant, and because this energy is only just above threshold for inelastic scattering such channels will not be important and all $\eta = 1$. There are then six amplitudes:

$$A_{10+}, \quad A_{11-}, \quad A_{11+},$$
$$A_{30+}, \quad A_{31-}, \quad A_{31+}.$$

This notation is somewhat clumsy. We can put the l information in by replacing the $A_{l\pm}$ by S, P for $l = 0, 1$ and having subscripts $2t2j$, where $j = l \pm \frac{1}{2}$, except for $l = 0$, when we have only $2t$. Thus the six amplitudes are

$$S_1, \quad P_{11}, \quad P_{13},$$
$$S_3, \quad P_{31}, \quad P_{33}.$$

Then

$$d\sigma_+/d\Omega = |S_3 + (2P_{33} + P_{31})\cos\theta|^2 + |(P_{33} - P_{31})\sin\theta|^2, \tag{8.17}$$

$$d\sigma_-/d\Omega = \tfrac{1}{9}\{|(S_3 + 2S_1) + (2P_{33} + P_{31} + 4P_{13} + 2P_{11})\cos\theta|^2$$
$$+ |(P_{33} - P_{31} + 2P_{13} - 2P_{11})\sin\theta|^2\}, \tag{8.18}$$

$$d\sigma_0/d\Omega = \tfrac{2}{9}\{|(S_3 - S_1) + (2P_{33} + P_{31} - 2P_{13} - P_{11})\cos\theta|^2$$
$$+ |(P_{33} - P_{31} - P_{13} + P_{11})\sin\theta|^2\}, \tag{8.19}$$

where, for example,

$$S_1 = (e^{2i\delta_1} - 1)/2ik, \quad P_{11} = (e^{2i\delta_{11}} - 1)/2ik,$$

and so on. A few moments inspection will show that these rather involved expressions have just the structure expected from Eqs. (8.15) and (8.16).

The most important and obvious fact about our analysis is that it has permitted us to describe three differential cross-sections in terms of six phase shifts, whereas without the simplification of charge independence the number would be nine. Thus if we determine the differential cross sections at three angles for each of the three reactions, it should be possible to overdetermine the phase shifts and supply a check on the principle of charge independence. Unfortunately, the statistical accuracy of the experiments is insufficient to enable this to be done, and in practice, all the data at one energy are processed to find the phase shifts that best fit the data; this is done by a least-squares fit. We shall return to the subject of phase-shift analysis at higher energies in the next section. We devote the rest of this section to a discussion of the results of the low-energy phase-shift analysis.

The result indicated in Section 8.4 was that the $t = \tfrac{3}{2}$ state is particularly important. This is borne out by the strong peak in the $t = \tfrac{3}{2}$ total cross section. Let us connect this with the partial-wave analysis. Integrating Eq. (8.17) over all angles we obtain

$$\sigma = 4\pi\{|S_3|^2 + 2|P_{33}| + |P_{31}|^2\}. \quad (8.20)$$

Therefore

$$\sigma = (4\pi/k^2)\{\sin^2 \delta_3 + 2\sin^2 \delta_{33} + \sin^2 \delta_{31}\}. \quad (8.21)$$

The total cross-section is observed (Fig. 8.1) to reach just $8\pi/k^2$ at about 194 MeV, so that the dominant $t = \frac{3}{2}$ interaction is in the P_{33} state. Thus at 194 MeV (total center-of-mass energy 1236 MeV) the $t = \frac{3}{2}$, $j = \frac{3}{2}$ phase shift passes through 90° and there is said to be a resonance at that energy. The total cross section does not become appreciably greater than $8\pi/k^2$, so that the S_3 and P_{31} amplitudes must be relatively small.

This P_{33} resonance is now attributed to a $t = \frac{3}{2}$ unstable particle Δ, which has a place in the unitary symmetry classification of particles. The mass of this particle is such that only one strong-interaction channel is open to it, namely, pion plus nucleon. It should be possible to predict the parameters describing this resonance from a knowledge of the pion–nucleon interaction; that is, from a coupling constant for the pseudoscalar interaction of the pseudoscalar pion with a nucleon. The first attempt to do this was by Chew and Low (1956). Since that time there have been refinements to the theory, particularly by Hamilton and his co-workers (see Hamilton, 1967), who conclude that the pion–nucleon interaction can be considered in terms of four important contributions:

(1) Nucleon exchange;
(2) Δ (1236) exchange;
(3) ρ exchange (see Section 7.7);
(4) $(\pi\pi)_0$ exchange, the subscript meaning that the two-pion system has $t = 0$.

These contributions are relatively long range and leave the possibility of a short-range interaction. However, it does not contribute significantly, and it appears to be possible to explain the behavior of all the lower partial waves from the properties of these four long-range interactions. It may appear inconsistent that one contribution to the P_{33} resonance (i.e., the Δ) is due to the exchange of this particle. In fact, this contribution turns out to be small. In addition, it is very likely that the properties of all the partial waves and of the resonances that occur are intimately interlocked, so that a complete theory would have any one property depending on all, and the properties of a πN resonance depending upon the contribution to the basic interaction of the exchange of that resonance. This is the "bootstrap" concept.

The result of the Chew–Low theory, which essentially included only interaction (1), is an effective range formula for the P-wave phase shifts; it reads

$$\frac{\eta^3}{\omega}\cot\delta = \frac{\lambda}{f^2}\left(1 - \frac{\omega}{\omega_0}\right), \tag{8.22}$$

where η is the center-of-mass momentum and ω is the center-of-mass energy less the nucleon mass, in units of $m_\pi c$ and $m_\pi c^2$, respectively; f^2 is the pion–nucleon coupling constant (Section 10.11); ω_0 is a parameter; and λ depends upon the state and is

$$\begin{array}{ccccc} 2t2j & 33 & 31 & 13 & 11 \\ \lambda = & \tfrac{3}{4} & -\tfrac{3}{2} & -\tfrac{3}{2} & -\tfrac{3}{8} \,. \end{array}$$

For P_{33} this leads to a positive phase shift of 90° at $\omega = \omega_0$, so that ω_0 is the resonance energy. Plotting $(\eta^3/\omega)\cot\delta$ against ω should give a straight line, which extrapolated to $\omega = 0$ will give a value of f^2. Hamilton's analysis (1967) shows that the value of f^2 so obtained is inconsistent with other determinations, a result due to a contribution to the interaction from $(\pi\pi)_0$ exchange. The theory also predicts negative phase shifts for the remaining P-waves. This is qualitatively correct for all but P_{11} and indicates that other interactions must be important, particularly for the P_{11}-wave.

A theoretical treatment of S-wave scattering is more difficult than for P-wave waves because this wave explores the shorter-range interactions, which are less amenable to calculation. This short-range interaction is repulsive and equal in both $t = \tfrac{1}{2}$ and $t = \tfrac{3}{2}$, so it contributes negatively to the phase shifts. The addition of $(\pi\pi)_0$ exchange lessens the negative effect of the short-range interaction but leaves it degenerate in the two states. The phases are split by the effect of ρ exchange, which makes the S_1 interaction attractive and consequently gives it a positive phase shift, while the S_3 interaction becomes once more repulsive and has a negative phase shift. At low energies we can put the results in an effective-range form

$$k\cot\delta = 1/a + \cdots$$

and we have scattering lengths

$$a_1 = +0.180 \pm 0.008\,, \qquad a_3 = -0.091 \pm 0.005\,,$$

in units of $\hbar/m_\pi c$ (Hamilton, 1966).

8.6 Pion–Nucleon Scattering up to 2500 MeV

At incident energies above 200 MeV in π-N scattering an increasing number of inelastic channels open, which adds further to the complication of an increasing number of important partial waves. Nonetheless, phase-

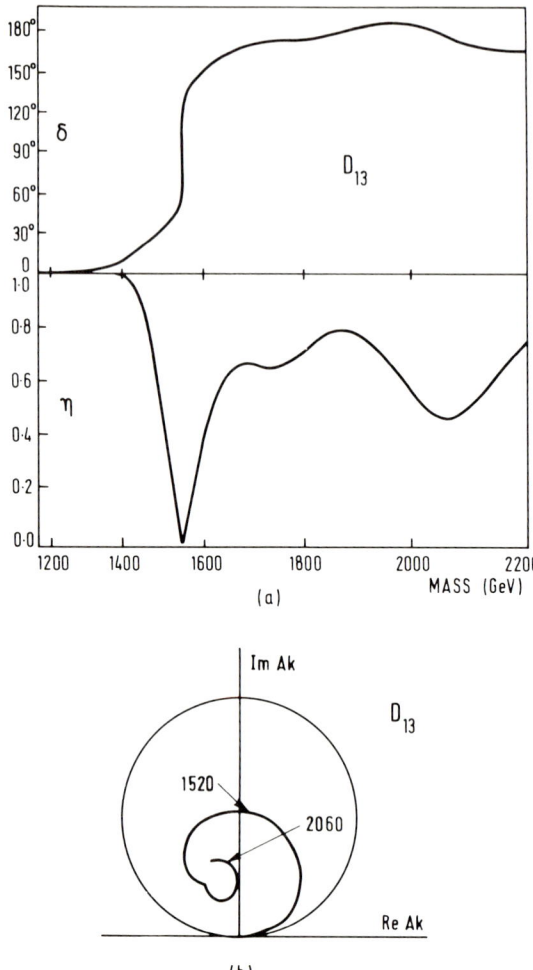

FIG. 8.3. Diagram showing the variation of the D_{13} pion–nucleon scattering amplitude with energy. (a) Plot of the phase shift δ and the absorption parameter η against the mass of the π-N system. (b) Trajectory of the amplitude on an Argand diagram. [Adapted from Donnachie et al. Physics Letters 26, B161(1968).]

shift analyses, as distinct from attempts to calculate and understand the interaction, have been made and will continue to be made (see, for example, Donnachie et al., 1968; Auvil et al., 1964; Bareyre et al., 1965a, b; Bransden et al., 1965; Roper et al., 1965; Roper and Wright, 1965). These analyses use the considerable amount of experimental data that now exists and often

8.6 Pion-Nucleon Scattering up to 2500 MeV

include theoretical data such as those obtained from the analytic properties of the partial-wave amplitudes and the need for continuity in the variation of phase shift and absorption parameters. The result is that analyses have been made up to 2190 MeV, and there is every possibility of extending them upward as more accurate data becomes available, in spite of the rapid increase in complexity (e.g., at least 18 amplitudes above 2000 MeV).

Let us look at some of the results. Figure 8.2 for the total cross sections in the $t = \frac{1}{2}$ and $\frac{3}{2}$ states shows that there is certainly a wealth of structure. The phase-shift analysis results are best shown as trajectories of the amplitude on an Argand diagram or as a plot of δ and η against energy. In Fig. 8.3 we sketch results of the analysis of Donnachie et al. (1968) for the D_{13} wave: (a) shows a simple plot of δ and η against total mass, while (b) shows the corresponding trajectory on the Argand diagram. It is evident that the well-known $N(1520)$ shows up clearly and is an example of a resonance where the phase shift passes very rapidly through 90° as the η nearly reaches 0. But the Donnachie solution shows another rapid decrease and rise in η that is characteristic of an inelastic resonance, and they attribute this to a D_{13} resonance at an energy of 2057 MeV with a width of 293 MeV. This proposed resonance, and others, require confirmation at the time of writing, but this example illustrates the method.

In some circumstances, a simpler approach can be used to obtain information. We have given in Eqs. (3.20) to (3.22), and (3.52) the formulae for the differential cross section and polarization in terms of the partial-wave amplitudes. The experimental results are usually analyzed by fitting the distribution to Legendre polynomials:

$$d\sigma/d\Omega = \sum_n B_n P_n(\cos\theta) \qquad (8.23)$$

and for the scattering asymmetry ϵ for a polarized target (polarization P)

$$(\epsilon/P)(d\sigma/d\Omega) = \sum_n C_n P_n^1(\cos\theta) . \qquad (8.24)$$

Alternatively, the polarization P of the nucleon recoiling from an unpolarized target is given by

$$P(d\sigma/d\Omega) = \sum_n C_n P_n^1(\cos\theta) \qquad (8.25)$$

It must be stressed that the coefficients C_n and B_n are not directly related to the partial-wave amplitudes, since the cross sections are intensities given by amplitudes squared. We can give examples of the connection. Using, S, P, D, with subscripts $2j$ for the partial-wave amplitudes (Tripp, 1966)

$$B_0 = |S_1|^2 + |P_1|^2 + 2|P_3|^2 + 2|D_3|^2 + \cdots ,$$

$$B_1 = \text{Re}\Big\{ 2S_1^*P_1 + 4S_1^*P_3 + 4P_1^*D_3$$

$$+ \frac{4}{5} P_3^*D_3 + \frac{36}{5}(P_3^*D_5 + D_3^*F_5) + \cdots \Big\},$$

$$B_2 = 2|P_3|^2 + 2|D_3|^2 + \frac{24}{7}|D_5|^2 + \frac{24}{7}|F_5|^2 + \cdots$$

$$+ \text{Re}\Big\{ 4S_1^*D_3 + 4P_1^*P_3 + 6S_1^*D_5 + 6P_1^*F_5$$

$$+ \frac{12}{7}(P_3^*F_5 + D_3^*D_5) + \frac{72}{7}(P_3^*F_7 + D_3^*G_7) + \cdots \Big\},$$

and so on, and

$$C_1 = \text{Im}\Big\{ 2S_1^*P_1 - 2S_1^*P_3 + 2P_1^*D_3$$

$$+ \frac{8}{5} P_3^*D_3 - \frac{18}{5}(P_3^*D_5 - D_3^*F_5) + \cdots \Big\},$$

$$C_2 = \text{Im}\Big\{ 2S_1^*D_3 - 2P_1^*P_3 - 2S_1^*D_5 + 2P_1^*F_5$$

$$+ \frac{10}{7}(P_3^*F_5 - D_3^*F_3) + \cdots \Big\},$$

and so on. It is worth noting some facts about these formulas.

(1) The squares of individual partial-wave amplitudes appear only in the angular-distribution coefficients (B_n) and then only in those for which $n < 2j$. Thus the D_3 wave can only generate up to $\cos^2 \theta$, although at first sight we might expect $\cos^4 \theta$. This is because the spin-nonflip amplitude $2D_3 P_2(\cos \theta)$ and the spin-flip amplitude $-D_3 P_2^1(\cos \theta)$ when squared and added do not have a $\cos^4 \theta$ term. This is more general, as can be seen from the Minami ambiguity: the angular distribution does not change if we interchange the parity of all waves, in which case the D_3 amplitude becomes the P_3 amplitude, which cannot have a complexity as great as $\cos^4 \theta$. Thus we see the greatest complexity of the wave with total angular momentum j is that of $\cos^{2j-1} \theta$.

(2) The "interference" terms are those due to the presence of two partial waves. We notice that interference between waves of even relative parity (e.g., S and D) contributes only to the even coefficients, while those of odd relative parity (e.g., S and P) contribute only to the odd coefficients.

(3) Although the Minami ambiguity requires that the angular distri-

8.6 Pion–Nucleon Scattering up to 2500 MeV

bution remains unchanged under the change of parity of all waves, the polarization changes sign. In this can be seen by interchanging the positions of the symbols thus:

$$S_1 \rightleftharpoons P_1, \qquad P_3 \rightleftharpoons D_3, \qquad D_5 \rightleftharpoons F_5, \quad \text{etc.}$$

This was proved more generally in Section 3.6.

(4) A particular interference term has the property of being a maximum when the amplitudes are in phase if it is of the form $\text{Re } L_{2j}^* L'_{2j}$, and when they are orthogonal if it is of the form $\text{Im } L_{2j}^* L'_{2j}$.

(5) The above formulas assume a single isotopic-spin channel is operating, for example, $\pi^+ p$. If two are operating as in $\pi^- p$ scattering, each amplitude is the sum of two amplitudes. Thus in the corresponding equations for $\pi^- p \to \pi^- p$ and $\pi^- p \to \pi^0 n$, each amplitude A_{2j} is replaced by $\frac{1}{3} A_{3,2j} + \frac{2}{3} A_{1,2j}$ and $\sqrt{\frac{2}{9}}(A_{3,2j} - A_{1,2j})$, respectively. The first subscript is $2t$, and the form of this replacement follows from Eq. (8.8) and is like that in Eqs. (8.15) and (8.16).

Let us consider the effect of these properties. A rapidly varying amplitude will show up in the even coefficients, and the maximum n in which it appears will fix its j but not its parity. If the amplitude is small but varying rapidly, as might happen in the case of a resonance weakly coupled to the πN system, then the square of the amplitude (in the even coefficients) may not show up clearly. However, if there is another, but slowly varying amplitude, the interference terms between the two may show up the variation more readily. If the parity of one amplitude is known, the appearance of the interference in even or odd coefficients will determine the parity of the other.

An example of this kind of argument is provided by the work of Duke et al. (1965) on π^+ and π^- scattering, in which they determined the coefficients B^+, B^-, C^+ and C^- over the incident momentum range 850–1700 MeV/c (mass range 1586–2024 MeV). This range contains several resonances, and we will concentrate on one, the $\Delta(1920)$, which has $t = \frac{3}{2}$ and is observed in both π^+ and $\pi^- p$ scattering. The coefficients up to $B_6{}^+$, but not $B_7{}^+$ or $B_8{}^+$, showed a marked peak at the mass 1920. The authors assume that this is due to a single resonance. Since it does not appear in $B_8{}^+$ it must have $j < \frac{7}{2}$, and since there is no structure in the small $B_9{}^+$, it is evident that no strong amplitudes with $j > \frac{7}{2}$ are to be found in this range. Again, since

$$B_7 = \text{Re}\left\{\frac{3150}{143}(D_5^* H_9 + F_5^* G_9) + \frac{9800}{429} F_7^* G_7 + \frac{2800}{429}(F_7^* H_9 + G_7^* H_9)\right.$$

$$\left. + \text{ terms in } G, H, \text{ and higher}\right\},$$

the only possible remaining contribution to B_7 is F_7-G_7 interference. However, B_7 is small, so that if there is a $j = \frac{7}{2}$ amplitude there is only one, F_7 or G_7. The coefficient B_6 is given by (neglecting waves of $j > \frac{7}{2}$)

$$B_6 = \frac{200}{11} \text{Re}\{D_5{}^*G_7 + F_5{}^*F_7\} + \frac{100}{33}\{|F_7|^2 + |G_7|^2\}.$$

Thus the peak is due to (a) F_7 or G_7 resonance, or (b) D_5 (or F_5) resonance interfering with a strong nonresonant G_7 (or F_7) amplitude, or (c) F_7 (or G_7) resonance with a strong nonresonant F_5 (or D_5) amplitude. Possibility (b) requires strong $j = \frac{5}{2}$ and $\frac{7}{2}$ nonresonant amplitudes to explain the peaks in $B_5{}^+$ and $B_6{}^+$ respectively (the $j = \frac{5}{2}$ squared amplitude only appears in $B_4{}^+$ and lower). The strengths of the amplitudes required in this case are sufficient to cause $B_0{}^+$ to be much larger than it is observed to be, and it follows that only (a) or (c) are possible, so that $j = \frac{7}{2}$ for the $\Delta(1920)$. To establish the parity of $\Delta(1920)$ it is necessary to consider the polarization data in a similar manner. The result is that the parity is even, i.e., the resonant wave is $F_{7/2}$.

This method of simple examination of coefficients obviously becomes more difficult at higher energies. Above about 2500 MeV the coupling of the πN system to any resonances that may exist has probably become very weak, and the description of elastic scattering by partial-wave analysis may cease to have any value.

8.7 The Kaon–Nucleon System

The four K-mesons are two isotopic-spin doublets with hypercharge $+1$ and -1. For both hypercharges the vector additions of the kaon isospin $\frac{1}{2}$ with the nucleon isospin $\frac{1}{2}$ will give isospin 0 or 1. Thus the kaon ($y = +1$) plus nucleon ($y = +1$) system can have $t = 0$ or 1 and $y = +2$, while the antikaon–nucleon system also has $t = 0$ or 1 but $y = 0$ (see Table 8.1). In Tables 8.5–8.7 we give the isotopic-spin vectors for the KN and $\bar{K}N$ systems. Neutral kaons as incident particles only come as K_2, so we have included the isotopic-spin decomposition of the K_2N system.

In the next section, we shall discuss low-energy kaon and antikaon scattering. The former appears to be rather uninteresting below 800 MeV/c. Above that momentum inelastic effects enter and there may be resonances in these $y = +2$ channels [see the review by Meyer (1968)]. However, this is not at present confirmed, and therefore we shall not discuss the KN system beyond the low-energy elastic scattering. The $\bar{K}N$ system is, in contrast, very rich in resonances. We shall discuss this system and in particular the $\Lambda(1520)$ in Section 8.9.

TABLE 8.5

Isotopic-Spin State Vectors for K^+N and K^-N Systems

Particle state	$\|t_a, t_{a3}, t_b, t_{b3}\rangle$	$\|t, t_3\rangle$	y
$\|K^+p\rangle$	$= \|\frac{1}{2}, +\frac{1}{2}, \frac{1}{2}, +\frac{1}{2}\rangle =$	$\|1, 1\rangle$	$+2$
$\|K^+n\rangle$	$= \|\frac{1}{2}, +\frac{1}{2}, \frac{1}{2}, -\frac{1}{2}\rangle =$	$\sqrt{\frac{1}{2}}(\|1,0\rangle + \|0,0\rangle)$	$+2$
$\|K^-p\rangle$	$= \|\frac{1}{2}, -\frac{1}{2}, \frac{1}{2}, +\frac{1}{2}\rangle =$	$\sqrt{\frac{1}{2}}(\|1,0\rangle - \|0,0\rangle)$	0
$\|K^-n\rangle$	$= \|\frac{1}{2}, -\frac{1}{2}, \frac{1}{2}, -\frac{1}{2}\rangle =$	$\|1, -1\rangle$	0

TABLE 8.6

Isotopic-Spin State Vectors for the KN and $\bar{K}N$ Systems

y	Total isotopic spin state $\|t, t_3\rangle$	Particle states
2	$\|1, +1\rangle$	$= \|K^+p\rangle$
	$\|1, \ 0\rangle$	$= \sqrt{\frac{1}{2}}(\|K^+n\rangle + \|K^0p\rangle)$
	$\|1, -1\rangle$	$= \|K^0n\rangle$
	$\|0, \ 0\rangle$	$= \sqrt{\frac{1}{2}}(\|K^+n\rangle - \|K^0p\rangle)$
0	$\|1, +1\rangle$	$= \|\bar{K}^0p\rangle$
	$\|1, \ 0\rangle$	$= \sqrt{\frac{1}{2}}(\|\bar{K}^0n\rangle + \|K^-p\rangle)$
	$\|1, -1\rangle$	$= \|K^-n\rangle$
	$\|0, \ 0\rangle$	$= \sqrt{\frac{1}{2}}(\|\bar{K}^0n\rangle - \|K^-p\rangle)$

TABLE 8.7

Isotopic-Spin State Vectors for the K_2N Systems

Particle state	Hypercharge and isotopic-spin state, $\|y, t, t_3\rangle$
$\|K_2p\rangle = \sqrt{\frac{1}{2}}(\|K^0p\rangle - \|\bar{K}^0p\rangle) = \frac{1}{2}\|2,1,0\rangle - \frac{1}{2}\|2,0,0\rangle - \sqrt{\frac{1}{2}}\|0,1,+1\rangle$	
$\|K_2n\rangle = \sqrt{\frac{1}{2}}(\|K^0n\rangle - \|\bar{K}^0n\rangle) = \sqrt{\frac{1}{2}}\|2,1,-1\rangle - \frac{1}{2}\|0,1,0\rangle - \frac{1}{2}\|0,0,0\rangle$.	

8.8 Low-Energy Kaon–Nucleon Scattering

By low energy we mean energies at which S-wave scattering is the only important effect. For kaons this is up to about 600 MeV/c and for antikaons it is up to about 300 MeV/c.

The scattering of kaons on protons at low energies is a very simple affair: below inelastic thresholds the reactions and corresponding amplitudes are

$$
\begin{aligned}
K^+ + p &\to K^+ + p & &A_1, \\
K^0 + p &\to K^0 + p & &\tfrac{1}{2}(A_1 + A_0), \\
&\to K^+ + n & &\tfrac{1}{2}(A_1 - A_0), \\
K^+ + n &\to K^+ + n & &\tfrac{1}{2}(A_1 + A_0), \\
&\to K^0 + p & &\tfrac{1}{2}(A_1 - A_0), \\
K^0 + n &\to K^0 + n & &A_1.
\end{aligned}
\qquad (8.26)
$$

Assuming charge independence, these reactions can all be analyzed by $t=0$ and $t=1$ amplitudes, A_0 and A_1, in each partial wave and the cross sections obtained as in π-N scattering. Observation of some these reactions is very difficult because neutron targets are available only in deuterium, which complicates interpretation, and neutral kaon beams always contain mixtures of K^0 and $\bar{K}^0$ (Chapter XIV). However, K^+p elastic scattering has been studied (Goldhaber et al., 1962) and, up to 640 MeV/c, analyzed in a finite-effective-range formalism. In Eq. (4.2) the $t=1$ scattering length is -0.29 ± 0.015 Fm ($=10^{-13}$ cm) and the effective range 0.5 ± 0.15 Fm. This negative scattering length means a repulsive interaction and a negative S-wave phase shift that has reached about $-30°$ at 600 MeV/c. An interaction that can explain the results is a repulsive core of radius 0.3 Fm. By 800 MeV/c the angular distribution is becoming nonisotropic and there is a rapid rise in the inelastic channels. A full analysis of this region does not yet exist, although, as we have mentioned, there are suggestions of a resonance.

The $t=0$ channel has been studied by Stenger et al. (1964). The S-wave phase shift is very small, corresponding to a scattering length of $+0.04 \pm 0.04$ Fm.

The antikaon scattering is more complicated because the presence of inelastic channels. We give these for the K^-p interaction below the three particle final state thresholds:

$$
\begin{aligned}
K^- + p &\to K^- + p & &\text{(a)} \\
&\to \bar{K}^0 + n & &\text{(b)} \\
&\to \Lambda + \pi^0 & &\text{(c)} \\
&\to \Sigma^0 + \pi^0 & &\text{(d)} \\
&\to \Sigma^+ + \pi^- & &\text{(e)} \\
&\to \Sigma^- + \pi^+ & &\text{(f)}
\end{aligned}
\qquad (8.27)
$$

All are exothermic, apart from (b) which has its threshold at 90 MeV/c laboratory incident momentum. Another two channels are open, namely, $\Lambda\pi^0\pi^0$ and $\Lambda\pi^+\pi^-$, but these are unimportant at low energies and will be neglected. A large number of other low-energy reactions can be written down for neutron targets and for $\bar{K}^0$ incident on proton and neutron targets.

8.8 Low-Energy Kaon–Nucleon Scattering

We shall analyze the K^-p, system but the results can easily be extended to the other reactions.

The $\bar{K}N$ system is a superposition of $t = 1$ and $t = 0$:

$$|K^-p\rangle = \sqrt{\tfrac{1}{2}}\{|1, 0\rangle - |0, 0\rangle\}, \qquad (8.28)$$
$$|\bar{K}^0n\rangle = \sqrt{\tfrac{1}{2}}\{|1, 0\rangle + |0, 0\rangle\},$$

so that the elastic (a) and charge exchange (b) scattering amplitudes are

$$\langle K^-p|T|K^-p\rangle = \tfrac{1}{2}(A_1 + A_0), \qquad (8.29)$$
$$\langle \bar{K}^0n|T|K^-p\rangle = \tfrac{1}{2}(A_1 - A_0),$$

where A_1 and A_0 are the S-wave scattering amplitudes in the $t = 1$ and $t = 0$ states, respectively. We assume zero effective range and analyze in terms of a complex scattering length [Eq. (4.8a)]

$$A_1 = \alpha_1/(1 - k\alpha_1), \qquad A_0 = \alpha_0/(1 - k\alpha_0),$$

where $\alpha_1 = a_1 + ib_1$ and $\alpha_0 = a_0 + ib_0$ are these complex scattering lengths. Then from Eqs. (4.9) and (8.29) we have

$$\sigma(K^-p \to K^-p) = \pi|\alpha_1/(1 - ik\alpha_1) + \alpha_0/(1 - ik\alpha_0)|^2, \qquad (8.30)$$
$$\sigma(K^-p \to \bar{K}^0n) = \pi|\alpha_1/(1 - ik\alpha_1) - \alpha_0/(1 - ik\alpha_0)|^2. \qquad (8.31)$$

The reaction cross section is

$$\sigma_r = (4\pi/k^2)\{\tfrac{1}{2}(1 - \eta_1^2) + \tfrac{1}{2}(1 - \eta_0^2)\},$$
$$= (2\pi/k^2)\{2 - |(1 + ik\alpha_1)/(1 - ik\alpha_1)|^2$$
$$\quad - |(1 + ik\alpha_0)/(1 - ik\alpha_0)|^2\}. \qquad (8.32)$$

The factor of $\tfrac{1}{2}$ enters because we have applied Eq. (4.10) to both isotopic-spin states, which are present equally in the initial state. Evidently σ_r must be equal to the sum of cross sections for the reactions (c) to (f) [Eq. (8.27)]. We must therefore consider these reactions, and start by defining three transition amplitudes:

$$t = 1, \quad \bar{K}p \to \Lambda\pi : \quad T_1^\Lambda;$$
$$t = 1, \quad \bar{K}p \to \Sigma\pi : \quad T_1^\Sigma;$$
$$t = 0, \quad \bar{K}p \to \Sigma\pi : \quad T_0^\Sigma.$$

Decomposing the initial and final states into the isotopic-spin states, we find

$$\langle \Lambda\pi^0|K^-p\rangle = \sqrt{\tfrac{1}{2}}\, T_1^\Lambda, \qquad (8.33)$$
$$\langle \Sigma^0\pi^0|K^-p\rangle = -\sqrt{\tfrac{1}{6}}\, T_0^\Sigma, \qquad (8.34)$$
$$\langle \Sigma^+\pi^-|K^-p\rangle = \tfrac{1}{2} T_1^\Sigma + \sqrt{\tfrac{1}{6}}\, T_0^\Sigma, \qquad (8.35)$$
$$\langle \Sigma^-\pi^+|K^-p\rangle = -\tfrac{1}{2} T_1^\Sigma + \sqrt{\tfrac{1}{6}}\, T_0^\Sigma, \qquad (8.36)$$

where the left-hand amplitudes are those that must be inserted into the
S-wave terms in Eqs. (3.20)–(3.22) to obtain cross sections. In addition,
we define the phase difference between T_1^Σ and T_0^Σ

$$\phi = \arg T_1^\Sigma - \arg T_0^\Sigma$$

and a $t = 1$ branching fraction

$$\epsilon = \frac{|T_1^\Lambda|^2}{|T_1^\Sigma|^2 + |T_1^\Lambda|^2} \tag{8.37}$$

Then, using Eq. (3.26), the cross sections are

$$\sigma(\text{c}) = (2\pi/k^2) |T_1^\Lambda|^2, \tag{8.38}$$

$$\sigma(\text{d}) = (2\pi/3k^2) |T_0^\Sigma|^2, \tag{8.39}$$

$$\sigma(\text{e}) = (\pi/k^2) |T_1^\Sigma + \sqrt{\tfrac{2}{3}}\, T_0^\Sigma|^2, \tag{8.40}$$

$$\sigma(\text{f}) = (\pi/k^2) |T_1^\Sigma - \sqrt{\tfrac{2}{3}}\, T_0^\Sigma|^2. \tag{8.41}$$

Hence

$$\sigma_r = (2\pi/k^2)\{|T_1^\Sigma|^2 + |T_0^\Sigma|^2 + |T_1^\Lambda|^2\}. \tag{8.42}$$

Breaking this into its $t = 1$ and $t = 0$ parts and equating to the corresponding parts of Eq. (8.32), we find

$$|T_0^\Sigma|^2 = 1 - \left|\frac{1 + ik\alpha_0}{1 - ik\alpha_0}\right|^2 = \frac{kb_0}{(1 + kb_0)^2 + k^2 a_0^2}, \tag{8.43}$$

$$|T_1^\Sigma|^2 + |T_1^\Lambda|^2 = 1 - \left|\frac{1 + ik\alpha_1}{1 - ik\alpha_1}\right|^2 = \frac{kb_1}{|1 + kb_1|^2 + k^2 a_1^2}. \tag{8.44}$$

So we obtain

$$\sigma(K^- p \to \Lambda \pi^0) = \frac{2\pi}{k} \frac{b_1 \epsilon}{(1 + kb_1)^2 + k^2 a_1^2}, \tag{8.45}$$

$$\sigma(K^- p \to \Sigma^0 \pi^0) = \frac{2\pi}{3k} \frac{b_0}{(1 + kb_0)^2 + k^2 a_0^2}. \tag{8.46}$$

and

$$\sigma(K^- p \to \Sigma^\pm \pi^\mp) = \frac{\pi}{k^2}\left[\frac{(1-\epsilon)b_1}{(1+kb_1)^2 + k^2 a_1^2} + \frac{2\epsilon b_0}{3\{(1+kb_0)^2 + k^2 a_0^2\}}\right.$$
$$\left. \pm 2\cos\phi\left(\frac{2(1-\epsilon)\epsilon b_0 b_1}{3\{(1+kb_0)^2 + k^2 a_0^2\}\{(1+kb_1)^2 + k^2 a_1^2\}}\right)^{1/2}\right]. \tag{8.47}$$

Equations (8.30), (8.31), and (8.45)–(8.47) have expressed a six-channel scattering problem in terms of six parameters $(a_0, b_0, a_1, b_1, \epsilon, \phi)$ of a zero-effective-range approximation for the S-wave scattering. Given that this approximation gives the correct energy variation, there can be no over-determination of the parameters nor a check of isotopic-spin conservation unless data from K^-n, $\bar{K}^0p$, and $\bar{K}^0n$, scattering is also put in (the cross sections are functions of the same parameters). This has not been done at the present time, but nonetheless this analysis is useful for two reasons. The first involves an extrapolation to high energies where a knowledge of the S-wave phase shifts is a useful part of the analysis of a resonance that occurs in the K^-p system (Section 8.9). This extrapolation can be done with some confidence if a good analysis of the low-energy behavior is available.

The second reason concerns the $t = 0$ channel. Consider the scattering length given by the favored Solution I of Kim (1965):

$$\alpha_0 = (-1.674 \pm 0.038) + i(-0.722 \pm 0.040) \quad \text{Fm}.$$

The real part is negative and larger than the expected range of the interaction, suggesting a bound K^-p state that, if it were not too strongly bound, would be unstable and decay into $\Sigma\pi$. Dalitz (1961) has given a formula for the mass and width of a bound state as a function of the scattering length, which, applied to this data, predicts a mass of 1410.7 ± 1.0 MeV and a width of 37.0 ± 3.2 MeV. Such a state is known with a mass 1405 MeV and width 35 MeV. The analysis of scattering helped to fix the quantum numbers of this resonance as those of the $t = 0$ S-wave: $TJ^P = 00^-$.

Our analysis has neglected the Coulomb interaction. This has several effects. The first, of course, is to add a Coulomb scattering amplitude into the nucleon scattering amplitudes for the elastic channel and a Coulomb barrier into the charged final states. A second effect is indirect: the Coulomb interaction breaks isotopic-spin symmetry and splits the $K^-\bar{K}^0$, K^+K^0, and np masses. This raises the threshold for charge-exchange scattering to an appreciable momentum and has a considerable effect on the analysis. We refer readers to the original zero-effective-range analysis of Dalitz and Tuan (1960) for details. Ross and Shaw (1961) have given a finite effective range treatment.

8.9 The Antikaon–Nucleon System

The measurement of total cross section of K^- mesons in hydrogen and deuterium can be used to determine the total cross sections effective in each of the two isospin states, $t = 0$, and 1, of the $y = 0$ system. The K^-n system is entirely $t = 1$; however, it cannot be studied directly but must be examined by analysis of K^- cross-sections in hydrogen and deuterium, a procedure that lends some uncertainty to the result. Both systems are rich in resonances, althongh, as in the case of the pion–nucleon system, these resonances become very inelastic and broad above 2 GeV. Such measurements have revealed several of these resonances, and a detailed examination of the system has been done in bubble chambers, mainly with K^-

mesons in liquid hydrogen. All the reactions of Eq. (8.27) are observed, as well as $\Lambda\pi\pi$ final states and others with more pions, which open as the incident energy increases. This wealth of reactions gives the investigator many means by which to study resonant systems, which may well decay into several of the available states. To illustrate this, we shall consider the $\Lambda(1520)$ discovered in the K^-p system (Ferro-Luzzi et al., 1962; Watson et al., 1963). These authors observed that at an incident momentum of 390 MeV/c there was a peak in the cross sections for the reactions

$$K^- + p \to \bar{K}^0 + n,$$
$$\to \Sigma + \pi,$$
$$\to \Lambda + \pi^+ + \pi^-,$$

but not in the reaction

$$K^- + p \to \Lambda + \pi^0.$$

The $\Lambda\pi^0$ system is a $t = 1$ state, so that the absence of the peak in the last reaction suggested that it was due to a $t = 0$ resonance. This is supported by the observation of the resonance in the $\Sigma^0\pi^0$ and $\Lambda\pi^0\pi^0$ systems, both of which, in this reaction, are entirely $t = 0$ states. In addition, later evidence shows the existence of this resonance in production but only in neutral states. A $t = 0$ assignment is indicated.

The mass and width from fitting of cross sections are 1518 MeV and 16 MeV, respectively.

The spin of this resonance was determined by a study of the angular distributions in the charge-exchange scattering. Ferro-Luzzi and co-workers expressed the differential cross section for charge-exchange and elastic scattering in the form

$$d\sigma/d\Omega = (1/4k^2)(A + B\cos\theta + C\cos^2\theta),$$

which is similar to our decomposition into Legendre polynomials in Eq. (8.23). At the resonance only the two C coefficients show any peak. No $\cos^3\theta$ terms or higher were required. This indicates $j = \frac{3}{2}$; i.e., a P or D resonant wave. We can attempt the simplest explanation in terms of non-resonant $t = 1$ and $t = 0$ S-wave scattering (amplitudes S_{11} and S_{01}, where the subscripts are t, $2j$) and either a resonant P_{03} or D_{03} amplitude. The differential cross-sections are then as follows:

$P_{3/2}$ resonance:

$$\frac{d\sigma}{d\Omega} = \frac{1}{4k^2}\{|S_{11} \pm S_{01}|^2 + |P_{03}|^2$$
$$+ 4\cos\theta \,\text{Re}[(S_{11} \pm S_{01})^*P_{03}] + 3|P_{03}|^2\cos^2\theta\}, \quad (8.48)$$

8.9 The Antikaon-Nucleon System

$D_{3/2}$ resonance:

$$\frac{d\sigma}{d\Omega} = \frac{1}{4k^2}\{|S_{11} \pm S_{01} - D_{03}|^2 \\ + 3\cos^2\theta\,[|D_{03}|^2 + 2\,\mathrm{Re}(S_{11} \pm S_{01})^* D_{03}]\}, \quad (8.49)$$

(plus for elastic and minus for charge-exchange scattering).

The absence of a strong $\cos\theta$ term influenced by the resonance indicates that it is $D_{3/2}$ under this simple assumption. However, the Minami ambiguity permits the same angular distribution to be fitted with P_{11}, P_{01}, P_{03} amplitudes instead of S_{11}, S_{01}, D_{03}. This is not an acceptable alternative because it would imply that the S-wave scattering observed at low energies had disappeared and been replaced by P-wave scattering. This cannot be the case because it would imply interesting changes in angular distributions in some transition region between low energies and 390 MeV/c. These are not observed, and the conclusion is that the resonance is $D_{3/2}$. Thus the TJ^P assignment for the $\Lambda(1518)$ is $0\tfrac{3}{2}^-$. This analysis shows how it is possible to determine the parity of a resonance by knowing that of another interfering wave, in this case from a knowledge of the low-energy behavior

The reaction

$$K^- + p \to \Sigma^+ + \pi^-$$

is also influenced by this resonance. A determination of the orbital angular momentum of the final state at resonance, that is, for the decay $\Lambda(1518) \to \Sigma^+ + \pi^-$, will give the $N\Sigma$ relative parity; it is even if $l = 2$, odd if $l = 1$ or 3. To do this it is necessary to measure Σ^+ polarization, which can be done by observing the decay $\Sigma^+ \to p\pi^0$ and by knowing the decay asymmetry parameters (Sections 6.4 and 11.14). This polarization is zero unless there are other partial waves present. The incident state contains S_{01}, S_{11}, and D_{03}. Even parity gives the same outgoing waves in $\Sigma\pi$ and and a polarization P given by

$$\frac{Pk^2}{\sin\theta}\frac{d\sigma}{d\Omega} \propto -\cos\theta\,\mathrm{Im}\left(\frac{1}{2}S_{11}^\Sigma + \frac{1}{\sqrt{6}}S_{01}^\Sigma\right)^* D_{03}^\Sigma$$

where the quantities S_{11}^Σ, S_{01}^Σ, and D_{03}^Σ are the transition amplitudes into the $\Sigma\pi$ system of $t = 1$, 0, and 0, respectively, with obvious notation for indicating l. This equation follows from Eqs. (8.25) and (8.35).

If the ΣN parity is odd, $\Sigma\pi$ outgoing amplitudes are P_{01}^Σ, P_{11}^Σ, P_{03}^Σ and the polarization is given by

$$\frac{Pk^2}{\sin\theta}\frac{d\sigma}{d\Omega} \propto +\cos\theta\,\mathrm{Im}\left(\frac{1}{2}P_{11}^\Sigma + \frac{1}{\sqrt{6}}P_{01}^\Sigma\right)^* P_{03}^\Sigma,$$

the reverse of the first case. (The Minami ambiguity shows that both these cases will give the same $\Sigma\pi$ angular distribution.)

FIG. 8.4. Argand diagram of the partial-wave transition amplitudes near the D-wave resonance in the reaction $K^- + p \to \Sigma^+ + \pi^-$. The tip of the D_{03} amplitude vector traverses the circle counterclockwise as the resonance is crossed. The numbers indicate the incident K^--meson momentum at three points on the way around. The other amplitudes will change only slowly. (Watson et al., 1963.)

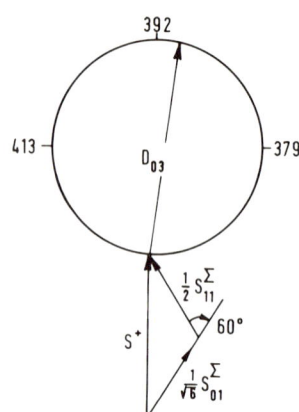

Let us suppose the ΣN parity is even; then we can predict the polarization. The branching ratios in the $\Sigma \pi$ modes of the $\Lambda(1518)$ suggested that $\arg S_{01}^{\Sigma} - \arg S_{11}^{\Sigma}$ was about $\pm 100°$, and the negative sign is favored by Kim's Solution I (Section 8.8). In the $\Sigma^+ \pi^-$ system the peak in the C coefficient of the angular distribution occurs at resonance. We can apply this to an equation like (8.49) to show that the S-wave amplitude S^+, in this case, $\frac{1}{2}S_{10}^{\Sigma} + \sqrt{\frac{1}{6}}S_{01}^{\Sigma}$, is nearly in phase with D_{03}^{Σ}. This gives an Argand diagram for these amplitudes near resonance as in Fig. 8.4. Now the polarization can be predicted: essentially it is given by $-2 \, \text{Im}(S^{+*}D_{03})$, which is positive below and negative above resonance if the S-wave amplitude varies slowly. The assumption that ΣN parity is odd makes the opposite prediction. The data show a polarization going through zero at resonance from having a positive value below resonance. The S-wave amplitudes certainly vary slowly, so this analysis shows that ΣN parity is even.

One possibility remains in this analysis. We have seen that neither the polarization nor the angular distribution changes if, at the same time as we interchange opposite parity amplitudes, we complex conjugate them (Section 3.6). We could then obtain the observed polarization for odd ΣN parity. However, the resonant amplitude would have to traverse the unitary circle in the clockwise direction, a variation disallowed by the Wigner condition (Section 4.4).

This analysis has been simplified considerably: its results could be considered as definite only if the results were consistent with all the data, such as the values of differential and total cross sections and polarization at various energies. Such a complete fit was made by Ferro-Luzzi and co-workers (Ferro-Luzzi et al., 1962; Watson et al., 1963). They also investigated the possibility of a $j = \frac{5}{2}$ resonance, but this hypothesis gave a poor fit to the data. The final conclusions do not differ from those we have given.

8.10 The Production of Baryon Resonances

This section will be devoted to discussing the methods of determining the spins and parities of resonances observed in production processes.

The Adair analysis, which was introduced in Section 8.2, can also be used on the baryon resonances. However, the sample of events must be selected for small-angle production, in fact, $\sin \theta \gtrsim 1/(l_{max} + 1)$, where θ is the center-of-mass production angle and l_{max} is the largest orbital angular momentum in the production process. This selection is very wasteful and may lead to such a small sample that the method becomes impossible. Thus it is desirable to look for methods that can make use of all events. In addition, it is desirable to include a parity determination. This usually involves measuring the polarization of the baryon produced in the decay of a baryon resonance. If the baryon produced is a Λ, Σ, or Ξ, the decay asymmetry of these particles will enable this to be done.

Let us consider the case of the $\Sigma(1385)$. This is observed in the reaction (see, for example, Alston et al., 1960; Ely et al., 1962)

$$K^- + p \to \Lambda + \pi^+ + \pi^-$$

for center-of-mass energies above the threshold 1515 MeV (380 MeV/c momentum). The Dalitz plot indicates that this reaction can go through two channels:

$$K^- + p \to \Sigma^+(1385) + \pi^-$$
$$\searrow$$
$$\Lambda + \pi^+$$

or

$$K^- + p \to \Sigma^-(1385) + \pi^+$$
$$\searrow$$
$$\Lambda + \pi^-.$$

The 1385 label is only a matter of convenience. The actual masses are 1383 and 1386 for the plus and minus, respectively. The width is 36 MeV; thus its decay is by strong interactions and the $\Lambda\pi$ decay product fixes $t = 1$.

Information can be obtained on the spin by observing the decay distribution of the $\Sigma(1385)$ in its own rest frame with respect to the direction (**n**), which is normal to its production plane. The Σ can be polarized in this direction and, if the spin is greater than $\frac{1}{2}$, this will generally give rise to an anisotropic decay distribution. The distribution was found to be of the form $1 + c \cos^2 \alpha$, where α is the angle between the direction of the decay Λ and **n**. This indicates $j \geqslant \frac{3}{2}$. To obtain the parity the Λ polarization is measured. The $\Lambda\pi$ state is $P_{3/2}$ or or $D_{3/2}$, depending on whether the $\Sigma(1385)$ is $\frac{3}{2}^+$ or $\frac{3}{2}^-$. These two cases give rise to different Λ polarizations.

The results indicated $\tfrac{3}{2}^+$, but full confirmation of this could only be obtained by an analysis that exhausts all the information available in the $\Sigma(1385)$ and the Λ decay. The way of doing this is due to Byers and Fenster (1963), and in fact is a general method of obtaining the spin and parity of a resonance from angular distributions and polarizations of the decay products when these consist of a spin-0 and spin-$\tfrac{1}{2}$ particle. In Section 2.15 we showed how to obtain the angular distribution and final polarization density matrix for the decay

$$\Delta(1236) \to \pi + N,$$

where the Δ has $J^P = \tfrac{3}{2}^+$, given the density matrix for the initial state. One obvious way of finding the J^P of a resonance decaying is to work out the decay distributions and polarizations for various J^P hypotheses and determine which fits the observations most closely. The Byers–Fenster method uses the observed distributions to obtain the parameters that specify the density matrix of a resonance. The number of parameters required and their interrelation give the spin and parity of the resonance. For convenience of description we apply the labels B^*, B, and π to the resonance and its spin-$\tfrac{1}{2}$ and spin-0 decay products; thus,

$$B^* \to B + \pi.$$

To describe the method we first consider the density matrix for the initial state B^*, of spin j. It is a Hermitian $(2j+1) \times (2j+1)$ matrix which, if normalized, requires for its specification $(2j+1)^2 - 1$ $[= 4j(j+1)]$ real parameters. Thus any such matrix can be constructed from the linear sum of $4j(j+1)$ basic Hermitian matrices. In a matrix representation of the operator $\mathbf{J}$ for spin j, there will be three such matrices. The remainder may be constructed from these by taking suitable products of these matrices. When this is done correctly, the basic matrices are a representation of spherical spin tensors T_L^M, where $-L \leqslant M \leqslant L$ and $L = 0, 1, \ldots, 2j$. These matrices are orthogonal,

$$\mathrm{Tr}(T_L^{M\dagger} T_{L'}^{M'}) = \frac{2j+1}{2L+1}\, \delta_{MM'}\delta_{LL'},$$

and transform under rotations as does $Y_L^M(\cos\theta)$ in each case. The procedure of building up these quantities is exactly analogous to constructing spherical harmonics from the position vector $\mathbf{r}$. Thus

$$Y_0 \propto 1,$$
$$Y_1^0 \propto (1/r)(z),$$
$$Y_1^{\pm 1} \propto \mp(1/r)(x \pm iy),$$
$$Y_2^0 \propto (1/r^2)(3z^2 - r^2), \quad \text{etc.,}$$

8.10 The Production of Baryon Resonances

and in fact the spherical spin tensors look formally the same:

$$T_0 \propto 1,$$
$$T_1^0 \propto J_z,$$
$$T_1^{\pm 1} \propto \mp(J_x \pm iJ_y),$$
$$T_2^0 \propto 3J_z^2 - \mathbf{J}^2.$$

The exact analogy does not continue, because J_x, J_y, and J_z do not commute whereas x, y, z do. However, the spirit is the same. The density matrix is given by

$$\rho = (2j+1)^{-1} \sum_{L=0}^{2j} (2L+1) \sum_{M=-L}^{M=L} t_L^{M*} T_L^M. \tag{8.50}$$

This defines the t_L^{M*}, which are called the multipole parameters and are complex numbers. If we were using this density matrix to find the expectation values of the spin, we would have

$$\langle J_z \rangle = \text{Tr}(\rho J_z)$$
$$= [j(j+1)]^{1/2} \text{Tr}(\rho T_1^0) = [j(j+1)]^{1/2} t_1^0. \tag{8.51}$$

Similarly, $\langle J_y \rangle$ and $\langle J_x \rangle$ are functions of $t_1^{\pm 1}$. If the production process is parity-conserving, then the density matrix ρ has some special properties. At present we are considering ρ defined so that the axis of quantization z is the normal to the production plane, and in the rest frame of the B^*. The ρ must be invariant under a parity transformation, and this is easily seen to be equivalent to a 180° rotation around the z axis. The operator responsible is $\exp(i\pi J_z/\hbar)$, and hence, from Eq. (2.98), we have

$$\exp(-i\pi J_z/\hbar) \, \rho \, \exp(+i\pi J_z/\hbar) = \rho. \tag{8.52}$$

From the definition of the density matrix this implies, for the element labeled $j_z j_z'$,

$$[\exp i\pi(j_z - j_z')] \rho_{pq} = \rho_{pq},$$

where j_z and j_z' are eigenvalues of the J_z operator for the base states in which the density matrix is expressed. It follows that $j_z - j_z'$ is zero or an even number, or that $\rho_{j_z j_z'} = 0$. This result means that the density matrix has a chessboard pattern of zeros. Returning to the expansion in terms of spin tensors, this has the effect that in Eq. (8.50) all t_L^{M*} with M odd are zero.

Byers and Fenster (1963) showed that the decay angular distribution of the B^* is given by

$$I(\theta, \phi) = \sum_{L,M} n_{L0} t_L^M Y_L^{M*}(\theta, \phi), \tag{8.53}$$

where $Y_L^{M*}(\theta, \phi)$ are the complex conjugates of the spherical harmonics and the n_{L0} are constants determined by L and the spin j of the resonance. The angles θ and ϕ are the spherical polar angles in the rest frame of the B^*. The x, y, z coordinate axes in this frame are those obtained by the following procedure. The coordinates set up in the production center of mass are along the directions of $\mathbf{k}_1 \times \mathbf{n}$, $\mathbf{k}_1$, and $\mathbf{n}$ (x, y, and z, respectively), where $\mathbf{k}_1$ is the direction of the incident particle and $\mathbf{n}$ is the normal to the production plane. Suppose the B^* is produced at angle ψ with respect to $\mathbf{k}_1$. Then the correct Lorentz transformation along its direction of motion ($\mathbf{k}_2$) will bring an observer into its rest frame from which he can see the center of mass moving. Thus he can remember both the direction in which the B^* is moving and the production plane, and hence set up the axes x, y, z with z perpendicular to $\mathbf{k}_2$ and the production plane, with y in the production plane so that it is at an angle ψ from the remembered B^* direction, and with $\mathbf{x} = \mathbf{y} \times \mathbf{z}$.

The terms in Eq. (8.53) are such that only even values of L are required, and it is obvious that the $t_L{}^M$ can be found by taking the moments of the spherical harmonics over the observed angular distribution. Thus from Eq. (8.53) we have

$$t_L{}^M = (1/n_{L0})\langle Y_L{}^M \rangle, \qquad L \text{ even}, \tag{8.54}$$

where the moment is

$$\langle Y_L{}^M \rangle = \frac{\int_0^{2\pi} \int_{-1}^{+1} Y_L{}^M(\theta, \phi) I(\theta, \phi) \, d\cos\theta \, d\phi}{\int_0^{2\pi} \int_{-1}^{+1} I(\theta, \phi) \, d\cos\theta \, d\phi}$$

If P is the polarization of the B, then its component along the direction of B in the B^* rest frame is $\mathbf{P} \cdot \mathbf{B}$ if $\mathbf{B}$ is the unit vector along this direction. Now

$$I(\theta, \phi) \mathbf{P} \cdot \mathbf{B} = \sum_{L,M} n_{L0} t_L{}^M Y_L^{M*}(\theta, \phi), \tag{8.55}$$

where only odd L values are required. Thus

$$t_L{}^M = (1/n_{L0})\langle \mathbf{P} \cdot \mathbf{B} Y_L{}^M \rangle, \tag{8.56}$$

where $\langle \mathbf{P} \cdot \mathbf{B} Y_L{}^M \rangle$ is the moment of $Y_L{}^M$ over the angular distribution of the longitudinal polarization.

We can now see that for a given j hypothesis we can find all the coefficients $t_L{}^M$. The correct j will in general yield all the corresponding $t_L{}^M$ ($L = 2j$) finite. If the hypothetical j is greater than the actual j, the values of the higher coefficients ($L > 2j$) will be zero. Thus a determination of the moments essentially yields j. To obtain the parity it is neces-

8.10 The Production of Baryon Resonances

sary to look at the transverse polarization. The angular distribution of the transverse components gives the odd L, $t_L{}^M$ multiplied by ± 1 for $l = j \mp \frac{1}{2}$, where l is the πB orbital angular-momentum states. Thus a comparison of the $t_L{}^M$ from transverse and longitudinal polarization gives the relative parity of the B^* and B.

To see this result in terms of physics is fairly straightforward. Consider a matrix M that, operating on the density matrix of B^*, gives the spin density matrix ρ for the B. The decay is assumed parity-conserving, so M has well-defined properties under the parity transformation (see Section 6.4). However, if, by hypothesis, we change the orbital parity of the final states, then it must reverse its properties under the transformation; i.e., a scalar M becomes pseudoscalar or vice versa. This can be done by multiplying from the left any assumed M by the only available pseudoscalar matrix $\boldsymbol{\sigma} \cdot \mathbf{B}$. So far as states are concerned this has the effect of transforming the final B spin state $|B\rangle$ by

$$|B\rangle \to \boldsymbol{\sigma} \cdot \mathbf{B} |B\rangle ,$$

where in a matrix representation $|B\rangle$ will be a two-component spinor. Now since

$$(\boldsymbol{\sigma} \cdot \mathbf{B})(\boldsymbol{\sigma} \cdot \mathbf{B}) = 1 ,$$

we have

$$\exp i \frac{\pi}{2} \boldsymbol{\sigma} \cdot \mathbf{B} = 1 + i \frac{\pi}{2} \boldsymbol{\sigma} \cdot \mathbf{B} - \left(\frac{\pi}{2}\right)^2 - \frac{i}{3!}\left(\frac{\pi}{2}\right)^3 (\boldsymbol{\sigma} \cdot \mathbf{B}) + \cdots$$

$$= \cos \frac{\pi}{2} + i \boldsymbol{\sigma} \cdot \mathbf{B} \sin \frac{\pi}{2} = i \boldsymbol{\sigma} \cdot \mathbf{B} .$$

The operator $i \boldsymbol{\sigma} \cdot \mathbf{B} = \exp i\pi(\boldsymbol{\sigma} \cdot \mathbf{B}/2)$ thus has the effect of rotating the physical system through $180°$ around the direction $\mathbf{B}$. It follows that whatever the transverse polarization is for one parity, it has the same magnitude but the reverse direction for the opposite parity.

The Byers–Fenster method is very powerful and has been used in several places. In any experimental situation the method will be limited by the statistical weight of the sample to be analyzed, and this must be taken into account in determining spins and parities. Shafer and Huwe (1964) used the method on a sample of 895 charged $\Sigma(1385)$ and were able to establish $J^P = \frac{3}{2}^+$ with considerable confidence. Schlein et al. (1963) used 80 examples of $\Xi(1530)$ to establish that the simplest J^P assignment was $\frac{3}{2}^+$.

The method, to be applicable, demands the satisfaction of several conditions. First the decaying resonance should be decaying as a free particle. Short-lived particles decaying partly within the interaction region will suffer from interference effects, which can invalidate the analysis.

Second, the production process must produce a sample of the resonance that has a considerable degree of polarization or alignment: unpolarized samples give trivial uniform angular distribution of all the interesting quantities. The third condition is that it must be possible to measure the polarization of the baryon produced in the decay of the resonance. This is straightforward if this particle is Ξ, Σ, or Λ, since these have an asymmetric decay distribution in most decay channels. This decay intensity distribution is of the form $(1 + \alpha \mathbf{P} \cdot \mathbf{k})\, d\Omega$, where $\mathbf{P}$ is the polarization and $\mathbf{k}$ is a unit vector along the direction of an outgoing decay product. The component of $\mathbf{P}$ along a direction $\mathbf{m}$ is given by the moment of $\langle \mathbf{m} \cdot \mathbf{k} \rangle$ over the observed angular distribution:

$$P_m = (3/\alpha)\langle \mathbf{m} \cdot \mathbf{k} \rangle\,.$$

In principle the determination of the parity does not require a knowledge of α, since only the sign of the transverse component relative to the longitudinal component is needed. However, it is important that α be large or the confidence level of the result will be low from a statistically limited sample.

Many baryon resonances are produced in reactions that are strongly peripheral. We shall discuss the peripheral reaction mechanisms in some detail in Chapter XII. It is worth noting here that it is sometimes possible to identify the nature of the exchanged particle and to make use of the restrictions this places upon the density matrix of the particles produced.

8.11 The Ω^-

One of the remarkable events in the physics of elementary particles was the successful prediction of the existence of an isotopic singlet baryon, with hypercharge -2, charge -1, and a predicted mass of about 1675 MeV.

The prediction was based on the assumption that the observed elementary particles were members of $SU(3)$ multiplets. The subject of unitary symmetry will be discussed in Chapter IX, and for the present it is sufficient to remark that the predicted Ω^- was the missing member of a family expected to number 10. One interesting aspect of the prediction was that the mass was just too low to allow decay by strong interactions and consequently that the Ω^- would be "stable" except against decay by weak interactions and would therefore have a mean life of above 10^{-10} sec. Another feature is that since it had $y = -2$, the most likely production reaction was of the kind

$$\bar{K} + N \to \Omega + K + K,$$

where the important feature is two $y = +1$ kaons in the final state. The first recognized example of the Ω^- was found by Barnes *et al.* (1964); the reaction observed was

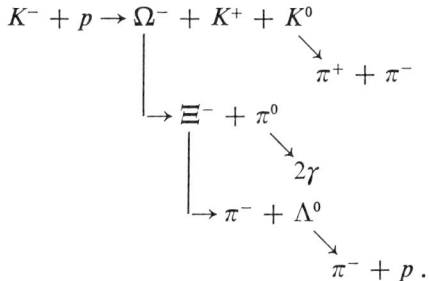

The world total of Ω^- particles is rising, but there is as yet an insufficient number to permit a determination of J^P. However, this is predicted to be $\tfrac{3}{2}^+$ and, from the manner of the prediction, it would be very surprising if it were not. The observed mass is about 1672 MeV and the mean life approximately 1.1×10^{-10} sec.

REFERENCES

Adair, R. K. (1955). *Phys. Rev.* **100**, 1540 (8.2).
Alston, M. *et al.* (1960). *Phys. Rev. Lett.* **5**, 520 (8.10).
Anderson, H. L., Fermi, E., Martin, R., and Nagle, D. E. (1953). *Phys. Rev.* **91**, 155 (8.4).
Auvil, P., Donnachie, A., Lea, A. T., and Lovelace, C. (1964). *Phys. Lett.* **12**, 76 (8.6).
Bareyre, P., Brickman, C., Stirling, A. V., and Villet, G. (1965a). *Phys. Lett.* **18**, 342 (8.6).
Bareyre, P., Brickman, C., and Villet, G. (1965b). *Phys. Rev. Lett.* **14**, 881 (8.6).
Barnes, V. E. *et al.* (1964). *Phys. Rev. Lett.* **12**, 204 (8.11).
Bransden, B. H., O'Donnell, P. J., and Moorhouse, R. G. (1965). *Phys. Rev. B* **139**, 1566 (8.6).
Byers, N., and Fenster, S. (1963). *Phys. Rev. Lett.* **11**, 52 (8.10).
Chew, G. F., and Low, F. E. (1956). *Phys. Rev.* **101**, 1570 (8.5).
Courant, H. *et al.* (1963). *Phys. Rev. Lett.* **10**, 409 (8.2).
Crawford, F. S., Cresti, M., Good, M. L., Stevenson, M. L., and Ticho, H. K. (1959). *Phys. Rev. Lett.* **2**, 114 (8.2).
Dalitz, R. H. (1962). *Rev. Mod. Phys.* **33**, 471 (8.8).
Dalitz, R. H., and Tuan, S. F. (1960). *Ann. Phys.* (*New York*) **10**, 307 (8.8).
Donnachie, A., Kirsopp, R. G., and Lovelace, C. (1968). *Phys. Lett. B* **26**, 161 (8.6).
Duke, P. J. *et al.* (1965). *Phys. Rev. Lett.* **15**, 468 (8.6).
Ely, R. P., Fung, S. Y., Gidal, G., Pan. Y. L., Powell, W. M., and White, H. S. (1962). *Phys. Rev. Lett.* **7**, 461; **8**, 48 (8.10).
Ferro-Luzzi, M., Tripp, R. D., and Watson, M. B., (1962). *Phys. Rev. Lett.* **8**, 28 (8.9).
Ferro-Luzzi, M. *et al.* (1965). *Nuovo Cimento* **36**, 1101 (8.3).
Goldhaber, S. *et al.* (1962). *Phys. Rev. Lett.* **9**, 135 (8.8).
Hamilton. J. (1966). *Phys. Lett.* **20**, 687 (8.5).

Hamiton, J. (1967). "High Energy Physics" (E. H. S. Burhop, ed.), Vol. I, p. 194. Academic Press, New York (8.5).
Kim, J. K., (1965). *Phys. Rev. Lett.* **14,** 29 (8.8).
Lee, T. D., and Yang, C. N. (1958). *Phys. Rev.* **109,** 1755 (8.2).
Meyer, J. (1968). *Proc. Int. Conf. Elem. Particles, Heidelburg, 1967,* North-Holland Publ., Amsterdam (8.7).
Roper, L. D., and Wright, R. M. (1965). *Phys. Rev. B* **138,** 921 (8.6).
Roper, L. D., Wright, R. M., and Feld, B. T. (1965). *Phys. Rev. B* **138,** 190 (8.6).
Ross, M., and Shaw, G. L. (1961). *Ann. Phys. (New York)* **13,** 147 (8.8).
Schlein, P. E., Carmony, D. D., Pjerrou, G. M., Slater, W. E., Stork, D. H., and Ticho, H. K. (1963). *Phys. Rev. Lett.* **11,** 167 (8.10).
Shafer, J. B., and Huwe, D. O. (1964). *Phys. Rev. B* **134,** 1372 (8.10).
Stenger, V. J., Slater, W. E., Stork, D. H., Ticho, H. K., Goldhaber, G., and Goldhaber, S. (1964). *Phys. Rev. B* **134,** 1111 (8.8).
Treiman, S. B. (1959). *Phys. Rev.* **113,** 355 (8.2).
Tripp, R, D. (1966). Baryon resonances. *Proc. Int. Sch. Phys. Enrico Fermi, Varenna, 1964.* **1,** 70 (8.6).
Watson, M. B., Ferro-Luzzi, M., and Tripp, R. D. (1963). *Phys. Rev.* **131,** 2248 (8.9).

IX

UNITARY SYMMETRY

9.1 Introduction

As our knowledge of the spectrum of elementary particles has improved, so attempts to classify the strongly interacting particles (hadrons) have been made. The first success came with the Sakata (1956) model, which in turn led to the eightfold way (Gell-Mann and Ne'eman 1964) and a realization of the importance of the symmetry groups $SU(2)$, $SU(3)$ and higher. In this chapter we shall introduce the idea of unitary symmetry and the theory of continuous groups, and apply these theories to the physics of the hadrons.

9.2 Symmetry and the Classification of States

One of the facts that emerges from the treatment of angular momentum in Chapter II is that the symmetry of physical systems under rotations in ordinary three-dimensional space leads to the conservation of angular momentum and to the classification of states according to their total and z component of angular momentum (j, j_z). In the absence of an external field the total energy of a system is independent of j_z, and thus the $2j + 1$ states associated with j are degenerate. Similarly, symmetry under rotations in isotopic-spin space is a property of the strong interactions, and as a consequence the hadrons can be classified according to the values of their total and third component of isotopic spin (t, t_3). These multiplets of t were recognized early in the extension of our knowledge of elementary particles; for example,

$$t = 0, \quad \Lambda;$$
$$t = \tfrac{1}{2}, \quad n, p;$$
$$t = 1, \quad \pi^+, \pi^0, \pi^-; \text{ etc.}$$

The question now arises whether strong interactions possess other symmetries that may lead to a classification of the particles into supermultiplets.

It is evident that the electromagnetic interaction is not symmetric under rotations in isotopic-spin space and that consequently the isotopic-spin multiplets are not precisely degenerate. This is another way of saying the mass differences within isospin multiplets is electromagnetic in origin. Since the known hadrons do not fall into supermultiplets as easily as they do into isospin multiplets, if there is a higher symmetry, it must be badly broken. Nonetheless it appears possible to classify particles into supermultiplets and to make considerable progress in relating masses, branching ratios, etc., although it is evident that this approach to hadron physics is only a beginning in one aspect of our understanding of elementary particles.

In order to develop our understanding, we must consider the theory of continuous groups. This may sound a forbidding subject, but in fact all we have to do is extend slightly and generalize the treatment we gave to angular momentum. The reader will find it helpful to consider the corresponding situation in angular momentum. In addition, we shall frequently draw attention to the group-theoretical aspects of angular momentum for clarification.

The literature on the subject of group theory and elementary particles is now considerable, and we refer readers to other works for a fuller treatment and enlightenment [see, among many, Behrends *et al.* (1962); de Swart (1963); Lipkin (1965); Carruthers (1966); Dyson (1966); Matthews (1967); Gasiorowicz (1966); Charap *et al.* (1967)]. The theory of groups, continuous groups, and their connection with quantum mechanics is an older subject. Readers interested in looking into these earlier aspects of the subject are referred to, for example, Weyl (1931, 1946); Hamermesh (1962); and Wigner (1959).

9.3 The Theory of Continuous Groups

Let us consider a complex vector space of n dimensions. This means that we can find n linearly independent vectors and that any arbitrary vector is a sum with complex coefficients of these n vectors. This set of base vectors is not necessarily unique. A transformation will generate from one vector $|\psi\rangle$ a second $|\psi'\rangle$, and if the length of the vector is preserved the connection is made via a unitary operator U (Section 1.6):

$$|\psi\rangle \to |\psi'\rangle = U|\psi\rangle,$$
$$\langle\psi| \to \langle\psi'| = \langle\psi|U^\dagger = \langle\psi|U^{-1}.$$

For example, a rotation of the physical system in ordinary space changes

9.3 The Theory of Continuous Groups

the eigenvector representing that state as in Eq. (2.6). The importance of the unitary transformation lies in the fact that it preserves the normalization; that is, the "length" of the vector. Thus, if

$$\langle \phi | \phi \rangle = 1$$

and

$$|\phi\rangle \to U|\phi\rangle = |\phi'\rangle,$$

then

$$\langle \phi' | \phi' \rangle = 1.$$

If we consider this in matrix notation, then U is an $n \times n$ unitary matrix, and $\langle \phi |$ and $|\phi \rangle$ are row and column vectors with n elements. (We shall frequently refer to matrices as a concrete representation of group theory.) Any unitary matrix U can be written $e^{i\phi}\mathcal{U}$, where $\mathcal{U}$ is again a unitary matrix but has the property Det $\mathcal{U} = 1$; that is, it is unimodular. If we take all the $n \times n$ matrices U that cause transformations of the vectors in our n-dimensional space, then a common phase can be removed so that all the remaining are unimodular. The $n \times n$ matrices U are a representtion of the unitary group $U(n)$, and the $n \times n$ matrices $\mathcal{U}$ of the special unitary group $SU(n)$.

If a physical system is invariant under the transformation represented by U, then the Hamiltonian

$$H \to H' = UHU^{-1} = \mathcal{U}e^{i\phi}He^{-i\phi}\mathcal{U}^{-1} = H.$$

If

$$H = e^{i\phi}He^{-i\phi},$$

then the system is invariant under a gauge transformation (see Section 9.4). Such an invariance will have an associated additive quantum number, which is conserved. We can, for example, associate this quantum number with baryon number. We do this and now consider the remaining transformations that are associated with the unitary operators $\mathcal{U}$. Invariance under these transformations will lead to the existence of further conserved quantum numbers that depend upon the group properties of $\mathcal{U}$.

As usual (cf. Section 1.6) we can write any one of the unitary operators $\mathcal{U}$ thus:

$$\mathcal{U} = \exp\left(i \sum_j \alpha_j G_j\right), \tag{9.1}$$

where the α_j are a set of real parameters and the G_j are a set of Hermitian operators called the generators of the transformation (J_x, J_y, and J_z in the case of angular momentum). All the $\mathcal{U}$ can be constructed out of finite a number of G's with different α_j. In matrix language, since the $\mathcal{U}$'s are unimodular, the G's are traceless.

All the properties of physical systems that we associate with the quantum theory of angular momentum, such as the classification of states, the vector addition of angular momenta, etc., follow from the commutation relations satisfied by the generators [Eq. (2.4)]. Invariance of physical systems under rotations means that the physical systems will be eigenstates of one of the operators (J_z) among the generators (**J**), and that conservation of the corresponding quantum number (j_z) is assured. In the more general case, the commutation relations satisfied by the generators determine completely the algebra of these operators. There will be some that are mutually commuting, and invariance under the corresponding transformations will mean that physical states can occur that are eigenstates of these operators and that the corresponding quantum numbers will be conserved. The number of generators required is called the order of the group s and the number of mutually commuting generators is called the rank r ($s = 3$, $r = 1$ for isotopic spin and angular momentum). Each physical state will then be characterized by r quantum numbers, in addition to others that appear later. The commutation relations satisfied by the generators are

$$[G_j, G_k] = C^l_{jk} G_l, \qquad (9.2)$$

where C^l_{jk} are called the structure constants and satisfy the conditions

$$C^l_{jk} = -C^l_{kj}, \qquad (9.3\text{a})$$

$$C^l_{jk} C^n_{lm} + C^l_{km} C^n_{lj} + C^l_{mj} C^n_{lk} = 0. \qquad (9.3\text{b})$$

Equation (9.3a) follows naturally from the form of a commutation relation and (9.3b) from the Jacobi relation. The usual convention of summing over repeated indices applies.

We can now give two examples of matrix representations: one is of the group $SU(2)$, which we have met before in isotopic spin, and the other is of $SU(3)$. The $SU(2)$ of isotopic spin has the same group algebra as does the group of rotations in three-dimensional space [$R(3)$], apart from the fact that rotation through 2π in $SU(2)$ is -1 and is the identity in $R(3)$. The 2 of $SU(2)$ means that, other than the trivial unit representation, the simplest matrix representation of the unitary operators $\mathscr{U}$ is a set of 2×2 matrices, and it follows in this case that the generators are 2×2 traceless Hermitian matrices. There are only three such matrices that are independent, and in one representation they are

$$T_1 = \frac{1}{2}\begin{bmatrix} 0 & 1 \\ 1 & 0 \end{bmatrix}, \quad T_2 = \frac{1}{2}\begin{bmatrix} 0 & -i \\ +i & 0 \end{bmatrix}, \quad T_3 = \frac{1}{2}\begin{bmatrix} 1 & 0 \\ 0 & -1 \end{bmatrix}, \qquad (9.4)$$

that is, half the Pauli matrices; these satisfy the commutation relations

$$[T_j, T_k] = i\epsilon_{jkl} T_l, \qquad (9.5)$$

where $j, k, l = 1, 2, 3$ and ϵ_{jkl} is entirely antisymmetric with $\epsilon_{123} = 1$,

9.3 The Theory of Continuous Groups

$\epsilon_{213} = -1$, etc. The next problem in $SU(2)$ is to find higher-dimensional matrices that satisfy Eq. (9.5). The representation in Eq. (9.4) we associate with isotopic spin $t = \frac{1}{2}$, and systems having isotopic spin $\frac{1}{2}$ are eigenstates of T_3 with eigenvalues $+\frac{1}{2}$ and $-\frac{1}{2}$. Representations corresponding to higher t exist: for example, for $t = 1$ we have

$$T_1 = \frac{1}{\sqrt{2}}\begin{bmatrix} 0 & 1 & 0 \\ 1 & 0 & 1 \\ 0 & 1 & 0 \end{bmatrix}, \quad T_2 = \frac{1}{\sqrt{2}}\begin{bmatrix} 0 & -i & 0 \\ i & 0 & -i \\ 0 & +i & 0 \end{bmatrix}, \quad T_3 = \begin{bmatrix} 1.0 & 0 & 0 \\ 0 & 0 & 0 \\ 0 & 0 & -1 \end{bmatrix},$$

which again satisfy Eq. (9.5). We know, by analogy with angular momentum, that representations corresponding to $t = \frac{3}{2}, 2, \frac{5}{2}, \ldots$, etc. also exist, and it is evident that the corresponding generating matrices are $2t + 1$ square. We have chosen the representations so that the generator T_3 is diagonalized, which corresponds to the normal convention in which the physical systems are classified as eigenstates of T_3. We know that these eigenstates are also eigenstates of the operator $T^2 = T_1^2 + T_2^2 + T_3^2$ with eigenvalues $t(t+1)$, where t is the dimension of the representation. This is a "Casimir" operator. Its eigenvalues differ only between representations of different dimension, but do not differ among the eigenstates of one dimension. T^2 has eigenvalues $\frac{3}{4}$ for order 2, 2 for order 3, $\frac{15}{4}$ for order 4, etc.

Now let us consider $SU(3)$. Other than the unit representation, the simplest $\mathscr{U}$ are 3×3 matrices. The lowest-order matrix representation of the generators will be all the eight independent 3×3 traceless Hermitian matrices; we now give one representation (Gell-Mann, 1962):

$$G_1 = \frac{1}{2}\begin{bmatrix} 0 & 1 & 0 \\ 1 & 0 & 0 \\ 0 & 0 & 0 \end{bmatrix}, \quad G_2 = \frac{1}{2}\begin{bmatrix} 0 & -i & 0 \\ +i & 0 & 0 \\ 0 & 0 & 0 \end{bmatrix}, \quad G_3 = \frac{1}{2}\begin{bmatrix} 1 & 0 & 0 \\ 0 & -1 & 0 \\ 0 & 0 & 0 \end{bmatrix},$$

$$G_4 = \frac{1}{2}\begin{bmatrix} 0 & 0 & 1 \\ 0 & 0 & 0 \\ 1 & 0 & 0 \end{bmatrix}, \quad G_5 = \frac{1}{2}\begin{bmatrix} 0 & 0 & -i \\ 0 & 0 & 0 \\ +i & 0 & 0 \end{bmatrix}, \quad G_6 = \frac{1}{2}\begin{bmatrix} 0 & 0 & 0 \\ 0 & 0 & 1 \\ 0 & 1 & 0 \end{bmatrix},$$

$$G_7 = \frac{1}{2}\begin{bmatrix} 0 & 0 & 0 \\ 0 & 0 & -i \\ 0 & +i & 0 \end{bmatrix}, \quad G_8 = \frac{1}{2\sqrt{3}}\begin{bmatrix} 1 & 0 & 0 \\ 0 & 1 & 0 \\ 0 & 0 & -2 \end{bmatrix}, \quad (9.6)$$

where

$$\mathscr{U} = \exp i\left(\sum_{j=1}^{8} \alpha_j G_j\right) = \exp i\boldsymbol{\alpha} \cdot \mathbf{G}.$$

The generators satisfy the commutation relations

$$[G_j, G_k] = i f_{jkl} G_l \qquad (9.7)$$

when the f_{jkl} are, as required, antisymmetric. The minimum number of f_{jkl} required to find all the nonzero commutations are given in Table 9.1.

TABLE 9.1

VALUES OF THE STRUCTURE CONSTANTS f_{jkl} FOR $SU(3)$

jkl	123	147	156	246	257	345	367	458	678
f_{jkl}	1	$\tfrac{1}{2}$	$-\tfrac{1}{2}$	$\tfrac{1}{2}$	$\tfrac{1}{2}$	$\tfrac{1}{2}$	$-\tfrac{1}{2}$	$\sqrt{\tfrac{3}{4}}$	$\sqrt{\tfrac{3}{4}}$

We have here a (three dimensional) representation of the group $SU(3)$; it is evident that the order of the group is 8 and, from an examination of Eqs. (9.6), that the rank is 2, since there are only two diagonal matrices. [In general, $SU(n)$ has rank $n - 1$ and order $n^2 - 1$.] If we have in mind an application to hadrons, then the rank 2 is what is required, for we have, apart from baryon number, two internal quantum numbers, the third component of isotopic spin and the hypercharge, to accommodate into a systematic scheme.

This three-dimensional representation is called the fundamental representation. Before considering representations of higher dimensions, we wish to introduce a graphical representation of the eigenvectors of the commuting generators. This is called a weight diagram. In general it consists of an r-dimensional space in which a point (or weight) represents an eigenstate and is plotted with coordinates that are the eigenvalues. Thus the weight of an eigenstate is an r-component vector $\mathbf{m}$ $(m_1, m_2, \ldots, m_r,$ the eigenvalues of the commuting generators). For $SU(3)$ the rank r is 2, so we have a simple two-dimensional weight diagram. Before considering weight diagrams further, we wish to introduce a new set of operators whose significance will become clear: we put

$$T_\pm = G_1 \pm iG_2,$$
$$U_\pm = G_6 \pm iG_7,$$
$$V_\pm = G_4 \pm iG_5, \qquad (9.8)$$
$$T_3 = G_3,$$
$$Y = (2/\sqrt{3}) G_8.$$

Notice that the commuting operators are T_3 and Y and, from an examination of the matrix representation, that their eigenvalues in the fundamental representation are (t_3, y):

9.3 The Theory of Continuous Groups

$(+\tfrac{1}{2}, +\tfrac{1}{3})$, eigenstate $\begin{bmatrix} 1 \\ 0 \\ 0 \end{bmatrix}$;

$(-\tfrac{1}{2}, +\tfrac{1}{3})$, eigenstate $\begin{bmatrix} 0 \\ 1 \\ 0 \end{bmatrix}$;

$(0, -\tfrac{2}{3})$, eigenstate $\begin{bmatrix} 0 \\ 0 \\ 1 \end{bmatrix}$.

The weight diagram for this representation is shown in Fig. 9.1. In accord with convention, we shall label this representation **3** (i.e., the dimension) or $D^3(1, 0)$ with a nomenclature that will become clear later.

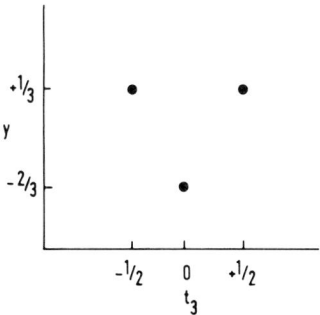

FIG. 9.1. Weight diagram for the $D^3(1, 0)$ representation of $SU(3)$.

We can now look at the effect of the operators of Eqs. (9.8). We find that these have an analogous effect to the lowering and raising operators of angular momentum $(J_\pm)$ and isotopic spin $(T_\pm)$. For example, the state $|t_3, y\rangle$ has

$$T_3 |t_3, y\rangle = t_3 |t_3, y\rangle,$$

and the states $T_\pm |t_3, y\rangle$ have

$$T_3(T_\pm |t_3, y\rangle) = (t_3 \pm 1) T_\pm |t_3, y\rangle.$$

This is the same as in isotopic spin. In general,

$T_\pm$ causes $\Delta y = 0$, $\Delta t_3 = \pm 1$,
$U_\pm$ causes $\Delta y = \pm 1$, $\Delta t_3 = \mp \tfrac{1}{2}$,
$V_\pm$ causes $\Delta y = \pm 1$, $\Delta t_3 = \pm \tfrac{1}{2}$.

See Fig. 9.2.

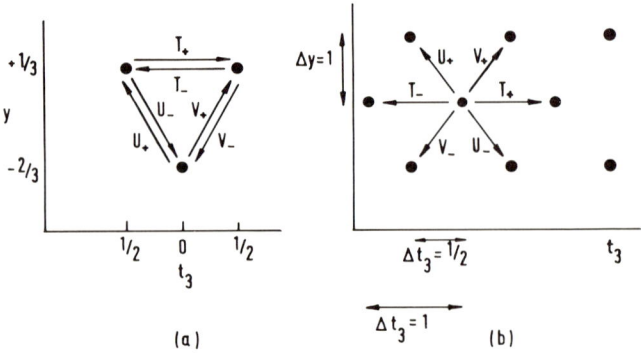

FIG. 9.2. The effect of the operators $T_\pm$, $U_\pm$, and $V_\pm$, (a) on the eigenstates of $D^3(1, 0)$ and (b) on the general eigenstate of $SU(3)$.

These operators we shall call the shift operators. There is a limit, of course, to repeated application of any of the shift operators to any eigenstate. For example, T_+ operating on the three states of $D^3(1, 0)$ is zero except for that having the eigenvector $|-\frac{1}{2}, \frac{1}{3}\rangle$. We shall find that the higher-dimensional representations have weight diagrams that contain more weights than $D^3(1, 0)$, and in these circumstances the shift operators change eigenstates into neighboring eigenstates if such exist in the representation; if they do not exist, the result is zero: for example, in $SU(2)$

$$T_+ |t, t_3 = t\rangle = 0.$$

Each weight diagram in $SU(2)$ is an array of points along a line. For the lowest-dimensional representation there are two points with $t_3 = \pm\frac{1}{2}$, for the next, three points with $t_3 = +1, 0, -1$, etc. This brings us to one property that has only been touched on so far. Both $SU(2)$ and $SU(3)$ have a one-dimensional representation; that is, a singlet representation whose matrix representation will have all matrices with one element that is zero. In $SU(2)$ this unitary singlet representation has $t_3 = 0$ ($t = 0$), and in $SU(3)$ it has $t_3 = 0$, $y = 0$ [label **1** or $D^1(0, 0)$]. As we have mentioned, the lowest-dimensional representation apart from the singlet is called the fundamental representation. There is another three-dimensional representation of $SU(3)$ called the complex-conjugate representation **3***. If we return to the transformation that was the starting point

$$\psi \to \psi' = \mathscr{U}\psi = \exp(i\,\boldsymbol{a}\cdot\mathbf{G})\,\psi,$$

we can evidently take the complex conjugate to obtain

$$\psi^* \to \psi'^* = \mathscr{U}^*\psi^* = \exp(i\,\boldsymbol{a}\cdot(-\mathbf{G}^*))\,\psi^*.$$

9.3 The Theory of Continuous Groups

Thus the generators $-\mathbf{G}^*$ are a new representation of the group. In the case of $SU(2)$ the generators produced in this way are not independent of the original; i.e., it is possible to find an S that satisfies

$$S\mathbf{G}S^{-1} = -\mathbf{G}^*,$$

so that the two representations have the same weight diagrams and are said to be equivalent. This is not the case for many of the representations of $SU(3)$, and in particular the complex conjugate of **3** is not equivalent. The weight diagram for **3*** $[D^3(0, 1)]$ is given in Fig. 9.3. [The commuting operators have been normalized as in Eq. (9.8).]

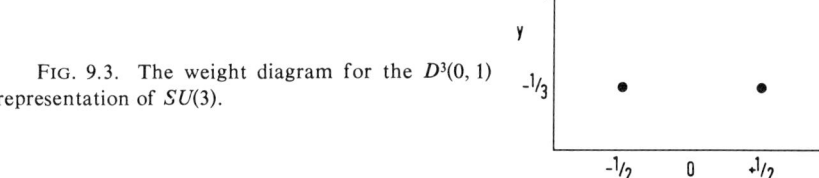

FIG. 9.3. The weight diagram for the $D^3(0, 1)$ representation of $SU(3)$.

In angular momentum and isotopic spin the higher-dimensional representations may be generated by the vector addition of two or more of the fundamental representations. Thus we know that the vector addition of two spin-½ systems can give rise to spin 1 and spin 0. We are familiar with this direct product of two representations; for example, in Eqs. (5.3)–(5.6), where it is done for spin ½ added to spin ½. There is a nomenclature for writing this, which is, for this example,

$$\mathbf{2} \otimes \mathbf{2} = \mathbf{3} \oplus \mathbf{1}. \tag{9.9}$$

The left-hand side implies the direct product of two two-dimensional representations; that is, the vector addition of two spin-½ systems. This is equal to the sum of a three-dimensional representation (spin 1) and a one-dimensional representation (spin 0). The reduction of a product as in Eq. (9.9) is called a Clebsch–Gordan series; the coefficients of this name give, of course, the amplitudes with which the eigenstates from each representation on the right occur for a given product of two eigenstates on the left. The product is said to be *reducible* if it is possible to reduce a product into a sum of representations and these representations are irreducible, since further decomposition would be possible were they not. For further information on reducible and irreducible representations we refer readers to the literature; for our present purpose we rely upon our knowledge of the

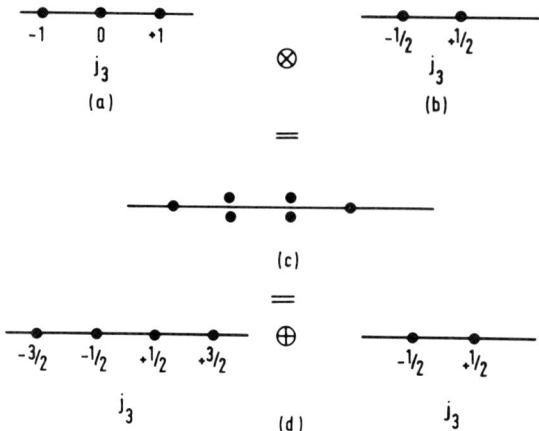

FIG. 9.4. The decomposition of the direct product $3 \otimes 2$ in $SU(2)$: (a) weight diagram for 3; (b) weight diagram for 2; (c) weight diagram for the product; (d) the weight diagrams in the Clebsch–Gordan series $3 \otimes 2 = 4 \oplus 2$.

vector addition of angular momentum to understand the decomposition of products of representations.

We can now consider the product of two representations, say $3 \otimes 2$ in $SU(2)$. We know that the weight diagrams allow occupied sites such that the weights are separated by the amount expected from properties of the shift operators J_+ and J_-, that is ± 1 (units of $\hbar$). We take the weight diagram for 3 (Fig. 9.4a) and 2 (Fig. 9.4b); the combined weight diagram is obtained by superimposing on each weight in one representation the center of the weight diagram of the second representation and replacing the first weight by all the weights of the second in the positions in which they are to be found in this superpositioning. Thus $j_z = -1$ becomes two weights $j_z = -\frac{3}{2}$ and $-\frac{1}{2}$. The combination is shown in Fig. 9.4c. The weights are $\frac{3}{2}$, $-\frac{3}{2}$ occupied once and $+\frac{1}{2}$, $-\frac{1}{2}$ each occupied twice (this is shown by displacing the heavy points from their coordinates). We know the reduction of this is into the two weight diagrams in Fig. 9.4d. This all corresponds to the vector addition of $j = 1$ with $j = \frac{1}{2}$ and is, in group-theory language,

$$3 \otimes 2 = 4 \oplus 2.$$

Now let us do this for $SU(3)$; we will start with $3 \otimes 3^*$. If we manipulate the weight diagrams of Figs. 9.1 and 9.3 as described in the previous paragraph, we obtain the combined weight diagram shown in Fig. 9.5. Again, when a site is occupied more than once, we show this by slightly

9.3 The Theory of Continuous Groups

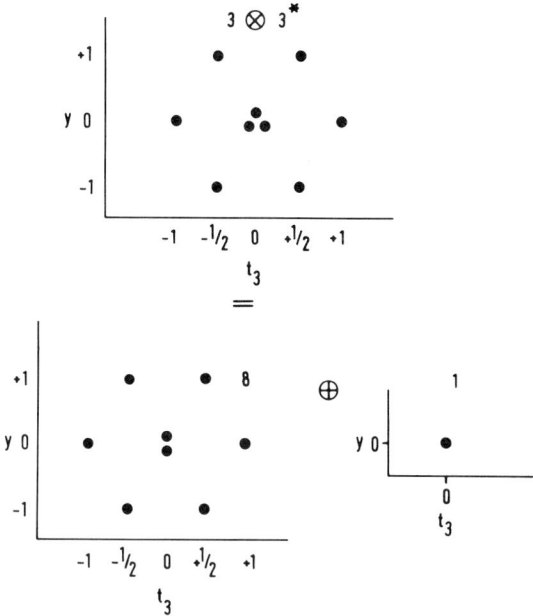

FIG. 9.5. The weight diagrams for the product $3 \otimes 3 = 8 \oplus 1$.

displaced points, one per weight. The decomposition of the product is not obvious but can be done once the rules about the structure of weight diagrams are known. These we shall give shortly, but at the moment state that the Clebsch–Gordan series is

$$3 \otimes 3^* = 8 \oplus 1 \qquad (9.10)$$

and the weight diagrams for the irreducible representations **8** and **1** are given in Fig. 9.5. Other representations may be found by taking other products of **3** and **3***. For example,

$$3 \otimes 3 = 6 \oplus 3^*,$$
$$3 \otimes 3 \otimes 3 = 1 \oplus 8 \oplus 8 \oplus 10. \qquad (9.11)$$

In Fig. 9.6 we give the weight diagrams for **10**, **10***, and **27**, which we have chosen from a list of the representations that begins

1, 3, 3*, 6, 6*, 8, 10, 10*, 15, 15*, 15', 15'*,

21, 21*, 24, 24*, 27, 28, 28*, ...

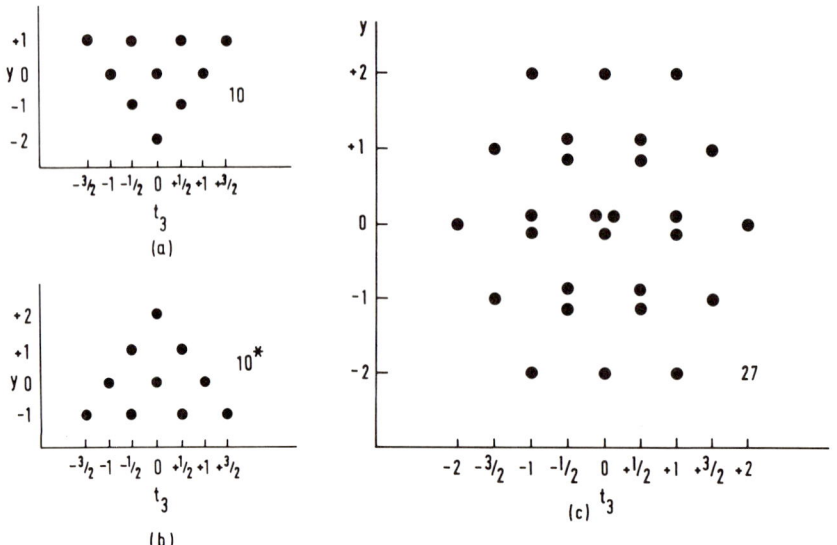

FIG. 9.6. The weight diagrams for **10** (a), **10*** (b), and **27** (c).

Once higher representations have been found, the process can be continued with these in the same manner: for example,

$$8 \otimes 8 = 1 \oplus 8 \oplus 8 \oplus 10 \oplus 10^* \oplus 27 . \tag{9.12}$$

We can now describe some properties of weight diagrams for irreducible representations of $SU(3)$. First we define some terms. The boundary is the series of straight lines that pass through all the outermost weights. The multiplicity of a given weight is the number of times a given weight occurs in each irreducible representation.

(1) The boundary is nonreentrant and generally six sided.

(2) The general boundary is shown in Fig. 9.7. It is symmetrical under rotation through 120°, and the lengths of the sides are specified by the integers p and q. Starting from the state with the maximum t_3, it takes p applications of the shift operator V_- to reach the next corner and q applications of the operator T_- to go from that corner to the next. Thus Fig. 9.7 is the boundary for $p = 3$, $q = 2$; that is, for the weight diagram of the irreducible representation **42** or $D^{42}(3, 2)$.

(3) There is only one weight with the maximum value of t_3, which is $\frac{1}{2}(p + q)$. The value for y of this weight is $\frac{1}{3}(p - q)$.

(4) All the weights on the boundary have multiplicity one.

(5) The multiplicity is two for all the weights one layer in from the boundary. It is three on the next layer and so on until the line joining the

9.3 The Theory of Continuous Groups

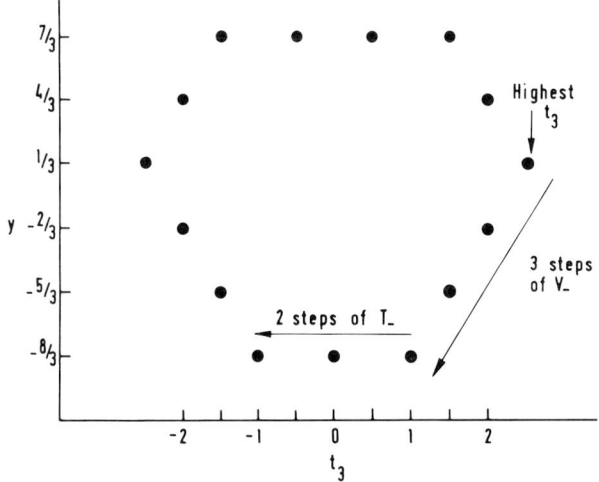

FIG. 9.7. The boundary of the irreducible representation $D^{42}(3, 2)$.

layer of weights in question becomes triangular, at which the multiplicity stops increasing (cf. Fig. 9.6c).

(6) It follows that weight diagrams that have triangular boundaries have multiplicity one throughout (cf. Fig. 9.6a).

(7) The dimension of the representation, which is equal to the total number of weights, is

$$\tfrac{1}{2}(2 + p + q)(1 + p)(1 + q).$$

It is evident that within a given representation the eigenvalues t_3, y do not completely specify a weight, since the multiplicity may be greater than one. If we consider all the weights that have the same y, then we have a set of weights that are several representations of the group $SU(2)$. For example, consider the $y = 0$ line of **27** (Fig. 9.6). The t_3 content is -2 once, -1 twice, 0 three times, $+1$ twice, and $+2$ once: This is obviously three $SU(2)$ representations of dimensions 5, 3, and 1. Anticipating the assignment of the familiar third component of isotopic spin to this component of these weights we see that we have three isotopic-spin multiplets with $t = 2$, 1, and 0, respectively. The operator $T^2 = \tfrac{1}{2}(T_+T_- + T_-T_+) + T_3^2 (= T_1^2 + T_2^2 + T_3^2$ of Chapter V) commutes with T_3 and with y, so it can be diagonalized and its eigenvalues $t(t + 1)$ can be used to specify the states. It also commutes with T_+ and T_-, so that these operators only shift between states having the same t. This is entirely analogous to $J^2 = J_x^2 + J_y^2 + J_z^2$ with eigenvalues $j(j + 1)$ in angular momentum. This brings us to the Casimir operators. These are functions of the generators that commute with all the generators and hence have eigenvalues that are the same throughout one representation. J^2 is a Casimir operator for angular momentum. There are two Casimir operators in $SU(3)$, one of which is obviously G^2; however, they are of no immediate interest, since we prefer to specify representations in $SU(3)$ not by their eigenvalues but by the dimension or the values of p and q. Readers interested in these operators are referred to the work of Carruthers (1966).

9.4 The Hadrons and $SU(3)$ Multiplets

The first model of elementary particles that used unitary symmetry was due to Sakata (1956), in which the triplet p, n, Λ was assumed to be the **3** representation and their antiparticles $\bar{p}, \bar{n}, \bar{\Lambda}$ the **3*** representation. If we compare the weight diagrams for **3** and **3*** (Figs. 9.1 and 9.3), we see that t_3 has the correct values for the isotopic spin, and $y + \frac{2}{3}B$, where B is baryon number, will give the correct hypercharge. The product

$$3 \otimes 3^* = 1 \oplus 8$$

implies that the states that can be formed from combination of baryon and antibaryon are a singlet with $y = 0$, $t_3 = 0$ and an octet with the isotopic-spin content

$y = 0,$ isospin singlet;
$y = 0,$ isospin triplet (π^+, π^0, π^-);
$y = +1,$ isospin doublet (K^+, K^0);
$y = -1,$ isospin doublet $(\bar{K}^0, K^-)$.

The identifications are in parentheses. At the time of the Sakata model no isospin singlets were known, but they have since been found (η and X^0). Thus, the Sakata model successfully predicted the multiplet structure of the pseudoscalar mesons. The $J^P = 0^-$ of the mesons was assumed to be due to the fact that the particle–antiparticle were in a $^1S_0$ state. The state $^3S_1$ would give rise to the nine vector mesons ($\omega, \phi, \rho, K^*, \bar{K}^*$). However, the Sakata model could not be made to accommodate the remaining $J^P = \frac{1}{2}^+$ baryons (Σ, Ξ) in a sensible manner.

The next step was made by Gell-Mann and separately by Ne'eman (see Gell-Mann and Ne'eman, 1964). The multiplet structure of the $\frac{1}{2}^+$ baryons (N, Λ, Σ, Ξ) was identical to that of the pseudoscalar mesons, and therefore they made the suggestion that they both belonged to the regular **8** representation of $SU(3)$ (the eightfold way). The **8** weight diagram is reproduced in Fig. 9.8 with labels indicating the particle identifications. Since these original identifications the ninth 0^- meson has been found, X^0, which is presumably a **1**. This structure of **1 + 8** is repeated in the vector mesons and in the 2^+ mesons (see Chapter VII). Among the baryon resonances the

9.4 The Hadrons and $SU(3)$ Multiplets

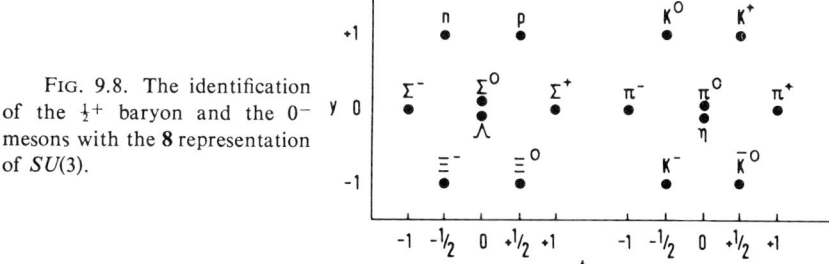

FIG. 9.8. The identification of the $\frac{1}{2}^+$ baryon and the 0^- mesons with the **8** representation of $SU(3)$.

situation is much less clear. Presumably the resonances in octet meson–nucleon systems will belong to the representations found in the direct product

$$8 \otimes 8 = 1 \oplus 8 \oplus 8 \oplus 10 \oplus 10^* \oplus 27 . \tag{9.13}$$

A large number of such resonances are now known, but assignment to multiplets is difficult, often because of lack of information about Ξ resonances. Several singlet candidates are known [$\Lambda(1405)$, $\Lambda(1520)$, etc.] and there are several suggested octets (Morales, 1968), but the real certainty is the existence of the **10**. The weight diagram and particle identification are shown in Fig. 9.9. The J^P assignments of $\frac{3}{2}^+$ are certain to the $\Delta(1236)$ and $\Sigma(1385)$ and are presumed for the $\Xi(1529)$ and the Ω^-. At the introduction of the eightfold way the tenth member of this decuplet, the Ω^-, was not known, and its existence was predicted from the nature of the weight diagram and its mass estimated from the Gell-Mann–Okubo mass formula. The discovery of this particle (Barnes et al., 1964) with the right mass was one of the great successes of unitary symmetry (see Section 8.11).

It is worth emphasizing one point: the unitary transformations, whose group structure has lead to those classification schemes, is expected to

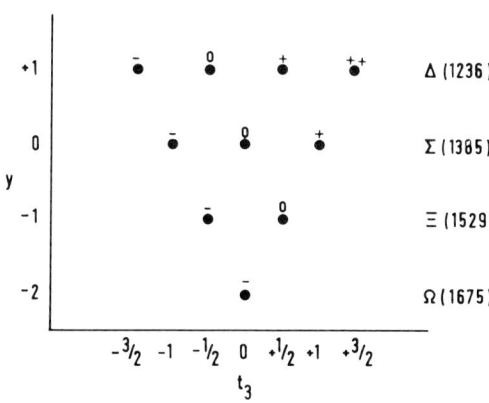

FIG. 9.9. The weight diagram for **10** and the particles identified with each weight; the sign above each weight is the particle charge.

commute with all the space-time transformations, so that within any multiplet the particles should have the same spin, parity, and mass. It seems certain that all but the last are satisfied. There appears to be an interaction that separates the masses within a multiplet but conserves parity and angular momentum. The analogy with the isotopic-spin multiplets is that electromagnetism splits the energy levels, but spin and parity are the same within the multiplet. We shall deal in more detail with the mass splitting in Section 9.7.

9.5 Properties of Representations

In this section, we shall briefly cover some topics that we shall need in considering the application of unitary symmetry. First, we shall consider the shift operators $T_\pm$, $U_\pm$, $V_\pm$: just as in angular momentum, these operators have matrix elements between the states that they connect. These matrix elements may be found by forming the eigenvectors of a representation from a direct product of eigenvectors of the fundamental representation and investigating the effects of the shift operators on the eigenvectors involved. This means knowing the matrix elements for the fundamental representations, and these may be found as follows. For example, consider the following matrix element of U_+U_-:

$$M = \langle -\tfrac{1}{2}, +\tfrac{1}{3} | U_+ U_- | -\tfrac{1}{2}, +\tfrac{1}{3} \rangle,$$

where the states belong to 3 are labeled $|t_3, y\rangle$ and are normalized. Since $U_+ | -\tfrac{1}{2}, +\tfrac{1}{3} \rangle = 0$ (there is no state to which U_+ can shift $|-\tfrac{1}{2}, +\tfrac{1}{3}\rangle$), we can put

$$M = \langle -\tfrac{1}{2}, +\tfrac{1}{3} | [U_+, U_-] | -\tfrac{1}{2}, +\tfrac{1}{3} \rangle.$$

Now the commutation relations give

$$[U_+, U_-] = \tfrac{3}{2} Y - T_3,$$

so that

$$M = 1.$$

However, using rule (d) (Section 1.2) for the matrix elements of products of operators, we can write

$$M = \langle -\tfrac{1}{2}, +\tfrac{1}{3} | U_+ U_- | -\tfrac{1}{2}, +\tfrac{1}{3} \rangle$$
$$= \sum_\alpha \langle -\tfrac{1}{2}, +\tfrac{1}{3} | U_+ | \alpha \rangle \langle \alpha | U_- | -\tfrac{1}{2}, +\tfrac{1}{3} \rangle.$$

An examination of Fig. 9.2 will show that there is only one state in the sum, that having $t_3 = 0$, $y = -\tfrac{2}{3}$, so we have

$$M = \langle -\tfrac{1}{2}, \tfrac{1}{3} | U_+ | 0, -\tfrac{2}{3} \rangle \langle 0, -\tfrac{2}{3} | U_- | -\tfrac{1}{2}, \tfrac{1}{3} \rangle = 1.$$

9.5 Properties of Representations

But $U_- = U_+^\dagger$; hence,

$$|\langle -\tfrac{1}{2}, \tfrac{1}{3} | U_+ | 0, -\tfrac{2}{3}\rangle|^2 = 1,$$

and

$$\langle -\tfrac{1}{2}, \tfrac{1}{3} | U_+ | 0, -\tfrac{2}{3}\rangle = e^{i\delta}.$$

Again, there is to be a choice of phase as in the case of angular momentum; the factors affecting the choice are discussed by Carruthers (1966). Conventionally the choice is made to correspond as closely as possible with the Condon and Shortley phases. In the representation **3** all the shift-operator matrix elements are $+1$. In **3*** they are all $+1$ except for $U_\pm$, which have matrix elements -1. The matrix elements in higher representations are then determined (Gasiorowicz, 1966).

The next important topic concerns the Clebsch–Gordan coefficients for $SU(3)$. In the direct product of two representations we can write out a Clebsch–Gordan series; for example,

$$\mathbf{3} \otimes \mathbf{3^*} = \mathbf{1} \oplus \mathbf{8}.$$

If we take the direct product of one state from **3** with one from **3***, it will in general be a linear superposition of the nine states from **1** and **8**. The coefficients of this linear superposition are the Clebsch–Gordan coefficients. They play the same role in $SU(3)$ as do the more familiar coefficients of the same name in angular momentum and isotopic spin $[SU(2)]$. Notation is now very important, as we have to be prepared to describe many parameters: using our bra-ket notation we use indices that are close to those used by de Swart (1963), who produced comprehensive tables of coefficients.

Let us consider the direct product of **a** and **b**. We are interested in that linear combination of products of pairs of states from **a** and **b** which is an eigenstate belonging to a particular irreducible representation (μ) appearing in the Clebsch–Gordan series for $\mathbf{a} \otimes \mathbf{b}$. We use μ, μ_a, and μ_b, as indices that specify the irreducible representations (e.g., they could be the eigenvalues of the Casimir operators). We use ν, ν_a, and ν_b, each symbol to cover all the quantum numbers t, t_3, y, which will specify completely one state within a representation [three numbers in $SU(3)$]. Then, by analogy with Eq. (2.40), the required linear combination is

$$|\mu_\gamma, \nu, \mu_a, \mu_b\rangle = \sum_{\nu_a, \nu_b} \langle \mu_a, \nu_a, \mu_b, \nu_b | \mu_\gamma, \nu, \mu_a, \mu_b\rangle |\mu_a, \nu_a, \mu_b, \nu_b\rangle. \quad (9.14)$$

The additional index γ distinguishes two representations that might otherwise appear to be the same; for example, the two **8**'s in $\mathbf{8} \otimes \mathbf{8}$ [Eq. (9.13)]. To be consistent with Carruthers (1966) and de Swart (1963) we write

$$\langle \mu_a, \nu_a, \mu_b, \nu_b | \mu_\gamma, \nu, \mu_a, \mu_b \rangle = \begin{pmatrix} \mu_a & \mu_b & \mu_\gamma \\ \nu_a & \nu_b & \nu \end{pmatrix}. \quad (9.15)$$

The indices ν, ν_a, and ν_b contain the additive quantum numbers t_3 and y, so that we have

$$t_3 = t_{a3} + t_{b3}$$

and

$$y = y_a + y_b,$$

with obvious notation. The total isotopic spins from a and b must add vectorially, as usual, so that we can form a linear sum of eigenstates of a and b to form eigenstates of

$$\mathbf{T} = \mathbf{T}_1 + \mathbf{T}_2, \qquad T_3 = T_{a3} + T_{b3},$$

and $Y = Y_a + Y_b$, using the well-known $SU(2)$ Clebsch–Gordan coefficients. So we put

$$|t, t_3, y, \mu_a, \mu_b\rangle$$
$$= \sum_{t_{a3}, t_{b3}} \langle t_a, t_{a3}, t_b, t_{b3} | t, t_3, t_a, t_b \rangle | \mu_a, \nu_a, \mu_b, \nu_b \rangle \quad (9.16)$$

exactly as in Eq. (2.40). We now form linear sums of the states $|t, t_3, y, \mu_a, \mu_b\rangle$, which are eigenstates belonging to the representation of interest

$$|\mu_\gamma, \nu, \mu_a, \mu_b\rangle = \sum_{\substack{t_a, t_b \\ y_a, y_b}} \begin{pmatrix} \mu_a & \mu_b & \mu_\gamma \\ t_a, y_a & t_b, y_b & t, y \end{pmatrix} |t, t_3, y, \mu_a, \mu_b\rangle. \quad (9.17)$$

The coefficients in this sum are called the isoscalar factors, since they are independent of the orientation in isotopic-spin space.

Thus the complete Clebsch–Gordan coefficient becomes

$$\begin{pmatrix} \mu_a & \mu_b & \mu_\gamma \\ \nu_a & \nu_b & \nu \end{pmatrix} = \langle t_a, t_{a3}, t_b, t_{b3} | t, t_3 \rangle \begin{pmatrix} \mu_a & \mu_b & \mu_\gamma \\ t_a, y_a & t_b, y_b & t, y \end{pmatrix}, \quad (9.18)$$

where we have contracted the isotopic-spin coefficient as was done just after Eq. (2.40). As an example let us consider the **10** in the product of $\mathbf{8} \otimes \mathbf{8}$. In particular let us pick out the $y = 0$, $t = 1$ states in **10** [particle $\Sigma(1385)$ if **a** is the baryon octet and **b** the pseudoscalar octet]. This will be a linear sum of the product states in $\mathbf{8} \otimes \mathbf{8}$ that have $y = 0 = y_a + y_b$ and $t = 1$ and $t_3 = t_{a3} + t_{b3}$. Particle identifications would make the final state a sum of the $t = 1$ states made from

9.5 Properties of Representations

	t_a	t_b	y_a	y_b	t
$\|\Xi K\rangle$	$\tfrac{1}{2}$	$\tfrac{1}{2}$	-1	$+1$	1
$\|N\bar{K}\rangle$	$\tfrac{1}{2}$	$\tfrac{1}{2}$	$+1$	-1	1
$\|\Sigma\eta\rangle$	1	0	0	0	1
$\|\Lambda\pi\rangle$	0	1	0	0	1
$\|\Sigma\pi\rangle$	1	1	0	0	1

Thus, using the isoscalar factors from de Swart,

$$
\begin{aligned}
|\Sigma(1385)\rangle = & \begin{pmatrix} 8 & 8 & | & 10 \\ \tfrac{1}{2}, -1 & \tfrac{1}{2}, +1 & | & 1, 0 \end{pmatrix} |\Xi K\rangle \quad = \sqrt{\tfrac{1}{6}}\,|\Xi K\rangle \\
& + \begin{pmatrix} 8 & 8 & | & 10 \\ \tfrac{1}{2}, +1 & \tfrac{1}{2}, -1 & | & 1, 0 \end{pmatrix} |N\bar{K}\rangle \quad -\sqrt{\tfrac{1}{6}}\,|N\bar{K}\rangle \\
& + \begin{pmatrix} 8 & 8 & | & 10 \\ 1, 0 & 0, 0 & | & 1, 0 \end{pmatrix} |\Sigma\eta\rangle \quad +\tfrac{1}{2}\,|\Sigma\eta\rangle \\
& + \begin{pmatrix} 8 & 8 & | & 10 \\ 0, 0 & 1, 0 & | & 1, 0 \end{pmatrix} |\Lambda\pi\rangle \quad -\tfrac{1}{2}\,|\Lambda\pi\rangle \\
& + \begin{pmatrix} 8 & 8 & | & 10 \\ 1, 0 & 1, 0 & | & 1, 0 \end{pmatrix} |\Sigma\pi\rangle \quad +\sqrt{\tfrac{1}{6}}\,|\Sigma\pi\rangle .
\end{aligned}
\quad (9.19)
$$

This is an example of Eq. (9.17). If now we were to choose a particular charge state of $\Sigma(1385)$, say the neutral, then the states on the right of Eq. (9.19) have to be expressed in their charge states. Thus those having $t = 1$, $t_3 = 0$ become, for example,

$$
\begin{aligned}
|\Xi K\rangle &= \sqrt{\tfrac{1}{2}}\,|\Xi^0 K^0\rangle + \sqrt{\tfrac{1}{2}}\,|\Xi^- K^+\rangle , \\
|\Sigma\eta\rangle &= |\Sigma^0 \eta\rangle , \\
|\Sigma\pi\rangle &= \sqrt{\tfrac{1}{2}}\,|\Sigma^+ \pi^-\rangle - \sqrt{\tfrac{1}{2}}\,|\Sigma^- \pi^+\rangle , \\
|\Lambda\pi\rangle &= |\Lambda\pi^0\rangle ,
\end{aligned}
\quad (9.20)
$$

and so on. These correspond to the decomposition of Eq. (9.18), and we have used the tables of Clebsch–Gordan coefficients in Appendix E. It is obviously easy to proceed to the complete expressions of Eq. (9.19) for various values of t_3.

The $SU(3)$ Clebsch–Gordan coefficients have various symmetry properties, which have been reviewed by de Swart (1963). As in $SU(2)$ the sum in Eq. (9.14) can be inverted to give

$$|\mu_a, \nu_a, \mu_b, \nu_b\rangle = \sum_{\mu,\tau} \begin{pmatrix} \mu_a & \mu_b & \mu_\tau \\ \nu_a & \nu_b & \nu \end{pmatrix} |\mu_\tau, \nu, \mu_a, \mu_b\rangle.$$

As in $SU(2)$ and angular momentum, there is a Wigner–Eckart theorem for $SU(3)$. To proceed let us consider an eigenstate $|\phi\rangle$ ($=|\mu, \nu\rangle$) of a representation μ. Under an $SU(3)$ transformation $\mathscr{U}$;

$$|\phi\rangle \rightarrow |\phi'\rangle = \mathscr{U} |\phi\rangle.$$

Now as in Eq. (2.63) we can express the state $\mathscr{U}|\mu, \nu\rangle$ as a sum of all the eigenstates of the representation:

$$\mathscr{U} |\mu, \nu\rangle = \sum_{\nu'} \mathscr{D}^\mu_{\nu\nu'} |\mu, \nu'\rangle, \tag{9.21}$$

where

$$\mathscr{D}^\mu_{\nu\nu'} = \langle \mu, \nu' | \mathscr{U} | \mu, \nu \rangle. \tag{9.22}$$

The matrix $\mathscr{D}^\mu_{\nu\nu'}$ corresponds to the rotation matrix of angular momentum. Now we consider an irreducible tensor operator: if it transforms according to the irreducible representation μ, then it is defined by

$$\mathscr{U} T(\mu, \nu) \mathscr{U}^{-1} = \sum \mathscr{D}^\mu_{\nu\nu'} T(\mu, \nu'). \tag{9.23}$$

As in our chapter on angular momentum we have left out the parameters α_j ($j = 1, \ldots, 8$) that specify the transformation: strictly, $\mathscr{U}$ and $\mathscr{D}$ are functions of the α_j. A tensor operator transforming according to a representation has the effect of producing from eigenstates of the representation μ_a states that belong to μ_b, where μ_b is contained in the reduction of $\mu \otimes \mu_a$. Thus $T(\mathbf{8}, \nu)|\mathbf{8}, \nu_a\rangle$ is a state that is a linear superposition of states from the irreducible representations in the reduction

$$\mathbf{8} \otimes \mathbf{8} = \mathbf{1} \oplus \mathbf{8} \oplus \mathbf{8} \oplus \mathbf{10} \oplus \mathbf{10^*} \oplus \mathbf{27}.$$

Obviously this corresponds to the description of tensor operators in angular momentum, which was given in Section 2.12. It is now possible to write down the Wigner–Eckart theorem for $SU(3)$. A proof is given by de Swart (1963). It states that

$$\langle \mu_b, \nu_b | T(\mu, \nu) | \mu_a, \nu_a \rangle = \sum_\tau \begin{pmatrix} \mu_a & \mu & \mu_b \\ \nu_a & \nu & \nu_b \end{pmatrix} \langle \mu_b \| T(\mu) \| \mu_a \rangle_\tau.$$

where the reduced matrix element is defined by this equation but is independent of the quantum numbers ν, ν_a, and ν_b. The number of terms is given by the index γ and depends on the number of times a representation μ_b appears in the reduction of $\mu \otimes \mu_a$. For example, there will be two terms if $\mu = \mu_a = \mu_b = \mathbf{8}$, one if $\mu = \mathbf{10}$, $\mu_a = \mu_b = \mathbf{8}$.

So far we have avoided the problem of distinguishing between representations of the same dimension when they occur in the reduction of a direct product. In, for example,

$$3 \otimes 3 \otimes 3 = 1 \oplus 8 \oplus 8 \oplus 10,$$

we could ask what distinguishes the two **8** representations. If the representations are constructed out of products of basic vectors belonging to **3**, then it is found that the two eights have different symmetry properties under the exchange of coordinates of two of the three vectors in the product. These symmetry properties do not change under $SU(3)$ transformations, so that they serve to distinguish between them. These statements apply quite generally.

9.6 Applications of $SU(3)$

The first application that we shall consider is that of obtaining $SU(3)$-invariant couplings between the pseudoscalar mesons (P) and the $\frac{1}{2}^+$ baryons (B). To introduce this subject let us consider the interaction between the nucleons and pions: the interaction Lagrangian density (Section 10.11) is of the form

$$\mathscr{L}_1 = \sum_{i=1}^{3} g_{\pi NN} \bar{\psi}_N \gamma_5 T_i \psi_N \phi_i.$$

Each of the four-component Dirac spinors $\bar{\psi}_N$ and ψ_N is also a two-component isotopic-spin spinor. ϕ is a three-component field describing the pions. We can lay aside the γ_5 structure and consider $\bar{\psi}_N T_i \psi_N$. This is a three-component quantity that evidently transforms like a vector in isotopic-spin space. ϕ_i is also a vector, so that the sum over i constitutes the scalar product and the result is an $SU(2)$ invariant. That it is possible to construct only one invariant becomes obvious when we consider the reduction of the direct product of three isotopic-spin states belonging to the $SU(2)$ representations **2**, **2**, and **3**. We have

$$2 \otimes 2 \otimes 3 = 1 \oplus 3 \oplus 3 \oplus 5.$$

Thus the single appearance of the scalar **1** shows that only one $SU(2)$-invariant Lagrangian can be constructed and only one coupling constant is required.

Now the case of the baryon-to-pseudoscalar meson Yukawa coupling is one of coupling three fields, each belonging to **8**. The relevant reduction is

$$\begin{aligned}
8 \otimes 8 \otimes 8 &= 8 \otimes (1 \oplus 8 \oplus 8 \oplus 10 \oplus 10^* \oplus 27) \\
&= 8 \oplus 1 \oplus 8 \oplus 8 \oplus 10 \oplus 10^* \oplus 27 \oplus 1 \oplus 8 \oplus 8 \oplus 10 \\
&\quad \oplus 10^* \oplus 27 \oplus 8 \otimes (10 \oplus 10^* \oplus 27).
\end{aligned}$$

There is no **1** in **8** ⊗ (**10** ⊕ **10*** ⊕ **27**), so that the reduction contains only two scalars. This indicates that we can construct two scalar invariants from the fields, and we will require two coupling constants to give their strength. These are usually called the D and F couplings. We put

$$\mathscr{L} = g\bar{B}\gamma_5[(1-\alpha)D_i + \alpha F_i]BP_i.$$

B and P have been used to represent the baryon and pseudoscalar meson fields, respectively. The quantities D_i and F_i in matrix notation are those combinations of the 8×8 matrix generators of the regular representation that make $\bar{B}DB$ and $\bar{B}FB$ representations of **8**, and are symmetric and antisymmetric, respectively, under baryon exchange. The strength of the D (symmetric) coupling is $(1-\alpha)g$, and that of the F (antisymmetric) coupling is αg. Unscrambling $\mathscr{L}$ into contributions from isotopic multiplets is best done in tensor notation (de Swart, 1963; Gourdin, 1967) and it gives, for example, that the pion–nucleon interaction contribution to the Lagrangian is $g\bar{N}\gamma_5 T N \cdot \boldsymbol{\pi}$. This is independent of α, so that g is just the pion–nucleon coupling constant (Section 10.11). Analysis using KN dispersion relations (Kim, 1967) has given $\alpha = 0.41 \pm 0.07$.

Given invariance under $SU(3)$, then we can predict relations between the decay widths for the strong decay of baryon or meson resonances. The difficulty occurs in comparing experimental widths. First, the particles involved in the decays may have very different masses, so that the energy release in decays to be compared may be very different. This effect is covered by including the phase-space and angular-momentum barrier-penetration factors, so that the matrix elements are compared. Second, it is obvious that $SU(3)$ is violated, and it is not known how this affects the coupling constants involved. However, let us look at a resonance decay—that of the $t=1$, $y=0$ member of **10** called $\Sigma(1385)$ into a pseudoscalar octet meson and a $\tfrac{1}{2}^+$ baryon. Since $\mathbf{8} \otimes \mathbf{8} \otimes \mathbf{10}$ contains **1** only once, there is only one coupling constant for all decays and we can calculate the ratio of matrix elements from the Clebsch–Gordan coefficents. Looking at Eq. (9.19), we see that

$$\langle \Xi\bar{K} | \Sigma(1385)\rangle : \langle NK | \Sigma(1385)\rangle : \langle \Sigma\eta | \Sigma(1385)\rangle :$$
$$\langle \Lambda\pi | \Sigma(1385)\rangle : \langle \Sigma\pi | \Sigma(1385)\rangle$$
$$= \sqrt{2} : -\sqrt{2} : \sqrt{3} : \sqrt{3} : \sqrt{2}. \tag{9.24}$$

However, only the decays

$$\Sigma(1385) \to \Sigma + \pi,$$
$$\Sigma(1385) \to \Lambda + \pi,$$

are kinematically allowed. From Eq. (9.24) we expect

$$\frac{|\langle \Sigma\pi | \Sigma(1385)\rangle|^2}{|\langle \Lambda\pi | \Sigma(1385)\rangle|^2} = \frac{2}{3}.$$

9.6 Applications of SU(3)

Putting in the kinematic factors Cutkosky (1963) predicted for the ratio of the decay widths

$$\frac{\Gamma(\Sigma(1385) \to \Sigma\pi)}{\Gamma(\Sigma(1385) \to \Lambda\pi)} = 0.16,$$

to be compared with the experimental value of about 0.1 (Particle Data Group, 1969). Since the coupling constant is the same, these predictions can be made about decays not only within the same isotopic-spin multiplet but also within the **10**. For example,

$$\frac{\langle \Xi\pi | \Xi(1529) \rangle}{\langle N\pi | \Delta(1236) \rangle} = \frac{1}{2}.$$

The analysis of all the decays within the $\frac{3}{2}^+$ decuplet has been reviewed by Ferro-Luzzi (1968). The observed resonance partial widths are consistent, within a factor of 2, with the $SU(3)$ predictions. This result is not bad, considering that $SU(3)$ symmetry is broken and that the partial widths vary from about 4 MeV for $\Sigma(1385) \to \Sigma\pi$ to 120 MeV for $\Delta(1236) \to N\pi$. Given this situation, it can be extended to other multiplets and used as a test of the assignment of baryons to multiplets. The position is complicated in the case of $\mathbf{8} \to \mathbf{8} + \mathbf{8}$ decays because again a coupling constant and an F/D ratio are involved. Ferro-Luzzi examined the 1968 assignments and showed that in general they are consistent [see also Tripp *et al.* (1967), Kernan and Smart (1966)].

The next application of $SU(3)$ is to scattering amplitudes. Consider, for example, pion–nucleon scattering: conservation of isotopic spin indicates relations between the scattering amplitudes for the observable reactions as discussed in Section 8.4. The π-mesons have $t = 1$ [**3** representations of $SU(2)$] and the nucleons have $t = \frac{1}{2}$ [**2** representation of $SU(2)$]. Since $\mathbf{3} \otimes \mathbf{2} = \mathbf{2} \oplus \mathbf{4}$, we expect two amplitudes to be operative, and they correspond to scattering in the isotopic-spin states $t = \frac{1}{2}$ (**2**) and $t = \frac{3}{2}$ (**4**). The relation between scattering amplitudes becomes very simple if one of these amplitudes is very strong compared to the other, as is the case for π–N scattering near 200 MeV incident pion energy. Even without this simplification, relations between amplitudes can be derived. Now consider the general reaction: scattering of pseudoscalar mesons by baryons into states consisting of 0^- meson $+ \frac{1}{2}^+$ baryon. A large variety of reactions come into this category: for example,

$$K^- + p \to \Lambda + \pi^0,$$
$$\pi^- + p \to \Lambda + K^0,$$
$$K^- + p \to \Xi^- + K^+,$$

and many others, many of which are not directly observable. If $SU(3)$ is not broken, then we expect that all the quantum numbers associated with this symmetry will be conserved, including those associated with the

Casimir operators. Any initial state is a superposition of states from the representations found in the usual reduction [Eq. (9.13)].

$$8 \otimes 8 = 1 \oplus 8 \oplus 8 \oplus 10 \oplus 10^* \oplus 27.$$

Unbroken $SU(3)$ means that transitions are not allowed between representations having different values of p and q $[D^n(p, q)]$. Thus seven amplitudes will occur

$$1 \rightleftharpoons 1, \quad 8_1 \rightleftharpoons 8_1, \quad 8_2 \rightleftharpoons 8_2, \quad 8_1 \rightleftharpoons 8_2, \quad 10 \rightleftharpoons 10,$$
$$10^* \rightleftharpoons 10^*, \quad \text{and} \quad 27 \rightleftharpoons 27.$$

Seven amplitudes means thirteen real numbers and one unobservable phase to describe all the two-body scattering reactions of the octet of 0^- mesons on the $\tfrac{1}{2}^+$ baryon octet at equivalent energies. As usual, it is not clear at what energies various reactions should be compared and how differing angular-momentum barrier penetrations in the contributing partial waves should be handled. If a resonant state is an obvious feature, then comparisons can be made. For example, the baryon decuplet would be an obvious feature of the scattering reactions

$$\pi + N \to \pi + N,$$
$$\pi + \Lambda \to \pi + \Lambda,$$
$$\pi + \Xi \to \pi + \Xi,$$

and their matrix elements at resonance peak should be simply related.

There is one simple technique that relates scattering and decay amplitudes and relies upon the symmetry of the weight diagram under rotation through 120°. We shall assume unbroken $SU(3)$ in this discussion. The isotopic-spin group $SU(2)$ is a subgroup of $SU(3)$, and we are now very accustomed to dealing with the conservation of isotopic spin. If, however, we look at weight diagrams rotated through 120°, then a new $SU(2)$ classification emerges involving what is called U-spin. In Fig. 9.10 the U-spin multiplets for the 0^- octet are ordered within each multiplet in increasing u_3. (We use u and u_3 to represent the eigenvalues.)

$$u = \tfrac{1}{2}: \quad K^-, \pi^-;$$
$$u = 1: \quad \bar{K}^0, \frac{\sqrt{3}\eta - \pi^0}{2}, K^0;$$
$$u = 0: \quad \frac{\sqrt{3}\pi^0 + \eta}{2};$$
$$u = \tfrac{1}{2}: \quad \pi^+, K^+.$$

Notice that the members of each U-spin multiplet have the same charge. [This is true in all representations of $SU(3)$.] Unbroken $SU(3)$ means con-

9.6 Applications of SU(3)

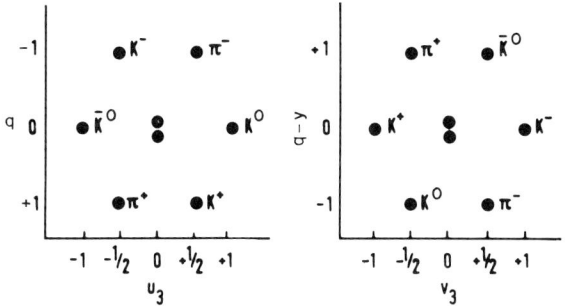

FIG. 9.10. The U- and V-spin multiplets in $SU(3)$. The central states have been left unlabeled: they are linear superpositions of π^0 and η.

servation of U-spin, and we can treat it in the same way as isotopic spin. Consider the reactions

$$
\begin{align}
\pi^- + p &\to \pi^+ + \Delta^-(1236), & \text{(a)} \\
\pi^- + p &\to K^+ + \Sigma^-(1385), & \text{(b)} \\
K^- + p &\to \pi^+ + \Sigma^-(1385), & \text{(c)} \\
K^- + p &\to K^+ + \Xi^-(1530). & \text{(d)}
\end{align}
\tag{9.25}
$$

Both particles involved on the left have U-spin $\frac{1}{2}$, so that the initial total U-spin is 0 or 1. On the right the mesons have $u = \frac{1}{2}$ and the baryons belong to the $u = \frac{3}{2}$ multiplet of **10**, so that the final state has $u = 1$ or 2. Transitions will only occur between u-states having $u = 1$. As usual we can use the $SU(2)$ Clebsch–Gordan coefficients to project out the relevant $u = 1$ amplitudes. Thus

$$
\begin{align}
|\pi^- p\rangle &= |1, 1\rangle, \\
|K^- p\rangle &= \sqrt{\tfrac{1}{2}}\,|1, 0\rangle + \sqrt{\tfrac{1}{2}}\,|0, 0\rangle, \\
\langle \pi^+ \Delta^-(1236)| &= \tfrac{1}{2}\langle 2, 1| + \sqrt{\tfrac{3}{4}}\,\langle 1, 1|, \\
\langle K^+ \Sigma^-(1385)| &= \sqrt{\tfrac{3}{4}}\,\langle 2, 1| - \tfrac{1}{2}\langle 1, 1|, \\
\langle \pi^+ \Sigma^-(1385)| &= \sqrt{\tfrac{1}{2}}\,\langle 2, 0| + \sqrt{\tfrac{1}{2}}\,\langle 1, 0|, \\
\langle K^+ \Xi^-(1530)| &= \sqrt{\tfrac{1}{2}}\,\langle 2, 0| - \sqrt{\tfrac{1}{2}}\,\langle 1, 0|,
\end{align}
$$

where the notation within the bras and kets is $|u, u_3\rangle$. We put

$$a_1 = \langle 1, u_3 | 1, u_3 \rangle$$

and find

$$
\langle \pi^+ \Delta^-(1236) | \pi^- p \rangle : \langle K^+ \Sigma^-(1385) | \pi^- p \rangle : \\
\langle \pi^+ \Sigma^-(1385) | K^- p \rangle : \langle K^+ \Xi^-(1530) | K^- p \rangle \\
= \sqrt{3}\, a_1 : -a_1 : a_1 : -a_1.
$$

This example serves to illustrate the method rather than to expose the conservation of U-spin, since these relations are not satisfied. Allowing for the factor of $\sqrt{3}$, the matrix elements for reactions (a) and (c) [Eq. (9.25)] are roughly equal over a wide range of energies. The matrix elements for reactions (b) and (d) are of the same order of magnitude, but very different from those of the other pair and also show a different energy behavior. Meshkov et al. (1964), who made this comparison, also showed that a symmetry-breaking interaction could reconcile these discrepancies, but this probably indicates the difficulties of applying $SU(3)$ symmetry to reaction cross sections. It may not be surprising that the symmetry that bears some relation to the internal structure of the particles does not apply directly to their scattering.

U-spin is also useful in obtaining relations between photoproduction reactions, since the form of the electromagnetic current is such that it appears to transform like the $u = 0$ member of an octet (Coleman and Glashow, 1961), so that the electromagnetic interaction only connects states having the same u and u_3. U-spin multiplets have the same charge, and thus rotations in U-spin space in the presence of an electromagnetic field do not change the energy; that is, the electromagnetic interaction is invariant under rotations in U-spin space and conserves U-spin rigorously. Let us apply these ideas to the reactions

$$\gamma + p \to \pi^+ + n, \quad (a)$$
$$\gamma + p \to K^+ + \Lambda, \quad (b) \quad (9.26)$$
$$\gamma + p \to K^+ + \Sigma^0. \quad (c)$$

The initial state has $u = \frac{1}{2}$, $u_3 = \frac{1}{2}$. The assignments on the right are

	π^+	K^+	n	$(\sqrt{3}\,\Lambda - \Sigma^0)/2$	$(\sqrt{3}\,\Sigma^0 + \Lambda)/2$
u	$\frac{1}{2}$	$\frac{1}{2}$	1	1	0
u_3	$-\frac{1}{2}$	$+\frac{1}{2}$	$+1$	0	0

There will be two amplitudes; one (a_1) couples the initial state to the final state that is the $u = \frac{1}{2}$ vector sum of $\mathbf{U}_a + \mathbf{U}_b$, where $u_a = \frac{1}{2}$ and $u_b = 1$; the second (a_0) couples the initial state to the final $u = \frac{1}{2}$ state that has $u_a = \frac{1}{2}$, $u_b = 0$. Now

$$|\pi^+ n\rangle = \sqrt{\tfrac{1}{3}}\,|\tfrac{3}{2}, \tfrac{1}{2}\rangle + \sqrt{\tfrac{2}{3}}\,|\tfrac{1}{2}, \tfrac{1}{2}\rangle,$$

where the notation within the ket refers to the total u and u_3 of the final state. Thus the transition matrix element

$$\langle \pi^+ n \,|\, \gamma p \rangle = \sqrt{\tfrac{2}{3}}\, a_1. \quad (9.27)$$

The state consisting of K^+ and the $u = 1$ combination of Σ^0 and Λ can be written

$$|K^+(\sqrt{\tfrac{3}{4}}\Lambda - \tfrac{1}{2}\Sigma^0)\rangle = \sqrt{\tfrac{3}{4}}|K^+\Lambda\rangle - \tfrac{1}{2}|K^+\Sigma^0\rangle$$
$$= \sqrt{\tfrac{2}{3}}|\tfrac{3}{2}, \tfrac{1}{2}\rangle - \sqrt{\tfrac{1}{3}}|\tfrac{1}{2}, \tfrac{1}{2}\rangle,$$

and taking its transition matrix element with the state $|\gamma p\rangle$ we obtain

$$\sqrt{\tfrac{3}{4}}\langle K^+\Lambda|\gamma p\rangle - \tfrac{1}{2}\langle K^+\Sigma^0|\gamma p\rangle = -\sqrt{\tfrac{1}{3}}a_1. \tag{9.28}$$

We can construct the a_0 amplitude in terms of the $u = 0$ combination of Σ^0 and Λ, but that is irrelevant because Eqs. (9.27) and (9.28) give immediately that

$$\langle \pi^+ n|\gamma p\rangle = \sqrt{\tfrac{1}{2}}\langle K^+\Sigma^0|\gamma p\rangle - \sqrt{\tfrac{3}{2}}\langle K^+\Lambda|\gamma p\rangle.$$

This relation is one between complex amplitudes, so that the square roots of the cross sections have to satisfy a triangular inequality. This has been tested by Boyarski *et al.* (1969) and by Elings *et al.* (1966) for high-energy photoproduction: the relation appears to fail at low-momentum transfers, and the reason is not known.

V-spin is associated with the last $SU(2)$ subgroup of $SU(3)$, and its eigenstates are found along the remaining axis of symmetry in the weight diagram, Fig. 9.10. Applications of U and V spin are discussed by Lipkin (1965).

9.7 Applications of Broken $SU(3)$

The inequality of masses within the $SU(3)$ multiplets makes it obvious that the symmetry is broken. Such symmetry breaking is familiar in atomic physics and within the isotopic-spin multiplets. In the latter the electromagnetic interaction is responsible, and we notice that if it has any effects other than breaking the degeneracy, such as producing new states, they are not obvious in strong interactions and we can continue to classify particles into isotopic-spin multiplets. The fact that the electromagnetic interaction conserves parity, angular momentum, and baryon number means that we expect all members of an isotopic multiplet to have the same spin and parity and baryon number. Looking at the $SU(3)$ multiplets we see that there must be another interaction, normally thought of as medium-strong, that splits the masses more obviously than does electromagnetism. The masses in the absence of these symmetry-breaking interactions would then be due to a strong, $SU(3)$-symmetric interaction. We can write down a mass operator M, which is made up of three terms:

$$M = M_S + M_{MS} + M_{EM},$$

with obvious notation. M_S is evidently a $SU(3)$ scalar, and therefore its expectation value for any member of a multiplet depends only on the representation and not upon t, t_3, or y. M_{MS} contains the effect of the medium-strong interaction which, although it conserves t and y, is not $SU(3)$-symmetric: in terms of tensor operators it will transform like the $t = 0$, $y = 0$ member of a representation. Postponing discussion of the even smaller effect of M_{EM}, we see that the mass of the particle labeled by ν within the representation μ is proportional to

$$\langle \mu, \nu | M | \mu, \nu \rangle = a_1(\mu) + \langle \mu, \nu | M_{MS} | \mu, \nu \rangle .$$

If $\mu = 8$, then only M_{MS} belonging to **1**, **8**, and **27** have nonzero expectation values (**10** and **10*** do not have any $t = 0$, $y = 0$ members). Following the notation of Section 9.5 we can write

$$M_{MS} = T(\mathbf{1}, \nu = 0) + T(\mathbf{8}, \nu = 0) + T(\mathbf{27}, \nu = 0) ,$$

where $\nu = 0$ means $t = t_3 = 0 = y$.

$T(\mathbf{1}, 0)$ has a constant matrix element, which we absorb into $a_1(\mu)$. $T(\mathbf{8}, 0)$ has, as usual, two matrix elements giving rise to two reduced matrix elements, $a_{8,1}(\mu)$ and $a_{8,2}(\mu)$. $T(\mathbf{27}, 0)$ gives one reduced matrix element $a_{27}(\mu)$. For details of this calculation readers are referred to de Swart (1963). Thus for the baryon octet we have four constants and four masses, and this serves no more than to give values for the four constants, which may themselves be meaningless. However, a_{27} is small, which justifies the assumption made by Gell-Mann (see Gell-Mann and Ne'eman, 1964) and by Okubo (1962) that M_{MS} transformed like the $t = y = 0$ member of an octet; i.e., $M_{MS} = T(\mathbf{8}, 0)$, so that the mass formula contains three constants: the result can be shown to be equivalent to the following formula, which holds for all representations although the constants may vary among representations (Carruthers, 1966)

$$M = a + by + c\{t(t+1) - \tfrac{1}{4} y^2\} . \tag{9.29}$$

For the baryon $\tfrac{1}{2}^+$ octet this leads to the relation

$$\tfrac{1}{2}(N + \Xi) = \tfrac{1}{4}(\Sigma + 3\Lambda) , \tag{9.30}$$

where we have used the particle signs to represent masses. Inserting the average isotopic multiplet masses in MeV gives 1128 for the left side and 1135 for the right. This can be considered a success, bearing in mind that the Σ- and Ξ-multiplet splitting due to electromagnetism is of the same order as the discrepancy.

For triangular representations (p or $q = 0$) we have $t = 1 + \tfrac{1}{2} y$ and Eq. (9.29) becomes

$$M = a' + b'y . \tag{9.31}$$

9.7 Applications of Broken $SU(3)$

This is the formula that predicts equal mass spacing between the isotopic-spin multiplets in the decuplet and permitted a prediction of the Ω^- mass before its discovery: the observed mass differences are

$$\Omega - \Xi(1530) = 142 \text{ MeV},$$
$$\Xi(1530) - \Sigma(1385) = 145 \text{ MeV},$$
$$\Sigma(1385) - \Delta(1236) = 149 \text{ MeV},$$

where the errors on the masses make a comparison in the last digit meaningless.

The Gell-Mann–Okubo mass formula also applies to mesons if we take the masses squared. There appears to be no good reason for this, apart from the fact that it is M that appears in field equations for fermions and M^2 in the field equations for bosons (cf. the Dirac and Klein–Gordon equations) and that it works. Since charge-conjugation invariance for bosons requires that there be no term linear in y, Eq. (9.29) becomes

$$M^2 = A + C\{t(t+1) - \tfrac{1}{4}y^2\}. \tag{9.32a}$$

This gives

$$4K^2 = \pi^2 + 3\eta^2,$$

which is correct to about 5%. The same formula should apply to the vector mesons:

$$4(K^*)^2 = \rho^2 + 3\omega^2, \tag{9.32b}$$

but this fails. The reason appears to be the following: there is a vector boson ϕ, mass 1019 MeV, that appears to be an $SU(3)$ singlet and has the same quantum numbers t, y as the isotopic singlet ω belonging to an octet. Since $SU(3)$ is broken, these two states will be mixed by the medium-strong interaction, so that the observed ϕ and ω will be superpositions of a unitary singlet ω_1 and an isospin singlet ω_8 belonging to an octet:

$$|\omega\rangle = |\omega_8\rangle \cos\theta + |\omega_1\rangle \sin\theta, \tag{9.33}$$

$$|\phi\rangle = -|\omega_8\rangle \sin\theta + |\omega_1\rangle \cos\theta, \tag{9.34}$$

The masses of the ω and ϕ [$m^2(\omega)$ and $m^2(\phi)$] are found from the mass (squared) matrix

$$\begin{bmatrix} m^2(\omega_8) & m^2(\omega_8\omega_1) \\ m^2(\omega_8\omega_1) & m^2(\omega_1) \end{bmatrix}.$$

Here the elements $m^2(\omega_8)$, $m^2(\omega_1)$ are the masses squared of the particles ω_8, ω_1 and the $m^2(\omega_8\omega_1)$ is the matrix element of the mass-squared operator between the states $|\omega_8\rangle$ and $|\omega_1\rangle$. This matrix has eigenvalues that are the required masses squared, and the corresponding eigenvectors are column matrices with elements that are the coefficients in Eqs. (9.33) and (9.34). The eigenvalues are

$$\lambda_{\pm} = \tfrac{1}{2}[m^2(\omega_8) + m^2(\omega_1) \pm \{[m^2(\omega_8) - m^2(\omega_1)]^2 + 4m^4(\omega_8\omega_1)\}^{1/2}] \quad (9.35)$$

and θ is given by

$$\tan 2\theta = \frac{2m^2(\omega_8\omega_1)}{m^2(\omega_8) - m^2(\omega_1)}. \quad (9.36)$$

Now

$$\lambda_+ + \lambda_- = m^2(\omega) + m^2(\phi) = m^2(\omega_8) + m^2(\omega_1) \quad (9.37)$$

and

$$\lambda_+\lambda_- = m^2(\omega)\, m^2(\phi) = m^2(\omega_8)\, m^2(\omega_1) - m^4(\omega_8\omega_1). \quad (9.38)$$

The mass of ω_8 is found from Eq. (9.32a) to be 930 MeV. Putting into Eq. (9.37) the masses $\phi = 1019$ MeV, $\omega = 783$ MeV we obtain 887 MeV for the mass of ω_1. From Eq. (9.38) the value of $m(\omega_8\omega_1)$ is found to be 403 MeV, and this gives a mixing angle $\theta \simeq 36°$. This means, of course, that the observed particles have nonzero probabilities, which are the squares of the amplitudes like $\sin\theta$ or $\cos\theta$, for being either in the $SU(3)$ singlet state or in the isosinglet octet state. The mass effects and other implications of ϕ-ω mixing have been discussed by Dashen and Sharp (1964). It is probable that the 0^- meson X^0 at 965 MeV is an $SU(3)$ singlet, which mixes slightly with the isosinglet η at 548 MeV and in this way causes the small discrepancy in the Gell-Mann–Okubo mass formula for the pseudoscalar mesons.

The next stage in the discussion of masses is to include the effect of $M_{\rm EM}$. This is expected to cause mass splittings of the order of 1/137 of the nucleon mass; that is, up to about 10 MeV. This should be compared to $M_{\rm MS}$, which causes mass splittings of the order of 10–15% of the nucleon mass. Cross terms can be neglected, as their effects will be 0.1%. Consider four members of a multiplet occupying the corner sites on a parallelogram in the weight diagram (Fig. 9.11). Now $M_{\rm EM}$ is a scalar in U-spin space (the electromagnetic interaction is the same for all members of a U-spin multiplet), so that in the absence of $M_{\rm MS}$ the effect of $M_{\rm EM}$ is the same, to

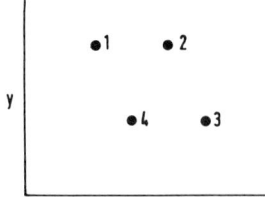

FIG. 9.11. The section of a weight diagram appropriate to the discussion of the parallelogram law.

9.7 Applications of Broken $SU(3)$

any order, for all members of a U-spin multiplet and thus, if M_{MS} is absent, $m_2 = m_3$, $m_4 = m_1$. M_{MS} is an isotopic-spin scalar, and in the absence of M_{EM} we have $m_1 = m_2$ and $m_4 = m_3$. The equation

$$m_1 - m_2 + m_3 - m_4 = 0 \tag{9.39}$$

is then the relation (the parallelogram law) expected to hold in the presence of both M_{MS} and M_{EM} if cross terms are neglected, since it is satisfied in the absence of either. This can be applied immediately to the baryon decuplet (Fig. 9.9) to obtain

$$\Delta^- - \Delta^0 = \Sigma^- - \Sigma^0 = \Xi^- - \Xi^0$$

and

$$\Delta^0 - \Delta^+ = \Sigma^0 - \Sigma^+ .$$

We have omitted the masses that label the states Σ and Ξ and identify them as the members of the decuplet. Data are not adequate at present to test these predictions. A complication occurs in the application of the parallelogram law to the baryon $\tfrac{1}{2}^+$ octet because one corner of any parallelogram will contain two particles, the Σ^0 and Λ. Consider the labeling $1 = n$, $2 = p$, $3 = \Sigma^+$, and $4 = \Sigma^0, \Lambda$. The superposition of Σ and Λ, which belongs to $u = 1$ (the other members are n and Ξ^0), is

$$\sqrt{\tfrac{3}{4}}|\Lambda\rangle - \tfrac{1}{2}|\Sigma^0\rangle,$$

so that the mass relation, applying the M_{EM} mass operator, is

$$\langle n | M_{EM} | n \rangle - \langle p | M_{EM} | p \rangle$$
$$= \langle \sqrt{\tfrac{3}{4}}\Lambda - \tfrac{1}{2}\Sigma^0 | M_{EM} | \sqrt{\tfrac{3}{4}}\Lambda - \tfrac{1}{2}\Sigma^0 \rangle - \langle \Sigma^+ | M_{EM} | \Sigma^+ \rangle .$$

This gives, writing particle symbols to stand for mass,

$$n - p + \Sigma^+ = \tfrac{1}{4}\langle \sqrt{3}\,\Lambda - \Sigma^0 | M_{EM} | \sqrt{3}\,\Lambda - \Sigma^0 \rangle .$$

Repeating this process for the opposite parallelogram (Ξ^-, Ξ^0, Σ^-, $\Sigma^0\Lambda$) we find

$$\Xi^- - \Xi^0 - \Sigma^- = -\tfrac{1}{4}\langle \sqrt{3}\,\Lambda - \Sigma^0 | M_{EM} | \sqrt{3}\,\Lambda - \Sigma^0 \rangle ;$$

hence we obtain the Coleman and Glashow (1961) relation

$$\Xi^- - \Xi^0 + n - p = \Sigma^- - \Sigma^+ . \tag{9.40}$$

The left-hand side is 7.8 ± 0.2 MeV and the right is 7.97 ± 0.11 MeV, which is excellent agreement.

To obtain information on magnetic moments, we must return first to $SU(2)$ to obtain a relation between the magnetic moments within the Σ triplet. In isotopic-spin space the electromagnetic interaction is a scalar plus the neutral component of an isovector. Using such an operator, the Wigner–Eckart theorem for $SU(2)$ shows that

$$2\mu_{\Sigma 0} = \mu_{\Sigma +} + \mu_{\Sigma -}. \tag{9.41}$$

$SU(3)$ can extend these predictions. The U-scalar nature of the electromagnetic interaction means that within a U-spin multiplet all members will have the same magnetic moment. Thus in the decuplet all the particles having the same charge should have the same magnetic moment. For the baryon octet the matter is again complicated by the U-spin singlet–triplet states at the center: however, we can write immediately, for the magnetic moments,

$$\mu_p = \mu_{\Sigma +}, \tag{9.42}$$

$$\mu_n = \mu_{\Xi 0}, \tag{9.43}$$

$$\mu_{\Xi -} = \mu_{\Sigma -}. \tag{9.44}$$

If μ also stands for the magnetic-moment operator and $|1, 0\rangle$ for the $u_3 = 0$ member of the U-spin triplet, we have

$$\mu_n = \mu_{\Xi 0} = \langle 1, 0 | \mu | 1, 0 \rangle$$
$$= \tfrac{3}{4}\mu_\Lambda + \tfrac{1}{4}\mu_{\Sigma 0} - \sqrt{\tfrac{3}{4}} \langle \Lambda | \mu | \Sigma^0 \rangle. \tag{9.45}$$

If $|0, 0\rangle$ is the U-spin singlet, then μ cannot mix $|1, 0\rangle$ and $|0, 0\rangle$ and we have

$$\langle 0, 0 | \mu | 1, 0 \rangle = 0$$

or

$$\langle \sqrt{\tfrac{3}{4}} \Sigma^0 + \tfrac{1}{2}\Lambda | \mu | \sqrt{\tfrac{3}{4}}\Lambda - \tfrac{1}{2}\Sigma^0 \rangle = 0 ;$$

that is,

$$\sqrt{\tfrac{3}{4}} \langle \Lambda | \mu | \Sigma^0 \rangle + \tfrac{3}{4}(\mu_\Lambda - \mu_{\Sigma 0}) = 0.$$

Substituting in Eq. (9.45) we find

$$\mu_n = \mu_{\Xi 0} = \tfrac{3}{2}\mu_\Lambda - \tfrac{1}{2}\mu_{\Sigma 0}. \tag{9.46a}$$

We can obtain one more relation, and that there is one more can be seen in the following way. The magnetic moment is the expectation value of an operator that has the transformation properties of the $u = 0$, $Q = 0$ member of an octet. Thus, in forming its matrix elements for an octet we expect all results to be expressed in terms of two constants, since **1** occurs twice in $\mathbf{8} \otimes \mathbf{8} \otimes \mathbf{8}$. So far the relations Eqs. (9.41)–(9.46a) do not express all magnetic moments in terms of two. To get the remaining relation we use the Wigner–Eckart theorem on the matrix elements $\langle 1, 0 | \mu | 1, 0 \rangle$ and $\langle 0, 0 | \mu | 0, 0 \rangle$. We have to remember that the notation is $|u, u_3\rangle$, so we are using the theorem from a skew viewpoint in the y–t_3 weight diagram, and

9.7 Applications of Broken $SU(3)$

we use the $SU(2)$ Clebsch–Gordan coefficients on U-spin and the isoscalar factors as if they were U-scalar factors. Then, remembering that μ has $u = 0$, $Q = 0$ and all vectors belong to **8**,

$$\langle 1, 0 | \mu | 1, 0 \rangle = \sqrt{\tfrac{1}{5}} \langle \mathbf{8} \| \mu \| \mathbf{8} \rangle_1 ,$$
$$\langle 0, 0 | \mu | 0, 0 \rangle = -\sqrt{\tfrac{1}{5}} \langle \mathbf{8} \| \mu \| \mathbf{8} \rangle_1 . \quad (9.46b)$$

Although we expect two, only one reduced matrix element appears, which makes it easy to find the last relation. Substituting

$$|1, 0\rangle = \sqrt{\tfrac{3}{4}} |\Lambda\rangle - \tfrac{1}{2} |\Sigma^0\rangle \quad \text{and} \quad |0, 0\rangle = \sqrt{\tfrac{3}{4}} |\Sigma^0\rangle + \tfrac{1}{2} |\Lambda\rangle$$

on the left and adding gives

$$\mu_\Lambda + \mu_{\Sigma^0} = 0 .$$

Collecting we have

$$\begin{aligned}
\mu_\Lambda &= -\mu_{\Sigma^0} = \tfrac{1}{2} \mu_n , \\
\mu_{\Sigma^+} &= \mu_p , \\
\mu_{\Sigma^-} &= \mu_{\Xi^-} = -\mu_p - \mu_n , \\
\mu_{\Xi^0} &= \mu_n .
\end{aligned} \quad (9.47)$$

These relations can be derived using the Wigner–Eckart theorem throughout, in which case all magnetic moments are given in terms of two reduced matrix elements. The fact that only one element appears in Eqs. (9.46b) makes the method used above simple.

The present experimental values for the magnetic moments are (Particle Data Group, 1969)

$$\mu_p = 2.792763 \pm 0.000030 ,$$
$$\mu_n = -1.913148 \pm 0.000066 ,$$
$$\mu_\Lambda = -0.73 \pm 0.16 ,$$
$$\mu_{\Sigma^+} = 2.4 \pm 0.6 ,$$

in units of nuclear magnetons. The relation between μ_{Σ^+} and μ_p is satisfied within the errors. There is a 1.4 standard-deviation discrepancy in the relation between μ_Λ and μ_n.

The relations of Eqs. (9.47) are also be satisfied by the charge, since it has the same kind of operator as the magnetic moment. These relations are not satisfied by the deviations of the masses from their values in the absence of electromagnetism, since the deviations are second-order effects.

The transition matrix elements $\langle \Lambda | \mu | \Sigma^0 \rangle$ is also given to be $-\sqrt{3}\, \mu_n / 2$. This is related to the decay rate $\Sigma^0 \to \Lambda \gamma$, since this must be a magnetic dipole transition. However, only a lower limit for the rate is known, and no check is possible.

9.8 Quarks

We know that the **3** and the **3*** representations of $SU(3)$ can be used to build up any higher representations by the direct product of a suitable number of these representations. Thus

$$\mathbf{3} \otimes \mathbf{3^*} = \mathbf{1} \oplus \mathbf{8}, \tag{9.48}$$

and

$$\mathbf{3} \otimes \mathbf{3} \otimes \mathbf{3} = \mathbf{1} \oplus \mathbf{8} \oplus \mathbf{8} \oplus \mathbf{10}. \tag{9.49}$$

This suggests that the observed multiplets may be the manifestation of the binding of particles carrying the quantum numbers of **3** and **3***, the so called quarks and antiquarks (Gell-Mann, 1964; Zweig, 1964). So far as hypercharge and isotopic spin are concerned, that this can be done is ensured by group theory; in Table 9.2 we give the relevant quark quantum

TABLE 9.2

QUANTUM NUMBERS OF THE QUARKS

	Quarks			Antiquarks		
	q_1	q_2	q_3	$\bar{q}_1$	$\bar{q}_2$	$\bar{q}_3$
t_3	$+\tfrac{1}{2}$	$-\tfrac{1}{2}$	0	$-\tfrac{1}{2}$	$+\tfrac{1}{2}$	0
y	$+\tfrac{1}{3}$	$+\tfrac{1}{3}$	$-\tfrac{2}{3}$	$-\tfrac{1}{3}$	$-\tfrac{1}{3}$	$+\tfrac{2}{3}$
B	$+\tfrac{1}{3}$	$+\tfrac{1}{3}$	$+\tfrac{1}{3}$	$-\tfrac{1}{3}$	$-\tfrac{1}{3}$	$-\tfrac{1}{3}$
Q	$+\tfrac{2}{3}$	$-\tfrac{1}{3}$	$-\tfrac{1}{3}$	$-\tfrac{2}{3}$	$+\tfrac{1}{3}$	$+\tfrac{1}{3}$

numbers, which can also be read from Figs. 9.1 and 9.3. All the mesons are made from a linear superposition of quark plus antiquark states: thus

$$|\pi^+\rangle = |q_1 \bar{q}_2\rangle,$$
$$|K^\circ\rangle = |q_3 \bar{q}_2\rangle, \quad \text{etc.}$$

If the quarks have spin $\tfrac{1}{2}$, then the pseudoscalar mesons would be the result of quark-antiquark in a state $^1S_0$, which has odd parity assuming odd relative parity for q and $\bar{q}$. The vector mesons would be the result of the state $^3S_1$.

Equation (9.49) shows that 27 baryons would be predicted from the binding of 3 quarks (and an identical number of antibaryons from binding 3 antiquarks). This implies that quarks have baryon number $+\tfrac{1}{3}$ and antiquarks baryon number $-\tfrac{1}{3}$. The 27 includes an **8** and a **10**, which can be identified with the well-known $\tfrac{1}{2}^+$ octet and $\tfrac{3}{2}^+$ decuplet. These multiplet spins and parities can be explained by assuming that the quarks have no relative orbital angular momentum and that the vector sum of the spins

9.9 Higher Symmetry Schemes

is $\frac{1}{2}$ and $\frac{3}{2}$, in the two cases. The observed facts about the baryons indicate that there are probably many more multiplets to be fully identified. Some of these may be Regge recurrences (see Chapter XII). However, there are many odd-parity baryons that require the quark model to allow orbital angular momentum. This adds very considerably to the spin and parity possibilities of the model. There is also the possibility of a $t = 0$, $y = +2$ resonance in the K^+N system (Cool et al., 1966). This cannot belong to a representation contained in $3 \otimes 3 \otimes 3$, and so in a quark model we require a $qqqq\bar{q}$ structure, and this possible resonance would belong to a **27** or **35** in $SU(3)$.

If we apply the usual formula $Q = t_3 + y/2$, then it appears that quarks will carry electric charge that is $\frac{1}{3}$ or $\frac{2}{3}$ that of the electron. This is quite a distinctive feature, which has been used in the search for quarks. So far they have not been observed, and the experiments can only put upper limits on production cross sections in conjunction with lower limits on mass. If quarks exist, they are probably heavier than 2 GeV. Thus, we have the interesting possibility of two particles of total mass greater than 4 GeV binding together to give mesons of mass as low as 0.14 GeV.

Quark models of elementary-particle processes have had some success (Dalitz, 1967; Lipkin, 1968; Kokkedee, 1969) but are so speculative that we do not pursue the subject here. There is one point, however; it is not clear why, if quarks exist, the bound states are only composed of combinations having total baryon number 1, 0, or -1:

$$q\bar{q}, \quad qqq, \quad \bar{q}\bar{q}\bar{q}, \quad \text{or} \quad qqqq\bar{q}, \quad \text{etc.}$$

This is perhaps the same question as asking why elementary particles only appear in the $SU(3)$ representations **1, 8, 10, 27** (?), and not in the others such as **6, 15**, etc. These others are just those that would appear from binding, for example, two quarks or two quarks and one antiquark. In fact nature appears to realize only representations of triality zero; that is, those having $p = q$(modulo 3). An alternative way is to say that only those representations appear that would be the result of binding n quarks and m antiquarks where $(n - m)$ is a multiple of 3.

9.9 Higher Symmetry Schemes

The success of $SU(3)$ in classifying some of the observed particles prompted an examination of the possibility that the hadrons and their strong interactions possess a higher symmetry and that the $SU(3)$ symmetry is a manifestation of a part of this symmetry. Another guiding fact was the existence of Wigner's supermultiplet theory of nuclear energy levels (Wigner, 1937; Dyson, 1966), which involves $SU(4)$ by combining the

$SU(2)$ of nucleon isotopic spin with the $SU(2)$ of ordinary spin. The first symmetry examined was that of the group $SU(6)$ (Gursey and Radicati, 1964; Sakita, 1964). Evidently this involves a six-dimensional complex vector space, so that the number of generators is 35 ($= n^2 - 1$) and the rank 5. There are two fundamental representations, **6** and **6***, and the regular representation is **35**. These are group-theory statements, and we must now examine the physical content of the use of $SU(6)$. This is most easily done by using the quark idea. If all quarks have spin $\frac{1}{2}$ and if we add s_z as another quantum number to the $SU(3)$ quantum numbers t_3 and y, then each quark becomes two possible states, one having spin up and the other spin down. Therefore the fundamental objects of $SU(6)$ are six quarks and six antiquarks. $SU(2)$ spin transformations will transform quarks of one ordinary spin state into another spin state; $SU(3)$ transformations will transform a quark in a given spin state into another quark in the same spin state. $SU(6)$ transformations will transform ordinary spin, isotopic spin, and hypercharge states. Postulating that the system is invariant under $SU(6)$ transformations requires a symmetry higher than symmetry under $SU(2)$ transformations of ordinary spin and $SU(3)$ transformations. It follows that $SU(6)$ will make more predictions about the properties of particles than either $SU(2)$ or $SU(3)$ separately.

We shall use the symbol $SU(2)_S$ to distinguish the group of $SU(2)$ transformations that involve ordinary spin from the $SU(2)_T$, which involve isotopic spin. $SU(3)$ will refer to the transformations we have come to associate with $SU(2)_T$ and hypercharge. The direct product $SU(2)_S \otimes SU(3)$ is a subgroup of $SU(6)$, and we can describe any representation of $SU(6)$ by its $SU(2)_S$ and $SU(3)$ content, using symbols that formally are of the kind (n, m); n will be the dimension of the $SU(2)_S$ representation, and m that of the $SU(3)$ representation. Thus the fundamental representations **6** and **6*** of $SU(6)$ are

$$\mathbf{6} = (\mathbf{2}, \mathbf{3}), \quad \mathbf{6^*} = (\mathbf{2}, \mathbf{3^*}).$$

That is, they are spin doublets and $SU(3)$ triplets. We expect that higher representations will contain observable particles, and to keep to integral charge and hypercharge we must use the usual quark–antiquark combination for mesons and three quarks for baryons. Thus we are interested in the Clebsch–Gordan series for the direct products

$$\mathbf{6} \otimes \mathbf{6^*} = \mathbf{1} \oplus \mathbf{35}, \tag{9.50}$$

$$\mathbf{6} \otimes \mathbf{6} \otimes \mathbf{6} = \mathbf{20} \oplus \mathbf{56} \oplus \mathbf{70} \oplus \mathbf{70}. \tag{9.51}$$

The reduction of **35** into its $SU(2)_S \otimes SU(3)$ components (Matthews, 1967) is as follows:

$$\mathbf{35} = (\mathbf{3}, \mathbf{8}) + (\mathbf{1}, \mathbf{8}) + (\mathbf{3}, \mathbf{1}); \tag{9.52}$$

9.9 Higher Symmetry Schemes

that is, an octet of spin-1 particles, an octet of spin-0, and a singlet spin-1 particle; with the $SU(6)$ singlet in Eq. (9.50) we have all the nine pseudoscalar and the nine vector particles. The decomposition of Eq. (9.51) is as follows:

$$20 = (4, 1) + (2, 8),$$
$$56 = (4, 10) + (2, 8),$$
$$70 = (4, 8) + (2, 10) + (2, 8) + (2, 1), \quad \text{twice.}$$

This contains a wealth of spin-$\frac{1}{2}$ and spin-$\frac{3}{2}$ multiplets: most noteworthy is **56**, which contains a spin-$\frac{3}{2}$ decuplet and a spin-$\frac{1}{2}$ octet, which we know can be identified immediately as the two lowest-energy $SU(3)$ baryon multiplets. The incompleteness of experimental information on baryon multiplets again prevents any other firm identifications. Since orbital angular momentum does not enter this quark model, all these predicted baryon multiplets would have even parity; many odd-parity baryons are known. This may also explain why the model has no place for the nine 2^+ mesons. Despite these difficulties, it is worth considering other predictions of $SU(6)$ symmetry.

$SU(6)$ predicts for the **56** the following formula (Gursey and Radicati, 1964) for the masses:

$$M = a_0 + b_0 y + c_0\{t(t+1) - y^2/4\} + d_0 j(j+1). \tag{9.53}$$

The important thing here is that the equal spacing in the decuplet requires $b_0 + 3c_0/2 = -146$ MeV, whereas the octet gives about -131 for this quantity. The agreement is satisfactory, considering that the discrepancies in the application of this kind of mass formula are of the order 10 MeV. The mass-squared form of Eq. (9.53) with $b_0 = 0$ applies to the mesons of **35** and gives

$$K^{*2} - \rho^2 = K^2 - \pi^2,$$

which is satisfied to within about 15%.

$SU(6)$ suggests the ϕ-ω mixing angle to be such that, in Eqs. (9.33) and (9.34)

$$\sin \theta = \sqrt{\tfrac{1}{3}} \ ;$$

that is $\theta = 35°$, in good agreement with the value already suggested, $36°$. Applied to meson–baryon interactions, $SU(6)$ gives the relative strength of the F and D interaction to be $F:D = 2:3$. This implies that $\alpha = 0.4$, in excellent agreement with the value 0.41 ± 0.07 found by Kim (1967). In addition, the assumption that the photon transforms as a member of an octet gives the remarkable result that the ratio of the total magnetic moments

$$\mu_p/\mu_n = -\tfrac{3}{2},$$

where the experimental value is -1.46.

It is remarkable that combining the internal symmetries that we associate with $SU(3)$ with an external symmetry property, namely spin and invariance under rotations, should have any success. The basic quark spin of $\tfrac{1}{2}$ has been treated nonrelativistically in the sense that it was assumed to be described by a two-component state vector (Pauli), whereas we know that a relativistic theory of spin-$\tfrac{1}{2}$ requires a four-component state vector. This means that the predicted F/D ratio is only correct for interaction between particles of zero momentum. If $SU(6)$ invariance is applied to obtaining relations between reaction amplitudes, it appears to fail, presumably because it is nonrelativistic but also, possibly, because orbital angular momentum couples to the spin. Johnson and Treiman (1965) assumed that in boson–baryon elastic scattering $SU(6)$ would hold best at high energies and at $0°$, where at least the baryon is at rest initially and finally. They obtained relations between forward scattering amplitudes that, by use of the optical theorem, become relations between total cross sections. The data agrees fairly well with these relations (Good and Xuong, 1965).

Attempts to extend $SU(6)$ to include relativistic invariance have been made. At the present these face problems that have not been solved. A discussion of this is beyond the scope of this book; interested readers are referred to the review by Matthews (1967).

9.10 Conclusion

Unitary symmetry is obviously a large subject, and this chapter cannot claim to do more than introduce the subject. We have avoided the use of many powerful techniques, such as tensor representations, in the hope that the physical content could be shown without them. References cited in Section 9.2 of this chapter will allow interested readers to find literature on these methods. Also we have omitted several topics from this chapter that are relevant to unitary symmetry and have discussed them elsewhere: for example, the $SU(3)$ applications to weak interactions are discussed in Chapter XI.

REFERENCES

Barnes, V. E. *et al.* (1964). *Phys. Rev. Lett.* **12**, 204 (9.4).
Behrends, R. E., Dreitlein, J. Fronsdal, C., and Lee, W. (1962). *Rev. Mod. Phys.* **34**, 1 (9.2).
Boyarski, A. *et al.* (1969). *Phys. Rev. Lett.* **22**, 1131 (9.6).
Carruthers, P. (1966). "Introduction to Unitary Symmetry." Wiley (Interscience), New York (9.2, 9.3, 9.5, 9.6, 9.7).

References

Charap, J. M., Jones, R. B., and Williams, P. G. (1967). *Rep. Progr. Phys.* **30**, 227 (9.2).
Coleman, S., and Glashow, S. L. (1961). *Phys. Rev. Lett.* **10**, 155 (9.6, 9.7).
Cool, R. L. *et al.* (1966). *Phys. Rev. Lett.* **17**, 102, (9.8).
Cutkosky, R. E. (1963). *Ann. Phys. (New York)* **23**, 415 (9.6).
Dalitz. R. H. (1967). *Proc. Int. Conf. High Energy Phys. 13th, Berkeley, 1966*, p. 215. Univ. of California Press, Berkeley, California (9.8).
Dashen, R. F., and Sharp, D. H. (1964). *Phys. Rev. B* **133**, 1585 (9.7).
de Swart, J. J. (1963). *Rev. Mod. Phys.* **35**, 916 (9.2, 9.5, 9.6, 9.7).
Dyson, F. (1966). "Symmetry Groups in Nuclear and Particle Physics." Benjamin, New York (9.2, 9.9).
Elings, U. B. *et al.* (1966). *Phys. Rev. Lett.* **16**, 474 (9.6).
Ferro-Luzzi, M. (1968). *Proc. CERN Summer School, 1968*, p. 431. CERN, Geneva (9.6).
Gasiorowicz, S. (1966). "Elementary Particle Physics." Wiley, New York (9.2, 9.5).
Gell-Mann. (1962). *Phys. Rev.* **125**, 1067 (9.3).
Gell-Mann, (1964). *Phys. Lett.* **8**, 214 (9.8).
Gell-Mann, M., and Ne'eman, Y. (1964). "The Eightfold Way." Benjamin, New York (9.1, 9.4, 9.7).
Good, R., and Xuong, N. (1965). *Phys. Rev. Lett.* **14**, 191 (9.9).
Gourdin, M. (1967). "Unitary Symmetries." North-Holland Publ., Amsterdam (9.6).
Gursey, F., and Radicati, L. A. (1964). *Phys. Rev. Lett.* **13**, 173 (9.9).
Hamermesh, M. (1962). "Group Theory." Addison-Wesley, Reading, Massachusetts (9.2).
Johnson, K., and Treiman, S. B. (1965). *Phys. Rev. Lett.* **14**, 189 (9.9).
Kernan, A., and Smart, W. M., (1966). *Phys. Rev. Lett.* **17**, 832 (9.6).
Kim, J. K. (1967). *Phys. Rev. Lett.* **19**, 1079 (9.6, 9.9).
Kokkedee, J. J. J. (1969). "The Quark Model." Benjamin, New York (9.8).
Lipkin, H. J. (1965). "Lie Groups for Pedestrians." North-Holland Publ., Amsterdam (9.2, 9.6).
Lipkin, H. J. (1968). *Proc. Int. Conf. on Elem. Particles, Heidelburg, 1967*, p. 253. North Holland Publ., Amsterdam (9.8).
Matthews, P. T. (1967). In "High Energy Physics" (E. H. S. Burhop, ed.). Academic Press, New York (9.2, 9.9).
Meshkov, S., Snow, G. A., and Yodh, G. B. (1964). *Phys. Rev. Lett.* **13**, 213 (9.6).
Morales, A. (1968). *Proc. CERN Summer School, 1968*, p. 235. CERN, Geneva (9.4).
Okubo, S. (1962). *Progr. Theor. Phys.* **27**, 949 (9.7).
Particle Data Group, (1969). *Rev. Mod. Phys.* **41**, 109 (9.6, 9.7).
Sakata, S. (1956). *Progr. Theor. Phys.* **16**, 686 (9.1, 9.4).
Sakita, B. (1964). *Phys. Rev. B* **136**, 1756 (9.9).
Tripp, R. D. *et al.* (1967). *Nucl. Phys. B* **3**, 10 (9.6).
Weyl, H. (1931). "The Theory of Groups and Quantum Mechanics." Methuen, London (9.2).
Weyl, H. (1946). "The Classical Groups." Princeton Univ. Press, Princeton, New Jersey (9.2).
Wigner, E. P. (1937). *Phys. Rev.* **51**, 106 (9.9).
Wigner, E. P. (1959). "Group Theory and its Application to the Quantum Mechanics of Atomic Spectra." Academic Press, New York (9.2).
Zweig, G. (1964). CERN Reps. TH 401, TH 402 (9.8).

X

FIELD THEORY

10.1 Introduction

This chapter is intended to introduce to readers the vocabulary of relativistic quantum mechanics and field theory. For a full discussion of these subjects, including the formalities of their development and the details of their application, we refer readers to more specialized works such as those of Bjorken and Drell (1964, 1965), of Jauch and Rohrlich (1955), of Schweber (1961), and of Bogoliubov and Shirkov (1955). We shall attempt to make familiar the ideas of field quantization, of annihilation and creation operators, and of Feynman diagrams without the full mathematical formalism. We shall illustrate our description by referring to quantum electrodynamics (QED) and its successes, and we shall end the chapter by extending the ideas to strong interactions.

10.2 First Quantization

The formulation of a simple wave mechanics corresponds to the process of first quantization, for such a process implies the incorporation of Planck's constant into the equations. The simplest wave mechanics was formulated in such a way that the equations of motion represented the motion of a single particle, free or under the influence of a field. There are three outstanding equations of this nature.

The Schrödinger equation:

(a) for a free particle

$$-\frac{\hbar^2}{2m}\nabla^2\psi = i\hbar\frac{\partial\psi}{\partial t} \, ;$$

(b) for a particle in an electromagnetic field

10.2 First Quantization

$$-\frac{\hbar^2}{2m}\sum_{l=1}^{3}\left(\frac{\partial}{\partial x_l} - \frac{ie}{\hbar c}A_l\right)^2\psi = i\hbar\frac{\partial\psi}{\partial t} - e c\phi.$$

The Dirac equation:

(a) for a free electron

$$-i\hbar c\sum_{l=1}^{3}\alpha_l\frac{\partial\psi}{\partial x_l} + mc^2\beta\psi = i\hbar\frac{\partial\psi}{\partial t};$$

(b) for an electron in an electromagnetic field

$$-i\hbar c\sum_{l=1}^{3}\alpha_l\left(\frac{\partial}{\partial x_l} - \frac{ie}{\hbar c}A_l\right)\psi + mc^2\beta\psi = i\hbar\frac{\partial\psi}{\partial t} - e c\phi.$$

The Klein–Gordon equation:

(a) for a free particle

$$-\hbar^2\nabla^2\psi + m^2c^2\psi = -\frac{\hbar^2}{c^2}\frac{\partial^2\psi}{\partial t^2};$$

(b) for a particle in an electromagnetic field

$$-\hbar^2\sum_{l=1}^{3}\left(\frac{\partial}{\partial x_l} - \frac{ie}{\hbar c}A_l\right)^2\psi + m^2c^2\psi = -\frac{\hbar^2}{c^2}\left(\frac{\partial}{\partial t} + \frac{iec\phi}{\hbar}\right)^2\psi.$$

$A_l (l = 1, 2, 3)$ are the three components of the vector potential and ϕ is the static electric potential. The quantities $\alpha_l (l = 1, 2, 3)$ and β that occur in the Dirac equations are 4×4 Hermitian matrices satisfying the following relations:

$$\alpha_l\alpha_m + \alpha_m\alpha_l = 2\delta_{lm} \qquad (l, m = 1, 2, 3), \tag{10.1}$$

$$\alpha_l\beta + \beta\alpha_l = 0, \tag{10.2}$$

$$\beta^2 = 1. \tag{10.3}$$

These equations are used by solving for the function ψ and by interpreting this function according to the postulates of quantum mechanics as described in Section 1.3. In particular there are functions of ψ that satisfy a continuity equation

$$(\partial\rho/\partial t) + \mathrm{div}\,\mathbf{j} = 0, \tag{10.4}$$

where ρ is a probability density and $\mathbf{j}$ a current associated with the particle. For example, in the case of the free-particle Schrödinger equation, this equation is satisfied by

$$\rho = \psi^*\psi, \tag{10.5}$$

$$\mathbf{j} = (\hbar/2mi)(\psi^*\,\nabla\psi - \psi\,\nabla\psi^*). \tag{10.6}$$

The corresponding quantities for the free-particle Dirac equation are

$$\rho = \phi^\dagger \phi, \tag{10.7}$$

$$j_l = c\phi^\dagger \alpha_l \phi \quad (l = 1, 2, 3). \tag{10.8}$$

Finding similar quantities for the free-particle Klein–Gordon equation creates difficulty; if **j** is taken as in Eq. (10.6), then to satisfy (10.4) we must have

$$\rho = \frac{i\hbar}{2mc^2}\left(\psi^* \frac{\partial \psi}{\partial t} - \frac{\partial \psi^*}{\partial t}\psi\right),$$

which can have negative values. This fact defeats any efforts to make a consistent one-particle theory from the Klein–Gordon equation. However, Pauli and Weisskopf (1934) showed how the equation could be reinterpreted as the equation of motion for a field that represented the behavior of many particles of all electric charges so that negative probability density is connected with an excess of negatively charged over positively charged particles; that is, we consider the probability density to be a charge density.

Although the Dirac theory is satisfactory as a one-particle theory, and as such successfully predicts the energy levels of the hydrogen atom, it fails to predict the fine structure. The trouble was recognized as being due to the difficulty of making the single-particle theory into a many-particle theory that remained Lorentz covariant. This difficulty can be overcome by interpreting the equation as being the equation of motion of a field that can contain many particles; that is, a many-particle theory, *ab initio*. This theory is very successful and correctly predicts the fine structure of the hydrogen-atom energy levels and the anomalous magnetic moment of the electron.

The process of interpreting these theories as many-particle theories is called second quantization.

10.3 Units and Notation

In this chapter we shall follow field-theory convention and use the natural system of units in which $\hbar = c = 1$. The unit of electric charge is defined in a rationalized system and we have therefore that

$$e \simeq (4\pi/137)^{1/2}.$$

We shall use the symbols p and k to represent the four-momenta of electrons or positrons and of photons or mesons, respectively. In this natural system all components of these four-vectors have the dimensions of cm^{-1}; that is, they are 2π times reciprocal wavelengths. It follows that m, the

10.4 The Lagrangian Formalism

symbol for mass, has the same quality.

For real particles we have

$$k^2 = k \cdot k = k_0^2 - \mathbf{k}^2 = \begin{cases} 0 & \text{for photons,} \\ m^2 & \text{for mesons,} \end{cases}$$

$$p^2 = p \cdot p = p_0^2 - \mathbf{p}^2 = m^2 \quad \text{for relevant spin-}\tfrac{1}{2}\text{ particle.}$$

We shall use the zeroth covariant component of the momentum four-vector for the total energy of a particle.

We now return to the covariant–contravariant notation introduced in Section 1.5. The covariant differential operator ∂_μ is

$$\partial_\mu \equiv \frac{\partial}{\partial x^\mu} \quad \left(= \frac{\partial}{\partial x^0}, \frac{\partial}{\partial x^1}, \frac{\partial}{\partial x^2}, \frac{\partial}{\partial x^3} \right)$$

and the contravariant ∂^μ is

$$\partial^\mu \equiv \frac{\partial}{\partial x_\mu} \quad \left(= \frac{\partial}{\partial x_0}, \frac{\partial}{\partial x_1}, \frac{\partial}{\partial x_2}, \frac{\partial}{\partial x_3} \right).$$

The usual replacement that occurs in wave mechanics is

$$\mathbf{p} \to -i\nabla \quad \text{and} \quad E \to i(\partial/\partial t).$$

The contravariant four-vector $p^\mu = (E, \mathbf{p})$ will therefore have the replacement

$$p^\mu \to i\partial^\mu$$

and the covariant four-vector $p_\mu = (E, -\mathbf{p})$ will be replaced by $i\partial_\mu$. The differential operator $\Box$ is defined by

$$\Box = \partial_\mu \partial^\mu = \frac{\partial}{\partial x^\mu} \frac{\partial}{\partial x_\mu} = \frac{\partial}{\partial t^2} - \nabla^2.$$

The space-volume element $dx\,dy\,dz$ is represented by $d\mathbf{x}$; the space-time element $dx^0\,dx^1\,dx^2\,dx^3$ by dx^4; and similarly for elements in momentum space.

10.4 The Lagrangian Formalism

A relativistic many-particle theory is set up by defining the dynamical field variables $\psi(x)$ that are operators and are functions of the space-time coordinates $x^\lambda (\lambda = 0, 1, 2, 3)$. The field variables exist at all points in space-time and can have more than one component at each point. The behavior of the field is described by the Lagrangian density, which is assumed to be a function of the field variables $\psi^\kappa(x)$ ($\kappa = 1, 2, \ldots, n$ for an n-component field) and of their derivatives $\phi_\lambda{}^\kappa(x) = \partial \phi^\kappa/\partial x^\lambda$ ($\lambda = 0, 1, 2, 3$). Thus

$$\mathscr{L} = \mathscr{L}(\phi^\kappa, \phi_\lambda{}^\kappa).$$

This implies that $\mathscr{L}$ is a function of the coordinates. The total Lagrangian or action is the density integrated over a suitable space–time volume Ω:

$$L = (1/i) \int_\Omega \mathscr{L}(\phi^\kappa, \phi_\lambda{}^\kappa)\, d^4x. \tag{10.9}$$

An action principle is used to predict the behavior of physically permissible systems. This principle requires that L be stationary for variations of the field variables at values they have when representing physical systems; it follows that the equations of motion are

$$\frac{\partial \mathscr{L}}{\partial \phi^\kappa} - \frac{\partial}{\partial x^\lambda} \frac{\partial \mathscr{L}}{\partial \phi_\lambda{}^\kappa} = 0.$$

Other information about the field may be found by making other variations, such as displacements of the boundary of Ω. The usual procedure is to use the $\mathscr{L}$ that gives the required equations of motion. We shall illustrate the information given in this section by giving the scalar field as example. This field has Lagrangian density given by

$$\mathscr{L} = \frac{1}{2}\left\{ \frac{\partial \phi}{\partial x^\lambda} \frac{\partial \phi}{\partial x_\lambda} - m^2\phi^2 \right\},$$

when ϕ is here the field variable for the scalar field. This leads to the equation of motion

$$\Box \phi + m^2 \phi = 0$$

which is the Klein–Gordon equation in covariant notation. Once the Lagrangian is known it is possible to find the commutators of the field variables, and it is this step that is the essential part of second quantization. We shall not quote the commutators at this stage but proceed to the next stage, which is to make a Fourier expansion of the field variables in terms of plane waves of fixed momentum. The Fourier coefficients must be operators, and their commutation relations can be derived from the commutation relations for the field variables. For the scalar field

$$\phi(x) = \frac{1}{(2\pi)^{3/2}} \int_{-\infty}^{+\infty} \frac{d\mathbf{k}}{(2k_0)^{1/2}} \{ a_k\, e^{-ik\cdot x} + a_\kappa{}^\dagger\, e^{ik\cdot x} \}, \tag{10.9a}$$

where $+(\mathbf{k}^2 + m^2)^{1/2} = k_0$. The volume of integration Ω has been taken to infinity. The operators a_k, $a_k{}^\dagger$ satisfy the commutation rules:

$$[a_k, a_{k'}] = [a_k{}^\dagger, a_{k'}^\dagger] = 0, \qquad [a_k, a_{k'}^\dagger] = \delta(k - k'). \tag{10.10}$$

10.4 The Lagrangian Formalism

The field variables obey the Heisenberg equation of motion

$$i\frac{d\phi}{dt} = [\phi, H],$$

where H is the Hamiltonian. This is an operator equation, and we can find its matrix elements between two states represented by state vectors $|1\rangle$ and $|2\rangle$, which are eigenfunctions of H with eigenvalues E_1 and E_2. Then

$$\left\langle 2 \left| i\frac{d\phi}{dt} \right| 1 \right\rangle = \langle 2 | \phi H - H\phi | 1 \rangle.$$

We can pick out one Fourier component $a_k e^{-ik \cdot x}$ and we have

$$\langle 2 | k_0 a_k e^{-ik \cdot x} | 1 \rangle = (E_1 - E_2) \langle 2 | a_k e^{-ik \cdot x} | 1 \rangle.$$

Therefore

$$k_0 = E_1 - E_2 \quad \text{if} \quad \langle 2 | a_k e^{-ik \cdot x} | 1 \rangle \neq 0.$$

This equation states that the total energy in the field represented by state vector $|1\rangle$ is greater than that in the field $|2\rangle$ by k_0 if the matrix element of $a_k e^{-ik \cdot x}$ between these states is nonzero. Thus this operator only has matrix elements between states that differ in energy by $k_0 = (\mathbf{k}^2 + m^2)^{1/2}$. Therefore the operator a_k operates on $|1\rangle$ to produce a state $|2\rangle$ that has an energy that is less by k_0; that is, it has the effect of removing one quantum of momentum k from the field; and as such it is called an annihilation operator. Similarly, $a_k{}^\dagger$ only has matrix elements between $\langle 2|$ and $|1\rangle$ if $E_2 - E_1 = k_0$; it follows that $a_k{}^\dagger$ has the effect of creating a quantum and is consequently called a creation operator.

We can define a vacuum state that contains no particle by the equation

$$a_k | 0 \rangle = 0 \quad \text{for all } k. \tag{10.11}$$

Let us put

$$a_k{}^\dagger | 0 \rangle = | 1 \rangle,$$

and

$$a_k{}^\dagger a_k{}^\dagger | 0 \rangle = a_k{}^\dagger | 1 \rangle = | 2 \rangle;$$

that is,

$$(a_k{}^\dagger)^n | 0 \rangle = | n \rangle,$$

We wish to consider the operator $a^\dagger a$, where we have dropped the subscript k, since all the following discussion is independent of k so long as the same subscript is implied throughout.

We have that

$$| n \rangle = a^\dagger | n - 1 \rangle.$$

Therefore
$$\begin{aligned}
a^\dagger a \,|n\rangle &= a^\dagger a a^\dagger \,|n-1\rangle \\
&= a^\dagger(1 + a^\dagger a)\,|n-1\rangle \\
&= |n\rangle + (a^\dagger)^2 a a^\dagger \,|n-2\rangle \\
&= |n\rangle + (a^\dagger)^2 (1 + a^\dagger a)\,|n-2\rangle \\
&= 2\,|n\rangle + (a^\dagger)^3 a \,|n-2\rangle \,.
\end{aligned}$$

Repeating the procedure in these steps, we find that
$$a^\dagger a\,|n\rangle = r\,|n\rangle + (a^\dagger)^{r+1} a\,|n-r\rangle\,;$$
when $r = n$, we have
$$a^\dagger a\,|n\rangle = n\,|n\rangle\,. \tag{10.12}$$

Now n is the number of quanta having the momentum of interest in the field and is the eigenvalue of the operator $a^\dagger a$. The Hamiltonian for this free scalar field can be found from the Lagrangian by the quantum analog of the classical formalism; this Hamiltonian can be expanded into a function of creation and annihilation operators. The result is

$$\begin{aligned}
H &= \int k_0 a_k^\dagger a_k \, d\mathbf{k}\,, \\
&= \int k_0 n_k \, d\mathbf{k}\,.
\end{aligned}$$

We expect the total energy to be the eigenvalue of the Hamiltonian, and this is satisfied by this equation.

If we require the action [Eq. (10.9)] to be invariant under Lorentz transformations, then there exist combinations of the field variables and their derivatives that do not vary with time and lead to conserved observables. These conservation laws are the same as the conservation laws discussed in Section 1.6, and the list of transformations with their equivalent conserved quantities (Table 1.1) applies in the Lagrangian formalism. In this formalism, the requirement of invariance under inhomogeneous Lorentz transformations without rotations draws attention to the importance of a tensor $T_{\lambda\mu}$ given by

$$T_{\lambda\mu} = \frac{\partial \mathscr{L}}{\partial \phi_\lambda^{\,\kappa}}\phi_\mu^{\,\kappa} - \mathscr{L} g_{\lambda\mu}\,.$$

The components with $\mu = 0$ can be integrated over all three-dimensional space to give a four-vector P_λ that is constant:

$$\frac{dP_\lambda}{dt} = \frac{d}{dt}\int T_{\lambda 0}\, d\mathbf{x} = 0\,.$$

10.4 The Lagrangian Formalism

P_λ can be identified as the total momentum–energy four-vector, and the conservation of energy and momentum is assured by the invariance of the Lagrangian density. Similarly, Lorentz rotations generate the angular-momentum tensor having components that can be connected with the components of orbital angular momentum and of spin. The conserved quantity is the total angular momentum. In the case of the one-component scalar field the spin is found to be zero, as expected.

There is one important transformation that can be applied to fields described by complex field variables (non-Hermitian operators)—the gauge transformation. Let us consider a complex scalar field for which the Lagrangian density is

$$\mathscr{L} = \left\{ \frac{\partial \phi^\dagger}{\partial x^\lambda} \frac{\partial \phi}{\partial x_\lambda} - m^2 \phi^\dagger \phi \right\}.$$

If ϕ and $\phi^\dagger$ are linearly independent, this gives two field equations:

$$\Box \phi + m^2 \phi = 0, \qquad \Box \phi^\dagger + m^2 \phi^\dagger = 0.$$

The gauge transformation changes the phase of the field variables

$$\phi \to \phi \, e^{-i\alpha} \quad \text{and} \quad \phi^\dagger \to \phi^\dagger \, e^{i\alpha}$$

and it is evident that the Lagrangian is invariant. In addition there is an associated tensor of unit rank

$$j^\lambda = ie \left(\phi^\dagger \frac{\partial \phi}{\partial x_\lambda} - \frac{\partial \phi^\dagger}{\partial x_\lambda} \phi \right), \tag{10.13}$$

which satisfies the continuity equation

$$\partial j^\mu / \partial x_\mu = 0$$

and has a zeroth component with a space integral that is conserved:

$$(d/dt) \int j_0 \, d\mathbf{x} = 0. \tag{10.14}$$

The tensor j^λ is a four-vector and is normally taken to be the four-vector of current and charge density. The conservation equation (10.14) expresses the overall conservation of electric charge. The same four-vector appears on the application of gauge transformation to Lagrangians describing other charged-particle fields. Equation (10.13) makes it evident that this four-vector does not exist for the real scalar field, and we deduce that such a field cannot describe charged particles. We expect that Lagrangians are invariant under other gauge transformations associated with the conservation of leptons (Section 11.2) and of baryons (Section 5.5). The Fourier expansion of the complex scalar field is

$$\phi(x) = \frac{1}{(2\pi)^{3/2}} \int \frac{d\mathbf{k}}{(2k_0)^{1/2}} \{a_k e^{-ik\cdot x} + b_k{}^\dagger e^{ik\cdot x}\},$$

$$\phi^\dagger(x) = \frac{1}{(2\pi)^{3/2}} \int \frac{d\mathbf{k}}{(2k_0)^{1/2}} \{a_k{}^\dagger e^{ik\cdot x} + b_k e^{-ik\cdot x}\},$$

where $k_0 = +(\mathbf{k}^2 + m^2)^{1/2}$ and the operators a_k, b_k, $a_k{}^\dagger$, and $b_k{}^\dagger$ satisfy commutation relations similar to those of Eq. (10.10). From our definition of the current operator [Eq. (10.13)] it follows that a and $a^\dagger$ annihilate and create particles of charge e while b and $b^\dagger$ annihilate and create particles of charge $-e$.

The scalar fields have been used in this section to exemplify the usefulness of the Lagrangian formalism in quantum field theory and to indicate how the insertion of commutation relations gives the field the required quantized structure. In the next two sections we discuss two rather more involved fields.

10.5 The Electromagnetic Field

It is usual to use the four-vector potential A^μ as the field variable, so that the observables **B** and **E** are given by

$$\mathbf{B} = \text{curl } \mathbf{A},$$

$$\mathbf{E} = -\frac{\partial \mathbf{A}}{\partial t} - \text{grad } \phi,$$

where $\phi = A^0$. The equations do not determine A^μ uniquely, for the observables are unaffected by the gauge transformation

$$A^\mu \to A^\mu + (d\chi/dx_\mu)$$

where χ is a scalar quantity. It is usual to apply the Lorentz subsidiary condition to the field, which in covariant notation reads

$$\partial A^\mu/\partial x^\mu = 0, \qquad (10.15)$$

in which case the field variables satisfy

$$\Box A^\mu = 0 \qquad (10.16)$$

and the uncertainty in A becomes due to the special gauge transformation

$$A^\mu \to A^\mu + (\partial \chi_0/\partial x_\mu),$$

where χ_0 satisfies

$$\Box \chi_0 = 0.$$

To quantize the electromagnetic field it is necessary to find a suitable Lagrangian. It turns out that there are several that are possible (see, for example, Muirhead, 1963; Schweber, 1961; Bjorken and Drell, 1965), all of which have difficulties of interpretation due to the fact that the field has quanta of zero mass and must be gauge invariant. Bjorken and Drell use

$$\mathscr{L} = -\frac{1}{2}\left(\frac{\partial A_\mu}{\partial x^\nu} - \frac{\partial A_\nu}{\partial x^\mu}\right)\frac{\partial A^\mu}{\partial x_\nu},$$

where the four components of the vector potential are treated as independent dynamical quantities. The usual procedure is followed to find the commutation relations. In fact, the four components are not independent, and by a suitable choice of gauge, namely that in which $\phi = 0$ and $\nabla \cdot \mathbf{A} = 0$, it is possible to describe the field in its simplest way. The two independent components then correspond to the two possible plane polarizations of light. The field satisfies Eq. (10.16), and the plane wave solution is

$$\mathbf{A} = \mathbf{A}_0 e^{-ik \cdot x}.$$

$\mathbf{A}_0$ must be at right angles to $\mathbf{k}$ to satisfy $\nabla \cdot \mathbf{A} = 0$. It follows from

$$\mathbf{E} = (\partial \mathbf{A}/\partial t) = ik_0 \mathbf{A}$$

and from

$$\mathbf{B} = \operatorname{curl} \mathbf{A} = i\mathbf{k} \times \mathbf{A}$$

that $\mathbf{E}$, $\mathbf{B}$, and $\mathbf{k}$ are mutually perpendicular. The general polarization will be a superposition of two mutually perpendicular polarizations $\mathbf{e}_1$ and $\mathbf{e}_2$ that with $\mathbf{k}$ form a mutually orthogonal set of directions. Then the general field can be expanded in terms of annihilation and creation operators

$$\mathbf{A}(x) = \frac{1}{(2\pi)^{3/2}} \int \frac{d\mathbf{k}}{(2k_0)^{1/2}} \sum_{j=1}^{2} (A_{kj}\mathbf{e}_j e^{-ik \cdot x} + A^\dagger_{kj}\mathbf{e}_j e^{+ik \cdot x}), \quad (10.17)$$

where $k_0 = |\mathbf{k}|$. The A_{kj} annihilates, and $A^\dagger_{kj}$ creates, a photon of momentum k and polarization j.

It is possible to expand the four components of the field separately. The four components are linearly dependent, and it turns out that in a calculation of any observable quantity the two superfluous polarizations always give canceling contributions.

10.6 The Dirac Field

The Dirac field is described by field variables that have four components and transform as do spinors. The field contains quanta that carry an intrinsic angular momentum of $\hbar/2$, and it appears that all such spin-$\frac{1}{2}$

particles in the free-field situation are described by a Dirac field.

We proceed by reforming the Dirac equation into covariant notation. We define four matrices by the equations

$$\gamma^k = \beta\alpha_k \quad (k = 1, 2, 3), \tag{10.18}$$

$$\gamma^0 = \beta, \tag{10.19}$$

where β and α_k are given by the relations of Eqs. (10.1)–(10.3). From these commutation relations for the matrices $\boldsymbol{\alpha}$ and β we find that

$$\gamma^\rho\gamma^\sigma + \gamma^\sigma\gamma^\rho = 2g^{\rho\sigma}I \quad (\rho, \sigma = 0, 1, 2, 3). \tag{10.20}$$

When defined as in Eqs. (10.18) and (10.19) three γ-matrices are anti-Hermitian,

$$\gamma^{\rho\dagger} = -\gamma^\rho \quad (\rho = 1, 2, 3),$$

and one is Hermitian,

$$\gamma^{0\dagger} = \gamma^0.$$

It is not normally necessary to use an explicit representation of the γ-matrices. Such a representation in its irreducible form will be a set of 4×4 matrices; see, for example, Bjorken and Drell (1964). Substituting Eqs. (10.18) and (10.19) into the free-particle Dirac equation (Section 10.2), we find that the free-field equation is

$$i\gamma^\rho(\partial\psi/\partial x^\rho) - m\psi = 0. \tag{10.21}$$

ψ is a four-component wavefunction, which is readily represented by a column matrix of four elements; therefore the Dirac equation, as expressed in Eq. (10.21), contains four equations. A second field variable, called the adjoint function $\bar{\psi}$, is given by

$$\bar{\psi} = \psi^\dagger \gamma_0$$

and satisfies

$$i(\partial\bar{\psi}/\partial x^\rho)\gamma^\rho + m\bar{\psi} = 0.$$

The Lagrangian density that generates these field equations is

$$\mathscr{L} = \bar{\psi}\left(i\gamma^\rho \frac{\partial}{\partial x^\rho} - m\right)\psi. \tag{10.22}$$

Let us now consider the plane-wave solutions to Eq. (10.21)

$$\psi(x) = u(p)\, e^{-ip\cdot x},$$

where

$$p \cdot x = p^0 x^0 - \mathbf{p} \cdot \mathbf{x}.$$

10.6 The Dirac Field

The spinor $u(p)$ satisfies the equation

$$(\gamma^\rho p_\rho - m)u(p) = 0, \tag{10.22a}$$

and the adjoint spinor, $\bar{u}(p) = u^\dagger(p)\gamma^0$, satisfies

$$\bar{u}(p)(p_\rho \gamma^\rho + m) = 0. \tag{10.22b}$$

There are four solutions for a given value of **p**, two having $p_0 = +(\mathbf{p}^2 + m^2)^{1/2}$ and the others having $p_0 = -(\mathbf{p}^2 + m^2)^{1/2}$. The four solutions form an orthogonal set. However, it is more convenient to make use of another orthogonal set, namely the four solutions $u_r(p)$ and $v_r(p)$, where $r = 1, 2$ that satisfy

$$(\gamma^\rho p_\rho - m)\, u_r(p) = 0,$$
$$(\gamma^\rho p_\rho + m)\, v_r(p) = 0.$$

The two solutions $u_r(p)$ go with the plane wave $e^{-ip\cdot x}$ and have momentum $\mathbf{p}(=p^1, p^2, p^3)$ and energy $p^0 > m$. The solutions $v_r(p)$ go with the plane wave $e^{+ip\cdot x}$, have momentum $-\mathbf{p}$ and energy $-p^0$.

We can now make a Fourier expansion of $\phi(x)$:

$$\phi(x) = \int \frac{d\mathbf{p}}{(2\pi)^{3/2}} \left(\frac{m}{p_0}\right)^{1/2} \sum_{r=1}^{2} \{a_{pr}\, u_r(p)\, e^{-ip\cdot x} + b_{pr}^\dagger\, v_r(p)\, e^{+ip\cdot x}\},$$

where $p_0 = +(\mathbf{p}^2 + m^2)^{1/2}$. There is a similar expansion for the adjoint function

$$\bar{\phi}(x) = \int \frac{d\mathbf{p}}{(2\pi)^{3/2}} \left(\frac{m}{p_0}\right)^{1/2} \sum_{r=1}^{2} \{a_{pr}^\dagger\, \bar{u}_r(p)\, e^{+ip\cdot x} + b_{pr}\, \bar{v}_r(p)\, e^{-ip\cdot x}\}.$$

The normal method of second quantization would be to derive the commutation relations for the field variables of this Dirac field; however, in this case commutation relations lead to physical inconsistencies, which can be avoided only by having the field variables obey anticommutation relations. The result is that the operators a, $a^\dagger$, b, and $b^\dagger$ also satisfy anticommutation relations. These are

$$\{a_{pr}, a_{p'r'}^\dagger\} = \delta(p - p')\delta_{rr'},$$
$$\{b_{pr}, b_{p'r'}^\dagger\} = \delta(p - p')\delta_{rr'},$$

where we have used the notation

$$\{A, B\} = AB + BA$$

All other combinations of these operators anticommute.

As in the case of the previous fields, these operators can be interpreted as creation and annihilation operators:

a_{pr} annihilates a particle of momentum $\mathbf{p}$ and energy p_0,
$a^\dagger_{pr}$ creates a particle of momentum $\mathbf{p}$ and energy p_0,
b_{pr} creates a particle of momentum $-\mathbf{p}$ and energy $-p_0$,
$b^\dagger_{pr}$ annihilates a particle of momentum $-\mathbf{p}$ and enegy $-p_0$,

The subscript r refers to the spin orientation of the particle concerned. Physically the creation of a particle of momentum $-\mathbf{p}$ and energy $-p_0$ is equivalent to the destruction of a particle of momentum $\mathbf{p}$ and energy p_0 and of opposite electric charge. Thus, applied to the electron–positron field, we have

a_{pr} annihilates electrons
$a^\dagger_{pr}$ creates electrons
b_{pr} annihilates positrons
$b^\dagger_{pr}$ creates positrons

all of momentum $\mathbf{p}$ and energy p_0

Thus, for this field,

$\psi(x)$ annihilates electrons or creates positrons at the point x
$\bar{\psi}(x)$ creates electrons or annihilates positrons at the point x.

Let us consider the properties of these creation and annihilation operators in more detail. We define the vacuum state $|0\rangle$ by

$$a|0\rangle = b|0\rangle = 0,$$

where we have dropped the subscripts on the operators; in the following argument, all operators refer to particles in the same state; that is, momentum $\mathbf{p}$ and spin r. Let us create a one-electron state $|1\rangle$,

$$|1\rangle = a^\dagger|0\rangle.$$

We now try to create a second electron having the same momentum and spin as the first:

$$|2\rangle = a^\dagger|1\rangle = a^\dagger a^\dagger|0\rangle = 0.$$

The last equality follows from the fact that $a^\dagger$ anticommutes with itself; that is,

$$\{a^\dagger, a^\dagger\} = 0.$$

This zero that appears when we try to create electrons of the same momentum and spin is interpreted to mean that it is impossible to have two electrons of the same momentum and spin in the same field. This result is gratifying, as it means that the implications of the Pauli exclusion principle

10.6 The Dirac Field

appear as a consequence of the anticommuting properties of the field variables and that the principle does not have to be put directly into the theory as a postulate.

It is possible to construct an operator that has eigenvalues equal to the number of quanta of specified momentum and spin in the field. It is $N_{pr} = a^\dagger_{pr} a_{pr}$. It has the property

$$N_{pr}(N_{pr} - 1) = 0,$$

so that

$$N_{pr} = 1 \text{ or } 0,$$

as required for a field that obeys the exclusion principle. $M_{pr} = b^\dagger_{pr} b_{pr}$ is the operator that has eigenvalues equal to the number of antiparticles in the field. The b operators behave in the same way as the a operators, and the antiparticles also obey the exclusion principle, but have charge opposite to that of the particle.

We can now discuss the other quantities that can be derived once the Lagrangian density is known. These are the current four-vector, the energy-momentum tensor, and the angular-momentum tensor. The current four-vector is

$$j^\rho = e\bar{\psi}\gamma^\rho\psi.$$

Taking the fourth component, integrating over space, and substituting the Fourier expansion, we obtain an expression for the conserved total charge of the field; it is

$$Q = e \sum_{r=1}^{2} \int d\mathbf{p}\, (a^\dagger_{pr} a_{pr} - b^\dagger_{pr} b_{pr}),$$

which confirms our assignment of one charge to the particle whose number operator is $a^\dagger a$ and opposite charge to the antiparticle whose number operator is $b^\dagger b$. The energy-momentum tensor leads to an expression for the total momentum four-vector

$$P^\mu = \int d\mathbf{p}\, p^\mu \sum_{r=1}^{2} \{a^\dagger_{pr} a_{pr} + b^\dagger_{pr} b_{pr}\}.$$

The fourth component is related to the energy and indicates that this quantity is a positive definite quantity equal to the total energy of particles and antiparticles in the field. The angular-momentum tensor can be used to find, for quanta of momentum $\mathbf{p}$, the component of spin angular momentum along the axis of quantization in the rest frame. It is

$$s = \tfrac{1}{2}\{a^\dagger_{p1} a_{p1} - a^\dagger_{p2} a_{p2} - b^\dagger_{p1} b_{p1} + b^\dagger_{p2} b_{p2}\}.$$

This result allows us to complete our interpretation of the a and b operators. All the particles created or annihilated have spin $\tfrac{1}{2}$; for individual quanta the component of this spin is $\pm\frac{1}{2}$ according to

$r = 1$ for particle or $r = 2$ for antiparticle with spin component $+\frac{1}{2}$,
$r = 2$ for particle or $r = 1$ for antiparticle with spin component $-\frac{1}{2}$,

It is worth noting at this point that we can form bilinear combinations of the field operators and the γ-matrices that have useful transformation properties. They are

$$\begin{array}{ll} \bar{\phi}\phi & \text{scalar (S),} \\ \bar{\phi}\gamma^\mu\phi & \text{vector(V),} \\ \bar{\phi}\gamma_5\phi & \text{pseudoscalar (P),} \\ \bar{\phi}\gamma_5\gamma^\mu\phi & \text{axial vector (A),} \\ (i/2)\bar{\phi}(\gamma^\mu\gamma^\nu - \gamma^\nu\gamma^\mu)\phi & \text{antisymmetric tensor (T),} \end{array}$$

where $\gamma_5 = i\gamma^0\gamma^1\gamma^2\gamma^3$. This is a complete list of the 16 independent 4×4 matrices that can be constructed from the γ-matrices. Thus no other independent quantities of this form can be found. Factors of i are included to make the bilinear covariants Hermitian when the field operators belong to the same particle.

10.7 Second Quantization and the Commutation Relations

The subjects described in the last section did not include a detailed consideration of the commutation relations, although these relations were stated to be the source of the commutation relations for the annihilation and creation operators. The commutation properties of the field variables are the real reason for the quantized nature of the field, and thus the essential part of second quantization is the determination of these properties. The commutation relations and their expectation values, which are also important in the theory, can be expressed as complex integrals that are solutions of the field equations with δ-function sources. A discussion of these so-called Green's functions is also beyond our scope, and readers are referred to the literature: for example, Bogoliubov and Shirkov (1959) and Jauch and Rohrlich (1955).

We shall take a brief look at one commutation relation—that for a scalar field:

$$[\phi(x), \phi(y)] = -iD(x - y),$$

where

$$D(x) = \frac{-i}{(2\pi)^3}\int_{-\infty}^{+\infty} e^{ik\cdot x}\,\delta(k^2 + m^2)\epsilon(k_0)\,d^4k, \qquad (10.23)$$

and where

$$\epsilon(k_0) = \begin{cases} +1 & \text{for} \quad k_0 > 0, \\ -1 & \text{for} \quad k_0 < 0. \end{cases}$$

If x and y are spacelike points, that is, if $\mathbf{x} - \mathbf{y} > (x^0 - y^0)$, which means x and y cannot be connected by a light signal, then $\phi(x)$ and $\phi(y)$ must commute and $D(x - y) = 0$. Hence $D(x)$ must be zero if x is a spacelike point with respect to the origin. We can test this by putting $x^0 = 0$ in Eq. (10.23), from which we find $D(x) = 0$. This particular equal time case can be transformed into any spacelike separation by a Lorentz transformation. This leaves $D(x)$ unchanged, and it is therefore zero for all spacelike points.

10.8 Interaction and the S-Matrix

In earlier sections of this chapter we have discussed free fields. In practice the fields are not free but interacting continuously, and this has several effects. One of these is to change "bare" particles into "dressed" particles; for example, an electron is in continuous interaction with the electromagnetic field, emitting and reabsorbing virtual photons, and this causes a change in mass, electric charge, and magnetic moment from the unobservable values for a bare electron to the observed values for a dressed electron. A second effect of the interaction is to cause an interchange of energy between interacting fields, so that particles are created, or destroyed, or change energy; these particle reactions take place only if energy and momentum are conserved. It is this second feature of interacting fields that interests us in this section, and we shall discuss quantum electrodynamics as an example of the interaction properties.

Let us consider two interacting fields, 1 and 2, which have free-field Lagrangian densities $\mathscr{L}_1$ and $\mathscr{L}_2$. The total Lagrangian density is given by

$$\mathscr{L} = \mathscr{L}_1 + \mathscr{L}_2 + \mathscr{L}_I,$$

where $\mathscr{L}_I$ is the contribution from the interaction energy. Many formal difficulties appear in the theory when this term is added, but some progress has been made in a perturbation theory in which solutions appear as a power series in the coefficient that occurs in $\mathscr{L}_I$. This perturbation theory must be set up in the "interaction representation" (see works cited in Section 10.1), which we do not discuss here.

$\mathscr{L}_I$ is able to cause transitions from an initial state represented by a state vector $|\alpha\rangle$ to a final state of the same energy and momemtum represented by $|\beta\rangle$. $|\alpha\rangle$ and $|\beta\rangle$ are defined at times distant in the past and future respectively; let us suppose that $t_\alpha = -\infty$ and $t_\beta = +\infty$. Then the probability amplitude for finding the state $|\beta\rangle$ at $t_\beta = +\infty$ is

$$\langle \beta | S | \alpha \rangle.$$

S is the unitary S-matrix, and it is this operator that we wish to relate to $\mathscr{L}_I$. There is no relation in closed form, but S can be expressed as a power series:

$$S = \sum_{n=0}^{\infty} \frac{(-i)^n}{n!} \int_{-\infty}^{+\infty} d^4x_1 \int_{-\infty}^{+\infty} d^4x_2 \cdots \int_{-\infty}^{+\infty} d^4x_n$$
$$\times P\{\mathscr{L}_1(x_1) \mathscr{L}_1(x_2) \cdots \mathscr{L}_1(x_n)\}. \qquad (10.24)$$

This is the necessary perturbation expansion, and will depend for its success upon the smallness of the coefficient contributed by $\mathscr{L}_1$. P is the chronological-ordering operator, which puts all operators following into order of increasing time from right to left.

Let us examine the effects of S for the case of quantum electrodynamics for which $\mathscr{L}_1$ and $\mathscr{L}_2$ are the free electron–positron and electromagnetic-field Lagrangians, respectively. In classical mechanics the electromagnetic field can be included by making the substitution

$$p^\mu \to p^\mu - eA^\mu(x),$$

which in quantum mechanics becomes

$$i(\partial/\partial x_\mu) \to i(\partial/\partial x_\mu) - eA^\mu(x). \qquad (10.25)$$

This can be realized by adding the interaction Lagrangian density

$$\mathscr{L}_1 = e\bar{\psi}(x)\gamma_\mu\psi(x)A^\mu(x), \qquad (10.26)$$

which leads to the field equation [cf. Eqs. (10.25) and (10.21)]

$$\gamma_\mu\left(i\frac{\partial}{\partial x_\mu} - eA^\mu(x)\right)\psi(x) = m\psi(x) \qquad (10.27)$$

and

$$\Box A^\mu(x) = e\bar{\psi}(x)\gamma^\mu\psi(x). \qquad (10.28)$$

We count terms in the expansion of S by the order n. The zeroth term is unity and of no immediate interest. Let us now consider the first term of the S-matrix, using $\mathscr{L}_1$ of Eq. (10.26):

$$S_1 = -e\int_{-\infty}^{+\infty} d^4x\{\bar{\psi}(x)\gamma_\mu\psi(x)A^\mu(x)\}.$$

It is evident that this operator is able to create and destroy photons, electrons, and positrons at the point x. The number of particles involved is one photon $[A^\mu(x)]$ and two particles of the electron–positron field $[\bar{\psi}(x)$ ond $\psi(x)]$ with the restriction that gauge invariance of $\mathscr{L}_1$ limits the effective operators in S_1 to those that conserve charge in the process occurring at the point x.

There are no reactions involving one photon and two particles of the electron–positron field alone that conserve energy and momentum. Therefore S_1 has no matrix elements except for electron or positron scattering by a fixed field such as that due to a heavy nucleus. In Coulomb scattering, for example, $\psi(x)$ [or $\bar{\psi}(x)$] destroys the incoming electron (or positron)

10.8 Interaction and the S-Matrix

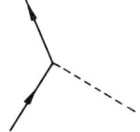

FIG. 10.1. The basic vertex of the Feynman graphs.

and $\bar{\psi}(x)$ [or $\psi(x)$] creates the outgoing electron (or positron); $A^\mu(x)$ describes the virtual photon exchanged between the scattered particle and the Coulomb field.

There is a graphical representation of the S_1-matrix as operating at the point x. In a diagram we draw a line from the edge of the diagram directed to the point x for the term $\psi(x)$. From the point x we draw a directed line toward the edge of the diagram for the term $\bar{\psi}(x)$. We draw a dotted line between x and the edge for the term $A^\mu(x)$. This basic diagram is shown in Fig. 10.1 and is called a vertex. If we place a time axis on the diagram, we can draw our lines with any orientation we please with respect to the axis: two examples are shown in Figs. 10.2a and 10.2b. We choose to interpret the lines pointing into the vertex as electron moving into or positrons moving away from x, and the lines pointing out of the vertex as electrons moving away from or positrons moving into x. It follows that our time axis immediately enables us to label the process in diagrams 10.2a and 10.2b as electron and positron scattering at a fixed field, respectively. We must note that the diagrams are really only representations of terms in the S-matrix; that we can put coordinates onto the diagrams and interpret them as pictures of what is actually happening is a concept due to Feynman (1949a,b). He showed that it was possible to formulate rules by which the matrix element of a given order of the S-matrix could be written down from an examination of all the diagrams of that order that could be drawn for the physical process concerned. The second-order term in the S-matrix is a convenient place to look at this method. This second-order term is

$$S_2 = \frac{e^2}{2} \int_{-\infty}^{+\infty} d^4x_1$$
$$\times \int_{-\infty}^{+\infty} d^4x_2 \, P\{\bar{\psi}(x_1)\gamma^\lambda \psi(x_1) A_\lambda(x_1) \bar{\psi}(x_2) \gamma^\mu \psi(x_2) A_\mu(x_2)\} . \quad (10.29)$$

It is evident that this contains terms corresponding to Feynman diagrams with two vertices (x_1 and x_2) and up to six incoming or outgoing particles.

FIG. 10.2. The Feynman Diagrams for (a) electron scattering by a fixed field, and (b) positron scattering by a fixed field.

The lines corresponding to the operators behind P do not all have to reach to the boundary of the diagram from a vertex. Some lines can start at one vertex, where they represent one operator, and end at the other vertex, where they represent another operator of the same field. When we interpret the diagrams containing one or more internal lines, then the number of ingoing and outcoming particles will be four or less. For example, of particular interest are the cases of two-particle initial-state and two-particle final-state reactions such as electron–electron scattering by the exchange of a virtual photon. It is evident that S_2 has all the necessary creation and annihilation operators and therefore that S_2 will have matrix elements for such processes.

It is possible to make a decomposition of S_2 into eight different terms, each of which is represented by a Feynman diagram. It is important to note that S_2 is not normally the only term in S that will contribute to the amplitude for a process that receives its lowest-order contribution from S_2. Higher-order contribution are the radiative corrections to be discussed in Section 10.9.

We refer readers to Schweber (1961) for a description of the mathematical details of the decomposition of the S-matrix into its terms. In Fig. 10.3, we show the Feynman diagrams O_1 to O_8 for the eight terms in S_2. The type of process for which term each has matrix elements can be found immediately from an examination of the corresponding graph. The external lines—that is, lines reaching the edge—continue to be interpreted as described for the diagrams we drew in connection with S_1 (Fig. 10.2). Thus solid lines directed into the diagrams are electons in the initial state or positrons in the final state, and vice versa for lines directed out of the diagram; external broken lines represent photons in the initial or final state. The internal lines connecting vertices represent virtual intermediate-state

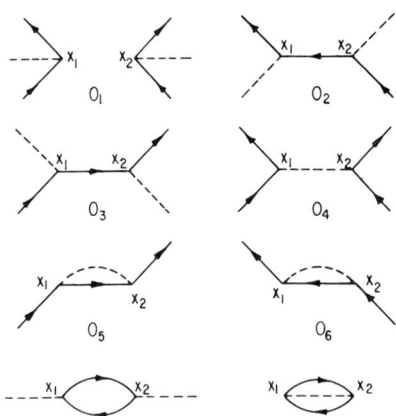

Fig. 10.3. The Feynman diagrams for the terms of S_2.

10.8 Interaction and the S-Matrix

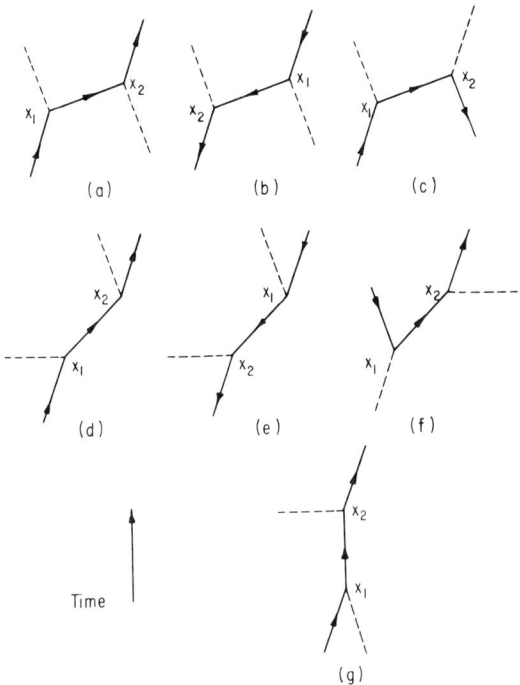

Fig. 10.4. Some Feynman diagrams of the O_3 type. (a) Compton scattering; (b) photon-positron scattering; (c) positron-electron annihilation; (d) electron bremsstrahlung; (e) positron brehmsstrahlung; (f) pair production; (g) photoelectric effect.

particles. With these properties in mind let us give seven redrawings of O_3 and identify them with physical processes. In Fig. 10.4 the orientation of the external lines with respect to the time axis fixes the physical process to which the diagram applies. Therefore the term O_3 has matrix elements for the processes:

> Compton effect;
> photon–positron scattering;
> positron–electon annihilation;
> electron bremsstrahlung;
> positron bremsstrahlung;
> pair production;
> photoelectric effect.

Wherever a nuclear electric field is involved, a photon line has been drawn perpendicular to the time axes; this is a virtual photon, which transfers the momentum imbalance to the participating nucleus. We note that other diagrams of second order may contribute to any of these physical processes.

As we have indicated, these diagrams have a straightforward physical interpretation. Thus in Fig. 10.4a, which contributes to Compton scattering, the incoming electron radiates a real photon at x_1 and propagates as a virtual particle to x_2, where it absorbs the incident photon to become the real final-state electron. Another diagram contributes: it appears from interchanging the photon lines so that the incoming electron absorbs the incident photon at x_1 and propagates to x_2, where it radiates a photon and becomes the real final-state electron. The fact that two diagrams can be drawn means that in the lowest-order calculation the total amplitude is the sum of two amplitudes, each associated with one diagram.

Feynman's rules enumerate the factors that must be written down in an amplitude for each diagram. These rules are most easily applied in momentum space: each external positron or electron line contributes a normalized spinor: each external photon line contributes a normalized polarization vector; each vertex a γ-matrix and a four-momentum δ-function. The internal lines contribute factors called propagators. Thus an internal photon line labeled with four-momentum k contributes a propagator proportional to $1/k^2$.

An internal electron or positron line labeled with four-momentum p contributes a propagator proportional to

$$\frac{\gamma^\mu p_\mu + m}{p^2 - m^2}.$$

We notice that these internal lines represent virtual states, so that $k^2 \neq 0$ for the internal photons and $p^2 - m^2 \neq 0$ for the virtual electrons. We also note that energy and momentum are conserved at each vertex, and since three particles are involved, one at least of these must be a virtual particle and be represented on a Feynman diagram by a line that ends at another vertex or on an external field.

We have attempted in this section to describe the meaning of the Feynman diagrams and to give the background to calculations in modern quantum electrodynamics. It is beyond the scope of this book to describe fully the Feynman rules. However, these are straightforward to use; they are described, with examples, by Bjorken and Drell (1964).

10.9 Renormalization and the Radiative Corrections

We have already noted that terms of the S-matrix higher than S_2 can contribute to processes that receive their simplest contribution from S_2. Thus the Compton effect has fourth-order contribution of the kind shown in Fig. 10.5; these are only a few of the contributions from this order.

10.9 Renormalization and the Radiative Corrections 315

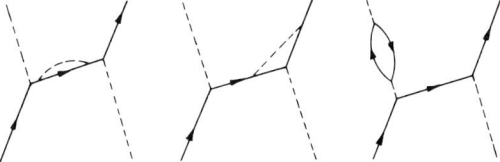

FIG. 10.5. The Feynman diagrams for three fourth-order contributions to Compton scattering.

Formal calculation of higher-order diagrams normally yields an infinite contribution, a result that is disturbing in view of the fact that the lowest-order calculation alone gives results that are finite and of the correct order of magnitude. A careful investigation shows that higher-order infinities are due to three basic types of diagram, which are shown in Fig. 10.6. Figure 10.6a shows a correction to an electron external line or propagator; 10.6b is a correction to a photon external line or propagator; 10.6c is a correction to a vertex.

Let us consider Fig. 10.6a, which is called the electron self-energy term because it has the effect of altering the electron energy. A formal calculation shows that the change in energy is infinite, and this difficulty is analogous to that which occurs in the classical theory of the electron. In that theory a point electron has infinite self-energy; if the self-energy is reduced to a finite value by giving the electron structure in the form of a spread-out charge, the theory leads to results that are inconsistent with the special theory of relativity. The change of energy in quantum theory is equivalent to a change in rest mass, and this gives rise to doubts as to the value of the mass that we should put into the Lagrangian density. We suppose that the observed mass m is the sum of the unobservable bare electron mass m_0 plus the unobservable self-energy δm:

$$m = m_0 + \delta m,$$

or

$$m_0 = m - \delta m.$$

We would expect the free-field Lagrangian to contain m_0 and not m, and in this spirit we can replace m [in Eq. (10.22), for example] by $m_0 = m - \delta m$; thus another term $+\delta m \bar{\psi}\psi$ appears in our stated Lagrangian density for the

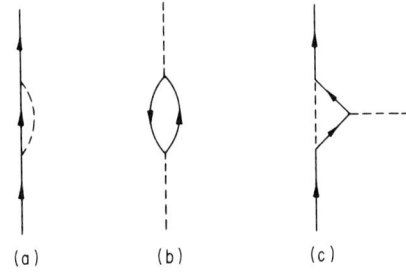

FIG. 10.6. The Feynman diagrams for the nonzero primitive divergent contributions.

(a)   (b)   (c)

electron–positron field. When we put this field into interaction with the electromagnetic field, we can shift this term into the interaction Lagrangian to give

$$\mathscr{L}_1 = e\bar{\psi}\gamma_\mu\psi A^\mu + \delta m \bar{\psi}\psi.$$

The theory is the same as before, apart from this extra term. When the electron self-energy term is calculated, we can cancel the normal infinity due to the first term of the interaction by the contribution from the $\delta m \bar{\psi}\psi$ if δm is given the correct value. This value is, of course, related to the divergent integral that produces the infinity, and is itself infinite, but this causes no difficulty because the canceling integrals are the same and δm is unobservable. This procedure is called mass renormalization; the essence of renormalization is the correct recognition of the divergent parts of the integrals involved. Once this has been done, the resulting infinity can be dropped and the renormalized theory give the correct result if everywhere that m occurs we use the value of the observed mass. By correct recognition of divergent integrals we mean those parts that disappear if the term $\delta m \bar{\psi}\psi$ included in the interaction Lagrangian.

Charge renormalization is required when we consider the effect of diagram b in Fig. 10.6. This effect is called vacuum polarization and increases the apparent coupling constant e by an infinite factor. Charge renormalization involves redefining the electric charge so as to absorb this infinite factor. We assume the electron has a bare charge e_0, which is unobservable; the infinities in the theory occur multiplied by some power of e_0, and we take this product to be the observable charge to the same power. Thus if the charge e occurring in the theory is taken to have the value observed and if we correctly factor out the infinities, we are left with a finite contribution. The polarization of the vacuum has no effect upon the mass of the photon, which must be zero by gauge invariance; however, it does have an effect upon physical processes involving photons in interaction with charged particles, and after charge renormalization the remaining finite contributions are a part of the radiative corrections.

We can now give an example to show how renormalization is applied to remove the divergences that occur in an actual problem. Let us consider the subtraction of infinities from the higher-order corrections to the scattering of an electron in the Coulomb field of a nucleus. The lowest order that has matrix elements for this process is S_1, and the Feynman diagram is shown in Fig. 10.7a. The radiative corrections that interest us are some of the processes that have S_3 matrix elements; those of interest are shown in Fig. 10.7b–10.7e. If we renormalize the mass by adding the term $\delta m \bar{\psi}\psi$ to the interaction Lagrangian, the effects of graphs 10.7b, 10.7c, and 10.7d are finite. Charge renormalization removes the effect of diagram 10.7e.

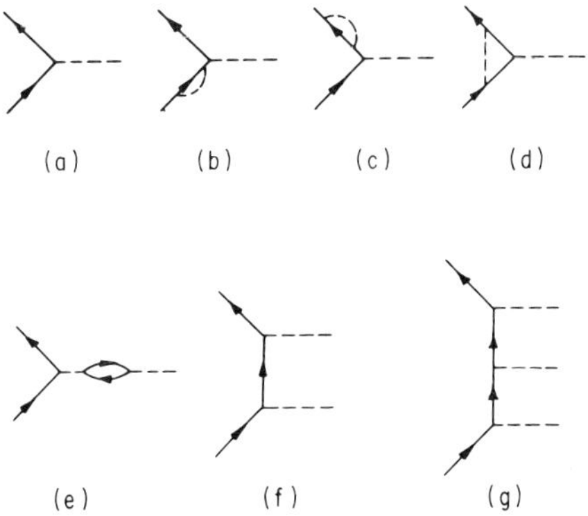

10.9 Renormalization and the Radiative Corrections

Fig. 10.7. The Feynman diagrams for the Coulomb scattering of electrons (a), with second-order radiative corrections (b)-(e). Parts (f) and (g) show the diagrams for second- and third-order scattering in the Coulomb field.

The result is a cross section for the scattering that includes the radiative corrections due to diagrams 10.7b to 10.7e. There are some other contributions of order 2 and 3 due to second- and third-order scattering in the Coulomb field—diagrams f and g of Fig. 10.7. These give a finite contribution only if the Coulomb field is modified for screening due to the atomic electrons (Jauch and Rohrlich, 1955). A final difficulty remains due to the "infrared" divergence, which is the contribution due to the radiation of virtual and real photons of very long wavelengths. It can be shown that if the calculations are done to all orders, these effects cancel and the result is a value for the only kind of cross section that can be measured, namely that for elastic plus inelastic scattering with the radiation of a photon of energy less than some value set by the resolution of the measuring equipment.

The foregoing discussion applies to scattering in the electric field of a nucleus; the same procedure can be applied to the behavior of an electron in a fixed magnetic field. After renormalization the radiative corrections indicate that the electron behaves as if it had an anomalous magnetic moment in addition to the magnetic moment of one Bohr magneton predicted by the one-particle Dirac theory. Unfortunately comparison between theory observation of the magnetic moment, and other quantities, is made difficult by uncertainties that remain in the value of the fine-structure constant (Drell, 1968). Thus the calculated value of the magnetic moment is in the range (Rich, 1968)

$$1.001159617 - 1.001159647 \quad \text{Bohr magnetons},$$

where extreme values are given by extreme values of the fine-structure constant and the calculational uncertainty is about 5 in the last figure. The experimental values is (Rich, 1968).

$$1.001159557 \ (\pm 27 \times 10^{-9}) \quad \text{Bohr magnetons}.$$

The discrepancy between this value and the nearest calculated value is one-and-a-half standard deviations (keeping in mind that the errors on the calculated values are only educated guesses).

The second problem to which the renormalization technique has been applied concerns the energy levels of the hydrogen atom. In the simple Dirac theory the $2S_{1/2}$ and $2P_{1/2}$ excited states are degenerate; in the many-particle quantum electrodynamics these levels split. The energy difference can be predicted by calculation of the radiative corrections to the motion of the bound electron. Such calculations have been made to fourth order. There are many contributions, but the most important are two: the first is due to renormalization of the electron self-energy diagrams, and it raises the $2S_{1/2}$ level relative to the $2P_{1/2}$ by 1079 MHz; the second is due to vacuum polarization and it depresses the $2S_{1/2}$ level by 27 MHz. The total calculated effect is a splitting of 1057.50 ± 0.11 MHz (Soto, 1966). The measured values of this energy difference are 1057.77 ± 0.10 (Triebwasser et al., 1953) and 1057.86 ± 0.10 MHz (Robiscoe, 1968). In this case the uncertainty in the value of the fine-structure constant has a small effect, and a real discrepancy may exist. Thus, although the theory has had results that give great confidence in its correctness, discrepancies remain to be cleared up in the low-energy region.

That renormalization can be carried out to all orders in the perturbation expansion for quantum electrodynamics was proved by Dyson (1949), and this means that the S-matrix has finite matrix elements between any two states involving photons, electrons, and positrons. However, the labor involved in performing the calculations prevents evaluation of any but a few of the lowest orders.

Throughout this discussion we have implied that, after renormalization, the finite contributions from each term of the series expansion of the S-matrix form a convergent series. That this appears to be the case is due to the smallness of the parameter $e = (4\pi/137)^{1/2}$ that occurs with increasing power in successive terms. However, it appears that the series may not be convergent (Dyson, 1952). No successful solution to this problem has been presented.

10.10 QED at High Energies

We have drawn attention to the great success of quantum electrodynamics in the last section, and its validity certainly extends over a very great range of energies. However, it is conceivable that at very large momentum transfers there will be a breakdown in the theory. For example, in electron–electron (Møller) scattering, the ability of the electrons to undertake a very large change in momentum may be inhibited by the existence of any structure that they may have, or altered by the effect of other interac-

10.10 QED at High Energies

tions. The normal quantum electrodynamics does not recognize the first possibility and assumes the interaction between electron and electromagnetic field takes place at a point. There exists no model or theory to predict deviations that might occur, so we represent these phenomenologically and place upper limits on the parameters describing the deviation.

We anticipate one feature of lepton physics; this is the electron-muon universality. The muon (Section 11.8) appears to be a heavy electron in its known interactions and thus normal quantum electrodynamics describes its low-energy electromagnetic behavior. Thus all that follows can be applied equally to muons or electrons, and many of the tests may be easier to apply to the former for experimental reasons. However, it does not mean that we expect that the deviations, if any, will be the same in both cases. This follows from the fact that the muon-electron mass difference indicates that there is some difference between these particles and suggests the possibility that muon and electron high-energy electromagnetic interactions will differ. We shall use the word *lepton* (Section 11.8) to mean electrons or positrons, positive or negative muons.

For any given Feynman diagram there are three types of contribution to the matrix element: the photon propagator, the lepton propagator, and the vertex factor. A breakdown in the theory can be represented by a modification to one or more of these, and various experiments can be devised that, taken together, test one at a time.

It is usual to modify a vertex by a form factor that is a function of the invariant four-momentum transferred to the lepton. Thus if, in Fig. 10.1, the lepton four-momentum changes from p_1 to p_2, the momentum transfer squared is

$$k^2 = (p_1 - p_2)^2 .$$

Unmodified, the vertex contributes to the matrix element a factor $e\gamma^\mu$. Modified by the form factor this becomes

$$e\gamma^\mu F(k^2) .$$

There is no reason to expect any particular form for $F(k^2)$, and it is usual to write it as

$$F(k^2) = (1 - k^2/\Lambda_v^2)^{-1} . \tag{10.30}$$

Any breakdown will have $\Lambda_v < \infty$. Now $\hbar k$ is a four-momentum, and thus $1/k^2$ has the dimensions of (length)^2. Thus $1/\Lambda_v$ is the "size" or distance that characterizes a breakdown in the vertex. $\hbar\Lambda_v/c$ has the dimensions of mass, and, for example, if its value is that of the nucleon, the size parameter is the reduced Compton wavelength, that is, 2.1×10^{-14} cm. For simplicity we shall call this quantity a cutoff.

In the vertex of Fig. 10.1, if the photon is real, $(p_1 - p_2)^2 = k^2 = 0$,

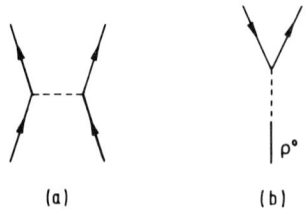

Fig. 10.8. Feynman diagrams for (a) Møller scattering and (b) ρ^0-meson decay into an electron–positron pair. The ρ-meson has the same spin and parity as the photon, and thus there is an amplitude for a ρ^0-meson becoming a timelike photon.

and one or both of the leptons is virtual. In this case, $F(k^2)$ is expected to be 1, and the suggested form [Eq. (10.30)] satisfies this. In Møller scattering (Fig. 10.8a) a virtual photon is exchanged between two electrons. A simple calculation shows that k^2 is zero for 0° scattering and becomes more negative (spacelike) as the scattering angle increases. We expect the scattering to become less than predicted from simple QED, and this is just the effect that $F(k^2)$ will have. As an example of a different kinematic region consider the ρ^0-meson decay into an electron and positron. The Feynman diagram is shown in Fig. 10.8b. Evidently k^2 is positive (timelike) and equal to m_ρ^2. As it stands, Eq. (10.30) would suggest an $F(k^2)$ greater than 1 for the vertex of timelike photons. This behavior is not expected, and Eq. (10.27) should be suitably modified in that region. This serves to show that there is nothing more in Eq. (10.27) than it being a convenient form for parametrizing any breakdown.

The photon propagator is $1/k^2$ in the unmodified theory, and the usual modification is to replace it by

$$\frac{1}{k^2} - \frac{1}{k^2 - \Lambda_\gamma^2} = \frac{1}{k^2}\left[\frac{1}{(1 - k^2/\Lambda_\gamma^2)}\right]. \tag{10.31}$$

Again $1/\Lambda_\gamma$ is a size characteristic of some breakdown, and $\hbar\Lambda_\gamma/c$ is a corresponding mass, sometimes called the regulator mass or the propagator momentum cutoff. The lepton propagators contain a factor $1/(p^2 - m^2)$, where p is the four-momentum caused by the virtual lepton. This is modified by replacement of this factor by

$$(p^2 - m^2)^{-1} - (p^2 - m^2 - \Lambda_l^2)^{-1}. \tag{10.32}$$

As before, a breakdown in the propagator shows up as Λ_l or $\Lambda_\gamma < \infty$, and it is again the case that these modifications cannot be correct for timelike photons ($k^2 > 0$) or leptons ($p^2 - m^2 > 0$).

It is interesting to note that the modification of Eq. (10.31) changes the Coulomb potential space variation thus:

$$1/r \to (1/r)[1 - \exp(-r\Lambda_\gamma)].$$

The proposed modifications are not independent, for it is impossible to satisfy various fundamental properties of S-matrix elements containing

10.10 QED at High Energies

closed loops if only one modification is made. Thus a modification of one propagator must at least involve a change in the other propagator or the vertex (Gatto, 1967). In addition, we expect strong-interaction effects to appear at some place. Thus timelike photons having $k^2 > (2m_\pi)^2$ are coupled to real neutral $J^P = 1^-$ states of hadrons having $(\text{mass})^2 = k^2$ (e.g., $\gamma \to \rho$-meson is probably an important process), and virtual states of these particles will effect the photon propagator.

A particle that is real is said to be on the "mass shell." A virtual particle $(k^2 \neq 0, p^2 - m^2 \neq 0)$ is said to be off the mass shell, and it is farther off the shell the greater the deviation of the magnitude of the four-momentum from its value for the real particle. The modifications introduced suggest that tests for the existence of the effects of a breakdown mean examining processes that involve virtual particles very far off the mass shell, and such processes always have very low cross sections. Thus the experiments we shall discuss are far from easy to perform. In addition, since every Feynman diagram that represents a physical process contains at least one virtual particle and one vertex, it is evident that determining where any breakdown is occurring can be difficult and may involve many experiments.

Testing the lepton propagator in the spacelike region has been done by measuring the cross section for the production of lepton pairs by γ-rays in the configurations in which the leptons emerge at equal angles left and right from the photon direction and with equal energy. Figure 10.9 shows the relevant Feynman diagrams. The process must occur in a nuclear electric field and involves momentum transfer (q) to the nucleus. However, the symmetric-pair arrangement reduces this momentum transfer to a minimum and allows the nuclear form-factor effects to be calculated from electron–nucleon scattering results. In addition, since this momentum transfer is small, the lepton vertex modification will be small and we can expect

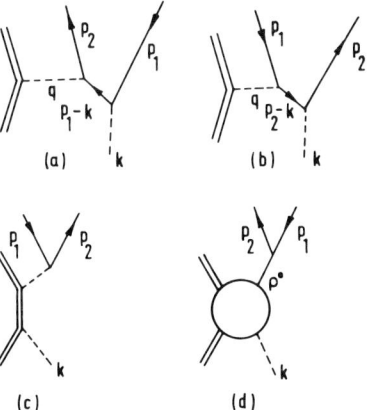

FIG. 10.9. Feynman diagrams for lepton-pair production by photons in the field of a nucleus. The nucleus is represented by the double line. The circle in (d) represents the process that leads to $\gamma \to \rho^0$ in a nuclear field.

any breakdown to be due to the virtual lepton. There are other complications. Figure 10.9c shows a so-called Compton contribution, which can be eliminated by averaging the data over the lepton charges. Figure 10.9d shows pair production by real or virtual ρ^0-mesons. They are copiously produced by photons and, although their branching ratio into lepton pairs is small, the effect on pair experiments can be considerable. Weinstein (1967) has reviewed the experimental data on symmetric-pair experiments. So far no deviations have be found, and the results are quoted as upper limits in fermis (10^{-13} cm) on the quantity Λ_l^{-1}. For the muon this limit is about 0.15 Fm ($\Lambda \cong 1200$ MeV/c) and for the electron 0.13 Fm (95% confidence). To probe to this limit, measurements have been made at squared momentum transfers (spacelike) carried by the virtual lepton up to about (700 MeV/c)^2. Wide-angle bremsstrahlung production has been used to test timelike momentum transfers by electrons up to 1000 MeV/c. No deviation was detected, and the upper limit on a momentum cutoff was put at 0.09 Fm (Siemann et al., 1969).

Electron–electron scattering has been studied at a center-of-mass energy of 600 MeV (Barber et al., 1966). This does not separate the effects of a photon propagator cutoff from those of a vertex form factor. If there is a momentum-transfer dependence described by a form factor $(1 - k^2/\Lambda^2)$ in the scattering amplitude, then this experiment places an upper limit on Λ^{-1} of 0.26 Fm (95% confidence). The elastic scattering of muons and electrons by protons can be compared at the same momentum transfer to obtain information on any cutoff in the muon electromagnetic interaction. If the assumption is made that the Barber result applies to the form factor at the γ-electron vertex, then a comparison yields an upper limit of 0.29 Fm on the cutoff for the γ-muon vertex (Weinstein, 1967).

The decays

$$\rho^0 \to e^+ + e^-,$$
$$\rho^0 \to \mu^+ + \mu^-$$

involve timelike virtual photons, and a comparison explores the difference between form factors at the γ-e and γ-μ vertices. If we put

$$\frac{1}{D^2} = \frac{1}{\Lambda_{v\mu}^2} - \frac{1}{\Lambda_{ve}^2},$$

then $D^{-1} < 0.17$ Fm. Other process, such as $\phi \to l^+ l^-$ and $p + \bar{p} \to l^+ + l^-$, can be used to make the same comparison at other timelike momentum transfers. The $^3S_1 - {}^1S_0$ splitting in positronium also involves timelike photons, but here the upper limit on the cutoff is about 2 Fm for an effect at the γ-e vertices or 2.8 Fm for an effect on the photon propagator (Weinstein, 1967).

10.11 Field Theory and Strong Interactions

Other processes such as e^+e^- or $\mu^+\mu^-$ pair production by muons or electrons can also be used to test quantum electrodynamics. There is also one other physical quantity that is very relevant. It is the anomalous part of the muon magnetic moment. If quantum electrodynamics applies to the muon as it does to the electron, then this anomalous part is calculable as a radiative correction. Deviations from the calculated value will indicate a breakdown in some part of the interaction. These processes involve photons in both timelike and spacelike regions. The experimental result given by the $g - 2$ experiments (Bailey et al., 1967) is, for the anomalous part, $(11,666 \pm 5) \times 10^{-7}$ muon magnetons, where theory predicts $11,656 \times 10^{-7}$. Form-factor effects should reduce the theoretical value, so the experimental result is in the unexpected direction. However, assuming that the experimental value is not less than theory by one part in 2000 gives an upper limit on a cutoff as 0.03 Fm. This is far lower than any other limit set experimentally.

10.11 Field Theory and Strong Interactions

An obvious feature of the physics of elementary particles is the strong interactions that exist between hadrons. The present-day view of these interactions is that they are basically due to the interactions of meson fields with baryons. Since baryons are conserved, the Feynman diagram for such an interaction is like the vertex in quantum electrodynamics, Fig. 10.1. The electron lines are now baryons, and the photon line is a boson. At the vertex, energy and momentum, hypercharge and charge, are conserved. Evidently we can write down the contribution of the interaction to the Lagrangian density. The principle normally followed in constructing these terms is that they are linear in each of the three fields concerned. And of course, the contribution to $\mathscr{L}$ must be scalar if parity is conserved. For example, we consider the coupling of an uncharged scalar or pseudoscalar field to a Dirac field (η-mesons to Λ-hyperons is an example of the second). If ϕ is the meson field, then ϕ and $\partial\phi/\partial x^\mu$ transform as scalar and vector or pseudoscalar and axial vector, since ϕ is a scalar or pseudoscalar. Thus the permitted interactions have $\mathscr{L}_1$ equal to

$g\bar{\psi}\psi\phi$	for scalar meson with scalar coupling (S),
$ig\bar{\psi}\gamma_5\psi\phi$	for pseudoscalar meson with pseudoscalar coupling (P),
$(f/m)\bar{\psi}\gamma^\mu\psi(\partial\phi/\partial x^\mu)$	for scalar meson and vector coupling (V),
$(f/m)\bar{\psi}\gamma^\mu\gamma_5\psi(\partial\phi/\partial x^\mu)$	for pseudoscalar meson and axial-vector coupling (A).

The coupling constants g and f have the same dimensions if the reciprocal mass factor is included in the derivative couplings.

The interaction terms written above have assumed that the baryon entering the vertex is the same as that which leaves it, as, for example, in

$$p \to p + \pi^0. \tag{10.33}$$

This need not be the case; as a first step consider

$$\Sigma^0 \to \Lambda + \pi^0. \tag{10.34}$$

The two baryon fields are associated with particles of different mass. If the Σ^0 and Λ had opposite parity, then the pseudoscalar π^0 would be emitted in a relative S-wave instead of a P-wave as it is in the process of reaction (10.33). To allow for this, an extra γ_5 must go into the baryon term of the interaction. Thus the P interaction $\bar{\psi}\gamma_5\psi\phi$ for reaction (10.34) would become $\bar{\psi}\psi\phi$ ($\gamma_5^2 = 1$). The next step is to consider charged-boson fields, for example,

$$p \to n + \pi^+.$$

Since it is believed that these interactions are charge independent, it is possible to describe all the pion–nucleon vertices by means of one interaction term, which must be invariant under transformations in isotopic-spin space. The pion field operators are redefined so as to have three components that transform like the components of vectors in isotopic-spin space ($\boldsymbol{\phi}$). The nucleon field has two components, which are formally similar to Pauli spinors that represent spin-$\frac{1}{2}$ states. Dropping the γ-matrices, the nucleon term becomes the isotopic-spin vector $\bar{\psi}\boldsymbol{\tau}\psi$, where $\boldsymbol{\tau}$ are three Pauli matrices (Section 2.9) operating on isotopic spinors. Then $\bar{\psi}\boldsymbol{\tau}\psi \cdot \boldsymbol{\phi}$ is a scalar in isotopic-spin space and describes a charge-independent interaction. Bjorken and Drell (1964) describe in detail the procedure for setting up such charge-independent interactions. In Chapter IX we considered the interactions that were not only charge independent but also $SU(3)$ invariant.

The known spin-0 mesons are pseudoscalar, so the interesting possibilities are the P or A couplings. However, for the pion–baryon interaction in first-order perturbation theory these couplings give the same result if $f = gm/(M_1 + M_2)$, where m is the pion mass and M_1, M_2 are masses of the two baryons involved at the vertex (Schweber, 1961).

Mesons of spin 1 and 2 are now known, and it is possible to construct interactions that are charge independent, conserve parity, etc. The electromagnetic interaction is formally similar to that which could exist between a spin-1 meson and the nucleons.

Strong interactions and electromagnetic interactions are both involved in photoproduction processes such as

$$\gamma + p \to \pi^+ + n \quad \text{and} \quad \gamma + p \to K^+ + \Lambda.$$

10.11 Field Theory and Strong Interactions

In addition to the interaction of fermions with the electromagnetic field and with the meson fields, this also involves the interaction of mesons with the electromagnetic field. This again is handled by the substitution

$$i(\partial/\partial x_\mu) \to i(\partial/\partial x_\mu) - eA^\mu(x),$$

but this does lead to a contribution to the interaction Lagrangian that is more complicated than it is in the case of quantum electrodynamics (Bjorken and Drell, 1964).

Once an interaction has been postulated, then it is formally possible to proceed to the calculation of transition rates and cross sections using the perturbation expansion. However, this leads to problems of renormalization and divergence. It has been shown that field theories of mesons interacting with nucleons are renormalizable for nonderivative couplings (Matthews and Salam, 1951). Despite this, the perturbation expansion diverges and does not give results that bear any relation to observation. In view of this the perturbation approach has been dropped. Instead a search has been made for nonperturbative solutions and calculations made using dispersion-relation techniques (Section 12.3).

We end this section by examining the coupling constants that occur in strong interactions; for orientation we return to quantum electrodynamics and recall that the coupling constant that occurs is e, the magnitude of the electron charge in units that are rationalized and have $\hbar = c = 1$. The value of e is approximately $(4\pi/137)^{1/2}$ and is dimensionless. Many readers will find these units unfamiliar, so let us return to the cgs and es units. We put e' to represent the electron charge in electrostatic units; then we have

$$e'^2/\hbar c \simeq 1/137.$$

This is the square of the constant that appears in the interaction Lagrangian density divided by 4π. It is the smallness of this quantity that makes the perturbation expansion an apparently convergent series after renormalization. We now put the coupling constants g and f into more familiar form; as they occur they are rationalized and dimensionless. If we rewrite our expressions in cgs units in which the coupling constants are g' and f', then we have to make the following replacements

$$g^2 \to \frac{4\pi g'^2}{\hbar c}, \quad f^2 \to \frac{4\pi f'^2}{\hbar c} \quad \left(\text{cf.} \quad e^2 \to \frac{4\pi e'^2}{\hbar c}\right).$$

In the pseudoscalar coupling the value of g' is such that

$$\frac{g'^2}{\hbar c} \simeq 14.5.$$

In view of the equivalence of axial-vector and pseudoscalar in first order it is possible to quote a value of f^2 instead of g^2:

$$f^2 = g^2(m_\pi/2m_N)^2 = 0.082 .$$

This is the coupling constant to which reference was made in Section 8.5. m_π and m_N are the pion, and nucleon masses, respectively.

10.12 Conclusion

It will be evident to the reader that this chapter has done no more than attempt to make field theory a more familiar subject. We have, for example, stopped short of attempting to describe how the S-matrix elements can be evaluated for particular processes, but we hope that readers will not take this to mean that they are to be deterred from discovering how this is done. In fact, the intention has been to reduce for the uninitiated the apparent formidableness of the subject and to encourage further reading.

REFERENCES

Bailey, J. et al. (1967). Proc. Int. Symp. Electron Photon Interactions High Energ. Stanford, 1967, p. 48. International Union of Pure and Applied Physics and U.S. Atomic Energy Commission (10.10).

Barber, W. C., Gittelman, B., O'Neill, G. K., and Richter, B. (1966). Phys. Rev. Lett 16, 1127 (10.10).

Bjorken, J. D., and Drell, S. D. (1964). "Relativistic Quantum Mechanics." McGraw-Hill, New York (10.1, 10.6, 10.8, 10.11).

Bjorken, J. D., and Drell, S. D. (1965). "Relativistic Quantum Fields." McGraw-Hill, New York (10.1, 10.5).

Bogoliubov, N. N., and Shirkov, D. V. (1959). "Introduction to the Theory of Quantised Fields." Wiley (Interscience), New York (10.1, 10.7).

Drell, S. D. (1968). Comments Nucl. Particle Phys. 2, 136 (10.9).

Dyson, F. J. (1949). Phys. Rev. 75, 486 (10.9).

Dyson, F. J. (1952). Phys. Rev. 85, 631 (10.9).

Feynman, R. P. (1949a). Phys. Rev. 76, 749 (10.8).

Feynman, R. P. (1949b). Phys. Rev. 76, 769 (10.8).

Gatto, R. (1967). In "High Energy Physics" (E. H. S. Burhop, ed.), Vol. II. Academic, Press, New York (10.10).

Jauch, J. M., and Rohrlich, F. (1955). "The Theory of Photons and Electrons." Addison-Wesley, Reading, Massachusetts (10.1, 10.7, 10.9).

Matthews, P. T., and Salam, A. (1951). Rev. Mod. Phys. 23, 311 (10.11).

Muirhead, H. (1963). "The Physics of Elementary Particles." Pergamon, New York (10.5).

Pauli, W., and Weisskopf, V. F. (1934). Helv. Phys. Acta 7, 769 (10.2).

Rich, A. (1968). Phys. Rev. Lett. 20, 967, 1221 (10.9).

Robiscoe, R. T. (1968). Phys. Rev. 168, 4 (10.9).

Schweber, S. S. (1961). "An Introduction to Relativistic Quantum Field Theory." Harper and Row, New York (10.1, 10.5, 10.8, 10.11).

Siemann, R. H., Ash, W. W., Berkelman, K., Hartill, D. L., Lichtenstein, C. A., and Littauer, R. M. (1969). Phys. Rev. Lett. 22, 421 (10.10).

References

Soto, M. F. (1966). *Phys. Rev. Lett.* **17,** 1153 (10.9).
Triebwasser, S., Dayhoff, E. S., and Lamb, W. E. (1953). *Phys. Rev.* **89,** 98 (10.9).
Weinstein, R. (1967). *Proc. Int. Symp. Electron Photon Interactions High Energ. Stanford, 1967,* p. 409, International Union of Pure and Applied Physics and U.S. Atomic Energy Commission (10.10).

XI

WEAK INTERACTIONS

11.1 Introduction

The weak interaction is responsible for a wide range of decay reactions. Many of the known particles decay by strong interactions into strongly interacting particles with mean lives of the order of 10^{-22} to 10^{-24} sec (e.g., $\rho \to 2\pi$, $\Sigma(1385) \to \Lambda\pi$). A few decay electromagnetically with mean lives of the order of 10^{-16} to 10^{-20} sec (e.g., $\pi^0 \to 2\gamma$, $\eta \to 3\pi$). If selection rules or accidents of mass difference prevent such decays, then the weak interaction, with its weaker selection rules, intervenes, and mean lives in the range 10^{-10} ($\Omega^- \to K^-\Lambda$) to 10^3 sec ($n \to pe\bar{\nu}$) are found. The only known particles finally stable against spontaneous free decay are the proton and antiproton, the electron and positron, and the neutrinos and photons.

Historically, the subject began with beta decay, and we shall deal with that aspect of weak interactions in the first part of this chapter, before proceeding to investigate the full properties of the weak interaction as revealed by the additional evidence available from the decays of strange particles.

11.2 The Description and Theory of Beta Decay

Beta decay is the name given to the manifestation of the following fundamental decay reactions:

$$n \to p + e^- + \bar{\nu}, \tag{11.1}$$

$$p \to n + e^+ + \nu. \tag{11.2}$$

For the time being we assume neutrino (ν) and antineutrino ($\bar{\nu}$) are different, and define them by these reactions. The second reaction is energetically forbidden for the free proton and can be observed only in certain complex

11.2 The Description and Theory of Beta Decay

nuclei. Other reactions connected with β-decay can be found by rearranging the particles in reactions (11.1) and (11.2) (i.e., by charge conjugating a particle and changing the side on which it appears); for example,

$$e^- + p \to n + \nu \tag{11.3}$$

is the reaction that takes place in nucleon capture of an electron from a K orbit, and

$$\bar{\nu} + p \to n + e^+ \tag{11.4}$$

is the antineutrino absorption reaction observed by Reines and Cowan (1959). Antinucleons are supposed to be able to undergo β-decay, although this has not been observed:

$$\bar{n} \to \bar{p} + e^+ + \nu, \quad \text{etc.} \tag{11.5}$$

The antiproton decay can occur only in complex antinuclei.

We note at this point that our definition of the neutrino and antineutrino is in line with the concept of the conservation of leptons. Lepton is the generic name for all the fermions having a mass less than that of the proton. The class is divided into leptons and antileptons; for example,

$$\text{leptons:} \quad e^-, \nu_e \quad \text{with lepton number } +1,$$

$$\text{antileptons:} \quad e^+, \bar{\nu}_e \quad \text{with lepton number } -1.$$

The law of the conservation of leptons states that the number of leptons, less the number of antileptons, is a constant.

In fact, there are also muons and two types of neutrinos, one associated with beta decay $(\nu_e, \bar{\nu}_e)$, the other associated with muon weak interactions $(\nu_\mu, \bar{\nu}_\mu)$. The law of lepton conservation becomes two laws, the first requiring the conservation of electron lepton number, and the second requiring the conservation of muon lepton number. We will discuss this situation in Section 11.8.

The mass of the electron neutrino has been established to be less than 60 eV (Bergkvist, 1969) and that of the muon neutrino less than 3.5 MeV (Barkas, 1956). However, all work on weak interactions assumes that these masses are zero. In addition, all evidence is consistent with spin $\frac{1}{2}$ for all neutrinos.

The simplest interaction Lagrangian density for four fermions interacting at a point that is invariant under proper Lorentz transformations and contains no derivatives of the field operators is

$$\mathscr{L}_i = \sum_r C_r (\bar{\psi}_p O_r \psi_n)(\bar{\psi}_e O_r' \psi_\nu) + \text{Hermitian conjugate}. \tag{11.6}$$

The C_r are constants that determine the strength of the interaction. The

O_r, O_r' are operators whose form we shall examine shortly. The remaining quantities are field operators that have the property of creating or destroying particles, thus:

ϕ_ν destroys a neutrino or creates an antineutrino,
$\bar{\phi}_e$ creates an electron or destroys a positron,
ϕ_n destroys a neutron or creates an antineutron,
$\bar{\phi}_p$ creates a proton or destroys an antiproton.

ϕ_ν refers, obviously, to the electron neutrino field as it occurs associated with ϕ_e. We shall not make the electron neutrino and muon neutrino distinction unless it is necessary to do so for clarity. Thus these four terms taken together will form a term in an S-matrix, which has matrix elements between a state consisting of neutron plus neutrino and a state consisting of proton plus electron, and which indicates that a transition from the first to the second state is possible; that is,

$$n + \nu \to p + e^-,$$

which is just reaction (11.1). This last statement can be checked by using the alternative property ϕ_ν has of creating antineutrinos. It is evident that such a term also has matrix elements between the states (antiproton plus positron) and (antineutron plus antineutrino); i.e., for the reactions

$$\bar{p} + e^+ \to \bar{n} + \bar{\nu} \quad \text{or} \quad \bar{p} \to \bar{n} + e^- + \bar{\nu}.$$

The Hermitian conjugate term is of the form

$$\sum_r \epsilon_r C_r (\bar{\phi}_n O_r^\dagger \phi_p)(\bar{\phi}_\nu O_r'^\dagger \phi_e), \tag{11.7}$$

where ϵ_r is ± 1, depending upon O_r and O_r' and

ϕ_e destroys an electron or creates a positron,
$\bar{\phi}_\nu$ creates a neutrino or destroys an antineutrino,
ϕ_p destroys a proton or creates an antiproton,
$\bar{\phi}_n$ creates a neutron or destroys an antineutron.

This term has matrix elements for the reactions

$$p \to e^+ + n + \nu,$$
$$\bar{n} \to \bar{p} + e^+ + \nu.$$

It is now obvious that the interaction of Eq. (11.6) covers all the reactions (11.1)–(11.5).

There are certain symmetry restrictions placed upon the operators O_r. At present our knowledge of β-decay is such that we believe that the interaction is invariant under proper Lorentz transformations, and, since it is

11.2 The Description and Theory of Beta Decay

local, under the *TCP* transformation; we know that it is not invariant under *C* or *P*. Let us examine the restrictions upon the O_r due to Lorentz invariance. The Lagrangian density must be scalar (under proper Lorentz transformations); the bilinear forms $(\bar{\psi}O_r\psi)$ in Eq. (11.6) are scalars, four-vectors, axial vectors, tensors, or pseudoscalars according to the operator O_r, and to make a scalar out of two such covariants requires that $O_r = O_r'$ and that a sum be performed over indices. In Fermi's original theory of the β-decay (Fermi, 1934) the O_r were chosen so that the covariants were four-vectors. We need not make such a restriction, and we will consider all the independent forms of O_r. We do not, however, consider interactions containing derivatives of the covariants because such interactions give obvious contradictions with experimental observation. The five independent O_r are

$$O_S = 1 \qquad \text{scalar,}$$
$$O_V = \gamma^\rho \qquad \text{vector,}$$
$$O_A = \gamma^\rho\gamma_5 \qquad \text{axial vector,} \qquad (11.8)$$
$$O_P = i\gamma_5 \qquad \text{pseudoscalar,}$$
$$O_T = (i/2)(\gamma^\rho\gamma^\sigma - \gamma^\sigma\gamma^\rho) \qquad \text{tensor.}$$

The covariants $\bar{\psi}O_r\psi$ transform as shown opposite each operator. In using the symbol O for one of there operators, we shall not distinguish between a covariant or contravariant form. Thus Fermi's original proposal for the interaction was

$$\mathscr{L}_1 = C_V(\bar{\psi}_p\gamma^\rho\psi_n)(\bar{\psi}_e\gamma_\rho\psi_\nu),$$

where, as always, a sum is implied over repeated indices, ρ in this case. This term is the scalar product of two four-vectors.

We know that the β-decay interaction must contain a part that is parity nonconserving; this, in turn, implies that there are terms in $\mathscr{L}_1$ that are pseudoscalar. We can construct a further five terms having this property by having $O_r' = O_r\gamma_5$. Thus for every term

$$C_r(\bar{\psi}_pO_r\psi_n)(\bar{\psi}_eO_r\psi_\nu) + \text{h.c.}$$

we can add

$$C_r'(\bar{\psi}_pO_r\psi_n)(\bar{\psi}_eO_r\gamma_5\psi_\nu) + \text{h.c.}[1]$$

These extra terms are pseudoscalar and have matrix elements between states of opposite parity. Notice that we have followed convention and put the extra γ_5 into the lepton covariant $(\bar{\psi}_eO_r\gamma_5\psi_\nu)$; the interaction is specified by the transformation properties of the nucleon covariant $(\bar{\psi}_pO_r\psi_n)$, which

---

[1] Where h.c. stands for Hermitian conjugate.

is scalar, vector, etc., as $O_r = O_S, O_V, \ldots$, etc. The complete interaction Lagrangian now reads

$$\mathscr{L}_1 = \sum_r \left(C_r(\bar{\phi}_p O_r \phi_n)(\bar{\phi}_e O_r \phi_\nu) + C_r'(\bar{\phi}_p O_r \phi_n)(\bar{\phi}_e O_r \gamma_5 \phi_\nu) \right) + \text{h.c.}$$
(11.9)

We can now state the values of ϵ_r of Eq. (11.7), which can be used to write out the Hermitian conjugate explicitly if required:

$\epsilon_r = 1$ if $O_r = O_r'$ i.e., for all the parity-conserving terms,

$\epsilon_r = 1$ if $O_r' = O_r \gamma_5$ and $r = $ V, A,

$\epsilon_r = -1$ if $O_r' = O_r \gamma_5$ and $r = $ S, T, or P.

Before proceeding, let us examine the effects on $\mathscr{L}_1$ if it is invariant under charge conjugation, time reversal, and the parity transformation separately.

P invariance requires $\mathscr{L}_1$ to be scalar, and in Eq. (11.9) all C_r' would have to be zero. T invariance requires $\mathscr{L}_1$ to be real, just as it required H to be real (Section 6.5); hence all the C_r and C_r' would have to be real, apart from a common phase factor, which has no effect. C invariance has the result that the parity-conserving matrix elements and the parity-nonconserving matrix elements have relative amplitudes that are imaginary, if there is no interaction in the final state. There is no such interaction in a first-order calculation with this simple Lagrangian; hence, if the C_r are real, the C_r' are imaginary, and vice versa. We do not impose these conditions, but keep them in mind so that we can observe the effect of any postulated invariance.

The object of all experimental investigation in β-decay is to confirm, or otherwise, this theory of β-decay and to ascertain the values of the coupling constants C_r and C_r'.

This theory is not renormalizable; however, the lowest order of an S-matrix perturbation expansion gives results that appear to agree closely with observation if suitable values of the coupling constants are employed. At present there is no solution to the problem of the divergences that remain in higher orders.

11.3 The Classification of Beta Decays

We will now briefly consider the evaluation of matrix elements of the interaction Lagrangian and show how this leads to a classification of the transitions by the changes in parity and angular momentum between the initial and final nucleus and by the degree of forbiddenness.

11.3 The Classification of Beta Decays

The first term of a perturbation expansion is found to give meaningful results; this term gives the transition rate ω in the form [Eq. (3.74)]

$$d\omega = 2\pi \sum |\langle\beta|H_I|\alpha\rangle|^2 \, d(dn/dE_\beta), \qquad (11.10)$$

where $\langle\beta|$ and $|\alpha\rangle$ are the final- and initial-state vectors and H_I is the interaction Hamiltonian. A sum is to be taken over all unobserved final particle spins and an average over all initial spin states. As shown, the transition rate is a differential quantity for the rate into an interval around the configuration given by β. The complete transition rate is found by an integration over all the independent dynamic variables, electron energy and direction, and neutrino direction. Källen (1964) and Okun (1965) describe in detail the evaluation of β-decay and other weak-interaction differential rates.

The matrix element $\langle\beta|H_I|\alpha\rangle$ includes integration over all the space variables of the particles involved; the outgoing leptons have wave functions of a simple form—plane wave for the neutrino, a Coulomb function (Blatt and Weisskopf, 1952) for the electron (or positron)—and their integration over space is in principle straightforward. In contrast to this, the integration over the nucleon wavefunctions can be a very complex matter, except in the case of free baryon decay and of some of the simpler nuclear β-decays. All other cases of β-decay take place in complex nuclei, and the integration requires a knowledge of nuclear structure; this problem occurs with varying degrees of difficulty according to the nucleus. Often the reverse procedure is applied: a knowledge of the properties of a given β-decay serves to give the nuclear matrix element and information on the structure of the nuclei involved.

The nuclear matrix element contains a factor $e^{-i\mathbf{p}\cdot\mathbf{r}}$, where $\mathbf{p}$ is the momentum of the recoiling nucleus. In performing the space integration of the nucleon wavefunctions, it is convenient to expand this exponential as a power series in $\mathbf{p}\cdot\mathbf{r}$. Each term in the expansion contributes to the matrix element an amount that is about $1/100$ of the previous term, so that only the first few terms have to be considered. In addition, each term imposes certain selection rules upon the nuclear transition. If the selection rules are such that a given transition allows the first term to give a contribution, the transition is said to be allowed; if the selection rules are such that the first nonzero contribution is that of $(\mathbf{p}\cdot\mathbf{r})^n$, the transition is said to be a forbidden transition of order n. In Table 11.1 we have listed the selection rules for the first two orders. These rules concern the change in parity and the change in angular momentum ($\Delta j = |j_\beta - j_\alpha|$) from the initial nucleus to the final nucleus and are derived from a nonrelativistic treatment of the nuclear matrix elements, an approximation justified for the low values of $\mathbf{p}$ found in ordinary β-decay. Thus a nucleus (Z, A) that

TABLE 11.1

CLASSIFICATION OF NUCLEAR β-DECAY

Order of forbiddenness	Change of parity	Change in angular momentum, Δj	Interaction
Allowed	No	0	S or V
	No	0 or 1 but not $0 \to 0$	T or A
First forbidden	Yes	1	S
	Yes	1	V
	Yes	0, 1, 2	T
	Yes	0, 1, 2	A
	Yes	0	P

is energetically able to undergo β-decay to the nucleus $(Z \pm 1, A)$ will do so by the lowest-order transition allowed by the required change in parity and nuclear angular momentum. If the lowest choice is anything but allowed, the number of interactions that can be involved is large and it becomes difficult to interpret the experimental data. In addition, the contribution of any pseudoscalar interaction is almost certainly very small (Konopinski, 1954), and we can therefore confine our discussion to the S, T, A, and V interactions in allowed transitions.

Fermi's original theory was based upon the V interaction alone, for which $\Delta j = 0$; however, there are many $\Delta j = 1$ transitions, indicating that there must also be some T and/or A interaction present. It is usual to classify the transitions into "Fermi" (S and V) and "Gamow–Teller" (T and A) after the authors who suggested these interactions (Fermi, 1934; Gamow and Teller, 1936). The nuclear parts of the matrix element are represented by M_F or M_{GT}; we can make this generalization because in the nonrelativistic limit, which applies to the nuclear matrix element, $|M_F|^2$ is independent of whether the interaction is S or V, and $|M_{GT}|^2$ is independent of whether the interaction is T or A (Källen, 1964).

11.4 Beta Decay: Pre-1956

We can now examine the type of measurements made before 1956 in an effort to determine what interactions are present and their strengths. These experiments were all interpreted under the assumption that the interaction was invariant under P, C, and T separately. The first experimental data concern the energy spectrum of β-particles (electrons or positrons) emitted from a radioactive source. Using the interaction of Eq. (11.9),

11.4 Beta Decay: Pre-1956

with all $C_r' = 0$, we expect the spectrum to be given by

$$N_\pm(E)\,dE = \underbrace{(dE/2\pi^3\hbar^7c^5)pE(E_0-E)^2}_{1}\,\underbrace{F(\mp Z,E)}_{2}$$
$$\times \underbrace{\{|M_F|^2|C_F|^2 + |M_{GT}|^2|C_{GT}|^2}_{}$$
$$\underbrace{\pm (m_ec^2/E)[1-(Z/137)^2]^{1/2}(b_F - b_{GT})\}}_{3}.\qquad(11.11)$$

$N_\pm(E)\,dE$ is the probability that a β-particle is emitted in one second with total energy in the range E to $E + dE$. E and p are the total energy and momentum of the emitted β-particle (upper sign for positron, lower for electron). E_0 is the maximum permissible energy of the β-particle, m_e is the electron mass, (1) is the density of states factor and other numerical factors, (2) is the factor that indicates the effect of the Coulomb distortion of the outgoing electron wave (as shown, it is a function of E and of the nuclear charge Z, of the final nucleus), (3) is the summed square of the matrix element, having M_F and M_{GT}, as already discussed, and

$$|C_F|^2 = |C_S|^2 + |C_V|^2, \qquad (11.12)$$
$$|C_{GT}|^2 = |C_T|^2 + |C_A|^2, \qquad (11.13)$$
$$b_F = 2\,\text{Re}\,C_S C_V^* |M_F|^2, \qquad (11.14)$$
$$b_{GT} = 2\,\text{Re}\,C_T C_A^* |M_{GT}|^2. \qquad (11.15)$$

We have written these formulas as if the C_r could be imaginary. This is not so if T invariance holds. The Fierz interference terms are b_F and b_{GT}; they are due to interference between the S and V interactions in Fermi transitions and between T and A in Gamow–Teller transitions.

A convenient way of plotting the spectrum is by the "Kurie plot." From Eq. (11.11) we have that

$$\left(\frac{N_\pm(E)}{pEF(\mp Z,E)}\right)^{1/2} = K(E_0 - E), \qquad (11.16)$$

where K contains the nuclear matrix elements and the Fierz interference terms. Graphs of the left-hand side of Eq (11.16) against E for allowed transitions yield straight lines indicating, within the experimental accuracy, that the nuclear matrix elements are constant (as expected) and that the Fierz interference terms are absent (Gerhart, 1958; Allen et al. 1955; Sherr and Miller, 1954).

It follows that the quantities $C_S C_V^*$ and $C_T C_A^*$ are purely imaginary or zero; they cannot be imaginary if T invariance holds, and we find that they must be zero. Thus the Fermi interaction is S or V but not both, and the Gamow–Teller interaction is T or A but not both.

We can integrate Eq. (11.11) to obtain the total transition rate $\omega = (\ln 2)/t$, where t is the half-life of the decaying nucleus:

$$\omega = (|C_F|^2 |M_F|^2 + |C_{GT}|^2 |M_{GT}|^2) \frac{fm_e^5 c^4}{2\pi^3 \hbar^3},$$

where

$$f = (1/m_e^5 c^9) \int_{E=m_e c^2}^{E=E_0} pE(E_0 - E)^2 F(\mp Z, E) \, dE$$

is dimensionless. This gives

$$ft = (2\pi^3 \hbar^7 \ln 2 / m_e^5 c^4)(|C_F|^2 |M_F|^2 + |C_{GT}|^2 |M_{GT}|^2)^{-1}. \quad (11.17)$$

The values of $\log_{10} ft$ fall very roughly into two groups; favored transitions have $\log_{10} ft$ about 2.9–3.7 and are transitions between corresponding states in mirror nuclei or between even–even and odd–odd nuclei; unfavored transitions have $\log_{10} ft$ values of about 5–6. Forbidden transitions have $\log_{10} ft$ values that are even higher.

The value of $|C_F|^2$ can be found from the ft values for $0^+ \to 0^+$ β-decays. Wu (1964) has reviewed the situation and discussed the corrections that have to be made; the important effects are due to nuclear size, screening by the atomic electrons, radiative corrections, and deviations from pure isotopic-spin states of the nuclear levels involved. The value found is (Freeman et al., 1964)

$$|C_F| = \left. \begin{array}{c} 1.4029 \pm 0.0022 \\ \text{or} \\ 1.3943 \pm 0.0017 \end{array} \right\} \times 10^{-49} \text{ erg cm}^3,$$

depending on a choice of screening correction.

The value of $|C_{GT}|$ can be found from an analysis of decays that are mixed Fermi and Gamow–Teller transitions. Using the value of $|C_F|$ found from the pure Fermi transitions, the value

$$\frac{|C_{GT}|}{|C_F|} = 1.231 \pm 0.010$$

is obtained (Particle Data Group, 1969).

We have given very recent experimental results of quantities known less accurately in 1956. At that time it was not known which interactions were contributing to the Fermi or to the Gamow–Teller transitions. In principle, electron–neutrino correlation experiments would have given the answer, but at that time these were insufficiently accurate to give decisive results.

11.5 Beta Decay: Post-1956

The nonconservation of parity was discovered late in 1956, and its experimental manifestation was first observed in the β-decay of ^{60}Co (Wu et al., 1957). This discovery entails a complete reexamination of the theory of β-decay. We relax the requirement that the Lagrangian be invariant under P and it becomes the full expression given in Eq. (11.9). In addition, Eqs. (11.12) to (11.15) become

$$|C_F|^2 = |C_V|^2 + |C_V'|^2 + |C_S|^2 + |C_S'|^2, \quad (11.18)$$

$$|C_{GT}|^2 = |C_A|^2 + |C_A'|^2 + |C_T|^2 + |C_T'|^2, \quad (11.19)$$

$$b_F = 2\,\mathrm{Re}(C_S C_V^* + C_S' C_V'^*)|M_F|^2, \quad (11.20)$$

$$b_{GT} = 2\,\mathrm{Re}(C_A C_T^* + C_A' C_T'^*)|M_{GT}|^2. \quad (11.21)$$

Equation (11.11) remains unchanged if the notation of Eqs. (11.18)–(11.21) replaces that of Eqs. (11.12)–(11.15). The subscripts S, V, A, T now refer to the Lorentz properties of the nuclear covariant. The absence of Fierz interference terms is no longer strong evidence that the Fermi interaction is S or V; b_F could easily be close to zero without requiring that $C_S = C_S' = 0$ or that $C_V = C_V' = 0$. The same argument applies to any conclusions regarding the Gamow–Teller interaction.

We can now examine some of the physically observable quantities associated with parity nonconservation. Let us consider the asymmetry in the emission of β-particles from polarized radioactive nuclei. The relevant factor in the angular distribution of β-particles is

$$\left\{1 + \frac{ac}{E}\langle\boldsymbol{\sigma}\rangle\cdot\mathbf{p}\right\},$$

where E is the electron total energy, $\mathbf{p}$ is the electron momentum, and $\langle\boldsymbol{\sigma}\rangle$ is the polarization of the decaying nucleus; the asymmetry factor a is given by

$$a = \frac{\mathrm{Re}[\mp|M_{GT}|^2\lambda_{jj'}(C_T^*C_T' - C_A^*C_A') - 2\delta_{jj'}\{j/(j+1)\}^{1/2} \times |M_F||M_{GT}|(C_T^*C_S' + C_T'^*C_S + C_A^*C_V' + C_A'^*C_V)]}{|M_F|^2|C_F|^2 + |M_{GT}|^2|C_{GT}|^2} \quad (11.22)$$

The sign is plus for positrons and minus for electrons, and we have neglected some small terms depending upon Z, the nuclear charge. The quantity $\lambda_{jj'}$ depends upon the nuclear angular momentum and its change

$$\begin{aligned}
\lambda_{jj'} &= +1 & \text{for} \quad & j \to j' = j - 1, \\
&= 1/(j+1) & & j \to j' = j, \\
&= -j/(j+1) & & j \to j' = j + 1,
\end{aligned}$$

where j is the angular momentum quantum number. We observe that if parity were conserved all $C_r' = 0$ and $a = 0$. Alternatively, if the interaction were charge-conjugation invariant, all C_r would be real and all C_r' would be imaginary, in which case $a = 0$. Thus the observation of an asymmetry indicates that the interaction is not invariant under C or P. The observations made upon ^{60}Co indicated that $a = -1$ for this transition, which is a pure Gamow–Teller with $j = 5$ and $j' = 4$. From Eq. (11.22) we see that this result is consistent with either

$$C_T = C_T' \quad \text{and} \quad C_A = C_A' = 0,$$

or

$$C_A = -C_A' \quad \text{and} \quad C_T = C_T' = 0.$$

The nonconservation of parity allows the emitted β-particles to be longitudinally polarized; that is, the decay rate depends upon $\boldsymbol{\sigma} \cdot \mathbf{p}$, where $\boldsymbol{\sigma}$ is the polarization vector of the β-particle and $\mathbf{p}$ is its momentum

$$\frac{\boldsymbol{\sigma} \cdot \mathbf{p}}{|\mathbf{p}|} = \mp \frac{v}{c} \frac{2 \operatorname{Re}[|M_F|^2 (C_S C_S'^* - C_V C_V'^*) + |M_{GT}|^2 (C_T C_T'^* - C_A C_A'^*)]}{|M_F|^2 |C_F|^2 + |M_{GT}|^2 |C_{GT}|^2},$$

v is the velocity of the β-particle. The sign is again plus for a positron and minus for an electron. The experimental results on the longitudinal polarization of electrons emitted in β-decay indicate that it is $-v/c$. The negative sign indicates that the spin is directed against the momentum (Deutsch et al., 1958; Alder et al., 1957). This result requires that either

$C_r = -C_r'$ and only V and A interactions are acting,

$C_r = +C_r'$ and only T and S interactions are acting.

Both these choices agree with the ^{60}Co asymmetry and the absence of Fierz interference terms.

The neutrinos or antineutrinos emitted in β-decay can also be longitudinally polarized. By considering Eqs. (11.6) and (11.7) and comparing the positions of the particle creation and destruction operators, we can expect that a neutrino [due to (11.7)] will be emitted with the same direction of polarization as an electron [due to (11.6)], unless $\epsilon_r = -1$. It follows that the neutrino has the same polarization direction as an electron if the interaction is V or A but opposite if the interaction is S or T. Thus a measurement of the polarization of neutrinos emitted from a β-active material that is emitting positrons or capturing K-electrons serves to decide between the V and A interaction and the S and T interaction. This measurement has been made in a most elegant manner by Goldhaber and collaborators (1958). The results indicate that the polarization directions are indeed the same. All the experimental results are therefore consistent with

$$C_V = -C_V' \quad \text{and} \quad C_A = -C_A',$$
$$C_S = C_S' = C_T = C_T' = C_P = C_P' = 0.$$

The nonzero coupling constants can be complex, but one of the phases involved may be removed as it is unobservable. Let us put C_V real; then C_A is also real if the interaction is invariant under time reversal. A measurement of the decay asymmetry for polarized neutrons determines, in principle, the relative phase between C_V and C_A and thus tests time-reversal invariance. Burgy and collaborators (1960) find that the experimental results are consistent with a relative phase between C_V and C_A of 180°. This is now assumed to be the case, and we have

$$C_A = -1.231 C_V$$

and, since $|C_F|^2 = (|C_V|^2 + |C_V'|^2)^{1/2}$ with $C_V' = C_V$,

$$\sqrt{2}\,|C_V| = |C_F| = g_V.$$

The β-decay interaction Lagrangian density is now written in the form in which it conventionally appears[2]

$$\mathscr{L}_I = \sqrt{\tfrac{1}{2}}\,(\bar{\phi}_p \gamma^\rho (g_V - g_A \gamma_5) \phi_n)(\bar{\phi}_e \gamma_\rho (1 - \gamma_5) \phi_\nu) + \text{h.c.}, \quad (11.23)$$

where $g_V/\sqrt{2} = C_V$ and $g_A/\sqrt{2} = |C_A|$ and $g_V = 1.4 \times 10^{-49}$ erg cm^3 (see Section 11.4). It appears that g_V and g_A owe their inequality to the effect of strong interactions of the nucleons involved in β-decay. In the absence of these interactions it is expected that $g_V = g_A$, and we have what is called the V–A interaction. We shall have more to say on the values of g_V and g_A in Sections 11.10 and 11.13.

11.6 The Two-Component Theory of the Neutrino

So far we have discussed the theory of β-decay as if the neutrino were a particle described by the Dirac equation as is the electron. Such a description requires a four-component wavefunction; however, if the neutrino is massless, it is possible to simplify the Dirac equation and describe the neutrino by a two-component wavefunction (Lee and Yang, 1957; Salam, 1957; Landau, 1957). This theory is not invariant under the C and P transformations separately.

Let us consider the Dirac equation in its covariant form [Eq. (10.21)]:

$$\left[i\gamma^\rho \frac{\partial}{\partial x^\rho} - \frac{mc}{\hbar} \right] \phi_\nu = 0.$$

---

[2] In many books the V–A theory appears with $1 + \gamma_5$ instead of $1 - \gamma_5$ in the currents. The difference is due to the choice of metric and is of no physical significance.

If this applies to a massless neutrino, it reduces to

$$\gamma^\rho \frac{\partial \psi_\nu}{\partial x^\rho} = 0$$

or

$$\gamma^0 \frac{\partial \psi_\nu}{\partial x^0} = -\gamma^l \frac{\partial \psi_\nu}{\partial x^l}. \tag{11.24}$$

In the original Dirac theory (Section 10.2) the matrices employed were α_l and β, from which the γ matrices are defined thus [Eqs. (10.18) and (10.19)]:

$$\gamma^l = \beta \alpha_l, \qquad \gamma^0 = \beta.$$

Thus Eq. (11.24) can be written

$$\frac{\partial \psi_\nu}{\partial x^0} = -\alpha_l \frac{\partial \psi_\nu}{\partial x^l}$$

or, in a more easily interpreted form,

$$i\hbar \frac{\partial \psi_\nu}{\partial t} = -i\hbar c \alpha_l \frac{\partial \psi_\nu}{\partial x^l}. \tag{11.25}$$

The α-matrices also have the following property:

$$\alpha_l = \gamma_5 \sigma_l = \sigma_l \gamma_5, \tag{11.26}$$

where σ_l is a 4×4 matrix formed from the Pauli matrix σ_l (Section 2.9): thus

$$\sigma_l = \begin{bmatrix} & & 0 & 0 \\ \sigma_l & & & \\ & & 0 & 0 \\ 0 & 0 & & \\ & & \sigma_l & \\ 0 & 0 & & \end{bmatrix}.$$

Substitute either of the equations implied in Eq. (11.26) into Eq. (11.25) and we find that

$$i\hbar \frac{\partial}{\partial t} \gamma_5 \psi_\nu = -i\hbar c \sigma_l \frac{\partial \psi_\nu}{\partial x^l}, \tag{11.27}$$

or

$$i\hbar \frac{\partial \psi_\nu}{\partial t} = -i\hbar c \sigma_l \frac{\partial}{\partial x^l} \gamma_5 \psi_\nu. \tag{11.28}$$

11.6 The Two-Component Theory of the Neutrino

By adding Eqs. (11.27) and (11.28) we obtain

$$i\hbar \frac{\partial}{\partial t}(1 + \gamma_5)\phi_\nu = -i\hbar c\sigma_l \frac{\partial}{\partial x^l}(1 + \gamma_5)\phi_\nu, \qquad (11.29)$$

and by subtracting we obtain

$$i\hbar \frac{\partial}{\partial t}(1 - \gamma_5)\phi_\nu = i\hbar c\sigma_l \frac{\partial}{\partial x^l}(1 - \gamma_5)\phi_\nu. \qquad (11.30)$$

If $(1 + \gamma_5)\phi_\nu$ and $(1 - \gamma_5)\phi_\nu$ represent plane waves, they are eigenfunctions of energy and momentum and we can write these equations:

$$E = +\boldsymbol{\sigma}\cdot\mathbf{p}c \quad \text{for the function} \quad (1 + \gamma_5)\phi_\nu,$$
$$E = -\boldsymbol{\sigma}\cdot\mathbf{p}c \quad \text{for the function} \quad (1 - \gamma_5)\phi_\nu.$$

These are massless particles, hence $E = pc$. Let us consider the $(1 - \gamma_5)\phi_\nu$ particle

$$p = -\boldsymbol{\sigma}\cdot\mathbf{p}.$$

If we call the positive energy state of this particle the neutrino, we observe that the neutrino spin vector must be directed antiparallel to the momentum. If we call the negative energy state the antineutrino, it follows that the antineutrino has its spin directed parallel to its momentum. The spin antiparallel corresponds to left-hand polarization for the photon and in this context is often called negative helicity; the spin-parallel case corresponds to right-hand polarization or positive helicity. In the language of field theory we find that

$(1 - \gamma_5)\phi_\nu$ destroys neutrinos with negative helicity or creates antineutrinos with positive helicity.

If we take the complex conjugate of Eq. (11.29), we find that

$\bar{\phi}_\nu(1 + \gamma_5)$ creates neutrinos with negative helicity or destroys antineutrinos with positive helicity.

We can make similar statements about the state function that obeys Eq. (11.30). If we now look at Eq. (11.23), we see that this β-interaction always creates negative-helicity neutrinos and positive-helicity antineutrinos. However, the interaction was suggested to be of the form of Eq. (11.23) in order that the theory agree with the experimental observation that neutrinos from β-decay have negative helicity. We have merely confirmed in this section that the $(1 - \gamma_5)$ operator is responsible for this.

To show that this is equivalent to describing the neutrinos by two component wavefunctions we choose the following representation of the γ-matrices.

$$\gamma^k = \begin{bmatrix} 0 & -\sigma_k \\ +\sigma_k & 0 \end{bmatrix}, \quad \gamma^0 = \begin{bmatrix} 0 & 1 \\ 1 & 0 \end{bmatrix}, \quad \gamma_5 = \begin{bmatrix} 1 & 0 \\ 0 & -1 \end{bmatrix}$$

$(k = 1, 2, 3)$.

Each matrix element is a 2×2 matrix, and the σ_k are the Pauli spin matrices. We put plane-wave solutions of the Dirac equation in the form $u(p)\, e^{-ip \cdot x}$, where $u(p)$ is a four-component spinor, which we express in terms of two two-component spinors, v and w:

$$u(p) = \begin{bmatrix} v \\ w \end{bmatrix}.$$

Then Eq. (11.24) becomes

$$\sum_{k=1}^{3} +p_k \begin{bmatrix} 0 & -\sigma_k \\ +\sigma_k & 0 \end{bmatrix} u + E \begin{bmatrix} 0 & 1 \\ 1 & 0 \end{bmatrix} u = 0$$

or

$$\begin{bmatrix} 0 & (\mathbf{p} \cdot \boldsymbol{\sigma} + E) \\ -(\mathbf{p} \cdot \boldsymbol{\sigma} - E) & 0 \end{bmatrix} \begin{bmatrix} v \\ w \end{bmatrix} = 0,$$

so that each spinor separately satisfies a matrix equation thus:

$$(\mathbf{p} \cdot \boldsymbol{\sigma} + E)w = 0,$$
$$(\mathbf{p} \cdot \boldsymbol{\sigma} - E)v = 0.$$

If we use w, then neutrinos have negative helicity and antineutrinos have positive helicity. If we use v, then the helicities are reversed. β-decay neutrinos and antineutrinos are of the w type. The operator $\frac{1}{2}(1 - \gamma_5)$ projects out from u the w component; i.e.,

$$\frac{1}{2}(1 - \gamma_5)\, u(p) = \begin{bmatrix} 0 \\ w \end{bmatrix},$$

and we can think of $(1 - \gamma_5)\varphi_\nu$ as the w-type of a two-component neutrino. Similarly $\frac{1}{2}(1 + \gamma_5)$ projects out the v-type.

It is interesting to ask why the interaction involves only one type of two-component neutrino. If we are willing to neglect the difference in the magnitudes of C_V and C_A, then there are several postulates that lead to V-A interaction and a two-component neutrino. For example, Marshak and Sudarshan (1958) postulated "chirality invariance," that is, invariance of the interaction Lagrangian under the transformation

$$\psi \to -\gamma_5 \psi \quad \text{and} \quad \bar{\psi} \to +\bar{\psi}\gamma_5$$

11.6 The Two-Component Theory of the Neutrino

of any one of the four fields involved. If we consider a covariant $(\bar{\psi}_2 O_r \psi_1)$ then this invariance requires

$$\gamma_5 O_r = -O_r \gamma_5 = O_r;$$

that is,

$$\{O_r, \gamma_5\} = 0.$$

Of the five possibilities [Eq. (11.8)] only V and A anticommute with γ_5, and hence

$$O_r = a\gamma^\rho + b\gamma^\rho \gamma_5.$$

But we require

$$-O_r \gamma_5 = O_r,$$

which gives

$$a = -b,$$
$$O_r = \gamma^\rho - \gamma^\rho \gamma_5,$$

leading to the V–A interaction with two-component neutrinos. The same result is obtained (Feynman and Gell-Mann, 1958) if one requires all the fields to appear as

$$(1 - \gamma_5)\psi \quad \text{and} \quad \bar{\psi}(1 + \gamma_5).$$

The Lagrangian density

$$\mathscr{L}_1 = \tfrac{1}{4} \sum_r C_r [\bar{\psi}_4(1 + \gamma_5) O_r (1 - \gamma_5)\psi_3][\bar{\psi}_2(1 + \gamma_5) O_r (1 - \gamma_5)\psi_1] + \text{h.c.}$$

will have nonzero contributions only from $r = $ V and A, which are the same because $\gamma_5(1 - \gamma_5) = -(1 - \gamma_5)$. Putting $O_r = \gamma^\rho$, γ_ρ gives Eq. (11.23) with $C_r = C_V = -C_A$.

We can show that the two-component theory can apply only to massless particles. Suppose we require that the neutrino always have negative helicity in all inertial frames. It follows that unless its velocity is that of light, we can always find a frame for which the helicity is opposite; for example, we would observe the momentum reversed but spin unchanged from a frame that is moving to overtake the neutrino. The neutrino mass must be zero if its velocity is to be that of light.

This theory is not invariant under the parity transformation or under charge conjugation separately. P changes the direction of the momentum of a neutrino, leaves the spin unchanged, and leaves the neutrino a neutrino; it therefore changes the allowed negative-helicity neutrino into a positive-helicity neutrino, which is not an allowed state. Similarly, C changes the allowed negative-helicity neutrino into a negative-helicity antineutrino, which is also not allowed. Thus the theory is not invariant under C or P;

however, the simultaneous operation CP connects allowed states, and the theory is invariant under the CP transformation (and consequently under time reversal).

There is a clear physical picture that shows immediately why the two-component theory leads to asymmetry in β-decay. Let us illustrate this by considering the decay of ^{60}Co:

$$^{60}\text{Co} \rightarrow {}^{60}\text{Ni} + e^- + \bar{\nu}.$$

This is an allowed Gamow-Teller reaction with $j \rightarrow j' = j - 1$, where $j = 5$ and $j' = 4$. We refer to Fig. 11.1, where we have indicated the

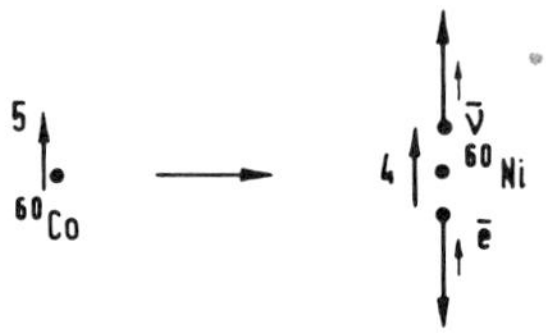

FIG. 11.1. Diagram showing a preferred configuration of spins and momenta in the β-decay of ^{60}Co. (The large arrows indicate momenta, the small arrows indicate spins. The unattached horizontal arrow indicates the direction of the decay reaction.)

angular-momentum vector of one ^{60}Co nucleus in a fully polarized specimen having $j_z = 5$. The final state consists of a ^{60}Ni nucleus with electron and antineutrino, which we have considered as ejected along the $+$ or $-z$ axis. If the antineutrino has positive helicity, the most likely configuration to conserve angular momentum will have the electron moving along the $-z$ axis with negative helicity (left-hand polarization) and the antineutrino moving along the $+z$ axis. It follows that the electrons ejected from a polarized specimen of ^{60}Co tend to emerge against the direction of the nuclear spin polarization. Next we consider an unpolarized specimen of ^{60}Co The observation of electrons in a particular direction picks out decays from nuclei that had nuclear spins preferentially aligned in the opposite direction; this gives the electrons a longitudinal polarization (in this case a negative helicity) that is in agreement with observation. Similarly, positron emitters give positrons that are longitudinally polarized with positive helicity.

These facts also follow from the properties of $(1 - \gamma_5)\psi_e$. We have seen that the V–A interaction is equivalent to all field operators appearing as $(1 - \gamma_5)\psi$ or $\bar{\psi}(1 + \gamma_5)$. As for neutrinos, this would cause the β-decay electrons or positrons to be 100% polarized with negative or positive helicity, respectively, if their mass were zero. In fact the nonzero mass leads to polarizations of magnitude v/c, v being the β-particle velocity.

In later parts of this chapter we shall use the label two-component neutrino to imply that the neutrino field always appears as $(1 - \gamma_5)\psi_\nu$ or $\bar{\psi}_\nu(1 + \gamma_5)$ in the Lagrangian interaction density, thus giving rise to neutrinos or antineutrinos with negative or positive helicity, respectively.

11.7 Conservation of Leptons in Beta Decay

So far in this chapter we have assumed conservation of lepton number, and it is worth examining the evidence for this. There are a few nuclei for which the following transition is energetically possible: two neutrons in the nucleus simultaneously undergo β-decay with the emission of two electrons but without emitting neutrinos:

$$n + n \to p + p + e^- + e^-.$$

For example,

$$^{124}\text{Sn}_{50} \to {}^{124}\text{Te}_{52} + 2e^-.$$

This process is called double β-decay and is supposed to proceed by the emission and reabsorption of a neutrino. The reactions are

$$n \to p + e^- + \bar{\nu},$$

followed by

$$n + \nu \to p + e^-.$$

These equations have been written maintaining the defined nature of the neutrinos involved. It is evident that reabsorption of the neutrino from the first reaction is possible only if neutrino and antineutrino are identical (Majorana theory: $\nu \equiv \bar{\nu}$). The nonobservation of double β-decay will be evidence that $\nu \neq \bar{\nu}$. At the present time the observations indicate that the double β-decay happens at a rate several orders of magnitude slower than that predicted by the Majorana theory, and the conclusion is that, as already defined, the neutrino is not the same as the antineutrino. This lends support to the hypothesis of lepton conservation. We notice that the two-component neutrino theory without lepton conservation would permit double β-decay. A full discussion of all the neutrino possibilities and of the double β-decay observations is given in the review by Primakoff and Rosen (1959).

The antineutrino-absorption experiment [Eq. (11.4)] yields a cross section that supports the two-component neutrino theory and lepton conservation (Reines and Cowan, 1959; Carter *et al.*, 1959). In addition, there is another neutrino-absorption experiment that supports lepton conservation: according to our lepton assignments fission products emit antineutrinos, which should be incapable of inducing the following neutrino absorption reaction:

$$^{37}\text{Cl}_{17} + \nu \to {}^{37}\text{A}_{18} + e^-.$$

Davies and Harmer (1959) have looked for but not observed this reaction caused by fission antineutrinos.

11.8 Muon and Pion Decay

Muons are produced by the decay of pions

$$\pi^+ \to \mu^+ + \nu_\mu, \tag{11.31}$$

$$\pi^- \to \mu^- + \bar{\nu}_\mu, \tag{11.32}$$

the mean pion life being 2.55×10^{-8} sec. Muons have a mean life of 2.20×10^{-6} sec and decay mainly by the mode

$$\mu^+ \to e^+ + \nu_e + \bar{\nu}_\mu, \tag{11.33}$$

$$\mu^- \to e^- + \bar{\nu}_e + \nu_\mu. \tag{11.34}$$

These lifetimes are consistent with these decays being due to weak interactions and, as we shall discuss, the interaction strength for muon decay is very nearly the same as g_V in β-decay.

The neutral particles in Eqs. (11.31)–(11.34) have been assumed to be neutrinos. This is justified by consistency and by the fact that conservation of energy, momentum, and angular momentum require them to be light, spin-$\frac{1}{2}$, neutral particles. The next important point is the neutrino labeling in Eqs. (11.31)–(11.34). The facts are that neutrinos and antineutrons from β-decay can cause inverse β-decay; for example, in the case of antineutrinos the sequence is

$$n \to p + e^- + \bar{\nu},$$
$$\bar{\nu} + p \to n + e^+,$$

with conservation of lepton number. However it is observed (Danby *et al.*, 1962) that the neutral particles from pion decay [Eqs. (11.31) and (11.32)] do not produce electrons or positrons but do produce muons, basically;

$$\nu_\mu + n \to p + \mu^-,$$
$$\bar{\nu}_\mu + p \to n + \mu^+.$$

Thus the neutrinos associated with muons are different from those associated with electrons. The law of lepton conservation now becomes two laws, the first of electron lepton conservation, the second of muon lepton conservation. The electron lepton quantum numbers have been given in Section 11.2. We shall assume the following muon lepton numbers:

μ^-, ν_μ: muon lepton number $+1$,

$\mu^+, \bar{\nu}_\mu$:   muon lepton number -1.

Equations (11.31)–(11.34) have been written so that both conservation laws

11.8 Muon and Pion Decay

are satisfied. These conservation laws explain the absence of decays and reactions that are otherwise expected to occur; for example,

$$\mu \to e + \gamma,$$
$$\mu^+ + e^- \to 2\gamma.$$

We shall now discuss the question of what is the interaction that is operating in muon decay. Muons brought to rest are found to decay asymmetrically with respect to their direction of motion, and this can happen only if parity is not conserved in muon decay and if the muons are polarized, which in turn can happen only if parity is not conserved in charged-pion decay.

Thus the μ-decay interaction is parity nonconserving and involves four fermion fields as in β-decay. We therefore use an interaction Lagrangian density that is the same in form as that in β-decay, thus:

$$\mathscr{L}_1 = \sum_r \{D_r(\bar{\phi}_\nu O_r \phi_\mu)(\bar{\phi}_e O_r \phi_\nu) + D_r'(\bar{\phi}_\nu O_r \phi_\mu)(\bar{\phi}_e O_r \gamma_5 \phi_\nu)\} + \text{h.c.} \quad (11.35)$$

Terms under the summation are responsible for μ^--decay, while the Hermitian conjugate terms are responsible for μ^+-decay. In subscripting the neutrino field operators we have not distinguished between muon and electron neutrinos: we mean that whenever ϕ_ν or $\bar{\phi}_\nu$ occurs the ν is of the kind that goes with the field in the same covariant; for example, $(\bar{\phi}_\nu \gamma^\rho \phi_\mu)$ means that the ϕ_ν field is the muon neutrino field. The interaction then conserves both lepton types.

We have the usual possible choices for the operators O_r, although we can expect that corresponding to the β-decay interaction to be correct, namely $D_V = -D_V'$ and $D_A = -D_A'$:

$$\mathscr{L}_1 = [\bar{\phi}_\nu \gamma^\rho (D_V + D_A \gamma_5) \phi_\mu][\bar{\phi}_e \gamma_\rho (1 - \gamma_5) \phi_\nu] + \text{h.c.} \quad (11.36)$$

We maintain this possibility (a) and the possibility of an unknown combination of interactions (b) and proceed to compare the predictions of the theory with observation.

The spectrum of electrons (or positrons) emitted from free muon decay, integrated over all directions, is given by

$$N(x)\,dx = \frac{12x^2\,dx}{\tau}\left\{1 - x + \frac{2}{9}\rho(4x - 3)\right\},$$

where $x = E/E_0$, the electron energy divided by the maximum possible electron energy, τ is the mean life of muon, and $N(x)\,dx$ is the number of electrons emitted per second in the energy range $x \to x + dx$. The

Michel parameter ρ (Michel, 1950) has the following value:

(a) $\rho = \frac{3}{4}$ or (b) $0 \leqslant \rho \leqslant 1$.

Cronin (1968), reviewing the weak-interaction data, gave

$$\rho = 0.752 \pm 0.003.$$

The second quantity of interest is the decay asymmetry parameter averaged over all electron energies:

$$N(\Omega)\, d\Omega = (1/\tau)(d\Omega/4\pi)\big(1 \pm |\sigma_\mu|(\xi/3) \cos \theta\big).$$

$N(\Omega)\, d\Omega$ is the number of electrons ($-$) or positrons ($+$) emitted per second into a solid angle $d\Omega$ about a direction Ω that makes an angle θ with the direction of muon polarization σ_μ. Theory predicts

(a) $\xi = 2 \operatorname{Re} D_V^* D_A / (|D_V|^2 + |D_A|^2)$, that is, $0 \leqslant |\xi| \leqslant 1$, or

(b) $0 \leqslant |\xi| \leqslant 3 - \frac{8}{3}\rho$.

The best value (Cronin, 1968) is $|\xi| = 0.973 \pm 0.014$. The helicity of positrons (electrons) from decay of an unpolarized source of muons is $-\xi$ ($+\xi$). The measurements (for example, Duclos et al., 1964) give $-\xi$ as positive and close to 1. Therefore we take $\xi = -1$ and hence $D_A = -D_V$.

The third quantity of interest is the parameter δ, which describes the variation of decay asymmetry with energy:

$$N(x, \Omega)\, dx\, d\Omega = (3x^2\, dx\, d\Omega/\tau)\{1 - x + (2\rho/9)(4x - 3) \\ \mp (\xi/3) \cos \theta (1 - x + 2\delta[4x - 3])\}.$$

This applies to fully polarized muons. The predicted values are

(a) $\delta = \frac{3}{4}$,

(b) $|\xi\delta| \leqslant \rho$.

The best value is $\delta = 0.752 \pm 0.009$ (Cronin, 1968).

It is evident that all the data are consistent with the V–A interaction, and it is aesthetically very acceptable that the form of the interaction in β-decay is the same as in muon decay. Assuming that this is the interaction, the mean life will be given by

$$\tau = 192\pi^3 \hbar^7 / g_\mu^2 m_\mu^5 c^4,$$

where $g_\mu^2 = |D_V|^2 + |D_A|^2 = 2|D_V|^2$. Using the most accurate value of $\tau = (2.198 \pm 0.001) \times 10^{-6}$ sec (Farley et al., 1962) and making the radiative corrections (Berman and Sirlin, 1962) it is found that (Freeman et al., 1964)

$$g_\mu = (1.4350 \pm 0.0003) \times 10^{-49} \text{ erg cm}^3.$$

11.8 Muon and Pion Decay

The interaction Lagrangian is therefore

$$\mathscr{L}_I = (g_\mu/\sqrt{2})(\bar{\phi}_\nu \gamma^\rho (1 - \gamma_5)\phi_\mu)(\bar{\phi}_e \gamma_\rho (1 - \gamma_5)\phi_\nu) + \text{h.c.} \quad (11.36a)$$

The remarkable equality between the coupling constant g_μ for muon decay and the g_V of β-decay (1.4029 or 1.3943 × 10^{-49} erg cm³, depending on screening correction) is to be noted. In fact, they are so close that the assumption is often made that they are basically equal and that the observed difference is due to the incomplete evaluation of the corrections that must be made to allow for nuclear effects etc., in β-decay (see Section 11.4). However, there is another suggestion for the reason for the discrepancy based upon the use of the unitary symmetry, which we shall discuss in Section 11.13. Nonetheless, the near equality of g_V and g_μ gives rise to the postulate of the conserved vector current, which we discuss in the next section.

The interaction Lagrangian density of Eq. (11.36a) is very similar to that of β-decay [Eq. (11.23)] if $g_V = g_A$ and is thus a manifestation of the V-A interaction of four fermions. If we rearrange the muon covariant thus,

$$(\bar{\phi}_\nu \gamma^\rho (1 - \gamma_5)\phi_\mu) = (\bar{\phi}_\nu (1 + \gamma_5)\gamma^\rho \phi_\mu),$$

we see that the muon neutrinos in this interaction have the same two-component properties as do the electron neutrinos. If this is the case for all muon neutrinos, then we expect the μ^+ from the decay of the spin-0 π^+ to have helicity the same as the muon neutrino, which is the other decay product; that is, negative helicity (Fig. 11.4). Likewise the μ^- from π^- should have the helicity of the muon antineutrino, which is positive. Several measurements have been made of muon helicity (Alichanov et al., 1960; Backenstoss et al., 1961; Bardon et al., 1961), and although the errors are large, the results are consistent with this assignment.

There is a further class of muon interactions, which is basically

$$\mu^- + p \rightarrow n + \nu.$$

This corresponds to K-capture of orbital electrons in β-decay. We shall discuss this process of μ-capture in more detail in Section 11.11. The experiments are insufficiently precise to permit an independent determination of the interaction type and strength, and at the present it is only possible to show that the results are consistant with a four-fermion interaction that is the same as in β-decay (Eq. 11.23) with the electron covariant $(\bar{\phi}_e \gamma^\rho (1 - \gamma_5)\phi_\nu)$ replaced by a muon covariant $(\bar{\phi}_\mu \gamma^\rho (1 - \gamma_5)\phi_\nu)$. The values of the coupling constants corresponding to g_V and g_A are certainly close to the values they have in β-decay.

From the evidence on β-decay and muon interactions a picture is emerging of a universal weak interaction that involves nucleons and muons

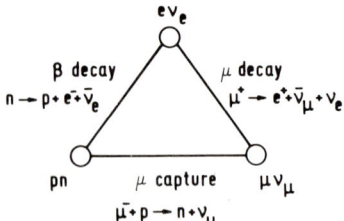

Fig. 11.2. The Puppi triangle, showing the three interactions that occur in weak interactions of nonstrange particles.

and electrons with their respective neutrinos. This is illustrated by the Puppi triangle (Fig. 11.2). The pair of particles at any vertex can interact with the pair of particles at either of the other two vertices with the same strength. The sides of the triangles in the figure are labeled with the resultant reaction type. An important aspect of this interaction is μ-e universality; all the evidence so far discussed is in favor of invariance under the following changes:

$$e \rightleftarrows \mu, \qquad \nu_e \rightleftarrows \nu_\mu, \qquad \bar{\nu}_e \rightleftarrows \bar{\nu}_\mu, \qquad m_e \rightleftarrows m_\mu.$$

This means that whenever an interaction occurs with $(\bar{\psi}_e \gamma^\rho (1 - \gamma_5) \psi_\nu)$, or its Hermitian conjugate, an equally valid interaction, having the same strength, will be one with the electron covariant replaced by a muon covariant, $(\bar{\psi}_\mu \gamma^\rho (1 - \gamma_5) \psi_\nu)$, or its Hermitian conjugate. Other evidence for μ-e universality will be given in this and later sections.

A universal weak interaction can cause other decays apart from those discussed so far in this chapter. For example, charged-pion decay [Eqs. (11.31) and (11.32)] can occur by the weak interaction in an intermediate state consisting of a virtual nucleon–antinucleon pair into which the pion can dissociate. The strong interactions leading to the nucleon pair cannot be calculated with present techniques, and instead of drawing Feynman diagrams showing routes for decay, we factor out the strong part and put it into a box that is entered by the π and indicate the weak part of the interaction as acting at a point on the surface of the box. At this point a four-fermion interaction is supposed to take place; for example, in π^--decay a virtual neutron and antiproton might be destroyed and a negative muon and a muon antineutrino created, although the diagram shows only the observed particles (Fig. 11.3). The effective part of the weak-interaction Lagrangian is presumably

$$\mathscr{L}_\mathrm{I} = (\bar{\psi}_p \gamma^\rho (g_\mathrm{V} - g_\mathrm{A} \gamma_5) \psi_n)(\bar{\psi}_\mu \gamma_\rho (1 - \gamma_5) \psi_\nu)$$
$$= g(J_\mathrm{V}{}^\rho + J_\mathrm{A}{}^\rho)^\dagger (M_{\mathrm{V}\rho} + M_{\mathrm{A}\rho}),$$

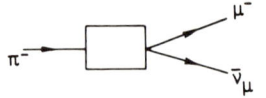

Fig. 11.3. Diagam for pion decay. The box contains all strong interaction effects.

11.8 Muon and Pion Decay

which has matrix elements $\langle \mu^- \nu | \mathscr{L}_1 | \pi^- \rangle$ between initial and final states. J_V and J_A are vector and axial-vector parts of the nucleon covariant operator, which we will now call "currents," and $M_V + M_A$ are the muon "currents." The matrix element can be factored thus:

$$\langle \mu^- \nu | \mathscr{L}_1 | \pi^- \rangle = \langle \mu^- \nu | M_V{}^\rho + M_A{}^\rho | 0 \rangle \langle 0 | J_{V\rho} + J_{A\rho} | \pi^- \rangle,$$

where $|0\rangle$ represents a vacuum state. This factoring is a symbolic separation of calculable and incalculable parts and can be performed because the lepton pair ($\mu^- \nu$) is produced at one space-time point, the consequence of a local theory. However, the part containing the J-currents must have certain symmetry properties. The whole matrix element is scalar or pseudoscalar, and since the first part is the component of a vector plus an axial vector, the second part must be the same. However, this second part must be a function of observable quantities, and the only independent one available is the four-vector of momentum and energy of the pion, p. So $\langle 0 | J_V{}^\rho + J_A{}^\rho | \pi^- \rangle$ is the component of a vector, and since the parity of the vacuum is even and of the pion odd, only the axial-vector part of the current has matrix elements between these states. Thus

$$\langle 0 | J_V{}^\rho + J_A{}^\rho | \pi^- \rangle = \langle 0 | J_A{}^\rho | \pi^- \rangle = p^\rho f(p^2).$$

f must be a scalar and therefore can only be a function of the sole independent scalar, namely $p^2 = +m_\pi^2$. The calculations of the free π^- decay rate will give a result containing $f^2(p^2)$ (see Section 11.11).

There is a rare alternative decay mode for the pion, thus

$$\pi^+ \to e^+ + \nu_e,$$
$$\pi^- \to e^- + \bar{\nu}_e.$$

Assuming μ–e universality, the calculation of the rate for this decay proceeds as for the muon decay with the replacement of muon leptons by electron leptons. The matrix element for J is exactly the same in the two cases, and although $f(p^2)$ cannot be calculated, it can be eliminated and the branching ratio calculated. The result is

$$\frac{R(\pi \to e\nu)}{R(\pi \to \mu\nu)} = 1.28 \times 10^{-4}.$$

The observed ratio is $(1.25 \pm 0.03) \times 10^{-4}$ (Di Capua et al., 1964). This is excellent evidence for μ–e universality.

It is possible to step back and make a more general choice of the four-fermion interaction. There are the usual possible choices, S, V, A, T, and P, for the structure of nucleon current, and arguments of the kind given above show that only A and P can contribute. We have given the result for A. If it were a P interaction, the branching ratio would be 5.5. This

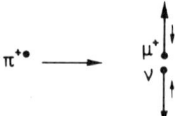

FIG. 11.4. Diagram showing the configuration of spins and momenta in π^+-decay. (The large arrows indicate momenta, and the small arrows indicate spins. The unattached horizontal arrow shows the direction of the decay reaction.)

is wrong by nearly five orders of magnitude and confirms our neglect of this interaction.

Källen (1964) has given a full discussion of the factoring of the matrix elements and the calculation of these decay rates and branching ratios.

At first sight it is remarkable that the electron decay is so rare. However, given two-component neutrinos we see that π^+-decay, for example, must lead to μ^+ or e^+ with negative helicity (see Fig. 11.4). The weak interaction prefers to create μ^+ and e^+ with positive helicity. The μ^+, with its greater mass and its relatively low energy in this decay, is much easier to create in this unpreferred state than is the very relativistic e^+.

11.9 The Universal Weak Interaction

In the last section we developed a picture of a universal weak interaction involving muons and electrons, with their respective neutrinos, and nucleons. The interaction is of the current–current type, where we have the choice of three so-called currents and their Hermitian conjugates:

$$M^\rho = (\bar{\phi}_\mu \gamma^\rho (1 - \gamma_5) \phi_\nu) = (\bar{\mu}\nu),$$
$$L^\rho = (\bar{\phi}_e \gamma^\rho (1 - \gamma_5) \phi_\nu) = (\bar{e}\nu),$$
$$J^\rho = (\bar{\phi}_n \gamma^\rho (1 - R\gamma_5) \phi_p) = (\bar{n}p),$$

where $R = |g_A/g_V|$. In the next section we shall see that it is likely that g_A is not equal to g_V because of strong interaction effects. Thus the basic nucleon current J^ρ may have $R = 1$, and it is the matrix elements of the axial-vector and vector parts at low momentum transfers that stand in the ratio R.

We have chosen our currents $(\bar{a}b)$ such that the particle created, a, has charge one unit less than that destroyed, b. The Hermitian conjugate induces the opposite change in charge. Evidently the interactions we have dealt with are such that

β-decay: $\mathscr{L}_I = (g_V/\sqrt{2})\{J^{\rho\dagger}L_\rho + \text{h.c.}\}$,

μ-decay: $\mathscr{L}_I = (g_\mu/\sqrt{2})\{M^{\rho\dagger}L_\rho + \text{h.c.}\}$,

μ-capture: $\mathscr{L}_I = (g/\sqrt{2})\{M^{\rho\dagger}J_\rho + \text{h.c.}\}$.

It is probable that g should be g_V (see Section 11.11). However, for the moment we will neglect the difference between the coupling constants and assume that there is one common to all currents. If we put

$$\mathscr{J}^\rho = J^\rho + M^\rho + L^\rho,$$

then the Lagrangian

$$\mathscr{L}_1 = \frac{g}{\sqrt{2}} \mathscr{J}^{\rho\dagger} \mathscr{J}_\rho$$

includes all the interactions as well as others not so far discussed. These are of the kind $M^\dagger M$, $L^\dagger L$, and $J^\dagger J$, which evidently describe muon–neutrino scattering, electron–neutrino scattering, and weak nucleon–nucleon scattering. The first two have not been observed, but the last will give rise to parity-nonconserving effects in nuclei that are just about detectable and of the expected magnitude [see data reviewed by Hamilton (1969)].

In latter sections of this chapter we shall extend the content of the current $\mathscr{J}$ to include a current that changes strangeness.

The currents described above all have $\Delta Q = -1$, while their Hermitian conjugates have $\Delta Q = +1$, Q being the electric charge. There appears to be no evidence for "neutral" currents, i.e., $\Delta Q = 0$. Such currents would cause, for example, the following decays:

$$K^0 \to \mu^+ + \mu^- \quad \text{or} \quad e^+ + e^-,$$

which have not been observed.

11.10 The Conserved Vector Current

In the discussion of previous sections we have found that the Fermi (g_V) coupling constant and the Gamow–Teller (g_A) coupling constant differ, and this is attributed to the differing effect of the strong interactions of the nucleons involved. In the language of field theory, we expect the coupling constant for the coupling of the basic nucleon vector and axial-vector currents to lepton currents to have a certain value that is renormalized by the virtual processes the nucleons undergo because of their strong interaction with the pion field. The renormalized, observed values of the vector and axial-vector coupling constants are not expected to be the same. It is fairly easy to see why renormalization is necessary by considering the free-neutron β-decay. An e^- and $\bar{\nu}$ are created and a neutron becomes a proton: however, a free neutron spends some of the time as a π^- and proton, and this virtual proton cannot undergo a decay that creates the e^-. Thus the time a neutron is in a position to β-decay is reduced by its strong interaction, and we expect the effective β-decay coupling to be reduced from the value for a nonstrongly interacting neutron. It is therefore surprising to find that g_V is almost equal to g_μ, the coupling constant in muon decay where there are no strongly interacting particles. It appears that the strong interactions

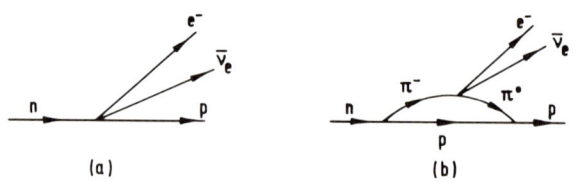

FIG. 11.5. Feynman diagrams for neutron decay, (a) directly and (b) indirectly via the virtual pion current interacting with the lepton current.

of nucleons have no effect on the coupling constant for Fermi transitions.

The coupling constant for the interaction between charged particles and the electromagnetic field is also unaffected by the presence of strong interactions. The proton charge is identical to the positron charge, although the former interacts strongly with other particles, and the latter does not. The reason for this lies in the fact that the electromagnetic vector potential is coupled to a conserved vector current. The result is that charge is conserved so long as we take into account all the fields that interact with the electromagnetic field. The physical reason is that although a proton spends a part of the time as a neutron and pion, its effective charge is not reduced because the pion is coupled equally to the electromagnetic field. This precedent suggests that the weak coupling of the vector nucleon current is unchanged because the virtual pions associated with it have a current that is equally coupled, with the result that neutron β-decay can take place even when the neutron is a virtual pion plus proton. See Fig. 11.5.

Let us examine the conserved-current idea. The electric current for the proton field is a four-vector

$$j^\rho = e\bar{\psi}_p \gamma^\rho \psi_p,$$

where e is the electric charge. We can generalize this to the nucleon field ψ_N by using the isotopic-spin operator T_3 (Section 5 4):

$$j^\rho = e\bar{\psi}_N(\tfrac{1}{2} + T_3)\gamma^\rho \psi_N,$$
$$= \frac{e}{2}\bar{\psi}_N \gamma^\rho \psi_N + e\bar{\psi}_N \gamma^\rho T_3 \psi_N = j_s^\rho + j_{v3}^\rho.$$

So far as rotations in isotopic-spin space are concerned, the first term transforms like a scalar and the second like the third component of a vector in isotopic-spin space; that is, they are isoscalar and isovector, respectively. In a theory that contains only nucleons, this equation defines the currents. However, other particles exist, and the isoscalar part of the current contains terms due to the isoscalar mesons such as ω, ϕ, etc., and to isoscalar bilinear combinations of the baryon fields. If the appropriate terms are

11.10 The Conserved Vector Current

added to give the complete isoscalar part of the current, $j_s{}^\rho$, it satisfies the the continuity equation

$$\partial_\rho j_s{}^\rho \equiv \partial j_s{}^\rho / \partial x^\rho = 0.$$

This four-dimensional divergence implies conservation of baryons. Similarly, isovector terms due to other fields must be added to make the complete isovector current which independently satisfies the same continuity equation, thus implying electric-charge conservation.

The weak-interaction vector current of nucleons has the form

$$J_V{}^\rho = (\bar{\psi}_p \gamma^\rho \psi_n) = (\bar{\psi}_N \gamma^\rho T_+ \psi_n),$$

where T_+ is the isotopic-spin operator, which raises the third component by one unit. Since T_+, T_-, and T_3 are the three components of an isovector, there is the possibility of a relation between J_V and j_V.

It was suggested by Feynman and Gell-Mann (1958) that to the weak current J_V must be added isovector currents due to other fields and that the resultant total current J_V was proportional to the current j_V. The consequences of this hypothesis are:

(1) The current J_V has zero divergence; that is,

$$\partial_\rho J_V{}^\rho = 0.$$

This continuity equation, by analogy with conservation of charge, gives rise to the name "conserved vector current" (CVC) hypothesis.

(2) The matrix elements of J_V are directly proportional to the matrix elements of the T_+ component (j_{V+}) of the isotopic-spin current, whose third component is j_{V3}.

(3) The transitions satisfy the selection rules for $\Delta t = 1$.

We must now look at the physical consequences of the CVC hypothesis. We first consider β-decay and μ-capture in which nucleons are involved in the weak interactions. The current $J_V + J_A$ induces transitions between physical states of the nucleons. The matrix elements of J_V namely $\langle 2 | J_V | 1 \rangle$, will be proportional to $\langle 2 | j_{V+} | 1 \rangle$. Now, the matrix element of the electric current between the two nucleon states is

$$\langle 2 | j_s{}^\rho + j_{V3}^\rho | 1 \rangle = \bar{u}(p_2) \{ e \gamma^\rho (\tfrac{1}{2} + T_3)$$
$$+ i(e/2M) \sigma^{\rho \lambda} q_\lambda [\mu_p(\tfrac{1}{2} + T_3) + \mu_n(\tfrac{1}{2} - T_3)] \}$$
$$\times u(p_1) e^{ip \cdot x}, \qquad (11.37)$$

where $\sigma^{\rho\lambda} = (i/2)(\gamma^\rho \gamma^\lambda - \gamma^\lambda \gamma^\rho)$, $\bar{u}(p_1)$ and $u(p_2)$ are eight-component spinors, that is spinors with two components in isotopic-spin space for each component in ordinary space. μ_p and μ_n are the anomalous nucleon magnetic moments, and M is the nucleon mass. q is the four-momentum transfer to the nucleon in the $1 \rightarrow 2$ transition. The first term is the conventional current contribution to the interaction of a Dirac particle with the electromagnetic field, the second is the contribution of the anomalous magnetic moment. This second part occurs because the nucleon interaction with meson fields gives rise to the anomalous moments, and therefore the observed values of the anomalous moments must be used in the physical matrix elements.

It is easy to pick out from Eq. (11.37) the part due to j_{V3} and by analogy construct the matrix element of $J_V{}^\rho$; thus,

$$\langle 2 | J_V{}^\rho | 1 \rangle = \bar{u}(p_2) \{\gamma^\rho T_+ + (i/2M)\sigma^{\rho\lambda}q_\lambda(\mu_p - \mu_n)T_+\} u(p_1) e^{iq\cdot x}.$$

This matrix element is coupled to the matrix element of the weak lepton currents with the coupling constant g_V. This coupling constant will be modified at large momentum transfers (q) in a manner we shall discuss later. We can now examine the consequences at low momentum transfer. The first term is a conventional vector term, but the second term corresponds to the anomalous nucleon magnetic moments and is called "weak magnetism" (Gell-Mann, 1958). It modifies the β-decay spectrum for $1^+ \rightarrow 0^+$ transitions by a factor, the important part of which is $(1 + \tfrac{8}{3}aE)$, where

$$a = \pm |g_V/g_A| [(1 + \delta)/2M]$$

(plus for β^- and minus for β^+). Such a term will always be present, but the CVC hypothesis prescribes the value of δ to be $\mu_p - \mu_n$. This term has been detected (Lee et al., 1963) in the β-decays $^{12}\text{B} \rightarrow {}^{12}\text{C} + e^-$ and $^{12}\text{N} \rightarrow {}^{12}\text{C} + e^+$ and the value of δ found to be correct within the experimental error of 15%.

Another consequence of the CVC hypothesis concerns the decay

$$\pi^+ \rightarrow \pi^0 + e^+ + \nu_e.$$

This is always expected to occur, since the β-decay interaction can occur in the virtual nucleon-plus-antinucleon states accessible from the π^+ state, but the rate cannot be calculated because of the usual difficulties of dealing with strong interactions. However, the CVC hypothesis implies that the vector current of the pion field itself is coupled to the lepton current with the same strength as is the nucleon vector current, so that direct β-decay of charged pions is possible. Consequently, CVC permits us to calculate, without any other assumptions, the rate for this decay. We can see that the axial vector, current J_A makes no contribution; since the matrix elements $\langle \pi^0 | J_V | \pi^+ \rangle$, $\langle \pi^0 | J_A | \pi^+ \rangle$ have the transformation properties of vector and axial vector respectively, and since it is impossible to make an axial vector

11.10 The Conserved Vector Current

from the physical vectors available, we must have $\langle \pi^0 | J_A | \pi^+ \rangle = 0$. The result is that the decay depends on g_V, and the predicted rate is 0.41 sec^{-1}, which is a branching ratio of

$$\frac{R(\pi^+ \to \pi^0 e^+ \nu)}{R(\pi^+ \to \mu^+ \nu)} = 1.07 \times 10^{-8}$$

(Lee and Wu, 1965). The average of the experimental results for this ratio is $(1.023 \pm 0.069) \times 10^{-8}$ (Particle Data Group, 1969). These tests and others involving nuclear β-decay, all of which give results in agreement with the CVC hypothesis, have been discussed in detail in a review article by Wu (1964). The evidence for the CVC hypothesis is strong, although it should be recognized that the hypothesis is making strong quantitative predictions about processes that will happen in any case and that full and complete confirmation will come only with precise quantitative agreement between the theory and experiment.

The CVC hypothesis also makes predictions about the effect of higher-momentum transfers in the case of the nucleon vector current. To discuss this we can consider a high-energy neutrino interaction:

$$\bar{\nu}_\mu + p \to \mu^+ + n.$$

As described so far the theory-predicts cross sections for this type of reaction that increase as the square of the center-of-mass momentum and that risk violating unitarity at sufficiently high energies (Section 11.16). In addition we expect that nucleon structure effects will begin to limit the cross section at energies of the order of a few GeV (Lee and Yang, 1960a). This can be described by the use of form factors, as in the case of electron–nucleon scattering (Section 13.6), and these factors will be functions of any variable Lorentz scalars. If we assume locality, there is only one independent variable of this kind, and it is conveniently chosen to be the square of the four-momentum (q) transferred to the nucleon between the initial (1) and final state (2). That is, $q = p_2 - p_1$, where p_2 and p_1 are the four-momenta of the states 2 and 1, respectively: as usual q^2 is an invariant. The matrix elements of J_V and J_A will therefore be functions of q^2 and, if we include the coupling constants, will have the form

$$\langle 2 | J_V^\rho | 1 \rangle = \sqrt{\tfrac{1}{2}}\, \bar{u}(p_2) [g_V(q^2) \gamma^\rho + i g_M(q^2) \sigma^{\rho\lambda} q_\lambda] u(p_1) e^{iq \cdot x}, \quad (11.38)$$

$$\langle 2 | J_A^\rho | 1 \rangle = \sqrt{\tfrac{1}{2}}\, \bar{u}(p_2) [g_A(q^2) \gamma^\rho \gamma_5 + g_P(q^2) \gamma_5 q^\rho] u(p_1) e^{iq \cdot x}. \quad (11.39)$$

Other terms can appear, but they have the wrong G-parity and are usually assumed to be zero (the second-class currents: Weinberg, 1958). The coupling constants are now functions of q^2. Equations (11.38) and (11.39) are in fact general enough to be the matrix elements between any two baryon states, but to apply the CVC hypothesis we limit the discussion to transitions between the nucleon states. In β-decay q^2 is very small and we have

$$g_V(0) = g_V \quad \text{and} \quad g_A(0) = g_A$$

in the notation used before. We note that it is possible that in the absence of strong interactions $g_A(0)$ could be equal to $g_V(0)$, and in this case the basic weakly interacting nucleon current

$$J_V{}^\rho + J_A{}^\rho = (g_V/\sqrt{2})\bar{\psi}\gamma^\rho(1-\gamma_5)\psi .$$

The CVC hypothesis requires

$$g_M(0) = g_V(0)[(\mu_p - \mu_n)/2M] \tag{11.40}$$

and the variation of g_V and g_M with q^2 to be the same as that of the corresponding nucleon electromagnetic isovector form factor (Section 13.6). To confirm such variations will require extensive neutrino interaction studies, although experiments so far done have given results roughly consistent with such a variation (Perkins, 1969).

The question of the properties of the axial-vector current naturally follows any discussion of the vector current. The inequality of g_u and g_A presumably shows that this current is not conserved. However, the hypothesis of a partially conserved axial-vector current (PCAC) appears to have some validity. This states that although the matrix elements $\langle 2|\partial_\rho V_A{}^\rho|1\rangle$ are not zero, they approach zero with increasing momentum transfer sufficiently quickly for any element to satisfy an unsubtracted dispersion relation (Dalitz, 1964; Bernstein et al., 1960). These relations give the matrix elements of $\partial_\rho J_A{}^\rho$ as functions of momentum transfer in the form of intergrals over intermediate states of various mass. These intermediate states have well-defined properties, which we can see by discussing Fig. 11.6. The nucleon axial-vector current can interact directly with a lepton current (Fig. 11.6a) or via some intermediate strongly interacting mesonic state M, so that the weak interaction is now between the lepton current and the mesonic field (Fig. 11.6b). Since parity is conserved between the nucleon current and the intermediate state, the interaction at the strong vertex will have to be between a nucleon axial-vector current and

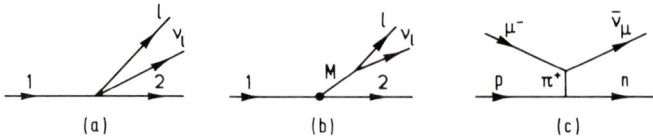

FIG. 11.6. Direct and intermediate states in baryon leptonic weak interactions.

11.10 The Conserved Vector Current

a pseudoscalar or axial-vector field. The field must also have isotopic spin 1 and odd G-parity. This last fact follows from noticing that this intermediate state is equivalent to a nucleon–antinucleon pair in a $J^P = 0^-$ or 1^+, $t = 1$ state, which has $G = -1$ (Section 6.8). The lowest-mass state with these quantum numbers is the pion, and the next lowest is one of three pions. For momentum transfers of interest the lowest-mass intermediate state is expected to be the most important in the dispersion relation, and since this is the pion, the diagram of Fig. 11.6b with M a pion shows that this contribution corresponds to a nucleon weak interaction actually taking place as the interaction of an emitted pion and the lepton current. Looking at Fig. 11.6c, which shows a contribution to μ^--absorption by a proton, we can see more clearly that this kind of effect is connected with the well-known decay of the pion. It is just this kind effect that is the cause of the term $g_P(q^2)\gamma_5 q^\sigma$ in Eq. (11.39); that is, the "induced pseudoscalar coupling." It can be calculated in terms of the amplitude for finding a virtual pion in the nucleon (i.e., the pion–nucleon coupling constant) and the known pion decay amplitude; the result is an effective coupling constant. For μ-capture the magnitude of the induced pseudoscalar coupling is about seven times g_A/m_μ. The effects of any induced pseudoscalar coupling in ordinary β-decay are too small to be detected.

Let us look at the form the dispersion relation argument takes: we have (Appendix D)

$$\left\langle 2 \left| \frac{\partial J_A^\rho}{\partial x^\rho} \right| 1 \right\rangle = \frac{i}{\sqrt{2}} \bar{u}(p_2) \gamma_5 u(p_1)$$
$$\times [(m_1 + m_2) g_A(q^2) + q^2 g_P(q^2)] e^{iq\cdot x}. \quad (11.41)$$

The PCAC hypothesis allows a dispersion relation to be written down for the second factor on the right-hand side, which has a pole term at $q^2 = m_\pi^2$ and a branch cut starting at $q^2 = (3m_\pi)^2$. The residue of the pole contains factors corresponding to the amplitude for the emission of a real charged pion ($\sqrt{2}\,G$) and to the amplitude for the decay of the pion into lepton plus neutrino (F). Taking the case of $m_1 = m_2 = M$, the nucleon mass

$$\frac{1}{\sqrt{2}}[2Mg_A(q^2) + q^2 g_P(q^2)] = \frac{+\sqrt{2}\,GFm_\pi^2}{m_\pi^2 - q^2} + \int_{(3m_\pi)^2}^\infty \frac{f(\sigma^2)\,d\sigma^2}{\sigma^2 - q^2}$$
$$= \frac{+\sqrt{2}\,GFm_\pi^2}{m_\pi^2 - q^2} \phi(q^2). \quad (11.42)$$

F is essentially proportional to the strong matrix element multiplied by the leptonic weak current in pion decay and is given by the decay rate; thus,

$$R(\pi^+ \to l^+\nu) = (F^2/4\pi)(m_l^2/m_\pi^3)(m_\pi^2 - m_l^2)^2, \quad (11.43)$$

with units $\hbar = c = 1$. From the discussion of Section 11.8 we can expect F to be independent of whether the leptons are $\mu\nu$ or $e\nu$. G is the pion-nucleon coupling constant ($G^2/4\pi = 14.5$).

In nuclear β-decay $q^2 \simeq 0$ and the nearest contribution to the dispersion integral is from the pole term: the higher-mass contribution should be small, so we put $\phi(0) = 1 + a$ with a expected to be small and obtain

$$Mg_A(0) = GF(1 + a). \tag{11.44}$$

This is the Goldberger–Treiman relation (1958) if $a = 0$. If we use units with $\hbar = c = 1$, then $g_A = 1.23\ g_V$, where $g_V = 1.5 \times 10^{-5}\,\text{GeV}^{-2}$. The width of the π^+-decay is $2.53 \times 10^{-17}\,\text{GeV}$, which gives $F = 1.06 \times 10^{-6}$ GeV^{-1}. The left-hand side of Eq. (11.44) is $1.33 \times 10^{-5}\,\text{GeV}^{-1}$ and the right-hand side is $1.43 \times 10^{-5}\,\text{GeV}^{-1}$, if $a = 0$. This is very satisfactory agreement, considering that this neglects the dispersion integral of Eq. (11.42). Thus the Goldberger–Treiman relation connects the weak axial-vector coupling constant with the strong interaction of pions.

If we put $a = 0$, Eq. (11.42) becomes

$$2Mg_A(q^2) + q^2 g_P(q^2) = +\frac{2M g_A(0) m_\pi^2}{m_\pi^2 - q^2}. \tag{11.45}$$

We want to apply this to find g_P for the μ-capture reaction

$$\mu^- + p \to n + \nu_\mu. \tag{11.46}$$

The momentum transfer is

$$q^2 = -m_\mu^2[1 + (m_\mu/M)]^{-1}$$
$$\simeq -0.9\, m_\mu^2. \tag{11.47}$$

If $g_A(q^2)$ varies with q^2 as does $g_V(q^2)$, which is expected, by CVC, to have the same variation as the electromagnetic isovector nucleon form factors, g_A in μ-capture will differ from $g_A(0)$ by less than 3%. Neglecting this difference, Eq. (11.45) gives

$$m_\mu g_P(q^2) = \frac{2Mm_\mu g_A(0)}{m_\pi^2 + 0.9\, m_\mu^2}$$
$$= 6.7\, g_A(0), \tag{11.48}$$

so that the axial-vector matrix element in μ-capture is

$$\langle 2|J_A{}^\rho|1\rangle = \sqrt{\tfrac{1}{2}}\, g_A(0)\, \bar{u}(p_2)[\gamma^\rho\gamma_5 + (6.7/m_\mu)q^\rho\gamma_5]u(p_1) e^{iq\cdot x}, \tag{11.49}$$

Weisberger (1965) and Adler (1965), using current commutation relations, have attempted to calculate the renormalization of the axial-vector coupling constant of β-decay. They obtained a sum rule involving neutrino cross sections. To evaluate the rule, it was necessary to use PCAC ideas and to use pion total cross sections. The different approximations lead to values of g_A/g_V that are 1.16 and 1.24, respectively, to be compared with the observed value of 1.23.

11.11 Muon Capture

We have already mentioned the muon-capture process, which is basically

$$\mu^- + p \to n + \nu_\mu.$$

This process is of interest because it gives information on the weak coupling constants when muons and nucleons are involved and because the induced pseudoscalar interaction can have a detectable effect. Muon absorption in hydrogen is difficult to observe directly because only one in two thousand stopping in hydrogen are absorbed in this way; the remainder decay. In addition, in liquid hydrogen 83% of the absorption takes place from a μ-molecule, so that the interpretation has uncertainties due to incomplete knowledge of the molecular physics of the initial state. The absorption rate has been measured, but this result alone is insufficient to find the coupling constants However, the absorption rate can be calculated if the following assumptions are made:

(1) e-μ universality holds, so that g_V and g_A in muon absorption have, apart from small form-factor effects, the same value as in β-decay,

(2) The CVC hypothesis is correct, so that the weak-magnetism term is well-defined.

(3) The dispersion-relation result for g_P [Eq. (11.48)] holds, so that $m_\mu g_P = 6.7 g_A$ at

$$q^2 = -m_\mu^2 \left(1 + \frac{m_\mu}{M}\right)^{-1},$$

as appropriate to the reaction.

There are several steps leading to absorption. In liquid hydrogen the first important occurrence after the μ^- has stopped is the formation of a $\mu^- p$ atom. There is then competition between absorption of the μ^- (17%) and transition to a $\mu^- pp$ molecule followed by absorption (83%). Thus there are two absorption rates, and what is observed is a combination of the two that depends upon the period of observation after stopping. The experimental results for absorption in liquid hydrogen are consistent with the predictions (Sens, 1966). Absorption in pure gaseous hydrogen is almost entirely from the singlet state of the $\mu^- p$ atom: the predicted rate is 626 sec^{-1} and that observed is 640 ± 70 sec^{-1} (Albergi Quaranta et al., 1967).

Muon capture by complex nuclei is complicated by the uncertainties in the nuclear effects. However, by choice of the absorption nuclear transitions, various coupling constants can be picked out. The evidence was reviewed by Ericson (1964) and, within the experimental error, is in accord with values of g_V and g_A in μ-capture being the same as in β-decay. The value of g_P can also be derived from absorption experiments and in particular from radiative absorption, since it is possible to reach $q^2 = m_\mu^2$, which is very much nearer the dispersion relation pole at $q^2 = m_\pi^2$ than is usual in μ-capture. The value of $m_\mu g_P = 6.7 g_A$ for proton absorption is not expected to hold: this induced coupling depends upon the presence of a cloud of virtual pions, and the structure of this cloud is likely to change in going from a free nucleon to a nucleon bound in nuclear matter; it follows that the induced effects will change just as the anomalous magnetic moments are changed. The change, however, seems to be very large: radiative capture in calcium (Conversi et al., 1964) gives for the magnitudes

$$m_\mu g_P = (13.3 \pm 2.7) g_A$$

and in copper (Chu et al., 1965) the factor is 20 ± 4.

At this point it is worth raising again the question of μ–e universality. All the evidence is in agreement with this idea, so that at our present state of knowledge the muon appears to be a heavy electron in all respects. Attempts have been made to detect muon properties that might indicate why it is so much heavier than the electron or whether it has some unexpected interaction. Measurements of the anomalous part of the magnetic moment do not disagree significantly with the predictions of quantum electrodynamics (Section 10.9), and muon–proton scattering (for example, Cool et al., 1965) has not yet revealed any unusual features. The reason for the existence of the muon is therefore a complete mystery.

11.12 The Leptonic Decays of Strange Particles

All the kaons, the strange baryons belonging to the same octet as the nucleons, and the Ω^--baryon have lifetimes that indicate that weak interactions are responsible for their decay. These decays and their different modes fall into two classes. The first we call leptonic decays because these have a lepton pair, $\mu \nu_\mu$ or $e \nu_e$, as or among the decay products. The second class, which will be discussed in Section 11.14, have no leptons among the decay products. It is important to note that parity is not conserved in any of these decays, which follows from observations of the following kinds: the τ–θ puzzle (Section 6.4), the longitudinal polarization of muons from $K_{\mu 2}$ decay (Coombes et al., 1957), and the decay asymmetry for polarized Λ hyperons (Crawford et al., 1958).

In discussing the leptonic decays we will frequently have no need to

11.12 The Leptonic Decays of Strange Particles

distinguish between a muon lepton pair ($\mu\nu_\mu$) and an electron lepton pair ($e\nu_e$) so we will use $l^-\bar{\nu}$ (charge conjugate $l^+\nu$) to cover both, and both types can be expected to appear if energy permits. The observed decay modes of interest in the context of this section and the next are

$$K_{l2}: \quad K \to l + \nu,$$
$$K_{l3}: \quad K \to \pi + l + \nu,$$
$$K_{l4}: \quad K \to \pi + \pi + l + \nu,$$
$$\Lambda: \quad \Lambda \to p + e^- + \bar{\nu},$$
$$\Sigma: \quad \Sigma^- \to n + l^- + \bar{\nu},$$
$$\Sigma: \quad \Sigma \to \Lambda + e + \nu,$$
$$\Xi: \quad \Xi^- \to \Lambda + l^- + \bar{\nu},$$
$$\Xi: \quad \Xi \to \Sigma + l + \nu.$$

In this list, charge signs have been omitted if there is a choice.

The leptonic modes have some very important properties, which are deduced from the observation of these modes and the absence of modes that might otherwise be expected. We shall discuss those decays having $|\Delta S| \geqslant 1$. Those having $\Delta S = 0$ have properties similar to those of nucleon β-decay and will be included in the discussion of baryon decays in Section 11.13. The $\Delta S \neq 0$ properties are

(1) The change in hypercharge and strangeness between the initial and final hadrons is 1 ($|\Delta y| = |\Delta S| = 1$). The possibility of $|\Delta S| = 2$ has, of course, been investigated. That an interaction leading to such a change is not present with comparable strength to that leading to $|\Delta S| = 1$ is established by the absence (Bingham, 1965) of any observed events of the kind

$$\Xi^- \to n + l^- + \bar{\nu}$$

and by the observed K_1, K_2 mass difference (Section 14.3).

(2) The changes in strangeness (ΔS) and ordinary charge (ΔQ) between the initial and final hadrons satisfy

$$\Delta S = \Delta Q.$$

Evidence that this is valid rests mainly upon nonobservation of the following decays [see Cronin (1968), Aubert (1969), Rubbia (1969), for discussion and references]:

$$\Sigma^+ \to n + l^+ + \nu,$$
$$K^+ \to \pi^+ + \pi^+ + l^- + \bar{\nu},$$
$$K^0 \to \pi^+ + l^- + \bar{\nu},$$

and
$$\bar{K}^0 \to \pi^- + l^+ + \nu,$$

all of which have $\Delta S = -\Delta Q$. Four Σ^+ events have been observed, one of which was interpreted to be $\Sigma^+ \to ne^+\nu$ and the remainder $\Sigma^+ \to n\mu^+\nu$, in a total number of 2.3 million charged Σ^+ decays. In K^+ decays no $\pi^+\pi^+e^-\bar{\nu}$ ($\Delta Q = -\Delta S$) final states were observed when the number of $\pi^+\pi^-e^+\nu$ ($\Delta Q = \Delta S$) was 264. It is usual to use X to represent the ratio of $\Delta Q = -\Delta S$ amplitude to the $\Delta Q = \Delta S$ amplitude in K_{l3}^0 decays. Cronin (1968) gives the average result

$$\operatorname{Re} X = +0.14 \pm 0.05,$$
$$\operatorname{Im} X = -0.12 \pm 0.05.$$

A nonzero Im X means violation of time-reversal invariance (see Section 14.8). This is the average of eight experiments, no one of which clearly indicates a nonzero Im X. More data are required before strong conclusions can be drawn. Thus, although the evidence is not overwhelming, it is possible to say that any $\Delta Q = -\Delta S$ amplitude in leptonic decays is certainly less than one fifth of the $\Delta Q = \Delta S$ amplitude.

(3) The changes in isotopic spin between the initial and final hadrons when $|\Delta S| = 1$ satisfy $\Delta t_3 = \tfrac{1}{2}$. This can occur as the result of $\Delta t = \tfrac{1}{2}$ or $\tfrac{3}{2}$. However, only $\Delta t = \tfrac{1}{2}$ is allowed in the K_{l2}, Λ, and $\Xi \to \Lambda$ modes because one strongly interacting state involved has $t = 0$. Contributions from $\Delta t = \tfrac{3}{2}$ are permitted in the remaining decays.

We must consider the consequences of these properties. Hadrons have quantum numbers that satisfy

$$Q = t_3 + (S + B)/2,$$

so that

$$\Delta Q = \Delta t_3 + \Delta S/2.$$

Therefore three possibilities exist:
(a) $\Delta Q = \Delta S = \pm 1$; hence, $\Delta t_3 = \pm\tfrac{1}{2}$ and $\Delta t = \tfrac{1}{2}$ or $\tfrac{3}{2}$;
(b) $\Delta Q = -\Delta S = \pm 1$; hence, $\Delta t_3 = \pm\tfrac{3}{2}$ and $\Delta t = \tfrac{3}{2}$;
(c) $\Delta t = \tfrac{1}{2}$, so that $\Delta t_3 = \pm\tfrac{1}{2}$, and since $\Delta S = \pm 1$, we have $\Delta Q = \Delta S$.

We have used the property (1) given above. Evidently property (2), the $\Delta S = \Delta Q$ rule, eliminates (b). The last, (c), is called the $\Delta t = \tfrac{1}{2}$ rule. We see that the $\Delta t = \tfrac{1}{2}$ rule gives the $\Delta S = \Delta Q$ rule, but not vice versa; it is therefore a stronger assumption to make. Evidence in its favor must show that there is no $\Delta t = \tfrac{3}{2}$ amplitude in the decays we are discussing; the most important field for making tests is in the K_{l3} decay modes of the K^+

11.12 The Leptonic Decays of Strange Particles

and K_L^0 mesons. The interaction Hamiltonian connecting initial and final states is the sum of two tensor operators H_1 and H_3 having the transformation properties of a $\frac{1}{2}$ and $\frac{3}{2}$ vector in isotopic-spin space. By the Wigner–Eckart theorem each matrix element can be set equal to the product of a Clebsch–Gordan coefficient and a reduced matrix element, so that

$$\langle t', t_3' | H_1 + H_3 | t, t_3 \rangle = -\langle t, t_3, \tfrac{1}{2}, \pm\tfrac{1}{2} | t', t_3' \rangle \frac{\langle t' \| H_1 \| t \rangle}{(2t'+1)^{1/2}}$$
$$- \langle t, t_3, \tfrac{3}{2}, t_3 - t_3' | t', t_3' \rangle \frac{\langle t' \| H_3 \| t \rangle}{(2t'+1)^{1/2}}.$$

We can call $\langle t' \| H_i \| t \rangle$ the amplitude A_i for the appropriate transition. This gives

$$\langle \pi^0 l^+ \nu | H_1 + H_3 | K^+ \rangle = -\sqrt{\tfrac{1}{6}} A_1 + \sqrt{\tfrac{1}{6}} A_3,$$
$$\langle \pi^- l^+ \nu | H_1 + H_3 | K^0 \rangle = -\sqrt{\tfrac{1}{3}} A_1 - \sqrt{\tfrac{1}{12}} A_3 = \langle \pi^+ l^- \bar{\nu} | H_1 + H_3 | \bar{K}^0 \rangle.$$

The last equality follows from CP invariance. We assume that the rule $\Delta S = \Delta Q$ is correct, so that there are no contributions from matrix elements $\langle \pi^+ l^- \bar{\nu} | H_3 | K^0 \rangle$ or $\langle \pi^- l^+ \nu | H_3 | \bar{K}^0 \rangle$. In reality the decays observed are not of K^0 and $\bar{K}^0$ but of K_S and K_L and since the K_S decays predominantly by nonleptonic modes it is the K_L decay we consider. Neglecting the small CP violation,

$$|K_L\rangle = \sqrt{\tfrac{1}{2}}(|K^0\rangle - |\bar{K}^0\rangle)$$

so that

$$\langle \pi^- l^+ \nu | H_1 + H_3 | K_L \rangle = \sqrt{\tfrac{1}{2}} \langle \pi^- l^+ \nu | H_1 + H_3 | K^0 \rangle$$
$$= -\sqrt{\tfrac{1}{6}} A_1 - \sqrt{\tfrac{1}{24}} A_3.$$

Similarly

$$\langle \pi^+ l^- \bar{\nu} | H_1 + H_3 | K_L \rangle = +\sqrt{\tfrac{1}{6}} A_1 + \sqrt{\tfrac{1}{24}} A_3.$$

If the $\Delta t = \tfrac{1}{2}$ rule holds, $A_3 = 0$ and we have a connection between decay rates:

$$R(K_L \to \pi^+ l^- \bar{\nu}) + R(K_L \to \pi^- l^+ \nu) = 2R(K^+ \to \pi^0 l^+ \nu), \quad (11.50)$$

and the particle spectra are expected to be the same in all decays, apart from the small effects of differing phase space due to the mass splittings in the multiplets. If there is any $\Delta t = \tfrac{3}{2}$ amplitude, the equality of Eq. (11.50) is modified by a factor on the right-hand side of

$$\left| \frac{A_1 + (A_3/2)}{A_1 - A_3} \right|^2.$$

The evidence that the $\Delta t = \frac{1}{2}$ rule is correct has been reviewed by Cronin (1968). The prediction of Eq. (11.50) holds separately for muon and electron decays so that the μ/e ratios in K^+ and K_L decays should be the same. The observed ratios are 0.66 ± 0.02 and 0.75 ± 0.04 (Particle Data Group, 1969). In addition, the total decay rate of K_L into pion-plus-lepton states will be twice the same mode rate for K^+; the observed ratio is 1.88 ± 0.08. These results come from averaging data from different experiments, which are not satisfactorily consistent, and they must therefore be treated with caution. If the last result can be trusted, it indicates that the A_3 amplitude is about 2% of A_1.

Equation (11.50) is correct, even if CP is not conserved (Section 14.7). If CP were strictly conserved, the two rates on the left-hand side would be equal.

The above discussion relates to the matrix elements connecting the hadron states, and so far nothing has been said about any basic four-fermion interaction that might cause the decay. In such a model there is a current-current interaction, and we see that it is necessary to have a baryon current ($\bar{b}a$) that satisfies the selection rules. The change in strangeness and charge must satisfy $\Delta S = \Delta Q$, and, if the $\Delta t = \frac{1}{2}$ rule is good, the covariant ($\bar{b}a$) must have the transformation properties of an isospinor in isotopic-spin space. Examples are ($\bar{\Lambda}p$), and $\sqrt{\frac{2}{3}}(\bar{\Sigma}^+ n) - \sqrt{\frac{1}{3}}(\bar{\Sigma}^0 p)$, a combination necessary to give the required transformation property. An obvious guess is that this baryon current is V–A as in β-decay, but this also requires investigation. Evidently it is possible to choose any combination of the usual five possible covariants with parity nonconservation. We shall consider first the leptonic K-decays: here the model requires a four-fermion weak interaction to occur in a virtual state of baryon–antibaryon. This cannot be calculated, but we can represent the action of this by operators F having matrix elements between the initial and final meson states. These operators have one of the usual transformation properties. Thus the interaction Lagrangian can be written

$$\mathscr{L}_1 = \sum_i F_i \bar{\psi}_e O_i (1 - \gamma_5) \psi_\nu + \text{h.c.}$$

where $i = $ S, V, A, T, P as usual. Let us see how this works for the K_{l3} decay modes; each F_i operates between a kaon initial state and a pion final state and thus contributes to the total matrix element a factor

$$\langle \pi | F_i | K \rangle .$$

Since we believe π and K have the same spin-parity 0^-, this quantity has the transformation property given by i. Thus $\langle \pi | F_S | K \rangle$ is a scalar, $\langle \pi | F_V | K \rangle$ a vector, etc. These matrix elements must be related to physically observable quantities connected with the π and K, and there are only

11.12 The Leptonic Decays of Strange Particles

two such, the kaon and pion four-momenta, p_K and p_π, respectively. From these we can construct two independent vectors, $p_K + p_\pi$ and $p_K - p_\pi$ for convenience, one independent scalar variable $(p_K - p_\pi)^2 = q^2$ (the square of the momentum transferred between the hadrons), and an antisymmetric tensor $(p_K{}^\rho p_\pi{}^\sigma - p_K{}^\sigma p_\pi{}^\rho)$. Thus

$$\langle \pi | F_S | K \rangle = F_S(q^2),$$
$$\langle \pi | F_V{}^\rho | K \rangle = (1/m_K)[(p_K + p_\pi)^\rho F_+(q^2) + (p_K - p_\pi)^\rho F_-(q^2)], \quad (11.50a)$$
$$\langle \pi | F_T{}^{\rho\sigma} | K \rangle = (1/2m_K{}^2) F_T(q^2)(p_K{}^\rho p_\pi{}^\sigma - p_K{}^\sigma p_\pi{}^\rho),$$

and the matrix elements of F_A and F_P are zero, since we cannot make an axial vector or pseudoscalar quantity out of the available quantities. F_S, F_+, F_-, and F_T are now form factors, which contain the effect of the strong interactions and which we cannot calculate. They will be functions of q^2. The factors $1/m_K$ and $1/m_K{}^2$ are included to give all these form factors the same dimension. From this point, the calculation of the decay rate and particle spectra can be easily done (Källen, 1964). Two facts are worth noting. The squared momentum transfer is

$$q^2 = (p_K - p_\pi)^2 = m_K{}^2 + m_\pi{}^2 - 2m_K E_\pi,$$

evaluated in the K rest frame; it therefore depends on the pion energy alone. The second concerns a vector interaction in K_{e3}; in this case the factor $F_-(q^2)(p_K - p_\pi)^\rho \bar{u}_e \gamma_\rho (1 - \gamma_5) u_\nu$ that occurs in the total matrix element reduces to $F_-(q^2) m_e \bar{u}_e (1 - \gamma_5) u_\nu$, which is small because m_e/m_K is small, and it follows that in this approximation the decay spectra are independent of F_-.

There is at present a certain amount of data on the K^+, K_L leptonic decays. Recent reviews have been given by Rubbia (1969) and by Cronin (1968). The spectra and correlations measured must be compared with theory, including the possibility of a quite violent variation of the form factors. However, all the K_{e3}^+ data are consistent with a vector interaction alone, without such variations. Assuming vector interaction alone, the variation of the form factor is small and it is sufficient to express it in the form

$$F(q^2) = F(0)\left[1 + (\lambda q^2/m_\pi{}^2)\right].$$

In K_{e3} decays only F_+ is effective, so that there is only one parameter λ_+ to find: for K^+, $\lambda_+ = 0.029 \pm 0.010$; for K_L, $\lambda_+ = 0.019 \pm 0.008$. The $\Delta t = \tfrac{1}{2}$ rule requires these two parameters to be equal.

The $K_{\mu 3}$ decay description requires both form factors, F_+ and F_-. There is no evidence for tensor or scalar interaction, and it is usual to assume μ-e universality and, by comparing $K_{\mu 3}$ and K_{e3} decays, to find values of $\xi = F_-/F_+$. The two form factors may not vary with q^2 in the same way, so that we put

$$\xi(q^2) = \xi(0)\,[1 + (\Lambda q^2/m_\pi^2)]\,,$$

where $\Lambda = \lambda_- - \lambda_+$. A nonzero imaginary part to ξ indicates a violation of time-reversal invariance. There are two principal ways of obtaining ξ. The first involves comparing the $K_{\mu3}$ and K_{e3} branching ratios, either over a limited region or over the whole of the Dalitz plot of the final state. For example, if $\Lambda = 0$, the branching ratio of the entire modes is given by

$$\frac{K \to \pi\mu\nu}{K \to \pi e\nu} = 0.649 + 0.127\,\mathrm{Re}\,\xi + 0.0193\,|\xi|^2\,.$$

This is quadratic and gives two values of ξ. Usually one value is near zero and the second is about -7. The latter value is inconsistent with spectra in the final state. The second method is to measure the muon polarization (Cabibbo and Maksymowicz, 1964). It is always 100%, but at an angle to the μ direction of motion that is a function of ξ and of the position of the decay on the Daliz plot. The μ-polarization is in the plane of decay for time-reversal invariance (Im $\xi = 0$) and has a component out of the plane if time-reversal invariance is violated. No such component has been observed.

The polarization measurements have given (assuming ξ constant) (Rubbia, 1969)

$$K^+: \quad \xi = -0.98 \pm 0.20\,,$$
$$K_L: \quad \xi = -1.45 \pm 0.26\,.$$

The branching-ratio results have, in general, given values of ξ nearer zero or positive. The origin of the discrepancy is not known, and more data are required. It is impossible at present to make strong statements about any variation of ξ with q^2.

The leptonic decays of the baryons have also been investigated in an effort to discover the form of the basic interaction. Given that it is a (V–A) current–current interaction, the strong-interaction effects, as in nuclear β-decay, will alter both coupling constants and the effective interaction will be of the form V–xA. We shall return to this point in Section 11.13. Two points, however, are worth noting. First, wherever hyperon leptonic decays can go into baryon plus $e\nu$ or $\mu\nu$, the branching ratios found are consistent with μ–e universality (Bernstein, 1968). The second point concerns rates. When the ideas of a universal Fermi interaction and of a conserved vector current were formulated, it became possible to predict hyperon leptonic decay rates. For the hypercharge-changing decays the predicted rates are about twenty times greater than observed. For example, the decay $\Sigma^- \to ne^-\bar{\nu}$ is predicted to have a decay rate of $1.1 \times 10^8\,\mathrm{sec}^{-1}$, while the observed rate is $(6.6 \pm 0.3) \times 10^6 \text{sec}^{-1}$ (Particle Data Group, 1969). A likely reason for the suppression of these rates is given by the theory of Gell-Mann and Cabibbo, which is discussed in the next section.

11.13 The Cabibbo Angle

The advent of $SU(3)$ symmetry has had consequences in weak-interaction theory. Gell-Mann suggested that the vector and axial-vector weak hadron currents transformed according to the eightfold representation of $SU(3)$. Thus the $\Delta Q = +1$, $\Delta S = 0$ current should transform like a π^+-meson and the $\Delta Q = +1$, $\Delta S = +1$ current like a K^+-meson. In fact, the total current contains both these possibilities, and so Cabibbo (1963) suggested it is a vector in $SU(3)$ space of unit length, with a component $\cos \theta$ along the π^+-vector and a component $\sin \theta$ along the K^+-vector. We represent the total current by J with vector and axial-vector parts J_V, J_A. Previously (cf. Section 11.9), we just had $\Delta S = 0$ currents in J, which we briefly represent by I, while we are now considering $\Delta S = 1$ currents, which we represent by K. Thus for the vector current

$$J_V = I_V \cos \theta_V + K_V \sin \theta_V.$$

The axial-vector current is also supposed to belong to an octet, so that

$$J_A = I_A \cos \theta_A + K_A \sin \theta_A.$$

An additional assumption is made that J_V belongs to the same octet as the electromagnetic current, which is, for the vector current $I_V(\Delta S = 0)$, the CVC hypothesis. If unitary symmetry were exact the $\Delta S = 1$ currents would also not be renormalized and would therefore couple with the same strength as do the $\Delta S = 0$ currents. This is not the case, and the couplings of the nonconserved currents are renormalized.

The decays $K^+ \to \mu^+ \nu$ and $\pi^+ \to \mu^+ \nu$ occur from the actions of the axial-vector parts of the currents, K_A and I_A, respectively (see Section 11.8 for arguments leading to this conclusion). Thus these decays have rates $R(K^+ \to \mu^+ + \nu)$, $R(\pi^+ \to \mu^+ + \nu)$ that are closely related. The hadronic matrix elements are

$$\langle 0 | J_A | K^+ \rangle = \sin \theta_A \langle 0 | K_A | K^+ \rangle$$

and

$$\langle 0 | J_A | \pi^+ \rangle = \cos \theta_A \langle 0 | I_A | \pi^+ \rangle.$$

If we ignore the possibility of any effects due to the difference in momentum transfer, we can show that the matrix elements of the basic currents are equal by applying the Wigner–Eckart theorem. I_A and K_A belong to an $SU(3)$ octet with the same properties as the octet containing the pions and the kaons. The vacuum is an $SU(3)$ singlet, so that

$$\langle 0|J_A|y, t, t_3\rangle \propto \langle 0, 0 | t_A, t_{A3}, t, t_3 \rangle \begin{pmatrix} 8 & 8 & 1 \\ y_A, t_A & y, t & 0, 0 \end{pmatrix} \langle 0 \| 8 \| 8 \rangle,$$

where y, t, t_3 are the quantum numbers of the initial meson and y_A, t_A, t_{A3} are those of the appropriate current and the 8's in the reduced matrix element indicate that an octet current interacts with an octet particle to produce the vacuum $\langle 0 |$. Thus for K^+-decay

and
$$y = -y_A = 1, \quad t = t_A = \tfrac{1}{2}, \quad t_3 = -t_{A3} = -\tfrac{1}{2},$$

$$\langle 0 | K_A | K^+ \rangle = -\sqrt{\tfrac{1}{2}}(-\tfrac{1}{2})\langle 0 \| 8 \| 8 \rangle$$
$$= \tfrac{1}{2}\sqrt{\tfrac{1}{2}}\langle 0 \| 8 \| 8 \rangle.$$

For π^+ decay the coefficients are $1/\sqrt{3}$ and $\sqrt{6}/4$, respectively, and we find

$$\langle 0 | I_A | \pi^+ \rangle = \langle 0 | K_A | K^+ \rangle;$$

it follows that

$$\frac{R(K^+ \to \mu^+ \nu)}{R(\pi^+ \to \mu^+ \nu)} = \tan^2 \theta_A \frac{m_K}{m_\pi} \frac{[1 - (m_\mu^2/m_K^2)]^2}{[1 - (m_\mu^2/m_\pi^2)]^2}.$$

This leads to $\theta_A = 0.2688 \pm 0.0006$ radians (Brene et al., 1966).

The decays $K^+ \to \pi^0 e^+ \nu$ and $\pi^+ \to \pi^0 e^+ \nu$ can be treated similarly and have the property that only the vector current is acting. Thus the ratio here gives information on θ_V. The result (Rubbia, 1969), assuming $\lambda_+ = 0.02$, is that

$$\theta_V = 0.247 \pm 0.008 \quad \text{radians}.$$

It is not clear whether θ_V would be the same as θ_A if momentum-dependent effects were to be correctly included in the calculation. There is a consequence to this theory as it stands; if we assume that the current $J = J_V + J_A$ is coupled with strength g_μ, the μ-decay coupling constant (Section 11.8), then the β-decay vector (Fermi) coupling constant should be $g_\mu \cos \theta_V$, since the $\Delta S = 0$ current that is operative is I_V: thus

$$g_\mu \cos \theta_V = 1.3915 \times 10^{-49} \quad \text{erg cm}^2 = (0.991 \pm 0.003) g_V$$

if we take $g_V = 1.4029 \times 10^{-49}$ erg cm³. With this value of g_V, the Cabibbo angle reduces but does not remove the discrepancy between g_μ and g_V. It is not surprising that discrepancies remain, since we know that $SU(3)$ symmetry is badly broken. In addition the existence of $SU(3)$ breaking causes

11.13 The Cabibbo Angle

difficulty in deciding the factors that must go into the formulas used to obtain the angles.

Cabibbo (1963) has also applied these ideas to the leptonic decays of hyperons. Here the matter is complicated by the fact that the matrix element of any member of an octet of currents between any two (1 and 2) baryon states also belonging to an octet contains two reduced matrix elements. To clarify the situation let us separate the matrix elements of the vector and axial-vector currents. We have, neglecting second-class currents,

$$\langle 2|J_V^\rho|1\rangle = \sqrt{\tfrac{1}{2}}\,\bar{u}(p_2)\{g_V(q^2)\gamma^\rho + i\sigma^{\rho\lambda}q_\lambda\, g_M(q^2)\}u(p_1)e^{iq\cdot x},$$

$$\langle 2|J_A^\rho|1\rangle = \sqrt{\tfrac{1}{2}}\,\bar{u}(p_2)\{g_A(q^2)\gamma^\rho\gamma_5 + \gamma_5 q^\rho g_P(q^2)\}u(p_1)e^{iq\cdot x}.$$

Let us consider the terms $g_V(q^2)$ and $g_A(q^2)$, which appear in the matrix element of the vector and axial-vector octet currents. The matrix element of any octet current between members of an octet will have two terms, corresponding to the F and D couplings of Section 9.6. We therefore consider four form-factors, $F_V(q^2)$, $D_V(q^2)$, $F_A(q^2)$, and $D_A(q^2)$ (we will drop the explicit q^2 dependence), which will appear in the matrix elements of J_V and J_A between various baryon states. The relative weights with which F and D appear are given by the Clebsch–Gordan coefficients of the two 8's in the series.

$$\mathbf{8}\otimes\mathbf{8} = \mathbf{1}\oplus\mathbf{8}\oplus\mathbf{8}\oplus\mathbf{10}\oplus\mathbf{10}\oplus\mathbf{27}.$$

If $\Delta S = 0$, the F and D appear multiplied by $\cos\theta$, and if $|\Delta S| = 1$ they appear multiplied by $\sin\theta$. An overall coupling constant g_μ will be put in finally, and we remove factors from the Clebsch–Gordan coefficients to agree with the known matrix elements in neutron decay. Thus by a procedure that is essentially the use of the Wigner–Eckart theorem we find the g_A of Eq. (11.39) as given in Table 11.2. Consider now the vector-current matrix elements, which will involve D_V and F_V. The CVC hypothesis allows us to find F_V and D_V by considering the matrix elements of the electromagnetic current j between neutron states. The electric current behaves like a U-spin singlet and thus has $SU(3)$ transformation properties that are the same as

$$\tfrac{1}{2}(\sqrt{3}\,\rho^0 + \omega).$$

Thus the matrix element of j between neutron states is given by

$$\langle n|j|n\rangle \propto \langle n|\sqrt{3}\,\rho^0 + \omega|n\rangle,$$

which gives no F-like term, and hence

$$\langle n|j^\lambda|n\rangle \propto \bar{u}_n(p_2)\gamma^\lambda D_V(q^2)u_n(p_1)e^{iq\cdot x},$$

TABLE 11.2

THE MATRIX ELEMENTS OF THE CURRENT OCTET

Decay	Matrix element contributions		
	Vector: $g_V/g_\mu =$	Axial vector: $g_A/g_\mu =$	Weak magnetism:[a] $g_M/g_\mu =$
$n \to p e^- \bar{\nu}$	$\cos \theta_V$	$(D_A + F_A) \cos \theta_A$	$\dfrac{\mu_p - \mu_n}{m_n + m_p} \cos \theta_V$
$\Sigma^- \to \Sigma^0 e^- \bar{\nu}$	$-\sqrt{2} \cos \theta_V$	$-\sqrt{2}\, F_A \cos \theta_A$	$-\dfrac{1}{\sqrt{2}} \left(\dfrac{2\mu_p + \mu_n}{2m_\Sigma} \right) \cos \theta_V$
$\Sigma^\pm \to \Lambda e^\pm \nu$	0	$+\sqrt{\dfrac{2}{3}}\, D_A \cos \theta_A$	$-\sqrt{\dfrac{3}{2}} \dfrac{\mu_n}{m_\Lambda + m_\Sigma} \cos \theta_V$
$\Xi^- \to \Xi^0 e^- \bar{\nu}$	$-\cos \theta_V$	$(D_A - F_A) \cos \theta_A$	$-\dfrac{\mu_p + 2\mu_n}{2m_\Xi} \cos \theta_V$
$\Lambda \to p l^- \bar{\nu}$	$\sqrt{\dfrac{3}{2}} \sin \theta_V$	$+\sqrt{\dfrac{3}{2}} \left(\dfrac{D_A}{3} + F_A \right) \sin \theta_A$	$\sqrt{\dfrac{3}{2}} \dfrac{\mu_p}{m_\Lambda + m_p} \sin \theta_V$
$\Sigma^- \to n l^- \bar{\nu}$	$\sin \theta_V$	$(F_A - D_A) \sin \theta_A$	$\dfrac{\mu_p + 2\mu_n}{m_\Sigma + m_n} \sin \theta_V$
$\Xi^- \to \Lambda l^- \bar{\nu}$	$\sqrt{\dfrac{3}{2}} \sin \theta_V$	$\sqrt{\dfrac{3}{2}} \left(F_A - \dfrac{D_A}{3} \right) \sin \theta_A$	$\sqrt{\dfrac{3}{2}} \dfrac{\mu_p + \mu_n}{m_\Xi + m_\Lambda} \sin \theta_V$
$\Xi^- \to \Sigma^0 l^- \bar{\nu}$	$\dfrac{1}{\sqrt{2}} \sin \theta_V$	$\dfrac{1}{\sqrt{2}} (F_A + D_A) \sin \theta_A$	$\dfrac{1}{\sqrt{2}} \dfrac{\mu_p - \mu_n}{m_\Xi + m_\Sigma} \sin \theta_V$

[a] μ_p and μ_n are the anomalous nucleon magnetic moments.

where $q = p_2 - p_1$. At $q^2 = 0$ this matrix element must be zero, so we have $D_V(q^2 = 0) = 0$. The neutron β-decay involves the combination $D_V + F_V$ in the vector matrix element, which by CVC is 1. Thus

$$F_V(q^2 = 0) = 1 \,.$$

We are dealing with hyperon leptonic decays in which q^2 is small, and so F_V and D_V are not expected to vary significantly from their $q^2 = 0$ values. So we put $F_V(q^2) = 1$, $D_V(q^2) = 0$. The entries in the vector column of Table 11.2 follow from these results. Without data other than $g_A/g_V = 1.23$ we could deduce that $F_A + D_A = 1.23$, but, as we shall discover, all the information on hyperon leptonic decays has been used to obtain values of F_A and D_A.

The next contribution to be considered is the weak magnetism, which is the second term in the vector matrix element. The CVC hypothesis requires this to be, for the neutron decay [Eq. (11.40)].

$$g_M(q^2) = \frac{g_V(q^2)(\mu_p - \mu_n)}{2M},$$

11.13 The Cabibbo Angle

where μ_p and μ_n are the anomalous magnetic moments. Since

$$\frac{g_V(q^2)}{g_\mu(q^2)} = F_V(q^2) = 1,$$

we have

$$g_M(q^2) = g_\mu(q^2) \frac{\mu_p - \mu_n}{2M}.$$

To obtain the remaining weak-magnetism terms we use the following $SU(3)$-based prescription. Since baryon octet magnetic moments are found from two reduced matrix elements of an octet operator between the octet states [Eq. (9.44)], it means that there are two contributions (D-like and F-like) to the weak magnetism in the same proportion as in the axial-vector current. Thus, using Table 11.2, we see that $n \to pe^-\bar{\nu}$ decay and CVC gives

$$(D_M + F_M) = \frac{\mu_p - \mu_n}{2M} \cos\theta_V$$

where subscript "M" refers to weak magnetism. In the $\Delta S = 0$ decay $\Sigma^\pm \to \Lambda e^\pm \nu$, CVC requires the weak magnetism to be proportional to $\sqrt{2}$ times the $\Sigma^0 \to \Lambda\gamma$ transition magnetic moment [Eq. (9.43)] which is, according to $SU(3)$, $-\sqrt{3}\,\mu_n/2$. Hence

$$+\sqrt{\tfrac{2}{3}}\,D_M = -\sqrt{\tfrac{3}{2}}\,(\mu_n/2M)\cos\theta_V.$$

Solving these equations we have

$$D_M = -\tfrac{3}{2}(\mu_n/2M)\cos\theta_V,$$

$$F_M = \frac{\mu_p + (\mu_n/2)}{2M}\cos\theta_V.$$

We have deliberately left the mass uncertain. The prescription now says put $2M = m_1 + m_2$, where m_1 and m_2 are the initial and final baryon masses. In addition, if $|\Delta S| = 1$, change $\cos\theta_V$ to $\sin\theta_V$. All the weak-magnetism contributions to the matrix elements follow by combining D_M and F_M as in the axial-vector current.

The induced pseudoscalar contribution $g_P(q^2)$ in the axial-vector current is often neglected, since it contributes only of the order of 1% in the decays that give a muon. Nambu (1960) gives

$$g_P(q^2) = g_A(q)^2 \frac{m_2 + m_1}{q^2 - m^2},$$

where m is the mass of the pion if $\Delta S = 0$ and of the kaon if $|\Delta S| = 1$. m_1 and m_2 are the masses of the baryons involved. This result is based on an extension of the Goldberger–Treiman relation for the induced pseudoscalar coupling [see Eq. (11.45)]. $g_A(q^2)$ is, of course, given by the appropriate F_A and D_A combination from Table 11.2.

Filthuth (1969) reported on the most recent values of the form factors and angles, using data on $\Lambda \to pe^-\bar{\nu}$, $\Sigma^- \to ne^-\bar{\nu}$, $\Sigma^- \to n\mu^-\bar{\nu}$, $\Sigma \to \Lambda e\nu$, $\Xi^- \to \Lambda e^-\bar{\nu}$. Since g_μ is known from muon decay, the unknowns are F_A, D_A, θ_V, and θ_A. A very satisfactory fit is obtained assuming $\theta_V = \theta_A$:

$$\theta = \theta_V = \theta_A = 0.235 \pm 0.006 \quad \text{radians,}$$
$$F_A = +0.49 \pm 0.02,$$
$$D_A = +0.74 \pm 0.02.$$

If θ_V is not assumed to be equal to θ_A, the analysis gives similar values differing by much less than their errors.

This analysis neglects the effects of corrections due to the electromagnetic interaction. It also takes advantage of the Ademello–Gatto theorem (1964), which states that there are no first-order corrections due to $SU(3)$ breaking. Higher-order effects are absorbed by the Cabibbo angle. Thus many data on the baryon decays are well fitted by the coupling constant from μ-decay and three other parameters, which gives strong support for this theory of baryon leptonic decays.

11.14 The Nonleptonic Decays of Strange Particles

In this section we will discuss the decay modes of the long-lived strange particles that do not yield leptons, only hadrons. In spite of the fact that all the particles involved are strongly interacting, the lifetimes involved indicate that the weak interactions are responsible for the decays. The decays do not conserve strangeness and are of the type

$$K \to \pi + \pi,$$
$$K \to \pi + \pi + \pi,$$
$$\Lambda \to \pi + N,$$
$$\Sigma \to \pi + N,$$
$$\Xi \to \pi + \Lambda.$$

We know already that parity and C are not conserved in these decays (Section 11.12). Until 1964, it was thought that CP was conserved, but the observation of $K_2^0 \to \pi^+\pi^-$ indicated that there is a very weak CP-violating interaction. Since it is weak (see below), we shall assume CP conservation in this section and defer discussion of CP violation to Chapter XIV.

11.14 The Nonleptonic Decays of Strange Particles

It is convenient to consider first the decays of the neutral kaons, K_1 and K_2 (Section 7.5), which are eigenstates of CP with eigenvalue $+1$ and -1, respectively. CP conservation gives selection rules for decay into pions. A $\pi^+\pi^-$ system has C eigenvalue $(-1)^l$ and P eigenvalue $(-1)^l$, so that $CP = +1$; a $\pi^0\pi^0$ system has l even for reasons of symmetry, so it also has $CP = +1$. Therefore, decay into two pions is allowed for K_1 and forbidden for K_2. As mentioned, $K_2 \to \pi^+\pi^-$ has been observed with a rate ratio (see Steinberger, 1969)

$$\frac{R(K_2 \to \pi^+\pi^-)}{R(K_1 \to \pi^+\pi^-)} \simeq 3.6 \times 10^{-6}.$$

It is the smallness of this number that justifies the neglect of CP violation in this section.

The three-pion states have various CP eigenvalues depending upon charge states and wavefunctions. From Section 7.8 we can see that a state of three pions, two of which are identical, cannot be in a state of zero total angular momentum and even parity. Since the neutral kaons have zero spin, the parity of the $3\pi^0$ decay state must be odd. Such a state of three π^0-mesons is even under C and hence odd under CP. Therefore

$$K_2 \to 3\pi^0 \quad \text{is allowed},$$
$$K_1 \to 3\pi^0 \quad \text{is forbidden}.$$

The remaining accessible charge state is $\pi^+ + \pi^- + \pi^0$. Evidence from the decay

$$K^+ \to \pi^+ + \pi^- + \pi^+$$

suggests that the relative orbital angular momenta in these three-pion decays are both zero (Section 7.9) and hence the state has odd parity. This charge state is even under C and hence is odd under CP. Thus the decay

$$K_1 \to \pi^+ + \pi^- + \pi^0$$

is forbidden into meson states of zero relative angular momentum. The decay into states of higher angular momenta will be inhibited by the angular-momentum barrier by a factor of at least 100 (Gell-Mann and Rosenfeld, 1957).

An examination of actual rates reveals the striking fact that $K^+ \to \pi^+\pi^0$ is 700 times slower than $K_1^0 \to \pi\pi$. If we assume pseudoscalar kaons, then angular-momentum conservation requires that the two pions in $K^+ \to \pi^+\pi^0$ or $K_1^0 \to \pi^+\pi^-$, or $\pi^0\pi^0$ be in an S-state. By Bose statistics we know that the overall wavefunction describing the two-pion state must be symmetric, so that the isotopic-spin wavefunction of the final state must also be symmetric. The possible total isotopic spins of the final state are $t = 0$, 1, or 2 with $t_3 = +1$ for $\pi^+\pi^0$ and $t_3 = 0$ for $\pi^+\pi^-$ and $\pi^0\pi^0$. The relevant state vectors $|t, t_3\rangle$ for the vector addition of two isotopic spins of 1 are

$\pi^+\pi^-, \pi^0\pi^0$
$$|0, 0\rangle = \sqrt{\tfrac{1}{3}}\,|\pi^+\pi^-\rangle - \sqrt{\tfrac{1}{3}}\,|\pi^0\pi^0\rangle + \sqrt{\tfrac{1}{3}}\,|\pi^-\pi^+\rangle,$$
$$|1, 0\rangle = \sqrt{\tfrac{1}{2}}\,|\pi^+\pi^-\rangle - \sqrt{\tfrac{1}{2}}\,|\pi^-\pi^+\rangle,$$
$$|2, 0\rangle = \sqrt{\tfrac{1}{6}}\,|\pi^+\pi^-\rangle + \sqrt{\tfrac{2}{3}}\,|\pi^0\pi^0\rangle + \sqrt{\tfrac{1}{6}}\,|\pi^-\pi^+\rangle,$$

$\pi^+\pi^0$
$$|1, 1\rangle = \sqrt{\tfrac{1}{2}}\,|\pi^+\pi^0\rangle - \sqrt{\tfrac{1}{2}}\,|\pi^0\pi^+\rangle,$$
$$|2, 1\rangle = \sqrt{\tfrac{1}{2}}\,|\pi^+\pi^0\rangle + \sqrt{\tfrac{1}{2}}\,|\pi^0\pi^+\rangle.$$

The only functions symmetric under pion exchange are $|0, 0\rangle$, $|2, 0\rangle$ for K_1^0 decay and $|2, 1\rangle$ for K^+ decay. The initial states have $t = \tfrac{1}{2}$, so that if the transition is limited to $\Delta t = \tfrac{1}{2}$, then $K^+ \to \pi^+\pi^0$ would be entirely forbidden and K_1^0 would decay into the $|0, 0\rangle$ state, giving an expected branching ratio of

$$\mathcal{R} = \frac{R(K_1 \to \pi^0\pi^0)}{R(K_1 \to \pi^0\pi^0 \text{ and } K_1 \to \pi^+\pi^-)} = \frac{1}{3},$$

with a slight correction for the different phase space in the two modes. In fact, the observed decay $K^+ \to \pi^+\pi^0$ indicates that there must be some $\Delta t = \tfrac{3}{2}$ or $\tfrac{5}{2}$ transition. Assuming the first alone is present and taking the known $K^+ \to \pi^+\pi^0$ decay rate, $\mathcal{R}$ can be corrected for the presence of the corresponding $\Delta t = \tfrac{3}{2}$ amplitude. The result is (Gell-Mann and Rosenfeld, 1957)

$$0.28 \leqslant \mathcal{R} \leqslant 0.38.$$

The experimental value is 0.300 ± 0.005 (Gobbi et al., 1968). The source of the $\Delta t = \tfrac{3}{2}$ amplitude is not at present clear. If the interaction is a virtual four-fermion interaction, then it will be between two currents, one having $\Delta S = 0, \Delta t = 1$ the other having $\Delta S = 1, \Delta t = \tfrac{1}{2}$. This means that the overall changes will be $\Delta S = 1$ and $\Delta t = \tfrac{1}{2}$ or $\tfrac{3}{2}$ with comparable strength. However, the $\Delta t = \tfrac{3}{2}$ seems almost absent. Electromagnetic interactions can give rise to a $\Delta t = \tfrac{3}{2}$ transition when only $\Delta t = \tfrac{1}{2}$ is present in the weak interaction, but the intensity factor would be roughly $(1/137)^2$. The problem is, therefore, why is the $\Delta t = \tfrac{3}{2}$ amplitude what it is and not as large as the $\Delta t = \tfrac{1}{2}$ amplitude or as small as expected from electromagnetic effects alone? Cabibbo (1964) and Gell-Mann (1964) have extended the idea of the weak current belonging to one octet of $SU(3)$ to the nonleptonic decays and show that $K_1^0 \to 2\pi$ decay would be forbidden if $SU(3)$ were a perfect symmetry and therefore must go by symmetry-breaking effects. This would depress a natural, symmetry-absent, rate by

11.14 The Nonleptonic Decays of Strange Particles

TABLE 11.3

THE $\Delta t = \frac{1}{2}$ RULE AND THE THREE-PION DECAY RATES OF KAONS

Rates	($\Delta t = \frac{1}{2}$) × (Phase space) × (EM correction) = Prediction						Experimental results
$\dfrac{K^+ \to \pi^+\pi^+\pi^-}{K^+ \to \pi^0\pi^0\pi^+}$	4	×	$\dfrac{1}{1.21}$	×	1.035	= 3.42	3.30 ± 0.10
$\dfrac{K_L \to \pi^0\pi^0\pi^0}{K_L \to \pi^0\pi^+\pi^-}$	$\frac{3}{2}$	×	$\dfrac{1.48}{1.31}$	×	0.965	= 1.63	1.70 ± 0.12
$\dfrac{K_L \to \pi^0\pi^+\pi^-}{K^+ \to \pi^0\pi^0\pi^+}$	2	×	$\dfrac{1.31}{1.21}$	×	1.02	= 2.22	1.73 ± 0.10

a factor of 10–20 to its observed value and, considering what the relative rates of $K^+ \to \pi^+\pi^0$ and $K_1^0 \to 2\pi$ would be in the absence of this damping, suggests that the K^+ is decaying by an electromagnetically induced $\Delta t = \frac{3}{2}$ transition. This theory does not require there to be a nonelectromagnetic $\Delta t = \frac{3}{2}$ amplitude, although, as indicated before, it is to be expected.

The $\Delta t = \frac{1}{2}$ rule also predicts ratios between the three-pion decay rates of the kaons. These predictions and experimental results, shown in Table 11.3, are a summary of the data presented by Aubert (1969). In addition, the spectrum of π^0 in $K_L \to \pi^+\pi^-\pi^0$ and of π^- in $K^+ \to \pi^+\pi^+\pi^-$ are expected to deviate from phase space by a term linear in the meson kinetic energy, and the constant of proportionality a should be the same in the two cases. The results are (Aubert, 1969)

$$a^+ = -0.26 \pm 0.02,$$

$$a^0 = -0.20 \pm 0.014.$$

This linear term comes into the integrals that have to be done to obtain the phase-space corrections to the predicted ratios for the $\Delta t = \frac{1}{2}$ test. The different masses within the pion triplet make it impossible to do this with certainty, and the results could be incorrect by several percent. The discrepancies that exist in the tests might be due to the presence of a $\Delta t = \frac{3}{2}$ amplitude.

The $\Delta t = \frac{1}{2}$ rule also has consequences for the nonleptonic decay of hyperons. First, we consider the decay modes of the Λ:

$$\Lambda \to \pi^- + p, \quad \Lambda \to \pi^0 + n.$$

The initial state has $t = 0$, the final state $t = \frac{1}{2}$ or $\frac{3}{2}$. Thus there are two amplitudes a_1 and a_3 corresponding to $\Delta t = \frac{1}{2}$ and $\frac{3}{2}$ respectively, each consisting of a sum of S and P outgoing waves, giving four in all: a_{1S}, a_{1P}, a_{3S}, and a_{3P}. In the $\pi^- p$ mode the S- and P-wave amplitudes, using the Clebsch-Gordan coefficients (see Section 6.4), are given by

$$a_S = \sqrt{\tfrac{2}{3}}\, a_{1S} + \sqrt{\tfrac{1}{3}}\, a_{3S},$$
$$a_P = \sqrt{\tfrac{2}{3}}\, a_{1P} + \sqrt{\tfrac{1}{3}}\, a_{3P}. \tag{11.51}$$

If b_S and b_P are the corresponding amplitudes for $\pi^0 n$ decay,

$$b_S = -\sqrt{\tfrac{1}{3}}\, a_{1S} + \sqrt{\tfrac{2}{3}}\, a_{3S},$$
$$b_P = -\sqrt{\tfrac{1}{3}}\, a_{1P} + \sqrt{\tfrac{2}{3}}\, a_{3P}. \tag{11.52}$$

The decay rate into these modes is the intensity integrated over all directions, for which S–P interference disappears, so that we obtain for the total rate

$$R = |a_{1S} + a_{3S}|^2 + |a_{1P} + a_{3P}|^2.$$

The rate (R^-) into the $\pi^- p$ state is

$$R^- = |\sqrt{\tfrac{2}{3}}\, a_{1S} + \sqrt{\tfrac{1}{3}}\, a_{3S}|^2 + |\sqrt{\tfrac{2}{3}}\, a_{1P} + \sqrt{\tfrac{1}{3}}\, a_{3P}|^2.$$

If the $\Delta t = \tfrac{1}{2}$ rule is correct, $a_{3S} = a_{3P} = 0$ and we find

$$R^-/R = \tfrac{2}{3}.$$

The observed value is 0.653 ± 0.013 (Particle Data Group, 1969). The asymmetry parameter α^- in $\Lambda \to \pi^- p$ decay is

$$\alpha^- = \frac{-2\,\mathrm{Re}\, a_S^* a_P}{|a_S|^2 + |a_P|^2},$$

while the corresponding parameter α^0 in $\Lambda \to \pi^0 n$ is the same with the replacements $a_S \to b_S$ and $a_P \to b_P$. Substituting from Eq. (11.52) gives, for $\Delta t = \tfrac{1}{2}$,

$$\alpha^- = \alpha^0.$$

Cork et al. (1960) find

$$\alpha^0/\alpha^- = 1.10 \pm 0.27.$$

Thus the Λ decay properties are consistent with $\Delta t = \tfrac{1}{2}$.

If the decay interaction conserves CP and if there is no final-state interaction, the amplitudes a_{1S}, a_{3S} are real quantities. The effect of a final-state interaction can be taken into account by use of the Watson theorem (Section 6.6). In terms of real quantities

$$a_{1S} = |a_{1S}|\, e^{i\delta_1}, \qquad a_{3S} = \pm |a_{3S}|\, e^{i\delta_3},$$
$$a_{1P} = \pm |a_{1P}|\, e^{i\delta_{11}}, \qquad a_{3P} = \pm |a_{3P}|\, e^{i\delta_{31}},$$

11.14 The Nonleptonic Decays of Strange Particles

where δ are the pion–nucleon scattering phase shifts at the same center-of-mass energy. By allowing S- and P-waves we have put in parity non-conservation: if C were conserved the S- and P-waves would be relatively complex in the absence of final-state interactions and α^- and α^0 would be zero (Section 6.9). The final-state interactions given by the factors $e^{i\delta}$ are insufficient to cause the observed asymmetry, and C conservation must also be violated in the decay. In fact, measurements of polarizations in polarized Λ decay and study of $^4_\Lambda$He-decay branching ratios indicate $a_{1P}/a_{1S} \simeq 0.35$.

The Σ-decay also contributes to the information on the isotopic-spin selection rule in the decay modes

$$\Sigma^+ \to \pi^+ + n,$$
$$\Sigma^+ \to \pi^0 + p,$$
$$\Sigma^- \to \pi^- + n.$$

Amplitudes having $\Delta t = \frac{1}{2}, \frac{3}{2}$, or $\frac{5}{2}$ could operate here, but we shall look at the consequences of $\Delta t = \frac{1}{2}$ alone. We are therefore interested in the matrix elements of the tensor operator H_1: for example,

$$\langle \pi^+ n | H_1 | \Sigma^+ \rangle = A^+ = \sqrt{\tfrac{1}{3}} \langle \tfrac{3}{2}, \tfrac{1}{2} | H_1 | 1, 1 \rangle + \sqrt{\tfrac{2}{3}} \langle \tfrac{1}{2}, \tfrac{1}{2} | H_1 | 1, 1 \rangle,$$

where the quantum numbers are t, t_3. Similarly

$$\langle \pi^0 p | H_1 | \Sigma^+ \rangle = A^0 = \sqrt{\tfrac{2}{3}} \langle \tfrac{3}{2}, \tfrac{1}{2} | H_1 | 1, 1 \rangle - \sqrt{\tfrac{1}{3}} \langle \tfrac{1}{2}, \tfrac{1}{2} | H_1 | 1, 1 \rangle,$$
$$\langle \pi^- n | H_1 | \Sigma^- \rangle = A^- = \langle \tfrac{3}{2}, -\tfrac{3}{2} | H_1 | 1, -1 \rangle.$$

Using the Wigner–Eckart theorem to express these amplitudes in terms of reduced matrix elements we find

$$A^+ = -\tfrac{1}{6} \langle \tfrac{3}{2} \| H_1 \| 1 \rangle - \sqrt{\tfrac{2}{9}} \langle \tfrac{1}{2} \| H_1 \| 1 \rangle,$$
$$A^0 = -\tfrac{1}{3} \sqrt{\tfrac{1}{2}} \langle \tfrac{3}{2} \| H_1 \| 1 \rangle + \tfrac{1}{3} \langle \tfrac{1}{2} \| H_1 \| 1 \rangle,$$
$$A^- = -\tfrac{1}{2} \langle \tfrac{3}{2} \| H_1 \| 1 \rangle,$$

from which

$$A^+ + \sqrt{2}\, A^0 = A^-. \tag{11.53}$$

If we decompose the outgoing-wave amplitude into S- and P-wave parts, then this relation holds between the S- and P-wave amplitudes separately. As in the case of the Λ-decay these amplitudes are real quantities multiplied by $e^{i\delta}$, where δ is the phase shift in the corresponding pion–nucleon scattering state. At the energies of the Σ-decays the phase shifts are still small, and it is sufficient to neglect them. Each amplitude A in Eq. (11.53) can be represented as vector in a two-dimensional space whose projections on the two orthogonal axes are the S- and P-wave amplitudes. Equation

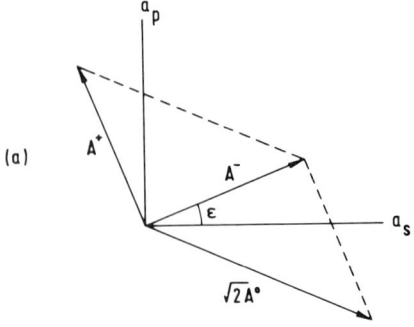

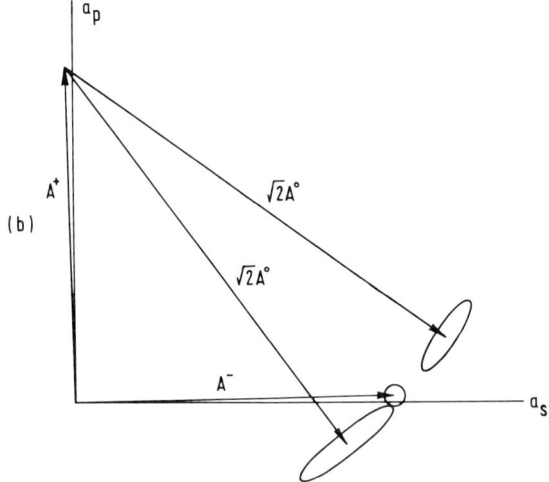

FIG. 11.7. Figures showing the relation between decay amplitudes in Σ-decay: (a) simplified diagram; (b) present experimental situation.

(11.53) is then a vector relation in this space. The square of the length of each vector is given by the corresponding decay rate; the channel decay rates are approximately equal, and Eq. (11.53) becomes the relation between the sides of a right-angle isosceles triangle; a possible arrangement of vectors is shown in Fig. 11.7a. The next step is to link this with the decay asymmetry; in any mode the decay asymmetry parameter is given by

$$\alpha = \frac{-2 \operatorname{Re} S^*P}{|S|^2 + |P|^2}.$$

All the quantities are real, so that

$$\alpha = -2 \frac{S}{(S^2 + P^2)^{1/2}} \frac{P}{(S^2 + P^2)^{1/2}}.$$

For the decay $\Sigma^- \to \pi^- n$ with asymmetry parameter α^- we find

$$\alpha^- = -2 \sin \epsilon \cos \epsilon = -\sin 2\epsilon,$$

11.14 The Nonleptonic Decays of Strange Particles

where ϵ is the angle between A^- and the S axis. Referring to Fig. 11.7a, we see that in this simple case of almost equal rates, the asymmetry parameters for the other modes would be simply related to ϵ; thus,

$$\alpha^+ \simeq +\sin 2\epsilon, \qquad \alpha^0 \simeq \pm\cos 2\epsilon.$$

The last sign ambiguity arises from the fact that the triangle can be reflected through A^- without failing Eq. (11.53). In fact, the asymmetry parameters are (Particle Data Group, 1969)

$$\alpha^+ = +0.017 \pm 0.037,$$
$$\alpha^0 = -0.955 \pm 0.070,$$
$$\alpha^- = -0.060 \pm 0.047,$$

so that ϵ is close to zero with A^+ and A^- lying close to an axis and perpendicular. The $\Delta t = \frac{1}{2}$ rule then predicts that A^0 lies at about 45° to the axes and that $\alpha^0 \simeq \pm 1$, consistent with the observed value. The remaining parameters β and γ (Section 6.4) are well enough known to show that $\Sigma^- \to \pi^- n$ is nearly pure S-wave and that $\Sigma^+ \to \pi^+ n$ is nearly pure P-wave. The remaining uncertainty is whether it is the S- or the P-wave that has the greater amplitude in $\Sigma^+ \to \pi^0 p$. The amplitudes are shown in Fig. 11.7b, which indicates the two possible solutions. It is clear that the triangle is nearly closing and that, within the errors, the P-wave amplitude is greater than the S-wave in $\Sigma^+ \to \pi^0 p$. Thus there is no positive evidence for violation of the $\Delta t = \frac{1}{2}$ rule in nonleptonic Σ-decay.

The cascade hyperon nonleptonic decays are

$$\Xi^0 \to \pi^0 + \Lambda, \qquad \Xi^- \to \pi^- + \Lambda.$$

The $\Delta t = \frac{1}{2}$ rule predicts equal decay asymmetries and decay rates in the ratio 1:2. The observed asymmetries are the same within 20%, and the rates ratio is 0.548 ± 0.036.

The $\Delta t = \frac{1}{2}$ rule for nonleptonic processes appears to be of considerable validity and leads back to the problems already mentioned. In these modes the weak interaction presumably occurs as a four-fermion current–current interaction with all particles in a virtual state. One current must have $\Delta S = 1$, $\Delta t = \frac{1}{2}$, the other can only have $\Delta S = 0$, $\Delta t = 1$; if, as Gell-Mann suggests, the currents are members of an octet, then the interaction $J^\dagger J$ can have transformation properties of the representations of the product of two identical octets. These are a singlet, an octet, and a 27-plet. The octet causes transitions that, if $|\Delta S| = 1$, have $\Delta t = \frac{1}{2}$. The 27-plet can cause transitions having $|\Delta S| = 1$ for which $\Delta t = \frac{1}{2}$ or $\frac{3}{2}$ and $|\Delta S| = 2$ for which $\Delta t = 1$. We have seen that there is no evidence that any transitions having $|\Delta S| = 2$ or $\Delta t = \frac{3}{2}$ are present. Various suggestions as to the reasons for this situation have been made, such as octet dominance (Coleman and Glashow, 1964) and a scheme of Lee and Yang (1960b) in which the observed selection rules are the consequence of properties assigned to the hypothetical weak intermediate vector boson (Section 11.15). None of these schemes provides more than a suggestion of a solution. In addition, the nonleptonic decays should also be damped by a factor $\cos\theta \sin\theta$, θ being the Cabibbo angle, in the same way as the leptonic decays are damped by $\sin\theta$. The actual rates indicate that this damping is essentially absent (Matthews, 1965). Thus the nonleptonic decays pose a considerable problem to theory.

11.15 The Intermediate Vector Boson

In fact, the interaction we have postulated is unlikely to have the form discussed, because such an interaction predicts cross sections for lepton-lepton scattering reactions such as

$$\bar{\nu}_e + e^- \to \mu^- + \bar{\nu}_\mu,$$
$$\bar{\nu}_e + e^- \to e^- + \bar{\nu}_e,$$

which increase indefinitely with energy, since there are no known form factors to alter the situation. The simplest modification is to change the interaction from the form in which four fermions interact at one point to one in which two fermion currents interact via the exchange of a vector boson (Fig. 11.8). Thus the total interaction Lagrangian

$$L_I = (g/\sqrt{2}) \int \big(\bar{\phi}_4(x)\gamma^\rho(1-\gamma_5)\phi_3(x)\big)\big(\bar{\phi}_2(x)\gamma_\rho(1-\gamma_5)\phi_1(x)\big) d^4x + \text{h.c.}$$

becomes

$$L_I = (g/\sqrt{2}) \iint \big(\bar{\phi}_4(x)\gamma^\rho(1-\gamma_5)\phi_3(x)\big) \\ \times F_\rho{}^\lambda(x-y)\big(\bar{\phi}_2(y)\gamma_\lambda(1-\gamma_5)\phi_1(y)\big) d^4x\, d^4y + \text{h.c.}$$

with

$$F_\rho{}^\lambda(x-y) = \left(\delta_\rho{}^\lambda + \frac{q_\rho q^\lambda}{W^2}\right)\frac{\exp iq\cdot(x-y)}{W^2 - q^2}.$$

q is the momentum transfer and W the mass of the exchanged spin-1 particle.

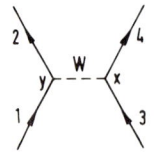

FIG. 11.8. Feynman diagram for the weak interaction via vector boson exchange.

11.15 The Intermediate Vector Boson

The basic interaction is now between a fermion vector plus axial-vector current with a spin-1 boson field ϕ with a coupling constant g_W:

$$\mathscr{L}_I = g_W\big(\bar{\phi}_4(x)\,\gamma^\rho(1-\gamma_5)\,\phi_3(x)\big)\,\phi_\rho(x) + \text{h.c.},$$

so that $g_V(0)/\sqrt{2} = g_W^2/W^2$. If this were the actual physical situation, the propagator factor for the boson would limit the predicted increase with energy of the neutrino–lepton reactions mentioned above.

The vector boson will have certain electric-charge properties; since it appears all the fermion currents are charge 1, i.e., $\Delta Q = \pm 1$ (Section 11.9), we expect the boson to have two charged states, $Q = \pm 1$. However, this leads to the possibility of $\Delta t = \frac{3}{2}$ in nonleptonic decays, so it is possible that there are also two neutral bosons (charge conjugate to one another) coupled so that the total effect is $\Delta t = \frac{1}{2}$. In another scheme due to Lee and Yang (1960b) the four boson states have isotopic-spin properties of a single and triplet in strangeness-conserving decays (as do, for example, the ρ and ω mesons) and the properties of two doublets $(K, \bar{K})$ in strangeness-changing decays. This is more restrictive than the $\Delta t = \frac{1}{2}$ rule and contains it.

Some of the consequences (as yet undetected) of the existence of the intermediate vector boson would be, for example:

(1) To increase the Michel parameter ρ to $\frac{3}{4} + \frac{1}{3}(m_\mu/W)^2$, which is a change much smaller than that due to radiative corrections.
(2) To increase the muon decay rate by $\frac{3}{5}(m_\mu/W)^2$.
(3) To add an amount $(2m_\mu M/W^2)g_A$ to the induced pseudoscalar coupling constant.

The charged boson would have at least the following decay modes:

$$W^+ \to \mu^+ + \nu_\mu,$$
$$\to e^+ + \nu_e,$$
$$\to \pi^+ + \pi^0, \quad \text{etc.}$$

with partial rates of the order of 10^{17} sec^{-1}.

At present the most likely place for detecting the boson is directly in high-energy neutrino interactions. W^+ production in the Coulomb field followed by a leptonic decay mode means that two charged leptons would be observed, e.g.,

$$\nu_\mu + \text{nucleus} \to \text{nucleus} + W^+ + \mu^-$$
$$\searrow$$
$$\mu^+ + \nu_\mu.$$

Neutrino experiments so far performed have not revealed any events that can be unambiguously assigned as involving W-production. However, it is possible to put a lower limit of 1.8 GeV on the W mass (Perkins, 1969).

The observation of an intermediate vector boson would be of very considerable importance to the development of our understanding of weak interactions.

11.16 Neutrino Interactions

Apart from the reactor neutrino-absorption experiments (Section 11.7), most experiments and measurements on weak interactions are performed on decaying systems, where the energy is naturally limited to several hundreds of MeV at the most. The expected rise of neutrino cross sections with energy and improvements in accelerator intensities have made technically possible the study of weak interactions at higher energies, and it is evident that neutrino interactions will be an important part of research on the next generation of high-energy accelerators. Thus far, such experiments have confirmed feasibility, have shown the existence of two neutrinos, and have given some crude information on high-energy weak interactions.

Let us examine the kind of information that has been or apparently will be obtained from a study of neutrino interactions (Perkins, 1969). First, it is likely that the intermediate vector boson, if it exists, will be observed in neutrino reactions; this is discussed in Section 11.15. Second, the elastic processes

$$\bar{\nu}_\mu + p \to \mu^+ + n,$$
$$\nu_\mu + n \to \mu^- + p$$

will give information on the high-momentum transfer behavior of the form factors in the weak interactions of nucleons. These are $g_V(q^2)$, $g_M(q^2)$, $g_A(q^2)$, and $g_P(q^2)$ of Eqs. (11.38) and (11.39). The conserved-vector-current hypothesis prescribes the q^2 behavior of $g_V(q^2)$ and $g_M(q^2)$ to be the same as that of the electromagnetic isovector form factors of the nucleons (Section 13.6). No such statements can be made about the axial-vector form factor $g_A(q^2)$ or the induced pseudoscalar form factor $g_P(q^2)$. Although the data are limited, the distribution of q^2 in the observed elastic events is consistent with the CVC hypothesis, and with the assumption that $g_A(q^2)$ varies with q^2 in the same way as does g_V, and that g_P is small.

A feature of the observations is the large inelastic cross-sections; for example,

$$\nu_\mu + N \to \mu^- + N + \pi + \cdots.$$

The number of such events and the energy observed indicate that the inelastic cross sections above 4 GeV are considerably greater than the elastic cross section. These reactions have been observed for nucleons in complex nuclei; further elucidation of these processes may require a study of neutrino interactions in liquid-hydrogen or deuterium bubble chambers. However, the production of Δ^{++} is already recognized as a prominent feature of single pion production:

$$\nu_\mu + p \to \mu^- + \Delta^{++}$$
$$\searrow$$
$$\pi^+ + p.$$

The production of strange particles by neutrinos in strangeness-nonconserving transitions will be of interest. The $\Delta Q = \Delta S$ rule allows the elastic (two-body) production of hyperons by antineutrinos but not by neutrinos. Thus antineutrinos can cause the reaction

$$\bar{\nu}_\mu + p \to \mu^+ + \Lambda^0,$$

whereas the strange-particle production by neutrinos might be of the form

$$\nu_\mu + n \to \mu^- + K^+ + n.$$

11.17 Conclusion

The appearance that there is a good theory of weak interactions is deceptive. We have seen that the intermediate vector-boson hypothesis has been introduced to limit the high-energy behavior. However, this still leads to a nonrenormalizable theory and to higher-order effects that can be greater than the first-order-calculation results known to agree with experiment. This situation can be arbitrarily avoided by a momentum cut-off in the theory. An upper limit to this cut-off can be calculated from the upper limits on the rates for second-order processes such as $K_L \to \mu^+ + \mu^-$. This gives about 36 GeV for the Fermi theory and about 100 GeV for the vector-boson theory. The $K_1 - K_2$ mass difference leads to an upper limit of about 5 GeV in the latter theory. These results were obtained by Ioffe and Shapalin (1967). How nature produces the cut-off, particularly if it is as low as 5 GeV, is a fascinating question (Low, 1969).

REFERENCES

Ademello, M., and Gatto, R. (1964). *Phys. Rev. Lett.* **13**, 264 (11.13).
Adler, S. L. (1965). *Phys. Rev. Lett.* **14**, 1051 (11.10).
Albergi Quaranta, A. *et al.* (1967). *Phys. Lett. B* **25**, 429 (11.11).
Alder, K., Steck, B., and Winther, A. (1957). *Phys. Rev.* **107**, 728 (11.5).
Alichanov, A. I., Galaktionov, Yu. V., Gorodkov, Yu. V., Eliseev, G. P., and Lyubimov, V. A. (1960). *Sov. Phys. JETP* **11**, 1380 (11.8).
Allen, R. A., Burcham, W. E., Chackett, K. F., Munday, G. L., and Reasbeck, P. (1955). *Proc. Phys. Soc. London Sect. A* **68**, 681 (11.4).

Aubert, B. (1969). *Proc. Topical Conf. Weak. Interactions*, CERN 69-7, p. 205. CERN, Geneva (11.12, 11.14).
Backenstoss, G., Hyams, B. D., Knop, G., Marin, P. C., and Stierlin, U. (1961). *Phys. Rev. Lett.* **6**, 415 (11.9).
Bardon, M., Franzini, P., and Lee, J. (1961). *Phys. Rev. Lett.* **7**, 23 (11.8).
Barkas, W. H. (1956). *Phys. Rev.* **101**, 778 (11.2).
Bergkvist, K. E. (1969). *Proc. Topical Conf. Weak Interactions*, CERN 69-7, p. 91. CERN, Geneva (11.2).
Berman, S. M., and Sirlin, A. (1962). *Ann. Phys. (New York)* **20**, 20 (11.8).
Bernstein, J. (1968). "Elementary Particles and Their Currents," p. 265. Freeman, San Francisco, California (11.12).
Bernstein, J., Fubini, S., Gell-Mann, M., and Thirring, W. (1960). *Nuovo Cimento* **17**, 757 (11.10).
Bingham, H. (1965). *Proc. Roy. Soc. Ser. A* **285**, 202 (11.12).
Blatt, J. M., and Weisskopf, V. F. (1952). "Theoretical Nuclear Physics." Wiley, New York (11.3).
Brene, N., Veji, L., Roos, M., and Cronstrom, C. (1966). *Phys. Rev.* **149**, 1288 (11.13).
Burgy, M. T., Krohn, V. E., Novey, T. B., Ringo, G. R., and Telegdi, V. L. (1960). *Phys. Rev.* **120**, 1829 (11.5).
Cabibbo, N. (1963). *Phys. Rev. Lett.* **10**, 531 (11.13).
Cabibbo, N. (1964). *Phys. Rev. Lett.* **12**, 62 (11.14).
Cabibbo, N., and Maksymowicz, A. (1964). *Phys. Lett.* **9**, 352 (11.12).
Carter, R. E., Reines, F., Wagner, J. J., and Wyman, M. E. (1959). *Phys. Rev.* **113**, 280 (11.7).
Chu, W. T., Nadelhaft, I., and Ashkin, J. (1965). *Phys. Rev. B* **137**, 352 (11.11).
Coleman, S., and Glashow, S. L. (1964). *Phys. Rev. B* **134**, 671 (11.14).
Conversi, M., Diebold, R., and Di Lella, L. (1964). *Phys. Rev. B* **136**, 1077 (11.11).
Cool, R. L., Lederman, L. M., and Tinlot, J. (1965). *Phys. Rev. Lett.* **14**, 724 (11.11).
Coombes, C. A., Cork, B., Galbraith, W., Lambertson, G. R., and Wenzel, W. A. (1957). *Phys. Rev.* **108**, 1348 (11.12).
Cork, B., Kerth, L., Wenzel, W. A., Cronin, J. W., and Cool, R. L. (1960). *Phys. Rev.* **120**, 1000 (11.14).
Crawford, F. S. *et al.* (1958). *Proc. Int. Conf. High Energy Phys. CERN, 1958*, p. 323. CERN, Geneva (11.12).
Cronin, J. W. (1968). *Proc. Int. Conf. High Energy Phys., 14th, Vienna, 1968*, p. 289. CERN, Geneva (11.8, 11.12).
Dalitz, R. H. (1966). *Proc. Int. Sch. Phys. Enrico Fermi, Varenna, 1964*. **32**, 206 (11.10).
Danby, G. *et al.* (1962). *Phys. Rev. Lett.* **9**, 36 (11.8).
Davies, R., and Harmer, D. S. (1959). *Bull. Amer. Phys. Soc.* **4**, 217 (11.7).
Deutsch, M., Gittelman, B., Bauer, R. W., Grodzins, L., and Sunyar, A. W. (1958). *Phys. Rev.* **107**, 1733 (11.5).
Di Capua, E., Garland, R., Pondrom, L., and Strelzoff, A. (1964). *Phys. Rev. B* **133**, 1333 (11.8).
Duclos, J., Heitze, J., de Rujula, A., and Soergel, V. (1964). *Phys. Lett.* **9**, 62 (11.8).
Ericson, T. E. O. (1964). Recent developments in mu capture. *CERN TH/474* (11.11).
Farley, F. J. M., Massam, T., Muller, T., and Zichichi, A. (1962). *Proc. Int. Conf. High Energy Phys. CERN, 1962*, p. 415. CERN, Geneva (11.8).
Fermi, E. (1934). *Z. Phys.* **88**, 161 (11.2, 11.3).
Feynman, R. P., and Gell-Mann, M. (1958). *Phys. Rev.* **109**, 193 (11.6, 11.10).

References

Filthuth, H. (1969). *Proc. Topical Conf. Weak Interaction, CERN 69-7*, p. 131. CERN, Geneva (11.13).
Freeman, J. M., White, R. E., Montagne, J. H., Murray, G., and Burcham, W. E. (1964). *Phys. Lett.* **8**, 115 (11.4, 11.8).
Gamow, G., and Teller, E. (1936). *Phys. Rev.* **49**, 895 (11.3).
Gell-Mann, M. (1964). *Phys. Rev. Lett.* **12**, 155 (11.14).
Gell-Mann, M. (1958). *Phys. Rev.* **111**, 362 (11.10).
Gell-Mann, M., and Rosenfeld, A. H. (1957). *Annu. Rev. Nucl. Sci.* **7**, 407 (11.14).
Gerhart, J. B. (1958). *Phys. Rev.* **109**, 897 (11.4).
Gobbi, B., Green, D., Hakel, W., Moffet, R., and Rosen, J. (1968). Reported by J. Cronin, *Proc. Int. Conf. High Energy Phys., 14th, Vienna, 1968.* p. 295. CERN, Geneva (11.14).
Goldberger, M. L., and Treiman, S. B. (1958). *Phys. Rev.* **111**, 358, **110**, 1178 (11.10).
Goldhaber, M., Grodzins, L., and Sunyar, A. W. (1958). *Phys. Rev.* **109**, 1015 (11.5).
Hamilton, W. D. (1969). *Progr. Nucl. Phys.* **10**, 1 (11.9).
Ioffe, B. L., and Shapalin, E. P. (1967). *Yad. Fiz.* **6**, 828; *Sov. J. Nucl. Phys.* **6**, 603 (11.17).
Källen, G. (1964). "Elementary Particle Physics." Addison-Wesley, Reading, Massachusetts (11.3, 11.8, 11.12).
Konopinski, E. J. (1954). *Phys. Rev.* **94**, 492 (11.3).
Landau, L. D. (1957). *Nucl. Phys.* **3**, 127 (11.6).
Lee, T. D., and Wu, C. S. (1965). *Annu. Rev. Nucl. Sci.* **15**, 381 (11.10).
Lee, T. D., and Yang, C. N. (1957). *Phys. Rev.* **105**, 1671 (11.6).
Lee, T. D., and Yang, C. N. (1960a). *Phys. Rev. Lett.* **4**, 307 (11.10).
Lee, T. D., and Yang, C. N. (1960b). *Phys. Rev.* **119**, 1410 (11.14, 11.15).
Low, F. E. (1969). *Comments Nucl. Particle Phys.* **3**, 36 (11.17).
Matthews, P. T. (1965). *Proc. Roy. Soc. Ser. A* **285**, 214 (11.14).
Marshak, R., and Sudarshan, G. (1958). *Phys. Rev.* **109**, 1860 (11.6).
Michel, L. (1950). *Proc. Phys. Soc. London Sect. A* **63**, 514 (11.8).
Nambu, Y. (1960). *Phys. Rev. Lett.* **4**, 380 (11.13).
Okun, L. B. (1965). "Weak Interaction of Elementary Particles." Pergamon Press, Oxford (11.3).
Particle Data Group, (1969). *Rev. Mod. Phys.* **41**, 109 (11.4, 11.10, 11.12, 11.14).
Perkins, D. H. (1969). *Proc. Topical Conf. Weak Interactions, CERN 69-7* p. 1. CERN, Geneva (11.10, 11.15, 11.16).
Primakoff, H., and Rosen, S. P. (1959). *Rep. Prog. Phys.* **22**, 121 (11.7).
Reines, F., and Cowan, C. L. (1959). *Phys. Rev.* **113**, 273 (11.2, 11.7).
Rubbia, C. (1969). *Proc. Topical Conf. Weak Interactions, CERN 69-7*, p. 227. CERN, Geneva (11.12, 11.13).
Salam, A. (1957). *Nuovo Cimento* [10], **5**, 299 (11.6).
Sens, J. C. (1966). *Proc. CERN Sch. Phys., 1966*, CERN 66-29 (11.11).
Sherr, R., and Miller, R. H. (1954). *Phys. Rev.* **93**, 1076 (11.4).
Steinberger, J. (1969). *Proc. Topical Conf. Weak Interactions, CERN 69-7*, p. 291. CERN, Geneva (11.14).
Weinberg, S., (1958). *Phys. Rev.* **112**, 1375 (11.10).
Weisberger, W. I. (1965). *Phys. Rev. Lett.* **14**, 1047 (11.10).
Wu, C. S. (1964). *Rev. Mov. Phys.* **36**, 618 (11.4, 11.10).
Wu, C. S., Ambler, E., Hayward, R. W., Hoppes, D. D., and Hudson, R. P. (1957). *Phys. Rev.* **105**, 1413 (11.5).

XII

STRONG INTERACTIONS

12.1 Introduction

It has not proved possible to find a theory of strong interactions that has had a quantitative success as great as that of quantum electrodynamics. Attempts using field theory and a perturbation expansion fail because of the strength of the coupling constants. Nonetheless, there has been success in the application of dispersion relations, and models, such as one-particle exchange and Regge poles, have given some insight into strong interactions. We shall discuss these subjects at an elementary level in this chapter.

12.2 The Mandelstam Variables

The variables introduced by Mandelstam (1959) to describe the kinematics of two-body collisions are the most convenient for discussions in this chapter. Therefore, we devote a section, at this point, to describing them and their properties. Consider the collision of two particles, masses m_1, m_2 with four-momenta p_1, p_2, to give particles m_3, m_4 with four-momenta p_3, p_4:

$$1 + 2 \rightarrow 3 + 4. \tag{12.1}$$

Conservation of energy and momentum requires

$$p_1 + p_2 = p_3 + p_4. \tag{12.2}$$

The Mandelstam variables are

$$s = (p_1 + p_2)^2, \tag{12.3}$$

$$t = (p_1 - p_3)^2, \tag{12.4}$$

$$u = (p_1 - p_4)^2. \tag{12.5}$$

12.2 The Mandelstam Variables

Now

$$s + t + u = p_1^2 + p_2^2 + p_3^2 + p_4^2 + (2p_1^2 + 2p_1p_2 - 2p_1p_3 - 2p_1p_4).$$

Eq. (12.2) shows that the term in parentheses is zero, and we have

$$s + t + u = m_1^2 + m_2^2 + m_3^2 + m_4^2. \tag{12.6}$$

We expect such a connection, since only two variables are required to describe the kinematics of a two-body collision, apart from its spatial orientation. The quantity s is the squared total center-of-mass energy of the system and t is the squared momentum transfer between particles 1 and 3 (or 4 and 2) and u that between particles 1 and 4 (or 3 and 2). The cosine of the center-of-mass scattering angle is given by

$$t = m_1^2 + m_3^2 - 2(E_1 E_3 - P_1 P_3 \cos \theta) \tag{12.7}$$

if E_1, E_3 and P_1, P_3 are total energies and three-momenta evaluated in the center-of-mass system. As s becomes very large $E_1 \simeq P_1$, $E_3 \simeq P_3$, and we have

$$t \simeq -2P_1P_3(1 - \cos \theta).$$

If E_{1L} is the incident laboratory total energy of particle 1, and particle 2 is at rest, we have

$$s = m_1^2 + m_2^2 + 2m_2 E_{1L}.$$

For large E_{1L} we have

$$s \simeq 2m_2 E_{1L}.$$

We now consider two other reactions related to the first by crossing the particles and changing them from particle to antiparticle. Obviously

$$1 + \bar{3} \to \bar{2} + 4 \quad \text{Channel II}$$

and

$$\bar{3} + 2 \to \bar{1} + 4 \quad \text{Channel III}$$

are related to the reaction of Eq. (12.1) (Channel I). For example,

$$\pi^- + p \to \pi^- + p$$
$$1 + 2 \to 3 + 4,$$

leads to

$$\pi^- + \pi^+ \to \bar{p} + p,$$

and to

$$\pi^+ + p \to \pi^+ + p.$$

To see how the variables for the three channels are related, consider continuing p_3 down from its physical region having $E_3 > m_3$ to $E_3 < -m_3$, where it now represents an antiparticle. We can define the four-momenta of the antiparticle so that it has positive energy and momentum

$$p_{\bar{3}} = -p_3.$$

Similarly we can continue p_2 down and put

$$p_{\bar{2}} = -p_2.$$

Now consider Channel II, in which particles 2 and 3 of Channel I reappear as antiparticles. The squared center-of-mass energy for Channel II is given by

$$s_{II} = (p_1 + p_{\bar{3}})^2$$
$$= (p_1 - p_3)^2$$
$$= t.$$

The momentum transfer in Channel II is given by

$$t_{II} = (p_1 - p_{\bar{2}})^2$$
$$= (p_1 + p_2)^2$$
$$= s.$$

Thus the variable s, which is the squared center-of-mass energy in Channel I, becomes the squared momentum transfer in Channel II; and t, which is the squared momentum transfer in Channel I, becomes the squared center-of-mass energy in Channel II. In Channel I, we will have

$$s \geqslant (m_1 + m_2)^2 \quad \text{and} \quad t \leqslant 0$$

and in Channel II

$$s \leqslant 0 \quad \text{and} \quad t \geqslant (m_1 + m_3)^2.$$

The variable u becomes the total center-of-mass energy for the Channel III. It has become usual to label reactions connected in this way by the Mandelstam variable that gives the total center-of-mass energy. Thus, if the s-channel is

$$\pi^- + p \to \pi^0 + n,$$

then the t-channel is

$$\pi^- + \pi^0 \to \bar{p} + n,$$

and the u-channel is

$$\pi^0 + p \to \pi^+ + n.$$

Since only two variables are required to specify completely the kinematics of any reaction, we can plot the kinematics configuration of any three reactions as a point in an s–t plane. Since $s + t + u = \sum m^2$ it can, in fact, be plotted with respect to the extended sides of an equilateral triangle of height $\sum m^2$. In Fig. 12.1 we have plotted such a triangle for the equal-

12.2 The Mandelstam Variables

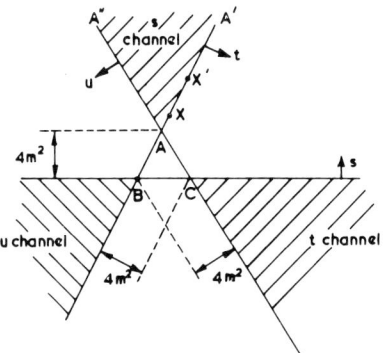

FIG. 12.1. Triangle plot of the values of the Mandelstam variables $s, t,$ and u for the equal-mass case: $s + t + u = 4m^2$. Shaded areas are the physically accessible regions in real scattering.

mass case. With the axes and directions correctly labeled, every point has $s + t + u = 4m^2$. The physically allowed regions are shaded. Consider the s-channel where the reaction is

$$1 + 2 \to 3 + 4.$$

At the point X, $s > 4m^2$, so this point represents a physically allowed scattering configuration. However, $t = 0$, so the center-of-mass angle from 1 to 3 is $0°$. Finally, u, the momentum transfer squared from 1 to 4, is negative. The same applies at X', but u is now more negative. Particle 4 is moving in the opposite direction to 1, and thus we expect an increasing momentum transfer between them as s increases. The line AA' evidently represents the kinematic condition in which the angle between $\mathbf{P}_1$ and $\mathbf{P}_3$ is zero, and the line AA'' the condition of zero angle between $\mathbf{P}_1$ and $\mathbf{P}_4$. When the masses become unequal, the physical-region boundary becomes less simple.

Equation (12.7) gives $\cos \theta$ in terms of t. For large s, evidently

$$t \simeq -2q^2(1 - \cos \theta), \qquad (12.8)$$

where q is the momentum in the center-of-mass system. Alternatively,

$$\cos \theta \simeq 1 + (t/2q^2). \qquad (12.9)$$

[The approximate equalities of Eqs. (12.8) and (12.9) become exact if the masses of all four particles involved are equal.] If now we are thinking of a continuation of the variables from one physical reaction to another, obviously t, which must be negative in the s-channel to give $\cos \theta$ between ± 1, will become positive in the t-channel and make $\cos \theta > 1$. The reason for considering this possibility follows from the idea that there is one function, $A(s, t, u)$, which is the scattering amplitude and which has an analytic continuation that describes the scattering in all three channels. If we have an expression for $A(s, t, u)$ in terms of s and $\cos \theta$ for the s-channel scattering, it is of interest to know how it behaves when continued to values of $\cos \theta > 1$ or even as $\cos \theta \to \infty$, as it may then describe the amplitude for t-channel scattering.

The property that the analytic continuation of the scattering amplitude will describe the three reactions related by crossing is called *crossing symmetry*.

It is important to be clear exactly what crossing symmetry says about scattering amplitudes. First, the s, t, and u variables do not specify spins or isotopic spins, so that the function $A(s, t, u)$, without other labels, is not a complete description of the scattering. Let us consider the elastic scattering process

$$\pi^+ + p \to \pi^+ + p.$$

Neglecting spins we use $T_{\pi^+ p}(p_1, q_1; p_2, q_2)$ to denote the amplitude for this process, where the pion four-momentum changes from q_1 to q_2 and the proton four-momentum from p_1 to p_2. Later we put $s = (p_1 + q_1)^2$, $t = (q_1 - q_2)^2$, $u = (q_1 - p_2)^2$, and we will have

$$s \geqslant (m_\pi + m_p)^2, \quad t \leqslant 0, \quad \text{and} \quad u \leqslant (m_p - m_\pi)^2 - t.$$

This amplitude does not change on crossing a particle from one side of the reaction to the other if, at the same time, we change it into its antiparticle and reverse the sign of all four components of momentum. If in this case we cross both mesons, we come to $\pi^- p$ elastic scattering and we have

$$T_{\pi^- p}(p_1, -q_2; p_2, -q_1) = T_{\pi^+ p}(p_1, q_1; p_2, q_2). \tag{12.10}$$

This does not tell us much about the amplitude for $\pi^- p$ scattering, since the left-hand side applies in a nonphysical region. This is because the total center-of-mass energy squared is the sum of the incoming momenta squared, $(p_1 - q_2)^2$, and this quantity is u for $\pi^+ p$ elastic scattering, which, from a figure for πp scattering such as that shown in Fig. 12.1, can never be greater than $(m_p + m_\pi)^2$ in the physical region for $\pi^+ p$ scattering.

We now rewrite the arguments of the amplitudes in terms of s, t, and u variables, and we see

$$T_{\pi^+ p}(s, t, u) = T_{\pi^- p}(u, t, s), \tag{12.11}$$

where the rule is that the first argument is the square of the sum of the incoming momenta, the second is the direct momentum transfer squared [the difference between first and third momenta in Eq. (12.4)], and the third is the crossed momentum transfer (the difference between first and fourth momenta).

If the amplitudes can be analytically continued, then the $T_{\pi^+ p}$ can be continued in the values of s, t, u until they are outside the physically attainable region for $\pi^+ p$ scattering and in the physically attainable region for π^-p scattering. Thus the amplitude T_{π^+p} by continuation also describes π^-p scattering.

Similar crossing between the q_1, p_2 variables in Eq. (12.10) shows that the analytic continuation of T_{π^+p} will also produce the amplitude for

$$\pi^+ + \pi^- \to p + \bar{p}.$$

All the reactions may be subjected to a *TCP* transformation, and we then obtain the amplitudes for pion–antiproton scattering. This, however, is trivial.

We are interested in amplitudes as functions of the variables s, t, and u, and in particular with the assumption that they are analytic in these variables considered as complex variables. This leads to dispersion relations.

12.3 The Analytic Properties of the *S*-Matrix

In Section 10.11, we discussed briefly the application of field theory to strong interactions, and indicated that the perturbation series solution could not be expected to apply. The theory of strong interactions has therefore developed by studying the properties of the transition amplitudes (which are related to elements of the *S*-matrix), in the hope that it would be possible to calculate their values without the use of field theory. Although this hope is not fully realized, the method has yielded a great deal of insight into the properties of strong interactions. We have previously stressed some of the properties satisfied by the strong-interaction *S*-matrix:

(1) symmetry under rotations in ordinary space and isotopic-spin space (conservation of angular momentum and isotopic spin),
(2) Lorentz invariance,
(3) invariance under charge conjugation and the parity transformation,
(4) unitarity,
(5) reciprocity,
(6) crossing symmetry.

Further progress is made by investigating the analytic properties of the transition amplitudes. In the early developments the analyticity was believed to depend upon the need for the *S*-matrix to satisfy the requirements of causality. However, it is, in fact, difficult to establish rigorously the connection between the two, and quite often one just assumes that the transition amplitudes are the real-axis boundary values of analytic functions (Eden *et al.*, 1966).

A transition amplitude is taken to be a function $T(s, t)$ of the two Mandelstam variables s and t. For t held constant, the assumption is made that T is an analytic function of the complex variable s, so that the physical values of this amplitude are the values of this function on the real axis of s. It is also possible to hold s constant and consider t as the complex variable, or to assume T is an analytic function of all three complex variables, s, t, and u (Mandelstam representation). We shall start by considering the first possibility for the two-body reaction with the s-channel:

$$1 + 2 \to 3 + 4.$$

For the moment we shall avoid various complications by assuming the particles are spinless and have zero isotopic spin. The experimentally accessible values of s in the s-channel are

$$s \geqslant s_0 = (m_1 + m_2)^2. \tag{12.12}$$

In the u-channel $\bar{3} + 2 \to \bar{1} + 4$, the accessible values of u are

$$u \geqslant u_0 = (m_2 + m_3)^2.$$

This means that, since $s + t + u = m_1^2 + m_2^2 + m_3^2 + m_4^2$ and t is fixed, values of $s \leqslant \bar{s}_0 = m_1^2 + m_2^2 + m_3^2 + m_4^2 - u_0 - t$ are also accessible. For elastic scattering $m_1 = m_3$, $m_2 = m_4$, and then $\bar{s}_0 = (m_1 - m_2)^2 - t$.

The unitarity of the S-matrix means that the T amplitude [Eq. (3.64)] satisfies the following relation (Eden, 1967):

$$-i(T_{\alpha\beta} - T_{\beta\alpha}^\dagger) = \sum_{\gamma} T_{\alpha\gamma} T_{\gamma\beta}^\dagger. \tag{12.13}$$

On the right-hand side the sum is over all real states that are accessible, allowing for conservation of energy and momentum and of all relevant quantum numbers. (This equation is a simplified one, since we have hidden in the summation integrals over phase space and δ-functions. Thus, terms giving the phase space for various states γ should be included.) As s increases, the threshold for various particle reactions will be passed (so-called normal thresholds); at each threshold, new terms appear on the right, and this leads to a singularity in T, which is a branch point with a discontinuity across the cut given by Eq. (12.13). The corresponding cut extends from the threshold to $s = \infty$ and is by convention taken to be along the real axis. If there is an exoergic reaction channel $1 + 2 \to 3 + 4$, then there will be a cut extending from a branch point with $s < s_0$. If the system $1 + 2$ has the same quantum numbers as a single-particle state of mass m_0 but $m_0 < m_1 + m_2$, then there is a singularity in T at $s = m_0^2$. This singularity is a pole. If this particle is unstable, then the corresponding pole moves away from the real axis. We will discuss this situation later.

In addition to singularities generated by thresholds and single-particle states in the s-channel, there will also be those caused by singularities in

12.3 The Analytic Properties of the S-Matrix

the u-channel (if t is fixed). A singularity at u' appears in s at a point $s' = \sum m^2 - u' - t$. In the case of cuts these extend from $s = s'$ to $s = -\infty$. These remarks about singularities apply equally to the complex planes of u with s fixed or of t with u fixed, if the appropriate interchanges of the three variables are made.

Since the scattering or transition amplitude is the value of the analytic function on the real axis and since there exist cuts along the real axis, it is important to know whether the amplitude is found by taking the limit of the amplitude from above or below the real axis. The answer to this question is given by perturbation theory (Eden et al., 1966) to be from above, so that the physical amplitude at total energy $\sqrt{s}$ is given by

$$T(s, t, u) = \lim_{\epsilon \to 0} T(s + i\epsilon, t, u). \tag{12.14}$$

Suppose now we consider crossing to the u-channel, keeping t fixed. The amplitude is now

$$\lim_{\epsilon \to 0} T(s, t, u + i\epsilon).$$

Since $s + t + u = \sum m^2$, the limiting procedure in the variable u is equivalent in this channel to approaching the values of s (which is negative in the u-channel) from the negative imaginary part of the plane. Thus, in Fig. 12.2, the crossing from the s-channel to the u-channel at constant t is equivalent to following a path in the complex variable s from the upper side of the positive real axis cut, through the real axis between s_0 and $\bar{s}_0$, and back toward the real s axis from below.

Before we examine dispersion relations for the transition amplitude, let us consider a simple dispersion relation for the scattering amplitude $f(\omega)$ for light of frequency ω. Applying Eq. (A.9) of Appendix A, we have

$$f(\omega) = \frac{1}{\pi i} P \int_{-\infty}^{+\infty} \frac{f(\omega') \, d\omega'}{\omega' - \omega}. \tag{12.15}$$

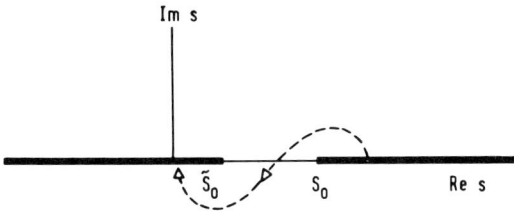

FIG. 12.2. Crossing from the s- to the u-channel means moving from real positive s-values, approached from above the real axis, to the real s-values, approached from below the real axis.

If we divide $f(\omega)$ into real and imaginary parts,

$$f(\omega) = D(\omega) + iA(\omega),$$

we obtain

$$D(\omega) = \frac{1}{\pi} P \int_{-\infty}^{+\infty} \frac{A(\omega') \, d\omega'}{\omega' - \omega}$$

and similarly for $A(\omega)$. Now it can be shown that

$$D(-\omega) = D(\omega) \quad \text{and} \quad A(-\omega) = -A(\omega),$$

so that

$$D(\omega) = \frac{1}{\pi} P \int_{-\infty}^{0} \frac{A(\omega') \, d\omega'}{\omega' - \omega} + \frac{1}{\pi} P \int_{0}^{+\infty} \frac{A(\omega') \, d\omega'}{\omega' - \omega}.$$

Changing the variable of integration in the first integral from ω' to $-\omega'$ and using the odd property of $A(\omega')$ gives

$$D(\omega) = \frac{1}{\pi} P \int_{0}^{\infty} \left[\frac{A(\omega') \, d\omega'}{\omega' + \omega} + \frac{A(\omega') \, d\omega'}{\omega' - \omega} \right]$$

$$= \frac{2}{\pi} P \int_{0}^{\infty} \frac{\omega' A(\omega') \, d\omega'}{\omega'^2 - \omega^2}.$$

We can use the optical theorem to relate $A(\omega')$ for forward scattering to the total cross section $\sigma(\omega)$:

$$\sigma(\omega) = \frac{4\pi}{\omega} A(\omega),$$

so that the real part of the forward-scattering amplitude is related to an integral over the total cross section:

$$D(\omega) = \frac{1}{2\pi^2} P \int_{0}^{\infty} \frac{\omega'^2 \sigma(\omega') \, d\omega'}{\omega'^2 - \omega^2}. \tag{12.16}$$

Let us consider now dispersion relations applied to the transition amplitude [a matrix element of T in Eq. (3.64)] for a strong interaction process (Pilkuhn, 1967). Figure 12.3 shows the form of the relevant contour. In the figure, it is shown as it is before expanding its circle to infinity, apart from small clockwise circles left around poles and the necessary journeys along cuts. The assumption is made that there are no other singularities, apart from those on the real axis. Then, using Eq. (A.3), we can write

$$T(z) = \frac{1}{2\pi i} \oint \frac{T(z') \, dz'}{z' - z}$$

12.3 The Analytic Properties of the S-Matrix

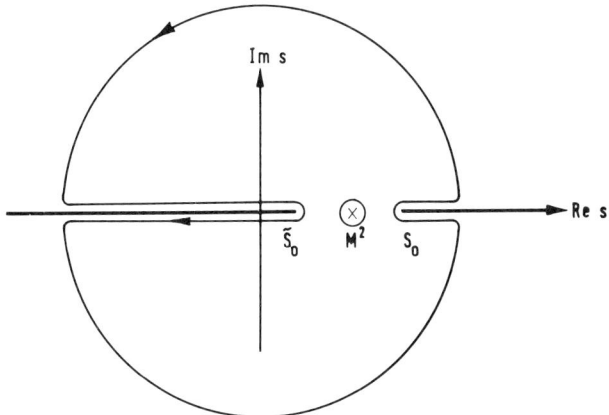

FIG. 12.3. Contour in the complex-s plane used to obtain a dispersion relation in the scattering amplitude. The heavy lines are cuts. The cross on the real axis is the position of a pole in the amplitude.

(z stands for a complex s). If $|T(z)| \to 0$ as $z \to \infty$, the contribution from the circle at infinity is zero and then the integral can be written

$$T(z) = \frac{1}{2\pi i}\left[\int_{-\infty}^{\bar{s}_0} + \int_{\bar{s}_0}^{-\infty} + \int_{+\infty}^{s_0} + \int_{s_0}^{\infty}\right]\frac{T(s')\,ds'}{s-s'} - R.$$

The integrals with limits that are the same but reversed are to be taken along different sides of the same cut. These can be combined, and the integrand becomes the discontinuity in $T(z)$ across the cut, namely $\Delta T(s')$. The last term follows from Eq. (A.5) and is the residue of the function $T(z')/(z'-z)$ at the pole in $T(z')$. This pole is only to be found in the amplitude given by the Born term, which is the name for the first nonzero term in a perturbation expansion. This term has

$$T(s) = c/(m^2 - s),$$

where c is a constant. Thus

$$\frac{T(z')}{z'-z} = \frac{c}{m^2-z'}\frac{1}{z'-z}.$$

Considered as a function of z' near the pole at $z' = m^2$, this has a residue $c/(m^2 - z)$. Finally, our dispersion relation reads

$$T(z) = \frac{-c}{m^2-z} + \frac{1}{2\pi i}\left(\int_{-\infty}^{\bar{s}_0} + \int_{s_0}^{+\infty}\right)\frac{\Delta T(s')\,ds'}{s'-z}.$$

If the theory is time-reversal invariant, the amplitude satisfies the following relation, for t real:
$$T(s + i\epsilon, t) = T^*(s - i\epsilon, t)$$
(Pilkuhn, 1967), which is the property called *Hermitian analyticity*. This means that $\Delta T(s') = 2i \operatorname{Im} T(s')$ and that $T(s)$ is real along the real axis between the branch points. Hence the relation becomes
$$T(z) = \frac{-c}{m^2 - z} + \frac{1}{\pi}\left[\int_{-\infty}^{s_0} + \int_{s_0}^{+\infty}\right]\frac{\operatorname{Im} T(s')\, ds'}{s' - z}.$$

The physical amplitude at energy $s(> s_0)$ is the limit of $T(s + i\epsilon)$ as $\epsilon \to 0$, and we have
$$T(s + i\epsilon) = \frac{-c}{m^2 - s - i\epsilon} + \frac{1}{\pi}\left[\int_{-\infty}^{s_0} + \int_{s_0}^{+\infty}\right]\frac{\operatorname{Im} T(s')\, ds'}{s' - s - i\epsilon}.$$

Taking the limit and using Eq. (A.7) we find
$$T(s) = \frac{-c}{m^2 - s} + \frac{1}{\pi}P\left[\int_{-\infty}^{s_0} + \int_{s_0}^{+\infty}\right]\frac{\operatorname{Im} T(s')\, ds'}{s' - s} + i\operatorname{Im} T(s).$$

This is now a dispersion relation for the real part of $T(s)$:
$$\operatorname{Re} T(s) = \frac{-c}{m^2 - s} + \frac{1}{\pi}P\left[\int_{-\infty}^{s_0} + \int_{s_0}^{+\infty}\right]\frac{\operatorname{Im} T(s')\, ds'}{s' - s}. \tag{12.17}$$

If, as $|z| \to \infty$, $T(z)$ does not go to zero, then it is necessary to make subtractions. Suppose that as $|z| \to \infty$,
$$T(z) \simeq |z|^{n-\alpha}$$
when n is a positive integer and $0 < \alpha \leqslant 1$. Then a dispersion relation with n subtractions is possible. To show how this works for $n = 1$, consider the Cauchy theorem [Eq. (A.3)] with the value of the amplitude at a point z_1 subtracted:
$$T(z) - T(z_1) = \frac{1}{2\pi i}\oint T(z')\left(\frac{1}{z' - z} - \frac{1}{z' - z_1}\right)dz'$$
$$= \frac{z - z_1}{2\pi i}\oint \frac{T(z')\, dz'}{(z' - z)(z' - z_1)}.$$

The integral now gives no contribution to the circle at infinity and converges. However, the cost is that one unknown constant has been added to the relation, namely $T(z_1)$. Every subtraction adds one more constant.

Let us apply now the dispersion relation to elastic scattering. To do this, we must consider the effect of crossing symmetry on the amplitudes for
$$1 + 2 \to 1 + 2$$

12.3 The Analytic Properties of the S-Matrix

and
$$\bar{1} + 2 \to \bar{1} + 2,$$

where the first is defined to be the s-channel, and the second the u-channel. We write the amplitude for the first $T_+(s + i\epsilon, t)$, where the $+i\epsilon$ means that the amplitude is found by approaching the real axis from above. The amplitude for the crossed reaction can be represented by $T_-(u - i\epsilon, t)$. However, T_+ must be the analytic continuation of T_- as discussed in Section 12.2 and expressed by Eq. (12.11). Hence

$$T_+(s' + i\epsilon, t) = T_-(u' - i\epsilon, t), \tag{12.18}$$

where

$$u' = \sum m^2 - s' - t. \tag{12.19}$$

Then we have, using Hermitian analyticity,

$$\int_{-\infty}^{s_0} \frac{ds'}{s' - s} \operatorname{Im} T_+(s') = \frac{1}{2i} \int_{-\infty}^{s_0} \frac{ds'}{s' - s} [T_+(s' + i\epsilon, t) - T_+(s' - i\epsilon, t)]$$

$$= \frac{1}{2i} \int_{s_0}^{-\infty} \frac{ds'}{s' - s} [T_-(u' + i\epsilon, t) - T_-(u' - i\epsilon, t)]$$

$$= \int_{s_0}^{-\infty} \frac{ds'}{s' - s} \operatorname{Im} T_-(u', t)$$

$$= \int_{u_0}^{\infty} \frac{du'}{u' + t + s - \sum m^2} \operatorname{Im} T_-(u', t).$$

Thus Eq. (12.17) becomes

$$\operatorname{Re} T_+(s, t) = \frac{-c}{m^2 - s} + \frac{1}{\pi} P\left[\int_{u_0}^{\infty} \frac{du'}{u' + t + s - \sum m^2} \operatorname{Im} T_-(u', t)\right.$$

$$\left. + \int_{s_0}^{\infty} \frac{ds'}{s' - s} \operatorname{Im} T_+(s', t)\right]. \tag{12.20}$$

The first integral is over values of u' that are physical for u-channel scattering, whereas the second integral is over values of s' that are physical for the s-channel. The difference is only that the incident particles are related by charge conjugation, while the target remains the same. Thus we can use s' to represent the squared center-of-mass energy in both cases, and since $u_0 = s_0 = (m_1 + m_2)^2$, we have

$$\operatorname{Re} T_+(s, t) = \frac{-c}{m^2 - s} + \frac{1}{\pi} P \int_{s_0}^{\infty} ds'$$

$$\times \left[\frac{\operatorname{Im} T_+(s', t)}{s' - s} + \frac{\operatorname{Im} T_-(s', t)}{s' + s + t - \sum m^2} \right]. \tag{12.21}$$

At $t = 0$ for elastic scattering the imaginary parts can be related to the total cross section by the optical theorem, so the dispersion relation connects the real part of the forward-scattering amplitude with the total cross sections. We shall apply this relation to pion–nucleon scattering in the next section.

The Mandelstam representation was mentioned earlier in this section (Mandelstam, 1958). This is a conjecture about the analytic properties of an invariant amplitude $T(s, t, u)$ in all three variables s, t, and u, where $s + t + u = \sum m^2$. If the variables are considered to be complex, then the relation is of the form

$$T(s, t, u) = \frac{1}{\pi} \int \frac{ds'}{s' - s} \rho_s(s')$$

$+$ similar terms in t and u

$$+ \frac{1}{\pi^2} \iint \frac{ds'\, dt'}{(s' - s)(t' - t)} \rho_{st}(s', t')$$

$+$ similar terms in t', u' and s', u',

where $s' + t' + u' = \sum m^2$. The functions ρ are called spectral functions. The properties of these functions are discussed by Chew (1962). The Mandelstam representation has not proved to be as useful as was first hoped.

12.4 Pion–Nucleon Scattering Dispersion Relations

The dispersion relation of Eq. (12.21) was derived for the condition of particles without spin or isotopic spin. This is not the case in pion–nucleon scattering, and we shall now take into account the spin of the nucleon. Since we are interested in applying Eq. (12.21), we shall consider $\pi^\pm$ scattering on protons.

The most general scattering amplitude $T_\pm$ that conserves parity is

$$T_\pm(s, t, u) = \bar{u}(p_2)[A_\pm(s, t, u) + \tfrac{1}{2} \gamma^\mu (q_1 + q_2)_\mu B(s, t, u)] u(p_1), \quad (12.22)$$

where the subscript "plus or minus" notation is the same as for Eq. (12.18). $\bar{u}(p_2)$ and $u(p_1)$ are the spinors for the final and initial proton states. A and B are called the Lorentz-invariant scattering amplitudes. Under crossing of the s- and u-channels, we have [see Eq. (12.10)]

$$s \rightleftarrows u,$$
$$T_+(s, t, u) = T_-(u, t, s). \quad (12.23)$$

Therefore

$$A_+(s, t, u) = A_-(u, t, s) \quad (12.24)$$

and

$$B_+(s, t, u) = -B_-(u, t, s), \quad (12.25)$$

12.4 Pion-Nucleon Scattering Dispersion Relations

because
$$q_1 \to -q_2 \quad \text{and} \quad q_2 \to -q_1.$$

There is a simpler variable E, given by
$$E = (1/2m_p)(s - m_p^2 - m_\pi^2 + \tfrac{1}{2}t) = (1/4m_p)(s - u), \quad (12.26)$$

so that $E \to -E$ under the crossing $s \rightleftharpoons u$. In the case of $t = 0$, E is the laboratory total energy of the incident pion (four-momentum q_1). We shall neglect the neutron-proton mass difference and put $m_p = m$.

The relation between the A and B amplitudes and the g and h amplitudes of Eq. (3.23) is given by the following: writing
$$g(\theta) - i h(\theta)\, \mathbf{n} \cdot \boldsymbol{\sigma} = f_1(\theta) + f_2(\theta)\, (\boldsymbol{\sigma} \cdot \hat{\mathbf{q}}_1)(\boldsymbol{\sigma} \cdot \hat{\mathbf{q}}_2),$$

where $\hat{\mathbf{q}}_1$ and $\hat{\mathbf{q}}_2$ are unit vectors along the direction of incident and scattered pion, we have
$$g(\theta) = f_1(\theta) + f_2(\theta) \cos \theta,$$
$$-h(\theta) = f_2(\theta) \sin \theta.$$

Then f_1 and f_2 are given by
$$f_1(\theta) = \frac{E^* + m}{16\pi m s^{1/2}} [A(s, t) + (s^{1/2} - m) B(s, t)],$$
$$f_2(\theta) = \frac{E^* - m}{16\pi m s^{1/2}} [-A(s, t) + (s^{1/2} + m) B(s, t)].$$

where E^* is the nucleon center-of-mass energy and θ is, of course, related to t, so that the independent variables are s and t.

Before writing down dispersion relations for any amplitudes, it is necessary to investigate the poles that may exist in these amplitudes. This is best done by examining the first nonzero terms in the perturbation expansion, the so-called Born terms. There are two contributions to the Born term, which have Feynman diagrams shown in Fig. 12.4. Contribution (a) has an amplitude
$$G^2 \, \bar{u}(p_2) \, \gamma_5 \, \frac{\gamma^\mu p_\mu + m}{p^2 - m^2} \, \gamma_5 \, u(p_1), \quad (12.27)$$

where $p = p_1 + q_1 = p_2 + q_2$ and m is the mass of the nucleon intermediate state (neutron in the case of $\pi^- p$ scattering). The two γ_5-matrices come

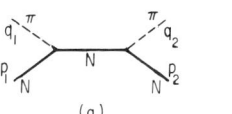
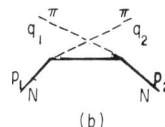

FIG. 12.4. Feynman diagrams for the Born terms in the amplitude for pion-nucleon scattering.

from the pseudoscalar coupling (Section 10.11). Using the properties of the γ-matrices and the Dirac equation [Eq. (10.22)], we find

$$\bar{u}(p_2)\gamma_5(\gamma^\mu p_\mu + m)\gamma_5 u(p_1) = -\bar{u}(p_2)\gamma^\mu(q_1 + q_2)_\mu u(p_1), \quad (12.28)$$

so that (a) does not contribute in first order to the invariant amplitude $A_\pm$. In addition, there is no contribution to B_+, since there can be no nucleon pole in the scattering amplitude [the $\Delta^{++}(1236)$ does not cause a pole on the real axis]. The contribution to B_- is $2G^2/(m^2 - s)$, where G is the renormalized pseudoscalar pion–nucleon coupling constant (Section 10.11). By similar arguments the Feynman diagram of Fig. 12.4b does not contribute to $A_\pm$, B_- and has a contribution $-2G^2/(m^2 - u)$ to B_+.

It is worth noting that we can define the pion–nucleon coupling constant, independently of field theory, as being given by the residue of the scattering amplitude at the nucleon pole.

It is possible to write down dispersion relations for the A and B amplitudes using Eq. (12.21) (Pilkuhn, 1967). However, we are most interested in a relation applying in the forward direction, since it is then possible to apply the optical theorem. We rewrite Eq. (12.22) to apply at 0°, where $q_1 = q_2$, $p_1 = p_2$, $t = 0$:

$$T_\pm(E, t = 0) = \bar{u}(p_1)[A_\pm(E, 0) + \gamma^\mu q_{1\mu} B_\pm(E, 0)]u(p_1)$$
$$= 2m[A_\pm(E, 0) + EB_\pm(E, 0)]. \quad (12.29)$$

This equation follows from the fact that if $\bar{u}(p_1) u(p_1) = 2m$, which is our normalization of the spinors, then

$$\bar{u}(p_1)\gamma^\mu u(p_1) = 2p_1^\mu,$$

and hence

$$\bar{u}(p_1)\gamma^\mu q_{1\mu} u(p_1) = 2p_1^\mu q_{1\mu}.$$

This is an invariant and, if evaluated in laboratory quantities, has the value $2mE$, since p_1^μ is the four-momentum of a stationary proton and $q_{10} = E$.

The imaginary part of the center-of-mass forward-scattering amplitude is related to the total cross section by the optical theorem [Eq. (3.84)], which becomes

$$\text{Im } T_\pm(E, 0) = 2ks^{1/2}\sigma_\pm(E)$$
$$= 2m(E^2 - m_\pi^2)^{1/2}\sigma_\pm(E). \quad (12.30)$$

We now apply the dispersion relation Eq. (12.21) to the forward-scattering amplitude of Eq. (12.29), using the known singularities in the $B_\pm$ amplitudes:

$$\text{Re } T_\pm(E, 0) = \frac{4G^2 Em}{-2mE \pm m_\pi^2} + \frac{1}{\pi} P\int_{m_\pi}^{\infty} dE'$$
$$\times \left[\frac{\text{Im } T_\pm(E', 0)}{E' - E} + \frac{\text{Im } T_\mp(E', 0)}{E' + E}\right].$$

12.4 Pion-Nucleon Scattering Dispersion Relations

Substituting from Eq. (12.30), we find

$$\frac{1}{2m} \text{Re } T_\pm(E, 0) = \frac{2G^2 E}{-2mE \pm m_\pi^2} + \frac{1}{\pi} P \int_{m_\pi}^{\infty} (E'^2 - m_\pi^2)^{1/2} dE'$$
$$\times \left[\frac{\sigma_\pm(E')}{E' - E} + \frac{\sigma_\mp(E')}{E' + E} \right]. \tag{12.31}$$

This relation is of no use because the integrals do not converge. However, a useful relation can be found from Eq. (12.31), namely

$$\frac{1}{2m} \text{Re } [T_-(E, 0) - T_+(E, 0)] = \frac{16\pi f^2}{E^2 - (m_\pi^2/2m)^2}$$
$$+ \frac{2E}{\pi} P \int_{m_\pi}^{\infty} \frac{dE' (E'^2 - m_\pi^2)^{1/2}}{E'^2 - E^2} [\sigma_+(E') - \sigma_-(E')], \tag{12.32}$$

where $f^2 = (G^2/4\pi)(m_\pi/2m)^2$. Relations of a similar kind were first derived by Goldberger *et al.* (1955) and applied to the available experimental data by Anderson *et al.* (1955). To obtain their relation we use a double subtraction by writing a dispersion relation for the function

$$\frac{(E + m_\pi)[T_\pm(E) - T_\pm(+m_\pi)] - (E - m_\pi)[T_\pm(E) - T_\pm(-m_\pi)]}{(E^2 - m_\pi^2)},$$

which has the same analyticity properties as $T(E)$. In addition we use crossing symmetry

$$T_\pm(+m_\pi) = T_\mp(-m_\pi)$$

to obtain

$$\text{Re } T_\pm(E) = \frac{1}{2}\left(1 + \frac{E}{m_\pi}\right) T_\pm(m_\pi) + \frac{1}{2}\left(1 - \frac{E}{m_\pi}\right) T_\mp(m_\pi)$$
$$\pm \frac{16\pi f^2 m(E^2 - m_\pi^2)}{m_\pi^2(1 \mp m_\pi^2/4m^2)} \frac{1}{(E \mp m_\pi^2/2m)}$$
$$+ \frac{2m(E^2 - m_\pi^2)}{\pi} P \int_{m_\pi}^{\infty} \frac{dE'}{(E'^2 - m_\pi^2)^{1/2}} \left[\frac{\sigma_\pm(E')}{E' - E} + \frac{\sigma_\mp(E')}{E' + E} \right]. \tag{12.33}$$

The $T_\pm(m_\pi)$ are related to the S-wave scattering lengths by

$$T_+(m_\pi) = 8\pi(m + m_\pi)a_3,$$
$$T_-(m_\pi) = \tfrac{8}{3}\pi(m + m_\pi)(2a_1 + a_3).$$

Anderson and co-workers showed that the values of the real part of the forward-scattering amplitude calculated from this dispersion relation were consistent over a wide range of energies with the values found from phase-shift analysis using $f^2 = 0.08$. More recently Samaranayake and

Woolcock (1965), using a related dispersion relation, have used more extensive data and fit to find the coupling constants and the scattering lengths:

$$f^2 = 0.0822 \pm 0.0018,$$
$$a_1 - a_3 = 0.292 \pm 0.020,$$
$$a_1 + 2a_3 = -0.035 \pm 0.012$$

(the lengths are in units of $h/m_\pi c$). Hamilton (1966) has found values (see Section 8.5) that are not quite consistent with these, but the methods are different and the source of the discrepancy is not clear. The high-energy cross-sections used may not be as accurate as believed. Dispersion relations have also been applied to forward charge-exchange scattering amplitudes (Höhler et al., 1966a).

Pion–nucleon scattering has also been studied by the use of partial-wave dispersion relations. These relations are found by considering the amplitudes as functions of the complex energy variable, as before. However, the singularities include the effects of thresholds in the t-channel and make the integration contours more complicated than in the case of the complete amplitude. However, the study (Hamilton, 1967) of these relations does lead to a description of low-energy pion–nucleon scattering in terms of only a few parameters and indicates the physical source of various effects (see Section 8.5).

12.5 Other Singularities

In our discussion of the analyticity of the S-matrix elements we have met two types of singularity, namely pole and branch point. A branch point is the manifestation of the threshold of a new particle reaction, while a pole is due to single-particle intermediate state. The thresholds mentioned are called normal thresholds. Another type of singularity does occur called an *anomalous threshold*, which can have a branch point below the lowest normal threshold. This type is discussed by Coleman and Norton (1964).

Other important singularities occur due to unstable particles: consider, for example

$$\pi^+ + p \to \Delta^{++}(1236) \to \pi^+ + p.$$

The amplitude for this kind of process is

$$\frac{m\Gamma}{m^2 - s - im\Gamma} \tag{12.34}$$

[compare with Eqs. (4.15) and (4.23)], where m is the mass of the unstable particle and Γ is its width. The pole in this amplitude is at $s = m^2 - im\Gamma$. Since m and Γ are real, this pole is at a value of s that is below the real axis.

12.6 Strong Interactions at High Energies

To reach this pole in the usual direction, i.e., from $s + i\epsilon$, means passing below the real axis. However, along the real axis we have a branch cut, and thus the pole is reached only by passing on to the second Riemann sheet. It follows that the "physical sheet"—that is, the one on which all contours were established—does not contain singularities in the amplitude that are due to the presence of unstable particles coupled to the reaction channel of interest. Of course if $\Gamma \to 0$, the pole appears on the real axis, and since $\Gamma = 0$ means stable it can appear only below the first normal threshold.

12.6 Strong Interactions at High Energies

In this section, we shall briefly review our present experimental knowledge of strong interactions at high energies. Accelerators have produced proton beams up to 70 GeV/c and secondary beams of an appropriately lower maximum momentum. It is these momenta that set the presently existing upper limit on energy in the systematic exploration of high-energy phenomena. We shall be primarily concerned with two-body collisions

$$a + b \to c + d.$$

The incident particle a will have to be stable in the sense of Section 9.3. The target particle b will normally be a proton or neutron (in deuterium). The possibility of using virtual targets such as pions or other mesons in the field of nucleons exists, but is beset with difficulties of interpretation. The two-body final state $c + d$ can be the same as the initial state and is not restricted to stable particles. Thus, although the reaction

$$\pi^+ + p \to \rho^0 + \Delta^{++}$$

appears as a final state of

$$\pi^+ + \pi^- + \pi^+ + p,$$

it can be considered as a two-body process, sometimes called quasi-two-body.

The quantities of interest in these reactions are the total cross section σ, the differential cross section $d\sigma/dt$, the polarization of any particles with spin in the final state, and the effect on these quantities of initial polarization. These quantities will be functions of s and, apart from σ, of t, the Mandelstam variables, and of the initial polarization.

The accessible energy range in any reaction extends from $s = (m_a + m_b)^2$ to ∞. This, at the present time, appears to divide itself roughly into two regions. The first is the "low-energy region," which extends from threshold to $s \simeq 10 \text{ GeV}^2$. There the behavior with energy of all quantities is more or less dominated by resonances. This is particularly true of the

$\bar{K}$-nucleon system, but it is not obviously true for the K-nucleon system, although the latter system is marked by fairly rapid cross-section changes with energy near $s = 2.3$ GeV². Rapid cross-section changes are not observed for any system in the high-energy region. There, all quantities vary smoothly with s, and although a slow or very rapid oscillatory behavior is not excluded, it seems likely that the presently accessible parts of the high-energy region are those where many quantities are beginning to approach some asymptotically characteristic behavior. A region sometimes called "Asymptopia" is supposed to exist in which the behavior as $s \to \infty$ is finally established. It is not clear at present where the boundary of this region occurs in the energy scale.

As an example of high-energy behavior we can mention the results of total cross-section measurements for nucleons, pions, kaons and antinucleons on nucleons. These appear to be decreasing with increasing s toward constant values, which are not the same. However, particle–nucleon and antiparticle–nucleon total cross sections are very close and tend to equality.

Elastic scattering $a + b \to a + b$ appears to be one of a few two-body reactions for which $d\sigma/dt$ for small t does not tend to zero as $s \to \infty$ (here, t is the squared momentum difference between the initial and final state of a). In fact, $d\sigma/dt$ at a fixed t appears to approach a nonvanishing limit. In the region of $0 \leqslant -t \leqslant 1$ GeV² the variation of the differential cross section is described by

$$d\sigma/dt = A \exp(Bt + Ct^2), \qquad (12.35)$$

with C almost negligible. B is in the range 4–12 GeV^{-2} and can vary with energy. For larger $-t$ the behavior often becomes less simple. For example, the differential cross section in proton–proton scattering decreases with increasing $-t$, but there are changes of slope and a flattening of the curve at 90° scattering, as expected (Allaby *et al.*, 1968), and in pion–proton scattering the cross section contains some structure at large $-t$ and a peak in the backward direction (Orear *et al.*, 1968). These latter features decrease as s increases.

Some two-body reactions also have the property that their channel total cross section and $d\sigma/dt$ do not decrease with energy. The feature common to all these reactions is that no changes in quantum numbers occur, apart from spin and parity, and when that occurs the change must be of the natural spin–parity series: 0^+, 1^-, 2^+, ... These reactions have the characteristics expected from the exchange of a particle with the quantum numbers of the vacuum. This exchanged "particle" appears as target to either of the incident particles. In each case, it can bring to the interaction orbital angular momentum and the corresponding parity. The orbital angular momentum need not be the same at each end of the exchange. Thus the following reactions are of this kind:

$$\pi + p \to \pi + N(1400),$$
$$\pi + p \to A_1(1070) + p.$$

The isovector and isoscalar parts of the photon have the quantum numbers of the neutral ρ and ω or ϕ meson, so the reactions

$$\gamma + p \to \rho^\circ + p,$$
$$\gamma + p \to \omega \text{ (or } \phi\text{)} + p,$$

are also among these reactions. The particle changes that are observed can occur in the field of complex nuclei, leaving the nucleus unchanged. They are therefore sometimes called coherent or diffraction production processes.

Most other two body or quasi-two-body reactions decrease with increasing energy, $d\sigma/dt$ varying between about s^{-1} and s^{-2}. Reactions involving double charge exchange, e.g., $K^-p \to \pi^+\Sigma^-$, vary approximately as s^{-4}. At low t the variation of the differential cross section is not as simple as it is in the case of elastic scattering, and, although the cross sections normally decrease with increasing momentum transfer, interesting structure is frequently observed. Care is required here on the values of the Mandelstam variables t and u, which interchange with a change of labeling in the final state. Conventionally in the case of a boson incident and produced, t is the square of the four-momentum difference between these two particles, so that small values of $-t$ mean small boson-production angles. Small values of u mean boson production angles near 180°. At the time of writing, the most recent compilation of data in this field is that edited by Kanada et al. (1967).

In Sections 12.8 and 12.9 we shall look at two models that go a part of the way to explaining some of the behavior that we have just briefly described.

12.7 Asymptotic Relations

A considerable amount of work has been done finding theoretical upper and lower limits on cross sections and amplitudes as $s \to \infty$. The assumptions made to obtain these relations vary: for example, using the Mandelstam representation, Froissart (1961) found that the total cross section $\sigma < c \log^2 s$, where c is a constant. These relations are not particularly restricting, and there is no experimental evidence for any breaking of the limits. They are described in some detail by Eden (1967).

Relations between cross sections, which are more directly testable, can also be derived. The best known is the Pomeranchuk theorem (1958), which states that as $s \to \infty$, the total cross section for particle–target collisions become asymptotically equal to that for antiparticle–target collisions. Interesting results also follow from the Okun–Pomeranchuk rule, which is the hypothesis that charge-exchange cross sections tend to zero as the energy increases. Given this, then the amplitudes and hence the differential cross sections for the elastic scattering of particles within an isotopic-spin multiplet on a given target become equal at high energies. The optical theorem then gives the same result for the total cross sections. The Okun–Pomeranchuk rule appears to generalize to all reactions involving the exchange of any quantum numbers other than those of the vacuum. Thus all nonelastic scattering reactions other than those that can go by a diffraction-like process will go asymptotically to zero.

The experimental evidence is that the particle and antiparticle cross-sections do appear to be converging but, in many cases, very slowly; and at the energies so far explored none has become equal within experimental error.

12.8 One-Particle-Exchange Mechanisms

The development of dispersion relations led naturally to a redefinition of coupling constants in terms of the residues at poles in scattering amplitudes. We have seen that there exist poles on the real energy (s) axis in the pion–nucleon scattering amplitude, which are due to single-particle intermediate states in the s- and u-channels at fixed t. It is evidently possible to have a pole in the amplitude due to a similar intermediate state in the t-channel. For example, the nucleon–nucleon scattering amplitude at constant s will be a function of t, and in the physical region as the center-of-mass scattering angle varies from 0 to 180°, t varies from zero into negative values: pion exchanges is expected to be important, and it follows there will be a pole in the amplitude at $t = m_\pi^2$, which is, of course, outside the physical region. The residue of the pole will contain the pion–nucleon coupling constant, which led Chew to suggest (1958) that an extrapolation of the differential cross section to this pole would measure this coupling constant. This procedure did not prove to be easy, because of the existence of second-order zeros and the exchange of other particles, but the ideas did generate other developments. We consider a reaction such as

$$\pi + N \rightarrow \pi + \pi + N \qquad (12.36)$$

and define t as the square of the momentum transfer to the nucleon. An important mechanism is expected to be pion exchange in the sense that the incident pion interacts with a virtual pion in the target nucleon; the Feynman diagram for this process is shown in Fig. 12.5a, and we see that it is the exchanged pion that carries the momentum transfer t. Evidently the amplitude for this process will contain a pole at the unphysical value of $t = m_\pi^2$. Considering now the residue at this point we see that it will contain the pion–nucleon coupling constant (f) and a term that describes the π–π interaction. In fact, the term depends, in this case, upon the $\pi\pi$ total cross section at a $\pi\pi$ center-of-mass energy (ω) equal to the effective mass of the two pions produced. Chew and Low (1959) proposed that this

12.8 One-Particle-Exchange Mechanisms

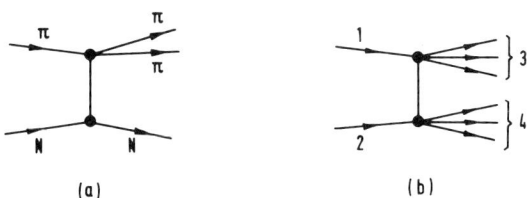

FIG. 12.5. (a) A one-pion-exchange diagram for inelastic pion–nucleon scattering. (b) The general one-particle-exchange diagram.

property be used as a means of measuring the $\pi\pi$ interaction cross section. The differential cross section $d^2\sigma/dt\, d\omega^2$ extrapolated to $t = m_\pi^2$ contains a factor $\sigma_{\pi\pi}(\omega)$:

$$\frac{d^2\sigma}{dt\, d\omega^2} \xrightarrow[t \to m_\pi^2]{} \frac{f^2}{2\pi m_\pi^2} \frac{t}{(t - m_\pi^2)^2} \frac{\omega}{P^2} \left(\frac{\omega^2}{4} - m_\pi^2\right)^{1/2} \sigma_{\pi\pi}(\omega), \quad (12.37)$$

where P is the incident-pion momentum in the laboratory. This formula can be used to obtain differential cross sections in the $\pi\pi$ system, once the relevant quantities are defined. Attempts to use this extrapolation technique were not outstandingly successful and gave results that were not free of ambiguities. The next development was essentially to postulate that, although other mechanisms were not absent, in the physical region closest to the pole the physical amplitude would be dominated by this pole. Thus Eq. (12.37) is taken to be true for values of t closest to m_π^2 instead of in the limit. An immediate prediction is that there will be strong preference for small-t (small-angle) events, which is just what is observed.

We see that small t means the exchange of almost real pions, and these are the ones associated with the long-range part of the interaction. Thus this model of the small-t production processes is sometimes known as the peripheral model. Going beyond one-pion exchange, we can apply the model to any reaction in which the exchange of one particle is expected to dominate; Fig. 12.5b shows the general one-particle exchange diagram with the possibility of the production of several particles at the vertices at both ends of the exchanged-particle line. It is now evident that if there is the possibility of a particularly strong interaction at one vertex at some value of the kinematic variables, then events having those values will be enhanced relative to those at other values. A well-known example of this is in two-pion production. The ρ-meson represents a region of strong π-π, isospin-1 interaction so that we expect, in reaction (12.36), to find a predominance of events in which the invariant mass of the two pions is that of the ρ-meson. This is just the case. In addition, the angular distribution of pions at this mass in the two-pion center-of-mass is expected to reflect that which would be observed in real π-π scattering at the mass of the ρ, namely $|P_1(\cos\theta)|^2$. This is nearly the case, but the distribution is evidently

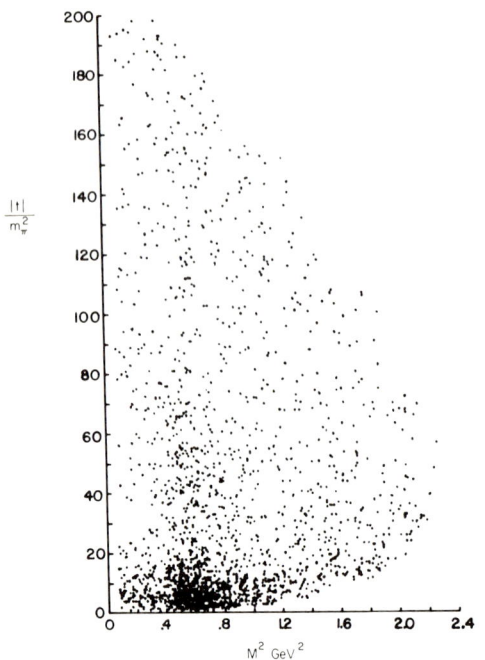

Fig. 12.6. A Chew-Low scatter plot of events in the reaction $\pi^- p \to \pi^- \pi^0 p^-$, where t is the squared momentum transfer to the proton and M^2 is the mass squared of the $\pi^- \pi^0$ system. The concentration of events shows the strong production of ρ^--mesons at low t. (Miller et al., 1967.)

influenced by other things. The application of these methods to the observation of bosons was discussed in Section 7.7.

Thus the predictions of the model can be summarized as (1) predominance of small-t events, (2) the strong production of resonances by quasi-two-body reactions. This is exemplified in an example of a Chew–Low plot (Fig. 12.6) for the reaction of Eq. (12.36). In this case we have a scatter plot of events according to the squared momentum transfer and to the invariant mass of the two final state pions. The strong ρ production at low t is clearly shown.

In fact, there is no reason why the exchanged particle should be a boson. If it is a baryon in the case of a boson incident on a proton, then the relevant pole is due to a single-particle state in the u-channel and at constant s this pole can be expected to dominate the angular distribution at u values close to the pole. The expected effect is a backward peak in the differential cross section: such peaks are observed. In general they are weaker and narrower than those in the forward direction in boson–nucleon interactions.

It is relevant to make some statements about the quantum numbers exchanged. In all one-particle-exchange processes, baryon number, charge, G-parity, strangeness, and isotopic spin are strictly conserved at both

12.8 One-Particle-Exchange Mechanisms

vertices. This immediately restricts the possible exchanges. For example, G-parity conservation forbids one-pion exchange contributing to the amplitude for elastic pion–nucleon scattering. The determination of the angular-momentum-parity factors is the same as if the exchanged particle were free. Thus in $\pi^+ + p \to \rho^+ + p$ the exchanged pion interacts with incoming π in a relative orbital angular-momentum state of $l = 1$ to give a 2π state of spin-parity 1^-. Finally, it is worth noting that the exchanged particle carries energy and momentum that satisfy not $E^2 - P^2 = m^2$ but $E^2 - P^2 = t$, and that energy and momentum conservation are satisfied separately at each vertex.

Consider a particle produced in a one-particle-exchange process. Its decay angular distributions and the polarization of any of its decay products are related to the spin state of the produced particle, which in turn is related to the production mechanism. Thus we should examine the restrictions that apply to the spin-density matrix of a produced particle. Let us consider

$$a + b \to c + d,$$

where d is produced by the interaction of an exchanged particle e with the incoming particle:

$$b \to c + e,$$
$$a + e \to d.$$

Let us set up Cartesian coordinates in the rest frame of d with the z axis along $\mathbf{p}_a$, with $\mathbf{p}_b$ and $\mathbf{p}_c$ in the xz plane (Fig. 12.7). $\mathbf{p}_e$ is then directed along $-z$. A parity transformation of this configuration (called A) reverses all momenta but leaves the spin states unchanged. A final rotation of 180° around the y axis restores the initial momenta but rotates the spin states,

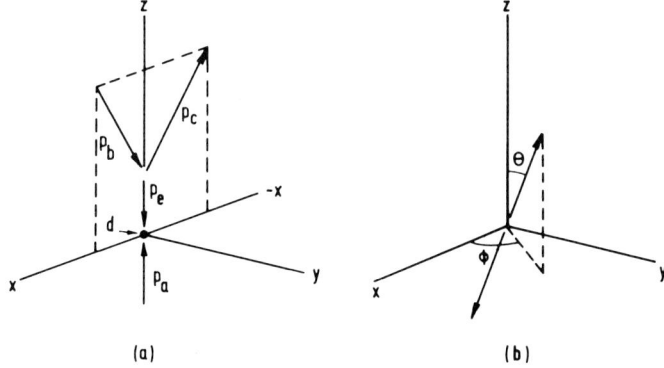

FIG. 12.7. (a) Diagram showing the Cartesian coordinates set up in the rest frame of d in the reaction $a + b \to c + d$. (b) Diagram showing the definition of the decay angles for the two-body decay of d.

giving us a system called B. If parity is conserved, the matrix element for the transition to B is the same (apart from a phase factor) as that to A. If the initial state is unpolarized, this means the final-state spin states are the same and the density matrix elements must be the same. Thus rotating the density matrix by 180° around the y axis should leave it unchanged.

The effect of this on the density matrix can be found by applying Eq. (2.98),

$$\rho = \mathscr{D}^{-1}\rho\mathscr{D},$$

where $\mathscr{D} = d^j_{j_z j_{z'}}(\beta)$ with $\beta = \pi$, from Eq. (2.68). Now

$$d^j_{j_z j_{z'}}(\pi) = (-1)^{j-j_z}\delta_{-j_z j_{z'}}.$$

So, using more convenient indices, we have

$$\sum_\beta d^j_{\alpha\beta}\rho_{\beta\gamma} = \sum_\epsilon \rho_{\alpha\epsilon}d^j_{\epsilon\gamma},$$

which becomes

$$(-1)^{j-\alpha}\rho_{-\alpha\gamma} = \rho_{\alpha-\gamma}(-1)^{j-\gamma},$$

or more succinctly,

$$\rho_{-\alpha-\beta} = \rho_{\alpha\beta}(-1)^{\alpha+\beta}.$$

This relation is true in general, not merely for the choice of z axis made in Fig. 12.7a (Gottfried and Jackson, 1964a). Combined with the Hermitian property of the density matrix

$$\rho_{\alpha\beta} = \rho^*_{\beta\alpha},$$

it leads to considerable constraints. Consider ρ-meson production: the density matrix can be written

$$\rho = \begin{bmatrix} \rho_{11} & \rho_{10} & \rho_{1-1} \\ \rho_{01} & \rho_{00} & \rho_{0-1} \\ \rho_{-11} & \rho_{-10} & \rho_{-1-1} \end{bmatrix} = \begin{bmatrix} \rho_{11} & \rho_{10} & \rho_{1-1} \\ \rho^*_{10} & \rho_{00} & -\rho^*_{10} \\ \rho_{1-1} & -\rho_{10} & \rho_{11} \end{bmatrix},$$

so that there are only four independent elements, ρ_{11}, ρ_{00}, ρ_{10}, ρ_{1-1}, with ρ_{1-1} real. The normalization Tr $\rho = 1$ would reduce this to three with $\rho_{11} = \frac{1}{2}(1 - \rho_{00})$ and the ρ angular decay distribution [using Eq. (2.95)] becomes

$$(3/4\pi)[\rho_{00}\cos^2\theta + \tfrac{1}{2}(1 - \rho_{00})\sin^2\theta - 2\rho_{1-1}\sin^2\theta\cos 2\phi + \sqrt{2}\,\text{Re}\,\rho_{10}\cos\phi\sin 2\theta],$$

where the angles are as shown in Fig. 12.7b.

The above analysis is a general one in that it is, in fact, independent of the one-particle-exchange model and depends only upon parity and

12.8 One-Particle-Exchange Mechanisms

angular-momentum conservation. Obviously, there will be more specific constraints if the exchanged particle is of a particular kind. For example, if it has spin 0, it can carry no information about any direction perpendicular to the direction of the momentum it contributes. Thus if the incident particle with which it interacts is unpolarized, we expect the decay distribution of the particle produced to be independent of ϕ (Fig. 12.7a). The density matrix in the spin-1 case, for example, will then have

$$\rho_{1\,-1} = \operatorname{Re} \rho_{10} = 0 \,.$$

The model has been widely applied to reactions involving incident pions or kaons; i.e., pseudoscalar particles. In these cases, if the exchanged particle is also pseudoscalar, it is impossible to feed any but the spin state with $j_z = 0$, and we have for $j = 1$

$$\rho_{11} = \rho_{-1\,-1} = \rho_{1\,-1} = \rho_{10} = 0 \,,$$
$$\rho_{00} = 1 \,.$$

Consider the case when the exchanged particle is a vector ($j^P = 1^-$) and interacts with an incident pseudoscalar meson to produce a vector particle ($s^P = 1^-$); then parity conservation requires that the interaction be in a state of odd orbital angular momentum. The state $s = 1$, $s_z = 0$ can only be fed from the state having $j_z = 0$, since $l_z = 0$. Since l, j, and s are all odd, the property of the Clebsch–Gordan coefficients [Eq. (2.53a)] means that this angular-momentum coupling cannot occur and the $s_z = 0$ state is not produced. Then $\rho_{10} = \rho_{00} = 0$, $\rho_{11} = \rho_{-1\,-1} = 0.5$, and the decay (into two pseudoscalar particles) is

$$\sin^2 \theta \, (1 - 2\rho_{1\,-1} \cos 2\phi) \,.$$

If the exchanged particle is axial vector, then this restriction on the density-matrix elements does not apply (see also Dalitz, 1966).

The azimuthal uniformity in one-pion exchange was suggested as a test of this process by Treiman and Yang (1962). It is satisfied for low-momentum transfers in many production reactions where pion exchange is an allowed contribution. It often fails for large-momentum transfers, but this is hardly surprising because the model is not expected to apply. There are also, however, reactions where one-pion exchange is allowed but the Treiman–Yang test fails. This is possibly because of the exchange of other particles or some other failure of the model.

The next important aspect of the peripheral model is a comparison of the t-distribution predicted and observed. At t large compared with m_π^2 in the one-pion-exchange model, we expect the factor containing t in Eq. (12.37) to become proportional to t^{-1}, so that the differential cross section does not fall very rapidly from its forward peak. This is in contradiction

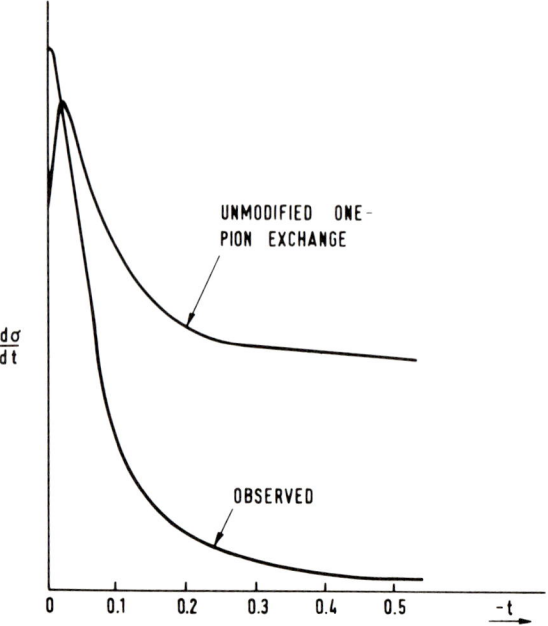

FIG. 12.8. Diagram showing the distribution in t of events in the reaction $\pi^- p \to \rho^- p$, as observed and as predicted by the unmodified one-pion-exchange model.

to experiment, where it is found that very many reactions satisfying the Treiman–Yang test have differential cross section varying as e^{Bt}, where B is between 4 and 12 GeV^{-2}. Thus, Fig. 12.8 shows how the t-distribution for $\pi^- p \to \rho^- p$ falls off very much more quickly than predicted by a simple one-pion-exchange model.

One of the simplest peripheral models is sometimes called the Born term model (BTM; see Pilkuhn, 1967). The amplitude is taken to be that given by the Feynman diagram for the particle exchange envisaged. The amplitude will contain two vertex terms, which will each contain appropriate coupling constants, and a propagator for the exchanged particle. Applying this to physical processes means assuming that the coupling constants defined at the pole when $t = m_\pi^2$ have the same value for kinematically accessible values of t, namely $t < 0$. This is the model that does not show a sufficiently rapid fall-off with t. The first modification of this model was due to Ferrari and Selleri (1961; 1963), who proposed and tried modifying the vertex function and the propagator by including a form factor that depended only upon t. Thus the form factor (Amaldi and Selleri, 1964)

$$F(t) = \frac{0.72}{1 - [(t - m_\pi^2)/4.73 m_\pi^2]} + \frac{0.28}{1 + [(t - m_\pi^2)/32 m_\pi^2]^2} \quad (12.38)$$

12.8 One-Particle-Exchange Mechanisms

had some success in fitting data supposedly involving a one-pion-exchange process. However, the real difficulty is in finding a reasonable basis for the use of form factors as distinct from the fitting of data. The next step came with the absorption model (Gottfried and Jackson, 1964b). The reasoning for this model goes as follows. The simple BTM gives an amplitude which violates unitarity, particularly by large contributions from low partial waves. These are just the ones that are expected to be strongly reduced by competition with other channels. Thus the model modifies each partial wave by an absorption factor that attenuates the lower waves more than the higher partial waves. This is normally done in the following way. The BTM amplitudes are decomposed with helicity amplitudes, each of which is analyzed into its partial-wave amplitude. Each incoming wave and each outgoing wave is shifted in phase by an angle δ—i.e., a normal phase-shift—so that we can consider the effect of interactions on the incoming and outgoing waves by replacing an amplitude T_j due to the BTM by

$$e^{i\delta_j} T_j e^{i\delta_j'}, \tag{12.39}$$

where j gives the partial-wave quantum number. The δ_j, δ_j' will, in general, be complex (to give absorption) and not the same, since the incoming and outgoing waves are not normally the same type of particle. The derivation of this result is discussed by Hearn and Drell (1967).

To proceed further, it is necessary to make assumptions about elastic scattering at high energies: first, that the scattering amplitude is purely imaginary and second, that the angular distribution is described by an exponential:

$$\frac{d\sigma_e}{dt} = \frac{\sigma^2}{16\pi^2} \exp\left(\frac{tR^2}{4}\right). \tag{12.40}$$

The terms in this equation follow from Eqs. (3.81) and (3.84) after changing $d\sigma_e/d\Omega$ into $(k^2/\pi) d\sigma_e/dt$. σ_e is the elastic-scattering cross section and σ the total cross section. R is the range parameter of the absorbing area of the target particle. From this it follows that

$$\exp i\delta_j = \left[1 - \frac{\sigma}{\pi R^2} \exp\left(\frac{-2j^2}{k^2 R^2}\right)\right]^{1/2}. \tag{12.41}$$

Thus numerical calculation is performed by absorbing each helicity partial wave by using the replacement

$$T_j \rightarrow T_j[1 - C\exp(-Aj^2)]. \tag{12.42}$$

If the incoming and outgoing particles are different, a fit may be obtained by having the square root of two similar factors with different C and A. Gottfried and Jackson (1964b), fitting $\pi^- p \rightarrow \rho^- p$, used $C = 0.765$ (to fit $\pi^- p$ elastic scattering) and $C = 1$ (complete absorption of the S-wave),

$A = 0.038$. Fits treating incoming and outgoing particles differently give R_f^2 of the final state to be two to three times larger than R_i^2 for the initial state if $\sigma_f/\pi R_f^2$ is taken to be 1 and $\sigma_i/\pi R_i^2$ to be 0.765, the value found from elastic scattering. This is hardly a reasonable result, since we do not expect that the total cross section for ρ-mesons on protons, for example, should be two to three times greater than that for pions. This indicates that the BTM amplitudes that are absorbed cannot be correct and therefore do not provide even a sound start for the absorption model. Nonetheless, it has been applied to a large number of processes involving both pseudo-scalar and vector meson exchange (see Hearn and Drell, 1967). Compared with the form-factor model there is no general improvement in the predicted t-dependence, but the absorption model does predict spin density matrix elements for particles produced, often in good agreement with the data and differing from those predicted by the form-factor model, which are identical to those of the unmodified BTM. However, many processes that are expected to be dominated by one-pion exchange, for example, show signs of the presence of the exchange of heavier particles, making it difficult to make meaningful fits.

It is evident that many of the difficulties in finding an adequate model come about because the full implications of unitarity are not included. Thus the effect of competing channels is not properly described by the absorption of Born amplitudes. The K-matrix model attempts to overcome this difficulty by constructing a real symmetric matrix, which by its definition gives a unitary S-matrix. The elements of the K-matrix are approximated by Born amplitudes (real) for each coupled channel, and the problem is one of finding these amplitudes, since there are normally a very large number of such channels. Considering the approximations required to obtain an answer, it is not surprising that this model does not greatly improve on the absorption model.

Another modification of the model comes by considering the meaning of factors that appear in the amplitude. To do this we consider the case of one-pion exchange in diagram (b) of Fig. 12.5. The amplitude is

$$V_{13}(t - m_\pi^2)^{-1} V_{24},$$

which consists of a propagator and two vertex factors. The pole model puts these vertex factors to the value they would have if the exchanged pion were real [$t = m_\pi^2$, $V_{13}^P = V_{13}(m_\pi^2)$, etc.], whereas in the Born term model the vertex terms depend upon t ($V_{13}^B = V_{13}(t), \ldots$). A comparison of the two vertex terms shows that

$$V_{13}^B = (p'/p)^l V_{13}^P$$

where p and p' are the momenta of the pion in the center of mass of the

12.8 One-Particle-Exchange Mechanisms

system 3 in the two cases (1) pion on the mass shell and (2) pion having mass $\sqrt{t}$.

$$p = (1/2m_3)\{[(m_1 - m_3)^2 - m_\pi^2][(m_1 + m_3)^2 - m_\pi^2]\}^{1/2},$$
$$p' = (1/2m_3)\{[(m_1 - m_3)^2 - t][(m_1 + m_3)^2 - t]\}^{1/2}.$$

l is the orbital angular momentum of the exchanged pion with respect to the incident particle 1. The difficulty that occurs in the BTM is that as $-t$ becomes large, $p' \simeq t$ and the effect of the propagator term is cancelled and the differential cross-section increases with increasing momentum transfer, contrary to observation. The BTM is thus worse than the pole model. Clegg (1964) and Durr and Pilkuhn (1965) suggested, by analogy with potential scattering theory, that the dynamical origin of the $(p'/p)^l$ factor is the angular-momentum barrier and consequently that the correct factor that changes the pole amplitude into a well-behaved amplitude is $[v_l(p'R)/v_l(pR)]^{1/2}$, where v_l are the angular-momentum barrier-penetration factors discussed in Section 4.5. These have the following properties:

$$v_l(x) = \left[\frac{x^l}{(2l-1)!!}\right]^2 \quad \text{for} \quad x \ll l$$

and

$$v_l(x) = \frac{1}{1 + l(l+1)/2x^2} \quad \text{for} \quad x \gg l.$$

Thus for small pR and $p'R$ we find

$$\left[\frac{v_l(p'R)}{v_l(pR)}\right]^{1/2} = \left[\frac{p'}{p}\right]^l$$

and the amplitude becomes the BTM amplitude. For large p' (large $-t$) the numerator becomes 1, and the behavior of the amplitude as a function of t is determined by the propagator as in the pole model. We have discussed only one vertex; similar factors will occur at the other vertex. If one of the vertices involves a baryon, it is necessary to use factors that include the effect of the spin structure of the interaction (see Durr and Pilkuhn, 1965). These authors call all these factors kinematical, as distinct from form factors, which are also functions of t, and show that without form factors or absorption this model predicts results close to those observed. The gap can be closed by including absorption or by form factors. This approach has been very successful in many reactions in which pion exchange is the predominant mechanism (Wolf, 1969). This model applies only to pseudoscalar exchange, and no way has been found to apply it to vector exchange. In any case all models of the exchange of spinning particles suffer from the appearance of a factor s^j, where j is the spin of the particle, thus giving incorrect high-energy behavior.

12.9 Regge Poles

We have seen in the last section that the one-pion-exchange model and other related peripheral mechanism models are capable of explaining a large amount of data. The difficulties begin when the exchanged particle has spin greater than 0, for in that case the Born contribution to the amplitude is proportional to

$$s^j/(t - m^2).$$

Thus, as s increases, the amplitudes and cross sections increase, in contradiction to experiment. This difficulty is connected with the fact that the field theories of particles of spin greater than 1 are not renormalizable.

Regge pole theory indicates a way out of this difficulty and does it by making the spin of the exchanged particle a variable in t. Thus j is replaced by $\alpha(t)$, where $\alpha(t = m^2) = j$ and $\partial \alpha/\partial t > 0$. We shall see that to prevent cross sections from diverging, all exchanges must have $\alpha(t = 0) \leqslant 1$.

Let us consider the partial-wave expansion of the scattering amplitude in the case of spinless particles:

$$f(\theta) = \sum_l (2l + 1) A_l P_l(\cos \theta). \tag{12.43}$$

We are going to reformulate the right-hand side as an integral over a suitable contour in the complex plane of the variable l', now considered to be complex. The contour C_1 will be one that comes from $+\infty$ on the real axis toward $l = 0$, passes around $l = 0$ in a counterclockwise manner, and goes back to $l = +\infty$ along the real axis. It misses the Re l integer points by circling above them on the way in and below them on the way out (Fig. 12.9). Consider now the function

$$1/\sin \pi l',$$

where l' is complex. Near the point $l' = l$, a real integer, $\sin \pi l'$ is approximately $\pi(l' - l)(-1)^l$. Thus this function of l' is analytic except when $l' = 0$ or a positive integer and it follows that, if we construct a contour C_2 circling one of these points,

$$\frac{1}{2\pi i} \oint_{C_2} \frac{dl'}{\sin \pi l'} = \frac{1}{2\pi i} \oint_{C_2} \frac{dl'}{\pi(-1)^l(l' - l)} = \frac{1}{\pi(-1)^l},$$

and that

$$\frac{1}{2\pi i} \oint_{C_2} \frac{(2l' + 1) P_{l'}(-\cos \theta) A_{l'}\, dl'}{\sin \pi l'} = \frac{(2l + 1)}{\pi} P_l(\cos \theta) A_l, \tag{12.44}$$

where we have used

$$P_l(-\cos \theta) = (-1)^l P_l(\cos \theta).$$

12.9 Regge Poles

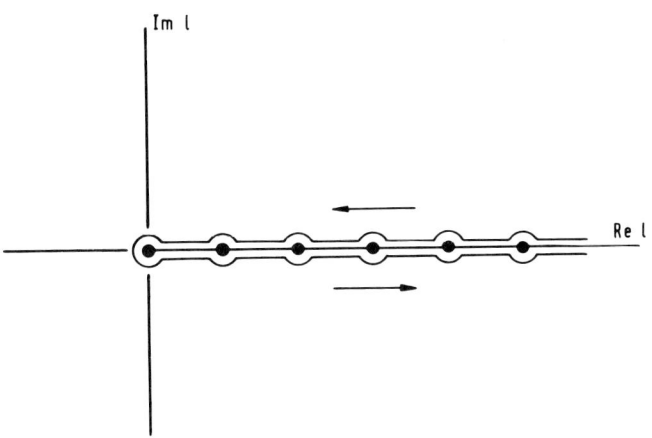

FIG. 12.9. The contour in the complex-l plane for the Sommerfeld–Watson transformation.

If now we change this contour, C_2, to the one first described, we shall encircle all the integer l points, and since other parts of the contour integral will give zero contribution, we have

$$\frac{1}{2i} \oint_{C_1} \frac{(2l' + 1) A_{l'} P_{l'}(-\cos\theta)\, dl'}{\sin \pi l'} = \sum_{l=0}^{\infty} (2l + 1) A_l P_l(\cos\theta)$$
$$= f(\theta). \qquad (12.45)$$

This is the Sommerfeld–Watson transformation. It relies upon the fact that $A_{l'}$ and $P_{l'}(-\cos\theta)$ can be continued to be analytic functions of the complex variable l' ($2l' + 1$ obviously is). The Legendre polynomial $P_{l'}(z)$ can be defined for complex l' and z and is analytic in these variables apart, from a cut in z from -1 to $-\infty$ (Morse and Feshbach, 1953). We will consider the analytic properties of $A_{l'}$ when we change the contour of integration. The reason for making a change is that we are interested in extending $\cos\theta$ out of the physical region. Remembering that [Eq. (12.7)]

$$t = m_1^2 + m_3^2 - 2(E_1 E_3 - P_1 P_3 \cos\theta) \qquad (12.46)$$

is negative for s-channel scattering but is positive in t-channel scattering, it is evident that we are interested in values of $\cos\theta$ outside the range -1 to $+1$. Therefore we write

$$f(z) = \frac{1}{2i} \oint_{C_1} \frac{(2l + 1) A_l P_l(-z)\, dl}{\sin \pi l}. \qquad (12.47)$$

We have dropped the prime on l. The function $f(z)$ then appears to diverge for large l as $z \to \infty$. Because $f(z)$ is a physical scattering amplitude it

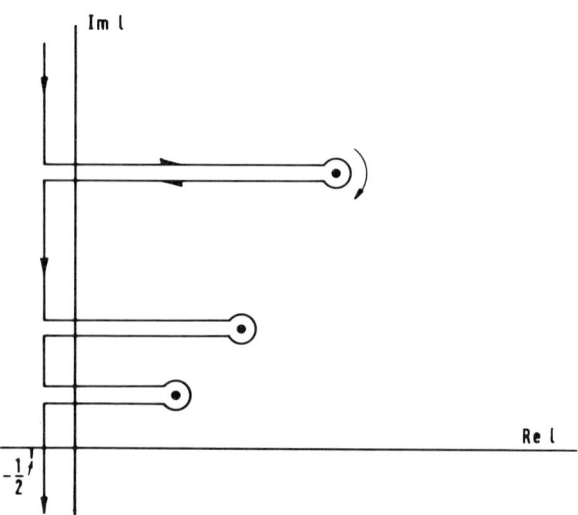

FIG. 12.10. The contour in the complex-l plane for the Regge representation of the scattering amplitude. Loops to singularities in A_l are shown.

cannot do this, as can be shown by shifting the contour to that shown in Fig. 12.10. It now moves from $-\tfrac{1}{2} + \epsilon + i\infty$ to $-\tfrac{1}{2} + \epsilon - i\infty$ and returns via a semicircle at infinity (ϵ is a small real positive quantity), apart from loops extending to singularities in A_l. The choice of contour is dictated by the fact that $P_l(z)$ for $z \to \infty$ has, for $\operatorname{Re} l \geqslant -\tfrac{1}{2}$, a behavior for large z that is z^l. Thus, for large z along the line $l = -\tfrac{1}{2} + \epsilon$ it oscillates with decreasing amplitude away from the real axes, and the result is that the contour integral, called the background term, can be neglected in the limit $z \to \infty$. Provided the singularities and the contributions from the semicircle are well-behaved, we now have a useful representation of the scattering amplitude.

Regge (1959) studied the amplitudes A_l and was able to show for a large class of potentials that the singularities in A_l were simple poles and that there were a finite number of them in the right half l-plane. This allows the separation of the poles, and we have

$$f(z) = \frac{1}{2i} \int_{-1/2+\epsilon+i\infty}^{-1/2+\epsilon-i\infty} \frac{(2l+1)P_l(-z)A_l\,dl}{\sin \pi l}$$
$$- \pi \sum_r \frac{(2\alpha_r + 1)\beta_r}{\sin \pi \alpha_r} P_{\alpha_r}(-z)\,, \qquad (12.48)$$

where the sum is over the poles in A_l, and α_r is the value of the complex angular momentum of the rth pole. The pole residues are written in this

12.9 Regge Poles

form to separate the angular-dependence effects $P_\alpha(z)$ from other terms. β_r is the residue of A_l at the pole.

So far, the possible variation with energy has not been included. Since we are considering the nonrelativistic case, we will use E for the energy rather than the Mandelstam variable s, and, for conciseness, will put $z = \cos\theta$. Then we can expose the possible energy variation by including E as an argument. Thus

$$f(E, z) = \frac{1}{2i} \int_{-1/2+\epsilon+i\infty}^{-1/2+\epsilon-i\infty} \frac{(2l+1)P_l(-z)A_l(E)\,dl}{\sin\pi l}$$
$$- \pi \sum_r \frac{[2\alpha_r(E) + 1]\beta_r(E)}{\sin\pi\,\alpha_r(E)} P_{\alpha_r(E)}(-z). \quad (12.49)$$

The parameters α_r and β_r are functions of E and give the position and residue of the rth Regge pole. As the energy varies, the position of the pole will move; the path it follows is called a Regge trajectory.

Let us consider the behavior of one of the Regge poles. For an energy E, less than threshold E_0, the $\alpha(E)$ is real, and it follows that there is a pole in the scattering amplitude when one $\alpha(E)$ is an integer. Such a pole corresponds to a bound state. As E increases above threshold, $\alpha(E)$ increases and, above E_0, becomes complex with $\text{Im}\,\alpha(E) > 0$. Consider now an energy E_R at which $\text{Re}\,\alpha(E_R)$ is an integer l. Close to E_R we can expand $\alpha(E)$ thus:

$$\alpha(E) = l + (E - E_R) \frac{\partial\,\text{Re}\,\alpha(E)}{\partial E}\bigg|_{E_R} + i\,\text{Im}\,\alpha(E). \quad (12.50)$$

Substituting into $\sin\pi\alpha$ we find that

$$\frac{\pi}{\sin\pi\alpha} = \frac{1}{\alpha'(-1)^l[E - E_R + i(\text{Im}\,\alpha/\alpha')]},$$

where

$$\alpha' = \frac{\partial\,\text{Re}\,\alpha}{\partial E}\bigg|_{E_R}.$$

Thus the contribution of this pole to the lth partial-wave amplitude is approximately,

$$\frac{\beta(E)/\alpha'}{E - E_R + i(\text{Im}\,\alpha/\alpha')} \quad (12.51)$$

This is just a Breit–Wigner amplitude for a resonance of width Γ, given by

$$\frac{\Gamma}{2} = \frac{\text{Im}\,\alpha(E_R)}{\dfrac{\partial\,\text{Re}\,\alpha(E)}{\partial E}\bigg|_{E_R}}. \quad (12.52)$$

Thus, the behavior of a Regge trajectory gives information on the physically observable bound and resonant states of a system of two particles. The formalism is based on the use of a potential presumed to exist between the particles. If this potential contains an exchange term V_E in addition to direct term V_D, then its value for angular momentum l is $V_D + V_E(-1)^l$. This means that there will be two Regge trajectories, one for even l and one for odd l, and that each can only give physical states at every other l. To allow for this possibility the Regge amplitude is written (Frautschi et al., 1962)

$$\frac{[2\alpha(E) + 1]\beta(E)}{\sin \pi \alpha(E)} \frac{\pi}{2} [P_{\alpha(E)}(+z) \pm P_{\alpha(E)}(-z)]. \tag{12.53}$$

A Regge trajectory that can have physical states at even l will have the $+$ sign (called even signature), and a trajectory that can only have odd-l physical states will have the $-$ sign (odd signature). If $V_E = 0$, then the two trajectories become one.

We have indicated our interest in extending the expression for an amplitude into nonphysical regions of the angle. In this case we wish to examine the effect of $z \to \infty$; now $P_\alpha(z) \to z^\alpha$ for $\mathrm{Re}\, \alpha \geqslant \frac{1}{2}$, so that the integrand in Eq. (12.48) is $z^{-1/2} z^{i\,\mathrm{Im}\, l}$, which means that for large z it is zero. Similarly the contribution from a single pole becomes

$$f(E, z) \xrightarrow[z \to \infty]{} \pi \frac{[2\alpha(E) + 1]\beta(E)}{2 \sin \pi \alpha(E)} [(+z)^{\alpha(E)} \pm (-z)^{\alpha(E)}].$$

Now $(-1)^\alpha = e^{-i\pi\alpha}$, and we then have the asymptotic contribution

$$f(E, z) \xrightarrow[z \to \infty]{} \frac{\pi}{2} \frac{[2\alpha(E) + 1]\beta(E)}{\sin \pi \alpha(E)} z^{\alpha(E)} (1 \pm e^{-i\pi\alpha(E)}).$$

Now let us consider relativistic scattering and the possibility of relating the scattering amplitudes in s-, t-, and u-channels with each other. We write down the contribution of a Regge pole to scattering in the t-channel. The energy now depends on t, and the scattering angle through the cosine, z_t, upon s. Thus, the rth Regge pole contributes

$$f_r(t, s) = \frac{\pi}{2} \frac{[2\alpha_r(t) + 1]\beta_r(t)}{\sin \pi \alpha_r(t)} [P_{\alpha_r(t)}(+z_t) \pm P_{\alpha_r(t)}(-z_t)]. \tag{12.54}$$

Now let us continue s, and hence z_t, up to large positive values and t downward to the negative values characteristic of s-channel scattering at high energy. From Eq. (12.9) applied to the t-channel we have

$$z_t = 1 + (s/2q^2), \tag{12.55}$$

which has the asymptotic behavior

$$z_t \xrightarrow[s \to \infty]{} s/2q^2.$$

12.9 Regge Poles

Thus the asymptotic contribution of the Regge pole to the s-channel scattering is

$$f_r(s, t) \xrightarrow[s \to \infty]{} \frac{\pi}{2} \frac{[2\alpha_r(t) + 1]\beta_r(t)}{\sin \pi \alpha_r(t)} \left(\frac{s}{2q^2}\right)^{\alpha_r(t)} (1 \pm e^{-i\pi\alpha_r(t)}). \quad (12.56)$$

The first argument in the amplitude f_r is used to indicate the channel in Eqs. (12.54) and (12.56).

That this procedure is possible was a conjecture originally developed by Chew *et al.* (1962), by Blankenbecler and Goldberger (1962), and by Frautschi *et al.* (1962). It has proved to be of considerable importance in the description of strong interactions.

To continue our discussion, we first consider elastic scattering. From Eq. (12.56) we can write the contribution of the rth pole

$$f_r(s, t) = \left(\frac{s}{s_0}\right)^{\alpha_r(t)} b_r(t) \frac{1 + \exp(-i\pi \alpha_r(t))}{2 \sin \pi \alpha_r(t)}, \quad (12.57)$$

where s_0 has been put in to avoid a variable dimension for $b_r(t)$. At $t = 0$ —that is, at zero degrees—

$$f_r(s, 0) = i(s/s_0)^{\alpha_r(0)} b_r(0). \quad (12.58)$$

Now the optical theorem gives the total cross section in terms of the imaginary part of the forward-scattering amplitude

$$\sigma = (4\pi/k) \operatorname{Im} f(0).$$

In this case, putting $k^2 = s/4$ and absorbing some factors, we have

$$\sigma \xrightarrow[s \to \infty]{} (1/s)(s/s_0)^{\alpha_r(0)} b_r(0). \quad (12.59)$$

If $\alpha_r(0) = 1$, then as $s \to \infty$ the cross section becomes constant. Equation (12.59) also shows that the trajectory with the highest $t = 0$ intercept will dominate the asymptotic total cross section. The pole that is supposed to do this is called the pomeron (P) and has $\alpha_p(0) = 1$. All other poles have $\alpha(0) < 1$ and do not affect asymptotic cross sections. The pomeron has the quantum numbers of the vacuum, and since we expect particle and particle and antiparticle to couple equally to such a pole, we predict in this model that particle and antiparticle total cross sections on any target will become equal and constant in the asymptotic region.

We have noted that Pomeranchuk proved that particle and antiparticle cross-sections should be equal from dispersion relations and the asumption of asymptotic but not, *a priori*, equal cross sections. This result was one motivation for the introduction of the pomeron (a vacuum trajectory) with $\alpha(0) = 1$ into Regge-pole theory.

For small t we may write for the pomeron

$$\alpha_P(t) = 1 + t \, \alpha_P'(t = 0), \quad (12.60)$$

then, for $s \to \infty$
$$d\sigma/dt = (s/s_0)^{2t\alpha_P'(0)} ; \qquad (12.61)$$
that is,
$$d\sigma/dt = \exp\{2t\alpha_P'(0) \ln(s/s_0)\} . \qquad (12.62)$$

We expect the trajectory slope $\alpha_P'(0)$ to be greater than zero, and since $t < 0$ in the physical region, this formula represents a forward or diffraction peak in the scattering cross section, with a width that decreases logarithmically with increasing s. This is the shrinkage of the diffraction peak predicted by Regge theory.

The prediction of a shrinking diffraction peak was an initial success of the Regge theory, since this was an observed effect in proton–proton scattering. However, no shrinkage was observed in $\pi^{\pm}p$ scattering, nor in K^-p scattering, and in $\bar{p}p$ scattering the diffraction peak was found to expand. These results apply to energies up to 20 GeV, and it can be argued that this is not the asymptotic region; in fact, it is possible to fit the data by including contributions from trajectories with quantum numbers other than those of the vacuum. Thus the asymptotic region is at present out of reach; by how much is not clear.

Let us now consider the problem of $\pi^+\pi^-$ scattering. The s-channel is
$$\pi^+ + \pi^- \to \pi^+ + \pi^- .$$

The t-channel scattering is the same and will have an amplitude to which various Regge poles will contribute. It is attractive to associate a pole with every state we can find of the $\pi^+\pi^-$ system, specified by quantum numbers that will be conserved, other than j. In this case only isotopic spin has to be considered. The states with isospin 0 or 2 can have only even angular momentum, and so we have two Regge trajectories of even signature: the physical manifestations of these poles, if they occur, will be in the spin-parity series 0^+, 2^+, 4^+, ... The isospin-1 trajectory will have odd signature and could have physically observable states in the spin-parity series 1^-, 3^-, 5^-, ... In view of these remarks it is interesting to examine the known meson spectrum for particles with these properties. The well-established candidates are

$$f(1260): \quad T^G J^P = 0^+ 2^+ ,$$
$$f'(1515): \quad = 0^+ 2^+ ,$$
$$\rho(765): \quad = 1^+ 1^- .$$

This suggests that $\pi^+\pi^-$ scattering will be dominated by these resonances in the energy range covered by these masses, and, as we have discussed in Section 7.7, indirect observation of $\pi\pi$ scattering indicates that this is the case. Turning now to the s-channel scattering in the asymptotic limit

12.9 Regge Poles

$s \to \infty$, we expect the scattering amplitude to involve these poles. Very little is known about the isospin-2 pole, and there are no known bosons with this isotopic spin, so we can afford to neglect it and consider the amplitude as being dominated by the exchange of the following Regge poles: one is the odd-signature ρ-pole, called this by its association with the ρ-meson; the others are even-signature isospin-0 poles. These latter poles have the feature that their lowest-spin physical manifestation would be "particles" with the quantum numbers of the vacuum, namely $J^P = 0^+$. The pomeron is one such pole and it appears to be necessary to include a second, P', also with the quantum numbers of the vacuum and having $\alpha_{P'}(0) = \frac{1}{2}$ (Igi, 1963). It is probable that the P' is on the same trajectory as the f-meson and so we shall discuss $\pi - \pi$ scattering assuming that there are three relevant Regge poles, ρ, P, and P'.

We wish to write down amplitudes in the s-channel in terms of the exchange of particles in the t-channel. For π-π scattering the relation is

$$A_T(s, t) = \sum_{T'} B_{TT'} A_{T'}(s, t),$$

where T and T' are the isospins (0, 1, or 2) in the s- and t-channels. B is an isospin crossing matrix which is, in this case (Chew, 1962),

$$B = \begin{bmatrix} \frac{1}{3} & 1 & \frac{5}{3} \\ \frac{1}{3} & \frac{1}{2} & -\frac{5}{6} \\ \frac{1}{3} & -\frac{1}{2} & \frac{1}{6} \end{bmatrix}.$$

Thus, for example, the exchange of a ρ-pole gives rise in the s-channel to isospin-0, -1, or -2 amplitudes of relative weight 1, $\frac{1}{2}$, $-\frac{1}{2}$.

We can now write the ρ-trajectory contribution as

$$\left(\frac{s}{s_0}\right)^{\alpha_\rho(0)} \frac{1 - \exp[-i\pi\alpha_\rho(t)]}{2 \sin \pi \alpha_\rho(t)} b_\rho(t) \qquad (12.63)$$

and similarly for the P and P' trajectories with appropriate $\alpha(t)$, $\beta(t)$, and signature. The differential cross section is given by

$$\frac{d\sigma}{dt} = \frac{\pi}{k^2} \frac{d\sigma}{d\Omega} = \frac{\pi}{k^4} |\sum_l (2l + 1) A_l P_l(\cos\theta)|^2. \qquad (12.64)$$

In the limit $s \to \infty$, $k^2 = s/4$, and under the Regge-pole hypothesis we have

$$\frac{d\sigma}{dt} = \frac{16\pi}{s^2} |\sum \text{Regge amplitudes}|^2.$$

The simplest place to apply this formalism is to the charge-exchange scattering:

$$\pi^+ + \pi^- \to \pi^0 + \pi^0.$$

This can only involve the ρ-trajectory exchange out of the three we are discussing. Its amplitude is the difference between the isospin-0 and -2 amplitudes ($\frac{3}{2}$ of ρ amplitude), so that we obtain

$$\frac{d\sigma}{dt}\text{(ch. ex.)} = \left(\frac{s}{s_0}\right)^{2\alpha_\rho(t)-2} \left| b_\rho(t) \frac{1 - \exp(-i\pi\alpha_\rho(t))}{2 \sin \pi\alpha_\rho(t)} \right|^2. \quad (12.65)$$

The various factors omitted are constants, and it is expected that $b_\rho(t)$ is a slowly varying real function of t whose value at $t = m_\rho^2$ may be deduced from the width of the ρ. This example serves to illustrate the application of Regge poles.

The theory has been applied successfully to measurable reactions where the quantum numbers exchanged are unique and thus only one trajectory is expected to contribute. Some of the reactions investigated are

(1) $\pi^- + p \to \pi^0 + n$,

(2) $\pi^- + p \to \eta + n$,

(3) $\pi^- + p \to p + \pi^-$ (backward scattering).

The exchanged pole in reaction (1) has natural spin–parity, isotopic spin 1, even G-parity, and baryon number and strangeness zero. The only possibility is the ρ-trajectory. A good fit to the cross section data in the range of s from 4 to 36 GeV^2 and $|t|$ from 0 to 1.5 GeV^2 has been obtained with $\alpha_\rho(0) = 0.57$ and slope $\alpha_\rho'(0) = 0.96$ GeV^{-2}. Note that $\alpha_\rho(t) = 0$ at $t = -0.59$ GeV^2, which gives, in the complete theory, the minimum observed in $d\sigma/dt$ near $t = -0.6$ (Höhler et al., 1966b). The second reaction requires the unique exchange $B = S = 0$, $T = 1$, $G = -1$, and $J^P = 0^+, 1^-, 2^+, \ldots$. An even-signature Regge pole could give rise to physically observable states at 0^+, 2^+, etc. There is one candidate, the $A_2(1315)$ meson, which has $T^G J^P = 1^- 2^+$. Phillips and Rarita (1965a, b) used KN and $\bar{K}N$ scattering data to find the parameters of this Regge trajectory, denoted by subscript "R," and found $\alpha_R(0) \simeq 0.4$, $\alpha_R'(0) \simeq 0.6$ GeV^{-2}, which also fitted data from reaction (2).

Reaction (3) is written to imply backward scattering; that is, the incoming π^- picks up an exchanged particle to become a proton. The exchanged particle must have $T = \frac{3}{2}$, $B = 1$, $S = 0$, and the physical candidate is the $\Delta(1236)$. A corresponding Regge trajectory that fits the data has $\alpha_\Delta(0) \simeq -0.3$ and $\alpha_\Delta'(0) = 1.1$ GeV^{-2}.

Regge-pole theory runs into difficulties in two places. It requires either conspiracy (see below) or Regge cuts in addition to poles to explain satisfactorily either the narrow peak in forward nucleon charge-exchange scattering

$$n + p \to p + n,$$

12.10 The Chew-Frautschi Plot

or the sharp forward peak observed in high-energy pion photoproduction (Boyarski *et al.*, 1968)

$$\gamma + p \to \pi^+ + n.$$

The π^+ photoproduction cross section should be zero in the forward direction if the process is due to Reggeized pion exchange alone.

On closer examination, the theory itself has many difficulties, of which we can mention only a few. If the interacting particles have spin, then at $t = 0$ (0° if the particles are the same in the final state as initially), angular-momentum conservation forces a connection between various scattering helicity amplitudes. This must be reflected in connections between the parameters of the Regge-pole amplitudes, which are otherwise unrelated (Van Hove, 1968). (It is not clear that the constraints imposed on one trajectory by one experimental circumstance are compatible with those imposed by another experimental circumstance.) The constraints can be satisfied in three ways. Conspiracy occurs when the constraints are satisfied by relations between the residues of different trajectories. Evasion occurs when the constraints are satisfied by the vanishing of certain of the residue functions at $t = 0$. A third way is by requiring the existence of a sequence of Regge poles (daughters) that have the same quantum number as a parent and in conjunction with it are capable of satisfying the constraint equations. Other difficulties occur, such as the need for fermion trajectories to occur in pairs of opposite parity (Volkov and Gribov, 1963; Barger and Cline, 1967). In addition, the amplitudes that enter integrals of the kind in Eq. (12.49) may have not only pole singularities but also cuts. For a full discussion of these difficulties see, for example, Van Hove (1968).

Recently (1968) a new look has been given to the Regge-pole model. The idea of *duality* comes from the success of Regge poles at lower energies than might be expected, and it suggests the Regge amplitude is an average of the resonant behavior shown in many *s*-channels at low energies. In addition, amplitude models have been proposed that satisfy all the necessary Regge properties and that have been applied with some success. Readers are referred to reviews in recent conference proceedings for details (for example, Jackson, 1969; Jacob, 1969).

12.10 The Chew-Frautschi Plot

In the last section, we have seen that high-energy scattering is, in some cases, well described by the exchange of Regge poles. The important quantity is $\alpha(t)$, the complex angular momentum exchanged at a momentum transfer t. For scattering we have $t < 0$; but by continuation $\alpha(t)$, $t > 0$ will have some meaning, and where Re $\alpha(t)$ is an integer or half-integer, the pole manifests itself as a real particle with mass $\sqrt{t}$. Chew and

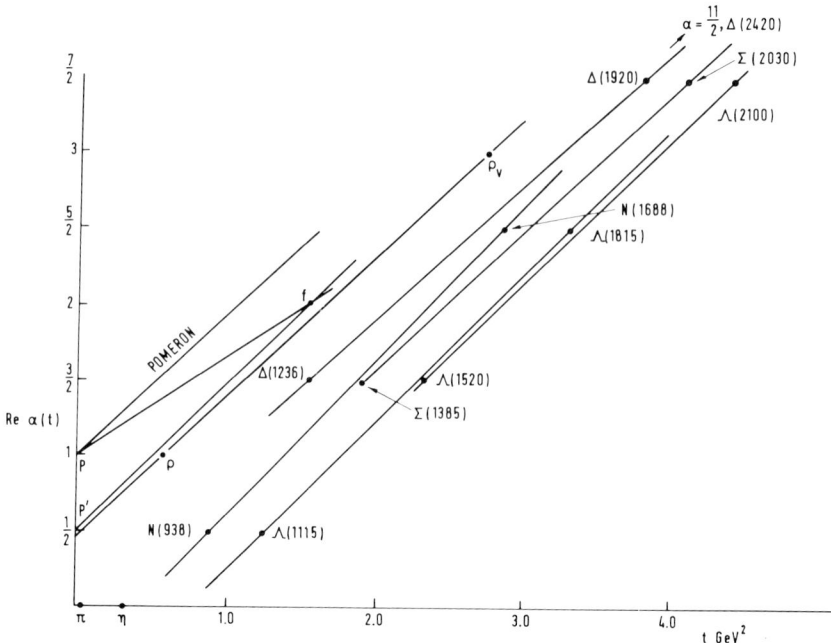

FIG. 12.11. A Chew–Frautschi Plot.

Frautschi (1962) suggested that all hadrons, including resonances, lay on Regge trajectories. Since we associate a signature with every trajectory, which gives zero at every other integral or half-integral value of $\alpha(t)$, we will expect to find particles separated by two units of angular momentum on each trajectory. Thus the ρ-meson trajectory, if it exists, will have the real ρ at $t = 0.585$ GeV^2 with Re $\alpha(t) = 1$, and we expect to find its first recurrence at Re $\alpha(t) = 3$. The $\rho_V(1665)$ (Particle Data Group, 1969) is a possible candidate ($t = 2.76$ GeV^2). However, better examples of recurrences are known, and in Fig. 12.11 we show some of the more convincing trajectories with their hadrons. An interesting point emerges: the slopes of the trajectories are all about 1.0 GeV^{-2}. This is an interesting result because in Regge's original work (1960) it was shown that the slope was connected with the range of the scattering potential. Thus the equal slopes suggest that the "potential" range for all hadrons is the same. In a bootstrap theory all hadrons are bound states of one another: for example, a ρ is partly a bound state of $\pi\pi$ and partly of $N\bar{N}$ and so on, while N is a bound state of πN, $\pi\Delta$, etc. The radius of interaction is expected to be approximately the same in all cases, and this is manifested in nearly equal Regge slopes.

It is difficult to find convincing recurrences for the mesons, although their trajectories are frequently used in theoretical work. In addition, many baryon states do not have or are not obvious recurrences. One reason may be that the assumption of a common and a constant slope is incorrect and makes it difficult to identify the recurrences.

The $SU(3)$ classification of particles would suggest that the hadron multiplets would recur so that the $J^P = \frac{1}{2}^+$ octet (N, Λ, Σ, Ξ) has a recurrence with $J^P = \frac{5}{2}^+$ and that the particles are $N(1688)$, $\Lambda(1815)$, $\Sigma(1910)$, and $\Xi(2030)$. The identification of high-mass multiplets is uncertain, and it is not completely clear at present whether they are Regge recurrences or other multiplets to be accommodated in some higher symmetry scheme.

REFERENCES

Allaby, J. V. et al. (1968). Phys. Lett. B **27**, 49 (12.6).
Anderson, H. L., Davidon, W. C., and Kruse, U. E. (1955). Phys. Rev. **100**, 339 (12.4).
Amaldi, U., and Selleri, F. (1964). Nuovo Cimento **31**, 360, (12.8).
Barger, V., and Cline, D. (1967). Phys. Rev. Lett. **19**, 1504 (12.9).
Blankenbecler, R. B., and Goldberger, M. L. (1962). Phys. Rev. **126**, 766 (12.9).
Boyarski, A. M. et al. (1968). Phys. Rev. Lett. **20**, 300 (12.9).
Chew, G. F. (1962). "S-Matrix Theory of Strong Interactions." Benjamin, New York (12.3, 12.9).
Chew, G. F. (1968). Phys. Rev. **112**, 1380 (12.8).
Chew, G. F., Frautschi, S. C., and Mandelstam, S. (1962). Phys. Rev. **126**, 1202 (12.9).
Chew, G. F., and Frautschi, S. C. (1962). Phys. Rev. Lett. **8**, 41 (12.10).
Chew, G. F., and Low, F. E. (1959). Phys. Rev. **113**, 1640 (12.8).
Clegg, A. B. (1964). Nuovo Cimento **34**, 244 (12.8).
Coleman, S., and Norton, R. E. (1964). Nuovo Cimento **38**, 438 (12.5).
Dalitz, R. H. (1966). Proc. Int. Sch. Phys. Enrico Fermi. **33**, 141 (12.8).
Durr, H. P., and Pilkuhn, H. (1965). Nuovo Cimento A **40**, 899 (12.8).
Eden, R. J. (1967). "High Energy Collisions of Elementary Particles," pp. 41, 211. Cambridge Univ. Press, London and New York (12.3, 12.7).
Eden, R. J., Landshoff, P. V., Olive, D. I., and Polkinghorne, J. C. (1966). "The Analytic S-Matrix," pp. 3, 16, Cambridge Univ. Press, London and New York (12.3).
Ferrari, E., and Selleri, F. (1961). Phys. Rev. Lett. **7**, 387 (12.8).
Ferrari, E., and Selleri, F. (1963). Nuovo Cimento **27**, 1450 (12.8).
Frautschi, S. C., Gell-Mann, M., and Zachariasen, F. (1962). Phys. Rev. **126**, 2204 (12.9).
Froissart, M. (1961). Phys. Rev. **123**, 1053 (12.7).
Goldberger, M. L., Miyazawa, H., and Oehme, R. (1955). Phys. Rev. **99**, 986 (12.4).
Gottfried, K., and Jackson, J. D., (1964a). Nuovo Cimento **33**, 309 (12.8).
Gottfried, K., and Jackson, J. D. (1964b). Nuovo Cimento **34**, 735, 1843 (12.8).
Hamilton, J. (1966). Phys. Lett. **20**, 687 (12.4).
Hamilton, J. (1967). In "High Energy Physics" (E. H. S. Burhop, ed.), Vol. I, p. 193. Academic Press, New York (12.4).
Hearn, A. C., and Drell, S. D., (1967). In "High Energy Physics" (E. H. S. Burhop, ed.), Vol. II, p. 219. Academic Press, New York (12.8).
Höhler, G., Baacke, J., and Strauss, R. (1966a). Phys. Lett. **21**, 223 (12.4).
Höhler, G., Baacke, J., Schaille, H., and Sonderegger, P. (1966b). Phys. Lett. **22**, 203 (12.9).
gi, K. (1963). Phys. Rev. **130**, 820 (12.9).

Jackson, J. D. (1969). *Proc. Int. Conf. Elementary Particle Physics, Lund, 1969.* p. 61. Institute of Physics, Lund (12.9).
Jacob, M. (1969). *Proc. Int. Conf. Elementary Particle Physics, Lund, 1969.* p. 125. Institute of Physics, Lund (12.9).
Kanada, H., Kobayashi, T., and Sumi, Y. (1967). *Progr. Theor. Phys. Suppl.* **41, 42** (12.6).
Mandelstam, S. (1958). *Phys. Rev.* **112**, 1344 (12.3).
Mandelstam, S. (1959). *Phys. Rev.* **115**, 1741 (12.2).
Miller, D. H. *et al.* (1967). *Phys. Rev.* **153**, 1430.
Morse, P. M., and Feshbach, H. (1953). "The Methods of Theoretical Physics," McGraw-Hill, New York (12.9).
Orear, J. *et al.* (1968). *Phys. Rev. Lett.* **21**, 389 (12.6).
Particle Data Group, (1969). *Rev. Mod. Phys.* **41**, 109 (12.10).
Phillips, R. J. N., and Rarita, W. (1965a). *Phys. Rev. B* **139**, 1336 (12.9).
Phillips, R. J. N., and Rarita, W. (1965b). *Phys. Rev. B* **140**, 200 (12.9).
Pilkuhn, H. (1967). "The Interactions of Hadrons," pp. 177, 277, North-Holland Publ., Amsterdam (12.3, 12.4, 12.8)
Pomeranchuk, I. Ya. (1956). *Sov. Phys. JETP* **3**, 306 (12.9).
Pomeranchuk, I. Ya. (1958). *Sov. Phys. JETP* **7**, 499 (12.7).
Regge, T. (1959). *Nuovo Cimento* **14**, 951 (12.9).
Regge, T. (1960). *Nuovo Cimento* **18**, 947 (12.10).
Samaranayake, V. K., and Woolcock, W. S. (1965). *Phys. Rev. Lett.* **15**, 936 (12.4).
Treiman, S. B., and Yang, C. N. (1962). *Phys. Rev. Lett.* **8**, 140 (12.8).
Van Hove, L. (1968). CERN 68–31 (12.9).
Volkov, D. V., and Gribov, V. N. (1963). *Sov. Phys. JETP* **17**, 720 (12.9).
Wolf, G. (1969). *Phys. Rev.* **182**, 1538 (12.8).

XIII

THE ELECTROMAGNETIC INTERACTION OF HADRONS

13.1 Introduction

Quantum electrodynamics, which was described in Chapter X, concerns the interaction between the electromagnetic field and electrons and positrons. Since many of the strongly interacting particles also carry electric charge or magnetic moment, they also interact with the electromagnetic field. However, the situation is complicated by the strong interaction, and there is no quantitatively precise theory. There are nevertheless many interesting and significant phenomena, which we shall describe in this chapter.

13.2 The Electromagnetic Interaction

In Chapter X, the basic electromagnetic interaction was found by making the replacement

$$i(\partial/\partial x_\mu) \to i(\partial/\partial x_\mu) - eA^\mu, \tag{13.1}$$

which gives rise to a term in the Lagrangian density

$$e\bar{\psi}\gamma^\mu\psi A_\mu \tag{13.2}$$

for the case of quantum electrodynamics. If the Lagrangian is gauge invariant (Section 10.4) then the theory defines a current

$$j^\mu = e\bar{\psi}\gamma^\mu\psi \tag{13.3}$$

that is conserved,

$$\partial j^\mu/\partial x^\mu = 0, \tag{13.4}$$

and has associated with it a charge that is conserved. We know that the

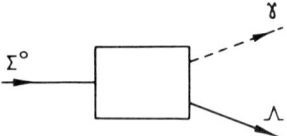

FIG. 13.1. A diagram indicating the lack of knowledge of the internal dynamics of a photon-hadron interaction. The photon interacts with some current generated by the strong interactions and causes the $\Sigma^0 \to \Lambda$ transition.

magnitude of the electric charge of the strongly interacting particles is always the same as that of the electrons, and this means that a Lagrangian describing the total interaction of all particles and the electromagnetic field is invariant under the gauge transformation connected with electric charge. The presence of strong interactions does not change this coupling constant. This situation is exactly as in the case of the vector weak coupling constant (Section 11.10). For a full discussion see Bernstein (1968).

We can associate a Feynman-diagram vertex with the interaction of Eq. (13.2), as, for example, Fig. 10.1. If the particle that enters the vertex has the same mass as that which leaves, then at least one of the three particles, including the photon, must be virtual to conserve energy and momentum at the vertex. A possible exception is when the particles do not have the same mass, in which case all the particles can be real. For example, the decay

$$\Sigma^0 \to \Lambda + \gamma \qquad (13.5)$$

conserves energy and momentum. However, this decay illustrates another point. The two particles are not charged, and thus the interaction cannot be described by the interaction of Eq. (13.2). It is possible to write down an interaction term that describes the interaction of a magnetic dipole with the electromagnetic field, but this would involve an extra unknown parameter (a coupling constant). The principle of minimal electromagnetic interaction (Watson, 1952; Bernstein, 1968) requires that the basic interaction be that of the electromagnetic field with the charged current. The anomalous moments of the electron and muon can be calculated correctly (Section 10.9) using this minimal interaction. In the case of hadrons, this cannot be done, but it is assumed that the interaction is minimal and that properties such as anomalous magnetic moments are a consequence of strong-interaction effects. Thus, instead of placing the outgoing photon as emerging from a vertex, it can be indicated as coming from a box (Fig. 13.1). The box hides our ignorance of the dynamic details.

Essentially there are two kinds of processes that will figure in our discussion. The first kind involves a real photon involved in a hadronic transition (Fig. 13.2a)

$$A \rightleftharpoons B + \gamma. \qquad (13.6)$$

For example, there are electromagnetic decays such as

$$\omega \to \pi^0 + \gamma, \qquad (13.7)$$

$$\Delta^+ \to p + \gamma \qquad (13.8)$$

13.2 The Electromagnetic Interaction

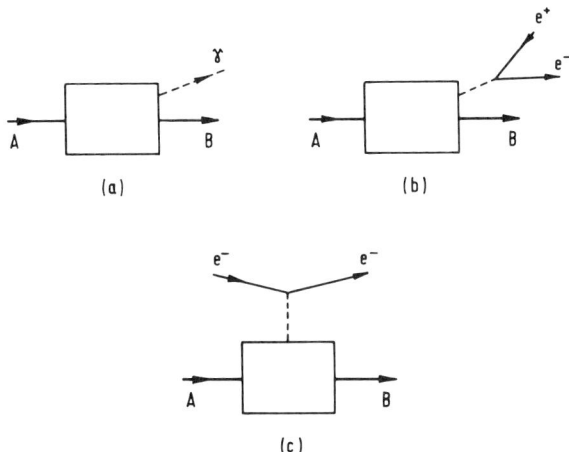

FIG. 13.2. Hadronic transitions involving (a) real photons and (b) and (c) virtual photons. The hadronic state B can be more than one particle.

and photoproduction processes such as

$$\gamma + p \to \Delta^+ \to \pi^+ + n. \tag{13.9}$$

The second kind of process is one in which the photon is virtual. It must then have a charged current source (or sink), and the most convenient current is provided by electrons or positrons. Charged hadrons are also possible sources, but the interpretation is doubly complicated and may be masked by the strong-interaction effects. (There is, however, one exception, which we shall meet shortly.) Thus the basic process (13.6) becomes (Fig. 13.2b)

$$A \to B + e^+ + e^-, \tag{13.10}$$

or, for example (Fig. 13.2c):

$$e^- + A \rightleftharpoons B + e^-, \tag{13.11}$$

We see that the real photon is replaced by a virtual photon, which interacts with the electron current. As examples we can give corresponding to Eqs. (13.7)–(13.9),

$$\omega \to \pi^0 + e^+ + e^-, \tag{13.12}$$

$$\Delta^+ \to p + e^+ + e^-, \tag{13.13}$$

$$e^- + p \to \Delta^+ + e^- \searrow \pi^+ + n. \tag{13.14}$$

The exception we mentioned is the use of nuclear electric fields as the source of virtual photons. The decay

$$\pi^0 \to \gamma + \gamma$$

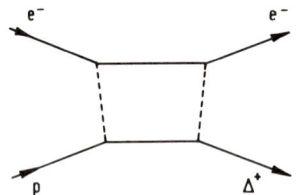

FIG. 13.3. Double photon exchange in inelastic electron-proton scattering.

has been reversed using one photon incident on the nuclear electric field Z, which provided the other (now virtual) photon

$$\gamma + Z \to \pi^0 + Z.$$

This is the Primakoff effect (Section 7.2). The nucleus can remain unchanged during this transition.

It is possible that all processes of the second kind can involve not only one but more than one virtual photon. This multiple-photon exchange becomes increasingly unlikely with increasing number; nonetheless, it has been considered. Figure 13.3 shows a diagram for double-photon exchange in the transition of Eq. (13.14). So far, such processes have not shown any of the effects expected from multiple photon exchange.

It is obvious that the processes of Eqs. (13.10) and (13.11) are closely connected. From crossing symmetry and analyticity we expect that the same analytic function of the kinematic variables describes both processes. It is worth inquiring what kinematic regions are covered. If we use p_1 and p_2 to represent the four-momentum of the incident and scattered electron in (13.11), then the four-momentum squared, t, carried by the photon is given by

$$\begin{aligned} t &= (p_1 - p_2)^2 \\ &= p_1^2 + p_2^2 - 2p_1 p_2 \\ &= 2m_e^2 - 2E_1 E_2 + \mathbf{p}_1 \cdot \mathbf{p}_2, \end{aligned}$$

where E_1 and E_2 are the incident and scattered electron energies, respectively. For elastic scattering at zero degrees, $t = 0$. For larger angles, t becomes negative. For inelastic scattering, t is negative at $0°$, and becomes increasingly negative as the angle increases. Thus, the virtual photon exchanged in the transition of Eq. (13.11) is spacelike ($t < 0$) in analogy with intervals $(d\mathbf{r}, d\tau)$, which are said to be spacelike if $d\tau^2 - d\mathbf{r}^2$ is negative. In contrast, the transition of Eq. (13.10) involves timelike virtual photons ($t > 0$). To see this, use p_1 and p_2 to represent the four-momenta of the outgoing electron and positron. Now the four-momentum squared of the photon is given by

$$\begin{aligned} t &= (p_1 + p_2)^2 \\ &= 2m_e^2 + 2E_1 E_2 - 2\mathbf{p}_1 \cdot \mathbf{p}_2, \end{aligned}$$

which is greater than or equal to $4m_e^2$, as expected, since the virtual photon must create the e^+e^- pair.

One interesting example of Eq. (13.10) involves making B the vacuum;

$$A \to e^+ + e^- . \tag{13.15}$$

For example:

$$\rho^0 \to e^+ + e^- ,$$

which has been observed. The reverse of reaction (13.15) is also possible in electron–positron colliding-beam experiments. The processes

$$e^+ + e^- \to \rho^0, \omega, \text{ or } \phi$$

have all been observed (see Section 13.7). The virtual photons are timelike.

There is one further effect of the electromagnetic interaction, which we recall at this point. It is assumed that the mass splitting within isotopic-spin multiplets is due to the electromagnetic interaction. This subject has been treated in Section 9.7, and will not be discussed further, except to mention that efforts to calculate the mass splitting from theoretical models have not been successful.

Finally, all processes, including weak interactions, that have charged particles in the final state can give one or more photons in addition to the expected particles, and the exchange of virtual photons between the charged particles involved will change cross sections and rates from those expected in the absence of the electromagnetic interaction. Corrections can be made for these radiative effects, but sometimes the uncertainties in this procedure limit the precision that can be reached in the measurement of the corrected quantities. We shall not discuss radiative corrections further.

13.3 Isotopic-Spin Selection Rules

The selection rules for electromagnetic transitions involving hadrons that apply to isotopic spin, G-parity, and C-parity have been discussed in Section 6.8, and there is only one point to add. A virtual photon has the same isotopic-spin properties as a real photon, so that the hadronic transition selection rules in elastic or inelastic electron scattering are the same as those in the absorption or emission of a real photon. The $SU(3)$ properties of photons have been discussed in Section 9.6.

13.4 The Angular-Momentum Properties of the Electromagnetic Field

We can go some way toward understanding the rotational and angular-momentum properties of the electromagnetic field by classical means. We start by defining three orthogonal unit electric vectors $\mathbf{e}_x, \mathbf{e}_y, \mathbf{e}_z$, with directions along the x, y, and z axes, respectively. If we consider a plane electromagnetic wave moving in the $+z$ direction, then its electric vector **E** must be transverse to the direction of propagation. From classical physical optics we known we can express all pure states of beam polarization by a suitable superposition of two basic states, usually the two states of plane polarization. Thuse the basic states would be described by

$$\mathbf{E}_1(\mathbf{r}, t) = \mathbf{e}_x \exp i(kz - \omega t), \qquad (13.16)$$

$$\mathbf{E}_2(\mathbf{r}, t) = \mathbf{e}_y \exp i(kz - \omega t). \qquad (13.17)$$

These are not obviously eigenstates of spin angular momentum. To find these we rewrite the usual analytic $l = 1$ eigenstates of orbital angular momentum in Cartesian instead of spherical polar coordinates. Thus

$$Y_1^{+1}(\theta, \phi) = -\left(\frac{3}{8\pi}\right)^{1/2} \sin\theta\, e^{i\phi} = -\left(\frac{3}{8\pi}\right)^{1/2} \frac{(x + iy)}{r},$$

$$Y_1^0(\theta, \phi) = \left(\frac{3}{4\pi}\right)^{1/2} \cos\theta = \left(\frac{3}{4\pi}\right)^{1/2} \frac{z}{r},$$

$$Y_1^{-1}(\theta, \phi) = +\left(\frac{3}{8\pi}\right)^{1/2} \sin\theta\, e^{-i\phi} = \left(\frac{3}{8\pi}\right)^{1/2} \frac{(x - iy)}{r}.$$

These equations show that

$$\mathbf{e}_{+1} = -\sqrt{\tfrac{1}{2}}\, (\mathbf{e}_x + i\mathbf{e}_y), \qquad (13.18)$$

$$\mathbf{e}_0 = \mathbf{e}_z, \qquad (13.19)$$

$$\mathbf{e}_{-1} = \sqrt{\tfrac{1}{2}}\, (\mathbf{e}_x - i\mathbf{e}_y) \qquad (13.20)$$

have the rotational properties of a system of total angular momentum 1 and z-component $+1$, 0, and -1, respectively. These new polarization vectors have been written so that they are orthonormal,

$$\mathbf{e}_i{}^* \cdot \mathbf{e}_j = \delta_{ij},$$

and the choice of overall sign is made to be consistent with that made in Chapter II.

We now change our basic states for photons moving in the z direction to

$$\mathbf{E}_{+1}(\mathbf{r}, t) = \mathbf{e}_{+1} \exp i(kz - \omega t),$$

$$\mathbf{E}_{-1}(\mathbf{r}, t) = \mathbf{e}_{-1} \exp i(kz - \omega t).$$

From a knowledge of optics, we know that the superposition of plane polarization as in Eq. (13.18) gives circularly polarized light with the electric vector rotating clockwise around the z axis, and that of Eq. (13.20)

13.4 The Angular-Momentum Properties of the Electromagnetic Field

gives circularly polarized light having counterclockwise rotation. Thus right- (defined here as clockwise) circularly polarized photons have helicity $+1$ and left- (counterclockwise) polarized have helicity -1. If the photon is propagating in the $+z$ direction, these photons are in eigenstates of S^2 and S_z with quantum numbers s, s_z, which are 1, $+1$ and 1, -1 respectively.

Since no longitudinal field can be associated with a real photon, no $\mathbf{e}_0$ component is possible and no real photon can have zero-spin component along its direction of motion. This fact has an immediate consequence. Consider angular-momentum conservation in a transition in which a spin-0 state α emits a photon and becomes a second spin-0 state β. We orient our z axis along the direction of motion of the photon so that any relative orbital angular momentum l between photon and the system β has $l_z = 0$. The outgoing photon will be a superposition of the polarization state represented by $\mathbf{e}_{+1}$ and $\mathbf{e}_{-1}$. These states have $s_z = +1\,(-1)$ and must go with states of β having $j_z = -1\,(+1)$, which is impossible. It follows that a $0 \to 0$ transition with real photon emission is absolutely forbidden.

The transverse nature of the electromagnetic wave in free space is a consequence of the zero mass of the real photon. For virtual photons this restriction disappears, and such a photon can have a longitudinal component of field and helicity zero. It follows that the $0 \to 0$ transition is allowed if the photon is virtual. Thus the $0^+ \to 0^+$ electromagnetic transition from the 6.05-MeV excited state to the ground state of ^{16}O gives an electron–positron pair generated by a virtual photon. In addition, since electric charge is manifest by the interaction of single virtual photons, it is not impossible to detect electric charge on spin-0 particles.

Just as in the case of particle emission, the emission of a real photon involves the relative orbital angular momentum between the photon and the recoiling system. Thus the photon carries off total angular momentum j, which is the vector sum of l and spin $s = 1$. From the argument in the previous paragraphs, the photon carries off at least $j = 1$, and thus no possible addition of s and any l can be allowed to give j less than 1. To describe the angular distribution of vector radiation, we need the vector spherical harmonics that are eigenfunctions of J^2, J_z, L^2, and S^2. Since $s = 1$, there are three such functions for each j and j_z having $l = j, j \pm 1$. Each harmonic can be expressed as a sum of products of ordinary spherical harmonics and spin eigenfunctions

$$\mathbf{Y}(j, j_z, l, s) = \sum_{l_z=-l}^{+l} \sum_{s_z=-1}^{+1} \langle j, j_z | l, l_z, s, s_z \rangle Y(l, l_z) \chi(s, s_z),$$

where $s = 1$ and the sum is done under the restrictions $j_z = l_z + s_z$, $l - 1 \leqslant j \leqslant l + 1$. The $\chi(s, s_z)$ are used to represent the three polarization states of Eqs. (13.18)–(13.20). The state $\chi(1, 0) = \mathbf{e}_0$ is included because we shall be considering the general vector field, which can have a component in the z direction.

An arbitrary vector field can be expressed as a sum of vector spherical harmonics thus:

$$\mathbf{A}(\mathbf{r}) = \sum_{j=1}^{\infty} \sum_{j_z=-j}^{+j} \mathbf{A}(j, j_z, \mathbf{r}), \qquad (13.21)$$

where

$$\mathbf{A}(j, j_z, \mathbf{r}) = \sum_{l=j-1}^{j+1} g(l, l_z, s, s_z, r) \mathbf{Y}(j, j_z, l, s). \qquad (13.22)$$

g is a function of the radial distance $r = |\mathbf{r}|$. The parity of each term is the parity of $\mathbf{Y}(j, j_z, l, s)$, which has the parity of $Y(l, l_z)$, namely $(-1)^l$. Thus a vector field of total angular momentum j can have either parity, since, of the three terms in Eq. (13.22), two have parity $(-1)^{j\pm 1}$ and one has parity $(-1)^j$.

For each j we are particularly interested in the vector harmonic having $l = j$:

$$\mathbf{X}(j, j_z) = \mathbf{Y}(j, j_z, l = j, s = 1)$$

$$= \sum_{s_z=-1}^{+1} \langle j, j_z | l = j, l_z = j_z - s_z, s = 1, s_z \rangle Y(l, l_z) \chi(s, s_z).$$

Substituting the Clebsch–Gordan coefficients we find

$$\mathbf{X}(j, j_z) = \frac{j_z}{[j(j+1)]^{1/2}} Y(j, j_z) \mathbf{e}_0$$

$$- \left(\frac{(j+j_z)(j-j_z+1)}{2j(j+1)} \right)^{1/2} Y(j, j_z - 1) \mathbf{e}_{+1}$$

$$+ \left(\frac{(j-j_z)(j+j_z+1)}{2j(j+1)} \right)^{1/2} Y(j, j_z + 1) \mathbf{e}_{-1}.$$

The observable quantities are related to these amplitudes squared: because of the orthonormal properties of the polarization vectors we find

$$\mathbf{X}^*(j, j_z) \mathbf{X}(j, j_z) = \frac{j_z^2}{j(j+1)} |Y(j, j_z)|^2$$

$$+ \frac{(j+j_z)(j-j_z+1)}{2j(j+1)} |Y(j, j_z - 1)|^2$$

$$+ \frac{(j-j_z)(j+j_z+1)}{2j(j+1)} |Y(j, j_z + 1)|^2. \qquad (13.23)$$

13.4 The Angular-Momentum Properties of the Electromagnetic Field

These functions are normalized

$$\int \mathbf{X}^*(j, j_z) \mathbf{X}(j, j_z) \, d\Omega = 1 \, .$$

Returning to the electromagnetic field, we start by establishing some facts about the parity changes and relations. The relative parity of the fields **E** and **H** can be found by considering the instantaneous energy flux, which is given by the Poynting vector **P**:

$$\mathbf{P} = (1/4\pi) \, \mathbf{E} \times \mathbf{H} \, . \tag{13.24}$$

Since **P** is a vector, one of **E** and **H** is a vector and the other is an axial vector. The contribution to the Hamiltonian due to the interaction of the electromagnetic field with a system is given by

$$H = \int \mathbf{j} \cdot \mathbf{A} \, dx \, , \tag{13.25}$$

where **j** is the vector current density and **A** is the vector potential given by

$$\mathbf{B} = \text{curl } \mathbf{A} \, . \tag{13.26}$$

Consider the emission (or absorption) of a real photon described by the Hamiltonian of Eq. (13.25):

$$\alpha \rightleftharpoons \beta + \gamma \, .$$

The operator **j** causes the transition $\alpha \rightarrow \beta$, and its matrix element between these states is a vector if α and β have the same parity (no change) and an axial vector if they have opposite parity (change). Since H is scalar, **A** is vector or axial vector, respectively. Equation (13.26) indicates that **B** and **A** have opposite parity; hence, **B** is an axial vector for no change and vector for a change in parity. A vector has odd parity and an axial vector has even parity, and it follows that we can correctly follow parity changes if we assign to the photon the parity of its magnetic field.

If we consider the magnetic field of the emitted photon, it can be expanded into vector spherical harmonics; and if the field carries a given j, then from Eq. (13.22) we find that the parity can be $(-1)^j$ or $(-1)^{j+1}$. If the parity is $(-1)^j$, the field or the transition is called an electric multipole of order j (Ej). If the parity is $(-1)^{j+1}$ it is called an magnetic multipole (Mj). The first three, having $j = 1, 2, 3$, are called dipole, quadrupole, and octopole, respectively. Table 13.1 lists the parity carried off by the photon according to the multipolarity of the transition.

It is important to stress that the photon itself does not carry the parity assigned to it. This parity only indicates the change in parity between the initial and final states involved in the real photon emission (or absorption).

TABLE 13.1

ELECTROMAGNETIC MULTIPOLE PARITIES

Photon polarity	Electric multipole parity	Magnetic multipole parity
Dipole, $j = 1$	-1	$+1$
Quardrupole, $j = 2$	$+1$	-1
Octopole, $j = 3$	-1	$+1$
Multipole, j	$(-1)^j$	$(-1)^{j+1}$

By considering the time-averaged Poynting vector as a function of polar coordinates, it is possible to determine the angular distribution of the radiation emitted. If the transition is a pure electric or magnetic multipole, then the intensity $I(\theta, \phi)$ is given by

$$I(\theta, \phi) = |\mathbf{X}(j, j_z)|^2 .$$

This is independent of the magnetic or electric nature of the transition, so it is impossible to determine the change in parity in the transition from angular-distribution measurements alone. If the transition is not a pure multipole, then there is the possibility of interference between the amplitudes. Let us consider an example of a pure multipole transition. The decay

$$\omega \to \pi^0 + \gamma$$

is believed to be about 10% of the all ω decays. The transition is $1^- \to 0^-$, so that the lowest-order multipole transition is magnetic dipole. If the ω-mesons are unpolarized, there can only be uniform emission of radiation. Suppose we consider a sample of ω-meson, (spin 1) all having $s_z = +1$. Since the π^0 has $s = 0$, the outgoing radiation has $j = 1, j_z = +1$. Hence

$$I(\theta, \phi) = |\mathbf{X}(1, 1)|^2 .$$

From Eq. (13.23) we have

$$I(\theta, \phi) = (3/16\pi)(1 + \cos^2 \theta) .$$

Our treatment of the problem of the angular distribution has been very superficial, and we have not considered the problem of interference between multipoles or the problem of the polarization of outgoing radiation. In the next section, we shall consider briefly the photoproduction of mesons and deal with the reverse problem of the angular distribution of particles produced by the absorption of radiation. For full accounts of these problems we refer readers to Biedenharn (1960), to Frauenfelder and Steffen (1966), and to Devons and Goldfarb (1957).

We can summarize the selection rules in real photon emission or absorption. If the transition is

$$\alpha \rightleftharpoons \beta + \gamma$$

by a multipole of order j, j_z, then

$$|j_\beta - j| \leqslant j_\alpha \leqslant j_\beta + j,$$
$$j_{\alpha z} = j_{\beta z} + j_z$$

The relative parity of α and β is $(-1)^j$ for an electric and $(-1)^{j+1}$ for a magnetic multipole.

13.5 Photoproduction Processes

The photoproduction of mesons is an important aspect of the electromagnetic interaction of hadrons. In this section we shall discuss briefly the low-energy single meson photoproduction processes, such as

$$\gamma + p \to \pi^+ + n,$$
$$\gamma + p \to K^+ + \Lambda.$$

We shall restrict ourselves to low energies (< 1 GeV), since high-energy photoproduction, although it has been experimentally investigated, has not so far yielded to any theoretical analysis that gives insight into the processes occurring. The γ-p system is a channel to the $t = \frac{1}{2}$ and $t = \frac{3}{2}$ nonstrange baryon resonances. Thus the well-known pion–nucleon resonances can be involved as intermediate states in pion photoproduction and will strongly influence the cross section. For example, the $\Delta(1236)$ is important in single-pion production:

$$\gamma + p \to \Delta^+ \to \pi^0 + p.$$

The result is that there is a peak in the total cross section at a center-of-mass energy of 1236 MeV. None of these resonances has a large branching ratio into the $K\Lambda$ or $K\Sigma$ systems, so that they do not have an appreciable effect on the single-kaon photoproduction cross section.

Information about the photoproduction process can be obtained by considering the possible angular distributions. We consider, as an example, a magnetic-dipole transition. This means a change in J^P from proton to meson–nucleon system of $\frac{1}{2}^+ \to \frac{1}{2}^+$ or $\frac{1}{2}^+ \to \frac{3}{2}^+$. That is, a $P_{1/2}$ or $P_{3/2}$ final state. We will derive the angular distribution to be expected in the latter case. We set up axes with z along the direction of the incident photon beam. If it is unpolarized, then it can be considered as an incoherent sum of states having $j_{\gamma z} = +1$ and $j_{\gamma z} = -1$. (We use j_γ and not j to represent the angular momentum carried into the transition by the photon to avoid confusion with the total angular momentum, for which we also need a symbol. For a magnetic dipole $j_r = 1$, and the quantization axis ensures $j_{rz} = \pm 1$.) The proton target is an incoherent sum of states with $s_z = \pm \tfrac{1}{2}$. Let us represent by $|j_r, j_{rz}, s, s_z\rangle$ a given initial state. Then decomposing them into eigenstates of total angular momentum $|j, j_z\rangle$ we have

$$|1, +1, \tfrac{1}{2}, +\tfrac{1}{2}\rangle = |\tfrac{3}{2}, \tfrac{3}{2}\rangle, \qquad (13.27)$$

$$|1, -1, \tfrac{1}{2}, +\tfrac{1}{2}\rangle = \sqrt{\tfrac{1}{3}}\,|\tfrac{3}{2}, -\tfrac{1}{2}\rangle - \sqrt{\tfrac{2}{3}}\,|\tfrac{1}{2}, -\tfrac{1}{2}\rangle \qquad (13.28)$$

These two states occur with equal probability in the initial system. We need not consider the states having $s_z = -\tfrac{1}{2}$ since, by parity invariance, they give the same angular distribution. Since we are interested only in the $j = \tfrac{3}{2}$ final states we consider these decomposed into outgoing P-waves and proton spin states, $|l_l = 1, l_z, s, s_z\rangle$:

$$|\tfrac{3}{2}, \tfrac{3}{2}\rangle = |1, +1, \tfrac{1}{2}, +\tfrac{1}{2}\rangle, \qquad (13.29)$$

which gives an angular distribution

$$I_+(\theta) = |\langle \theta, \phi | 1, +1, \tfrac{1}{2}, +\tfrac{1}{2}\rangle|^2 = (3/8\pi)\sin^2\theta.$$

The other $j = \tfrac{3}{2}$ state decomposes

$$|\tfrac{3}{2}, -\tfrac{1}{2}\rangle = \sqrt{\tfrac{1}{3}}\,|1, -1, \tfrac{1}{2}, +\tfrac{1}{2}\rangle + \sqrt{\tfrac{2}{3}}\,|1, 0, \tfrac{1}{2}, -\tfrac{1}{2}\rangle, \qquad (13.30)$$

which gives an angular distribution

$$I_-(\theta) = (1/8\pi)\sin^2\theta + (1/2\pi)\cos^2\theta.$$

Because of the $\sqrt{\tfrac{1}{3}}$ in Eq. (13.28) this last state occurs with relative intensity of $\tfrac{1}{3}$, so that the complete angular distribution is proportional to

$$I = I_+(\theta) + \tfrac{1}{3} I_-(\theta) = 2 + 3\sin^2\theta.$$

In this way we can find the angular distribution, assuming a pure multipole transition. The results are given in Table 13.2 for the first four lowest multipoles.

It is worth noting how we can calculate the effect of plane-polarized photons. From Eqs. (13.16) and (13.17) we know a plane-polarized state will be a superposition of the two photon states implied in Eqs. (13.18) and (13.20). Since

$$\mathbf{e}_x = \sqrt{\tfrac{1}{2}}\,(\mathbf{e}_{+1} - \mathbf{e}_{-1}),$$

we have that the initial photon–nucleon state will be

$$\sqrt{\tfrac{1}{2}}\,|1, +1, \tfrac{1}{2}, -\tfrac{1}{2}\rangle - \sqrt{\tfrac{1}{2}}\,|1, -1, \tfrac{1}{2}, +\tfrac{1}{2}\rangle.$$

13.5 Photoproduction Processes

TABLE 13.2
Details of Possible Transitions in Single-Pion Photoproduction

Multipole	Meson-nucleon state	Angular distribution	Momentum dependence
$E1$	$S_{1/2}$	1	p
	$D_{3/2}$	$2 + 3\sin^2\theta$	p^5
$M1$	$P_{1/2}$	1	p^3
	$P_{3/2}$	$2 + 3\sin^2\theta$	p^3
$E2$	$P_{3/2}$	$1 + \cos^2\theta$	p^3
	$F_{5/2}$	$1 + 6\cos^2\theta - 5\cos^4\theta$	p^7
$M2$	$D_{3/2}$	$1 + \cos^2\theta$	p^5
	$D_{5/2}$	$1 + 6\cos^2\theta - 5\cos^4\theta$	p^5

If follows that the final $P_{3/2}$ meson-nucleon state will be the corresponding linear sum of the states of Eqs. (13.29) and (13.30).

$$\sqrt{\tfrac{1}{2}}\,|\tfrac{3}{2},\tfrac{3}{2}\rangle - \sqrt{\tfrac{1}{6}}\,|\tfrac{3}{2},-\tfrac{1}{2}\rangle = \sqrt{\tfrac{1}{2}}\,|1,+1,\tfrac{1}{2},+\tfrac{1}{2}\rangle$$
$$- \sqrt{\tfrac{1}{18}}\,|1,-1,\tfrac{1}{2},+\tfrac{1}{2}\rangle - \sqrt{\tfrac{1}{9}}\,|1,0,\tfrac{1}{2},-\tfrac{1}{2}\rangle.$$

The intensity of outgoing mesons is then

$$I(\theta,\phi) = |\sqrt{\tfrac{1}{2}}\,Y(1,1) - \sqrt{\tfrac{1}{18}}\,Y(1,-1)|^2 + |\sqrt{\tfrac{1}{9}}\,Y(1,0)|^2$$
$$= \tfrac{1}{2}(3/8\pi)\sin^2\theta\,|-e^{i\phi} - \tfrac{1}{3}e^{-i\phi}|^2 + (1/12\pi)\cos^2\theta$$
$$\propto \sin^2\theta\,(5 + 3\cos 2\phi) + 2\cos^2\theta.$$

As we expect, this intensity varies as $\cos 2\phi$, because photon plane-polarization defines a plane not a direction, and the distribution must be symmetric for reflections in the plane containing the z direction and the plane of photon polarization.

In the last column of Table 13.2, we have given the power law by which the photoproduction cross section is expected to vary with the center-of-mass momentum (p) of the photoproduced meson, near threshold. The outgoing angular-momentum barrier has a penetrability that varies as p^{2l} and the phase space available varies as p, so that the overall effect is p^{2l+1}.

Pion photoproduction at low energies (<400 MeV photon energy) has several distinct features. Just above threshold (about 150 MeV) charged-pion production from hydrogen

$$\gamma + p \to \pi^+ + n$$

has a much higher cross section than neutral-pion production

$$\gamma + p \to \pi^0 + p.$$

The former has a uniform angular distribution with total cross section varying as p, which suggests electric dipole transition to an $S_{1/2}$ final state. At energies above 200 MeV the charged-pion cross section begins to rise more rapidly, reaching a maximum near 330 MeV (center-of-mass energy of about 1240 MeV) with an angular distribution indicating the presence of at least two outgoing partial waves. The neutral-pion production increases as p^3, reaching a maximum at about the same energy, at which the angular distribution is very close to $2 + 3\sin^2\theta$. This indicates a magnetic dipole transition to a $P_{3/2}$ pion–nucleon state. All this is consistent with the importance in photoproduction of the $\Delta(1236)$, $\frac{3}{2}^+$ resonance in the pion–nucleon system. This can be reached from the $J^P = \frac{1}{2}^+$ nucleon state by a magnetic dipole or an electric-quadrupole transition. The angular distributions suggest the dominance of the former. In addition, there is some electric-dipole transition in the charged-pion production.

At higher energies photoproduction is influenced by the relevant higher-mass pion–nucleon resonances; in fact, evidence for the $t = \frac{1}{2}$ resonances $D_{3/2}$, $N(1518)$ and $F_{5/2}$, $N(1688)$ first came from photoproduction experiments. Walker (1969) has analyzed single-pion photoproduction up to 1.3 GeV into partial waves, taking into account the known resonances. There are several reasons for thoroughly investigating, both experimentally and theoretically, this region of photoproduction. Since the photon is a U-spin singlet, the observed selection rules for the photoproduction of resonances will provide additional aid in assigning resonances to multiplets. In addition, nucleon resonances weakly coupled to the pion–nucleon channel may well show up in the photoproduction of quasi-two-body final states, as in

$$\gamma + p \to \pi + \Delta$$

or

$$\gamma + p \to \rho + N.$$

There are also quark models that make predictions about photoproduction to be tested.

At energies where s-channel resonance effects should have become insignificant, Regge-pole analysis has been made of π^0 photoproduction with partial success (Bolon et al., 1967; Richter, 1968). However, charged-pion and -kaon photoproduction has not been convincingly Regge analyzed. In Section 13.7, we shall discuss pion photoproduction in the vector dominance model. For a review of the experimental and theoretical situation we refer readers to Harari (1967); Richter (1968); Rollnik (1965); and Höhler (1965).

13.6 Electromagnetic Form Factors

In this section, we shall consider the scattering of electrons by hadrons. We shall start with the most familiar, which is elastic electron–proton scattering. The most important amplitude is due to the one-photon-ex-

13.6 Electromagnetic Form Factors

change process. The electron–photon vertex is presumably well understood from quantum electrodynamics. If the proton were another nonstrongly interacting fermion, then presumably the same could be said about the photon–proton vertex. However, this is obviously not the case, since the proton has a large anomalous magnetic moment and strong interactions. Thus it is not possible to predict the scattering from quantum electrodynamics. Our lack of knowledge goes into form factors that are functions of the kinematic invariants.

Let us consider in detail the photon–nucleon vertex. We let p_1, p_2 be the initial and final four-momenta of the nucleon and represent by $|p_1\rangle$ and $\langle p_2|$, the corresponding state vectors. The current that interacts with the electromagnetic field is the matrix element of the current operator [Eq. (13.3)], $j^\mu(x)$, between these states, namely

$$J^\mu(x) = \langle p_2 | j^\mu(x) | p_1 \rangle .$$

Since $j^\mu(x)$ is a four-vector, the matrix element is also a four-vector. We therefore consider all possible independent four-vectors that can be constructed from the actual four-momenta present and from the bilinear covariants constructed from γ-matrices and the Dirac spinors representing the initial and final free nucleons, $u(p_1)$ and $u(p_2)$. The actual four-momenta are most conveniently combined thus:

$$q_+ = p_1 + p_2$$

and

$$q_- = p_1 - p_2 .$$

Other four-vectors that can be considered are of the kind

$$\bar{u}(p_2) \gamma^\mu u(p_1) ,$$
$$\bar{u}(p_2) \gamma^\mu \gamma^\nu q_{-\nu} u(p_1) , \quad \text{etc.}$$

However, the properties of the Dirac spinors [Eqs. (10.22a) and (10.22b)] reduce all such terms to a combination of three in this case, where the initial and final nucleons are both real particles (see Appendix D). Therefore the matrix element must be a linear combination of three, thus:

$$J^\mu(x) = \bar{u}(p_2) [G_1(t) q_+^\mu + G_2(t) q_-^\mu + G_3(t) \gamma^\mu] u(p_1) e^{i(p_2-p_1)\cdot x} , \quad (13.31)$$

where

$$t = (p_1 - p_2)^2 = q_-^2 .$$

The coefficients of the linear combination, G_1, G_2, and G_3, are called form factors. In general they will be functions of all the independent kinematic Lorentz-invariant variables involved. For this vertex, there is only one, chosen to be $t = (p_1 - p_2)^2$. [The quantity $(p_1 + p_2)^2$ is a function of t and of the masses.]

Current conservation places a strong restriction on the form factors.

It means
$$\partial J^\mu(x)/\partial x^\mu = 0 .$$

Apply this to Eq. (13.31) and we have
$$\bar{u}(p_2)q_{-\mu}[G_1(t)q_+^\mu + G_2(t)q_-^\mu + G_3(t)\gamma^\mu]u(p_1) = 0 .$$

Using again the properties of the Dirac equation this yields
$$G_1(t)(m_1^2 - m_2^2) + G_2(t)t + G_3(t)(m_1 - m_2) = 0 .$$

For elastic electron–nucleon scattering $m_1 = m_2$ and hence $t\,G_2(t) = 0$. This means that $G_2 = 0$. We are therefore left with two form factors:
$$\langle p_2 | j^\mu(x) | p_1 \rangle = \bar{u}(p_2)[G_1(t)q_+^\mu + G_3(t)\gamma^\mu]u(p_1)\,e^{i(p_2-p_1)\cdot x} . \tag{13.32}$$

It is possible to rearrange the matrix element into a more convenient form:
$$\langle p_2 | j^\mu(x) | p_1 \rangle = \bar{u}(p_2)[F_1(t)\gamma^\mu + i\sigma^{\mu\nu}q_{-\nu}F_2(t)]u(p_1)\,e^{i(p_2-p_1)\cdot x} , \tag{13.33}$$
where F_1 and F_2 are real.

We have discussed this matrix element as if it applied only to nucleons. In fact it also has the same form applied to the electron–photon vertex, and we can interpret these different form factors from what we know about the electron–photon interaction. The coupling of the current to the electromagnetic field is given by the observed electron charge, so that
$$\lim_{t=0} F_{e1}(t) = e . \tag{13.34}$$

This coupling correctly gives the Dirac magnetic moment of the electron ($g = 2$), so that the magnet-dipole term $\sigma^{\mu\nu}q_{-\nu}$ means that F_{e2} is left to describe the anomalous part of the electron's magnetic moment; hence
$$F_{e2}(0) = e\kappa/2m , \tag{13.35}$$
where κ is the anomalous part of the magnetic moment in Bohr magnetons. Thus the most general matrix element for the current operator between two electron states is given by Eq. (13.33), and the form factors can contain all modifications that may occur from radiative and higher-order corrections.

For protons the dominant corrections are due to strong interactions: current conservation means that Eqs. (13.32) and (13.34) apply and (13.35) is correct if we interpret κ as the anomalous part of the proton magnetic moment. For the neutron, $F_1(t = 0) = 0$ and κ will be its magnetic moment. This suggests a further redefinition of the form factors for the nucleons, so that
$$\langle p_2 | j^\mu(x) | p_1 \rangle$$
$$= e\,\bar{u}(p_2)[F_1(t)\gamma^\mu + i(\kappa/2M)F_2(t)\sigma^{\mu\nu}(p_2 - p_1)_\nu]u(p_1)\,e^{i(p_2-p_1)\cdot x} , \tag{13.36}$$

13.5 Electromagnetic Form Factors

where M is the nucleon mass. For the proton,

$$F_{p1}(0) = 1, \quad F_{p2}(0) = 1, \quad \kappa = +1.79,$$

and for the neutron

$$F_{n1}(0) = 0, \quad F_{n2}(0) = 1, \quad \kappa = -1.91.$$

The functions F_1 are sometimes called the Dirac form factors and the functions F_2, the Pauli form factors. They are studied experimentally by measuring electron–nucleon elastic scattering as a function of energy and momentum transfer. Using Eq. (13.36) for the nucleon vertex and assuming that the corresponding electron matrix element has $F_1 = 1$ and $F_2 = 0$ gives the Rosenbluth formula for the differential elastic-scattering cross section:

$$\frac{d\sigma}{dt} = 4\pi \left(\frac{\alpha}{t(s-M^2)} \right) \left[\frac{1}{2} t^2 \{F_1(t) + \kappa F_2(t)\}^2 \right.$$
$$\left. + \left\{ F_1^2(t) - \frac{t\kappa^2}{4M^2} F_2^2(t) \right\} ([s-M^2]^2 + ts) \right], \tag{13.37}$$

where α is the fine-structure constant, s is the squared total center-of-mass energy, and M is the nucleon mass. Expressed in terms of laboratory quantities, the differential cross section is

$$\frac{d\sigma}{d\Omega} = \left(\frac{e^2}{2E_0} \right)^2 \frac{\cos^2(\theta/2)}{\sin^4(\theta/2)} \frac{1}{1 + (2E_0/M)\sin^2(\theta/2)}$$
$$\times \left[F_1^2(t) - \frac{t}{4M^2} \left\{ 2[F_1(t) + \kappa F_2(t)]^2 \tan^2 \frac{\theta}{2} + \kappa^2 F_2^2(t) \right\} \right],$$
$$\tag{13.38}$$

where

$$t = -\frac{4E_0^2 \sin^2(\theta/2)}{1 + (2E_0/M)\sin^2(\theta/2)},$$

θ is the laboratory scattering angle, and E_0 is the incident-electron energy. This differential cross section decreases very rapidly with increasing angle (increasing $-t$). At larger angles the anomalous part of the magnetic moment contributes increasingly to the scattering through the term containing $\tan^2(\theta/2)$ and slows down the decrease in cross section with angle. The observed effect of the form factors is to decrease the cross section from the values expected if the form factors had their $t = 0$ values. Consider the Dirac form factor for the proton. If the proton had a pointlike structure, then we would expect this form factor to be constant. If, as we expect, the proton charge is spread out by strong-interaction effects, then the ability of this structure to absorb large momentum transfers by interaction with its charge is reduced and $F_1(t)$ should be a function decreasing with increasing $-t$. These arguments are analogous to those made in discussing X-ray scattering from the atomic electrons. In that situation it is possible to interpret the form factor as the Fourier transform of the radial charge distribution:

$$F(q^2) = \int \rho(\mathbf{r}) e^{i\mathbf{q}\cdot\mathbf{r}} d\mathbf{r}, \qquad (13.39)$$

where $\mathbf{q}$ is now a three-vector of momentum transfer.

The mean square radius $\langle r^2 \rangle$ is given by

$$\langle r^2 \rangle = \int \rho(\mathbf{r}) r^2 d\mathbf{r}.$$

To relate this to $F(q^2)$, expand the right-hand side of Eq. (13.39) thus:

$$F(q^2) = \int \rho(\mathbf{r}) d\mathbf{r} + i \int \rho(\mathbf{r}) \mathbf{q}\cdot\mathbf{r} \, d\mathbf{r} - \tfrac{1}{2} \int \rho(\mathbf{r}) (\mathbf{q}\cdot\mathbf{r})^2 d\mathbf{r} + \cdots.$$

For a spherically symmetric distribution we find

$$F(q^2) = \int \rho(\mathbf{r}) d\mathbf{r} - (q^2/6) \int \rho(\mathbf{r}) r^2 d\mathbf{r} + \cdots. \qquad (13.40)$$

Putting

$$F'(q^2) = \partial F(q^2)/\partial q^2$$

we have

$$\langle r^2 \rangle = -6F'(0). \qquad (13.41)$$

This treatment is not relativistic, but it can be applied to electron–nucleon scattering if the result is not interpreted too literally. It is found, for example, that the measured Dirac form factor of the proton corresponds to root mean square radius of about 0.8×10^{-13} cm, a plausible result, since we expect the nucleon to be spread out by approximately the pion Compton wavelength, 1.4×10^{-13} cm.

The early results on electron–proton scattering at momentum transfers up to 900 MeV/c [summarized by de Vries et al. (1962), see also Littauer et al. (1961)] indicated that the form factors were decreasing rapidly with increasing momentum transfer, F_{p2} more rapidly than F_{p1}. Electron–neutron scattering has to be investigated by a study of elastic and inelastic electron–deuteron scattering, and the interpretation is made very difficult by the bound state of the neutron. However, the early results on the Dirac form factor were consistent with zero for this quantity and showed that the Pauli form factor was varying in nearly the same way as that for the proton.

The form factors used above can be rearranged to define what are called the isotopic scalar and vector form factors. We start by considering

13.6 Electromagnetic Form Factors

the Gell-Mann–Nishijima formula for the charge, in units of the magnitude of the electron charge

$$Q = t_3 + (y/2).$$

For the nucleons $y = 1$, so we have

$$Q = \tfrac{1}{2} + t_3.$$

This suggests that the proton and neutron Dirac form factors are used to define scalar and vector form factors $F_{S1}(t)$ and $F_{V1}(t)$ thus:

$$F_{p1}(t) = \tfrac{1}{2}\{F_{S1}(t) + F_{V1}(t)\}, \tag{13.42}$$

$$F_{n1}(t) = \tfrac{1}{2}\{F_{S1}(t) - F_{V1}(t)\}. \tag{13.43}$$

The Pauli form factors can be similarly used to define Pauli isotopic scalar and vector form factors $F_{S2}(t)$, $F_{V2}(t)$:

$$1.79 F_{p2}(t) = \tfrac{1}{2}\{-0.12 F_{S2}(t) + 3.70 F_{V2}(t)\}, \tag{13.44}$$

$$-1.91 F_{n2}(t) = \tfrac{1}{2}\{-0.12 F_{S2}(t) - 3.70 F_{V2}(t)\}. \tag{13.45}$$

The isotopic form factors are defined so as to have the value $+1$ at $t = 0$. The physical interpretation of these form factors is simplified by considering reactions crossed from the ordinary elastic scattering; for example,

$$p + \bar{p} \to e^+ + e^-.$$

The initial state can be classified into isotopic singlet (isoscalar) and triplet (isovector) states. The former will involve only the F_{S1} and F_{S2} form factors and the latter only the F_{V1} and F_{V2} form factors. By crossing we expect the same isotopic decomposition to apply to electon scattering.

The data assembled by de Vries and collaborators (1962) showed that the behavior for $0 < |t| < 0.8$ (GeV/c)^2 of these isotopic form factors could be fitted with a formula

$$F(t) = \frac{a}{1 - (t/m^2)} + (1 - a), \tag{13.46}$$

where the constants a are close to 1 for all except F_{S2}, for which it is $\sim \tfrac{1}{2}$. The two isovector form factors were fitted with $m \simeq 600$ MeV and the two isoscalar form factors with $m \simeq 670$ MeV.

These fits were not very satisfactory, and this is understandable in view of the smooth behavior found for the Sachs form factors, which we define later. Nevertheless the empirical results on the form factors suggested that the dominating influence on the form factors were $J^P = 1^-$, 2π, and 3π states of a well-defined mass of about 600 MeV. The Feynman diagram of Fig. 13.4, where the virtual photon coupled to the electron current becomes

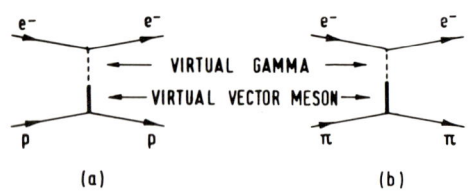

FIG. 13.4. The vector-meson interpretation of electromagnetic form factors of (a) nucleons and (b) pions.

a vector multipion state coupled to the nucleon, shows how this was expected to occur. The matrix element for the transition will contain a factor

$$\frac{1}{1 - (t/m_V^2)}, \qquad (13.47)$$

where m_V is the mass of the multipion state. In addition, since the photon has charge parity $C = -1$, these multipion states have $C = -1$ with $G = +1$ for the isovector and $G = -1$ for the isoscalar form factor. These were the facts that lead Nambu (1957) and later Frazer and Fulco (1959) to suggest the existence of a couple of pion resonances with $J^P = 1^-$ and $G = +1, G = -1$, respectively. These states were later observed and are now familiar to us as the ρ^0- and ω-mesons. The fact that $m_V = 600$ MeV, but $m_\rho = 765$, $m_\omega = 783$ MeV, possibly reflects the oversimplified nature of a model that attributes the form factors to the ρ^0 and ω but neglects other multipion states of the same quantum numbers. We can now explain partially the results expressed in Eq. (13.46). The equality of $F_{S1}(t)$ and $F_{V1}(t)$ means that the product of the coupling of photon to vector meson and of Dirac-like ($\bar{\psi}\gamma^\mu\phi\phi_\mu$) coupling of vector meson to nucleon is the same for ρ^0 and ω.

Experimentally it appears that all form factors have a large $-t$ behavior, which is

$$F(t) \simeq 1/t^2 \qquad \text{for} \qquad -t > 2 \quad \text{GeV}/c^2, \qquad (13.48)$$

where the model predicts t^{-1}. The t^{-2} dependence would appear if each form factor received contributions from two nearly equal-mass vector mesons ("dipole"). This may be the case for the isoscalar form factors, since there are two isosinglet vector mesons, the ω at 783 MeV and the ϕ at 1020 MeV. However, an another isotopic triplet partner to the ρ^0 at nearly the same mass has not been established.

Recently it has become conventional to express the experimental results in terms of the Sachs (1962) electric and magnetic form factors, defined by

$$G_E(t) = F_1(t) + (t\kappa/4M^2)F_2(t),$$
$$G_M(t) = F_1(t) + \kappa F_2(t).$$

13.6 Electromagnetic Form Factors

For the proton; $F_1(0) = 1$, $F_2(0) = 1$, $\kappa = 1.793$, so that
$$G_{pE}(0) = 1, \qquad G_{pM}(0) = 2.793.$$
For the neutron; $F_1(0) = 0$, $F_2(0) = 1$, $\kappa = -1.913$, so that
$$G_{nE}(0) = 0, \qquad G_{nM}(0) = -1.913.$$

As before, the G form factors can be decomposed into isotopic scalar and vector parts. Their advantage lies in the fact that they appear more directly connected with the observed electric and magnetic nuclear properties and that the Rosenbluth formula [Eq. (13.38)] simplifies considerably to

$$\frac{d\sigma}{d\Omega} = \sigma_0 \left[\frac{G_E^2 - (t/4M^2)G_M^2}{1 - (t/4M^2)} - \frac{t}{2M^2} G_M^2 \tan^2 \frac{\theta}{2} \right]$$

where

$$\sigma_0 = \left(\frac{e^2}{2E_0}\right)^2 \frac{\cos^2(\theta/2)}{\sin^4(\theta/2)} \frac{1}{1 + (2E_0/M)\sin^2(\theta/2)}.$$

There are no cross terms in the form factors, which makes analysis of experimental data more straightforward.

Experimental data (see Panofsky, 1968) are now available out to $-t = 25\,(\text{GeV}/c)^2$. Because G_M dominates at large momentum transfer, this form factor for the proton is the best known. Within errors up to $-t \simeq 4\,(\text{GeV}/c)^2$ the form factor G_{pE} satisfies

$$G_{pE} = G_{pM}(t)/\mu_p, \qquad (13.49)$$

where $\mu_p = 1 + \kappa_p$ and is the total proton magnetic moment. The neutron magnetic form factor is known to $-t \simeq 1.2\,(\text{GeV}/c)^2$ and satisfies

$$G_{nM}(t)/\mu_n = G_{pE}(t). \qquad (13.50)$$

It appears that the results expressed by Eqs. (13.49) and (13.50) are a correct scaling law for the form factors. Thus, all data nearly satisfy what is called the simple dipole fit:

$$G_{pE}(t) = \frac{G_{pM}(t)}{\mu_p} = \frac{G_{nM}(t)}{\mu_n} = \frac{1}{[1 - (t/0.71)]^2}. \qquad (13.51)$$

There are, however, known to be slight systematic deviations from this simple dipole fit. The neutron electric form factor is very difficult to obtain. Measurements of elastic electron–deuteron scattering and of thermal neutron scattering on atomic electrons show that G_{nE} is increasing from zero at $t = 0$ with $|\partial G_{nE}/\partial t| \simeq 0.5\,(\text{GeV}/c)^{-2}$. However, this increase is not maintained as $-t$ increases, and for $-t > 0.4\,(\text{GeV}/c)^2$ all experiment results are consistent with zero, although the errors are large. There is no satisfactory theoretical explanation of these data.

There is an interesting connection between electron–proton scattering and proton–proton scattering. It appears that for p–p scattering (Wu and Yang, 1965)

$$\frac{d\sigma}{dt} = \frac{d\sigma}{dt}\bigg|_{t=0} G_{pM}^4(t).$$

There is a simple argument that might explain this. If we think of G_{pM} as showing the effect of the matter distribution in the proton on an amplitude, then it appears as G_{pM}^2 in the e–p cross section. In p–p scattering the structure will appear twice, so the effect is G_{pM}^4 in the cross section.

So far, we have discussed electron–nucleon scattering, which explores the form factors for spacelike photons. As we have indicated, processes such as

$$e^+ + e^- \to p + \bar{p}$$

will involve timelike photons and thus explore an entirely new region of the form-factor behavior. The dipole fit of Eq. (13.51) probably has no meaning in this region. Experimental data will come from results obtained with electron–positron colliding-beam experiments at total energies greater than 2 GeV. Such energies are not yet available, and the alternative—measuring antiproton annihilation to electron–positron pair—has not yielded useful results.

The assumption has been made that all the results can be interpreted assuming only one-photon exchange; that is, that the Rosenbluth formula is correct. No deviations have been detected, and a comparison of electron and positron scattering and measurements of recoil proton polarization in e–p scattering do not show any of the effects expected if there were a two-photon exchange amplitude.

The next important form factor is that of the pion. Since this is a pseudoscalar particle, the matrix element of the electromagnetic current between two-pion states can contain only one term and there is only one form factor. Using the pole model we expect this form factor to be dominated by the ρ^0-meson. Until recently, the only data available were from experiments on pion–helium scattering and from pion electroproduction experiments. These results are hard to interpret and consequently have poor accuracy. The electron–positron storage rings have begun to give data on the reaction

$$e^+ + e^- \to \pi^+ + \pi^-,$$

and this measures the form factor in the timelike region (Augustin *et al.*, 1968; Auslander *et al.*, 1967). In the energy region around the ρ^0-meson,

13.7 The Vector-Dominance Model

F_π^2 appears to have a relativistic Breit–Wigner shape, as expected from a ρ-dominance model. We shall return to this subject in the next section.

The inelastic scattering of electrons by nucleons is a subject of growing interest. There are two aspects to this: the first concerns the production of baryon resonances (electroproduction):

$$e^- + p \to e^- + \Delta^+(1236),$$

for example. The second is inelastic scattering when the recoiling hadronic system is not in a resonant state; that is, continuum excitation. Considered from the electromagnetic point of view these processes are just an extension of photoproduction. If we use t to represent the squared four-momentum transferred by the virtual photon to the hadronic system in inelastic electron scattering, then t is always negative. In the corresponding photoproduction the real photon transfers four-momentum squared $t = 0$. Thus in the limit as $t \to 0$ the electroproduction cross section will be related to the photoproduction cross section. For t not zero the virtual photon is not limited to transverse polarization states, and a contribution to the cross section comes from a longitudinal polarization state of the virtual photon. Briefly, the experimental situation is that the resonance electroproduction cross-sections decrease with $-t$ as rapidly as does the elastic-scattering cross section. On the other hand, the continuum-excitation cross-sections decrease very much more slowly with increasing $-t$. This has caused speculation on the possibility that this may indicate the existence of pointlike charged bodies within the nucleon, possibly quarks (Panofsky, 1968).

The elastic and inelastic scattering of muons by nucleons has been studied experimentally. Unfortunately, available muon beams have very low intensity compared with available electron beams, so the accuracy and quantity of the data are that much less. However, the data do indicate that the scattering of muons is what is expected from electron–muon universality.

13.7 The Vector-Dominance Model

In the last section, we discussed electron–nucleon scattering and introduced the idea that the vector mesons ρ^0, ω, and ϕ played a part in determining the nucleon form factors. This implies a direct coupling of a photon to a vector meson, which in turn means that many other processes involving the interaction of hadrons with the electromagnetic field may be dominated by the effects of such a coupling, the so-called vector-dominance model (VDM). A common parameter in such a model will be the photon-to-vector-meson coupling. We shall first examine the definition of this coupling and discuss the values obtained for it in the following types of process: (1) the electromagnetic decay of vector mesons; (2) the photoproduction of vector mesons; (3) the relation between pion photoproduction and the production of vector mesons by pions. The simplest way to introduce the photon–vector-meson coupling is to consider the vector-meson dominance model for the charged pion form factor as shown by the Feynman diagram of Fig. 13.4b. We define the amplitude for a photon to become ρ^0-meson by g_ρ and use $f_{\rho\pi\pi}$ to represent the coupling constant of the ρ^0 to the pions. Then the charged-pion form factor $F_\pi(t)$ is given by

$$eF_\pi(t) = g_\rho f_{\rho\pi\pi}/(m_\rho^2 - t). \tag{13.52}$$

The terms on the right are the contributions to the scattering amplitude due to the photon–ρ^0 transition, the ρ^0 propagator, and the ρ^0 coupling to pions. The term on the left contains e, the electric charge, because, if we did not know anything about vector mesons, the corresponding contribution to the scattering amplitude would be of the form factor times the coupling of the electromagnetic field to the charged pion at $t = 0$. Now at $t = 0$, $F_\pi = 1$, so we have

$$g_\rho = em_\rho^2/f_{\rho\pi\pi}. \tag{13.53}$$

Unfortunately, we do not know whether this model is correct, so it is usual to start by a definition of the fields involved so that our photon–ρ-meson amplitude is written, by analogy with Eq. (13.53),

$$g_\rho = em_\rho^2/2\gamma_\rho, \tag{13.54}$$

and the photon–ρ^0 coupling is presented numerically by the value of the parameter $\gamma_\rho^2/4\pi$. For the treatment in field theory, which gives the precise definition of γ_ρ, and which is outside the scope of this book, we refer readers to the papers by Gell-Mann (1962), by Gell-Mann and Zachariasen (1961), and by Kroll et al. (1967). We shall have to be content with the prescription for the amplitude generalized for all vector mesons:

$$g_V = em_V^2/2\gamma_V. \tag{13.55}$$

We can immediately turn the argument around and, assuming that the ρ^0-meson dominates the charged-pion form factor, we have from Eq. (13.53)

$$\frac{\gamma_\rho^2}{4\pi} = \frac{1}{4}\frac{f_{\rho\pi\pi}^2}{4\pi}. \tag{13.56}$$

The coupling constant $f_{\rho\pi\pi}$ is given by the decay width for $\rho^0 \to \pi\pi$:

$$\Gamma(\rho^0 \to \pi^+\pi^-) = \frac{f_{\rho\pi\pi}^2}{4\pi}\frac{m_\rho}{12}\left[1 - \left(\frac{2m_\pi}{m_\rho}\right)^2\right]^{3/2}. \tag{13.57}$$

13.7 The Vector-Dominance Model

Using data on m_ρ and $\Gamma(\rho \to \pi^+\pi^-)$ from e^+e^- colliding-beam experiments Ting (1968) finds

$$\gamma_\rho^2/4\pi = 0.53 \pm 0.04 . \tag{13.58}$$

Using Eq. (13.41) we find that the rms charge radius on this model of the pion is expected to be about 0.63 Fm. As we have remarked in the previous section, there are very few data on the pion form factor for spacelike momentum transfers. From inelastic electron–proton scattering, Akerlof and co-workers (1966) found 0.8 ± 0.1 Fm. Electron–positron colliding-beam experiments permit the measurement of pion form factors in the timelike momentum-transfer region by observing

$$e^+ + e^- \to \pi^+ + \pi^- .$$

Near the mass of the ρ^0-meson we expect the reaction to be dominated by production of ρ^0, and this is the case. Using quantum electrodynamics and a charged-pion form factor, the total cross section is expected to be (Gatto, 1965)

$$\sigma(E) = \frac{\alpha^2 \pi}{12 E^2} \left(1 - \frac{m_e^2}{E^2}\right)^{3/2} |F_\pi(2E)|^2 , \tag{13.59}$$

where E is the energy of one of the colliding particles and m_e is the electron mass. Since the reaction is dominated by the physical ρ^0, we expect, assuming a nonrelativistic Breit–Wigner resonance, that

$$\sigma(E) = \frac{\pi \lambda^2}{4} (2j + 1) \frac{\Gamma(\rho^0 \to \pi^+\pi^-)\,\Gamma(\rho^0 \to e^+e^-)}{(2E - m_\rho)^2 + \Gamma^2/4} , \tag{13.60}$$

where j is the ρ^0 spin. Thus in the region of the ρ^0 the pion form factor is certainly dominated by this particle and $F_\pi(t = m_\rho^2)$ is observed to be about 40.

It is evident that colliding beams will be capable of exploring not only the charged-pion form factor but also charged- and neutral-kaon form factors through the reactions

$$e^+ + e^- \to K + \bar{K} .$$

The kaon form factor is expected to be dominated by the ϕ-meson at $t \simeq m_\phi^2$.

Equation (13.60) shows that at $2E = m_\rho$ the branching ratio for $\rho^0 \to e^+ + e^-$ is given by

$$\frac{\Gamma(\rho^0 \to e^+e^-)}{\Gamma(\rho^0 \to \text{all})} \simeq \frac{\Gamma(\rho^0 \to e^+e^-)}{\Gamma(\rho^0 \to \pi^+\pi^-)} = \frac{m_\rho^2 \sigma(m_\rho/2)}{12} .$$

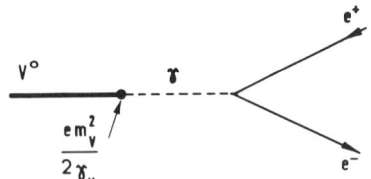

FIG. 13.5. Diagram for vector-meson decays.

This method has been used to obtain the $\rho^0 \to e^+e^-$ branching ratio and, by observation of

$$e^+ + e^- \to \phi \to K_1 + K_2,$$
$$e^+ + e^- \to \omega \to \pi^+ + \pi^- + \pi^0,$$

the branching ratios for $\phi \to e^+e^-$ and $\omega \to e^+e^-$. Measurements have also been made by producing these mesons in, for example, pion–proton interactions and observing the electron–positron decays. If these vector-meson decays are dominated by the vector-meson–virtual photon transition as in Fig. 13.5, then the partial decay widths are given by

$$\Gamma(V \to l^+l^-) = \frac{\alpha^2}{12}\left(\frac{\gamma_V^2}{4\pi}\right)^{-1} m_V \left(1 - \frac{4m_l^2}{m_V^2}\right)^{1/2}\left(1 + \frac{2m_l}{m_V^2}\right)$$

Thus the value of the appropriate $\gamma_V^2/4\pi$ may be found. Ting (1968) has reviewed the data available at that time and gives

$$\frac{\gamma_\rho^2}{4\pi} = 0.52 \begin{pmatrix} +0.07 \\ -0.06 \end{pmatrix},$$

$$\frac{\gamma_\omega^2}{4\pi} = 3.04 \begin{pmatrix} +1.07 \\ -0.66 \end{pmatrix}, \qquad (13.61)$$

$$\frac{\gamma_\phi^2}{4\pi} = 4.69 \begin{pmatrix} +1.24 \\ -0.81 \end{pmatrix}.$$

Note that as this parameter increases the amplitude for the photon $\rightleftharpoons$ vector-meson transition decreases.

The following decays can be connected if we assume vector dominance for the electromagnetic interaction:

$$\omega \to \pi^0 + \pi^+ + \pi^-,$$
$$\omega \to \pi^0 + \gamma,$$
$$\pi^0 \to \gamma + \gamma.$$

The appropriate Feynman diagrams under this assumption are shown in Fig. 13.6. The decay widths are given by (Gell-Mann et al., 1962):

$$\Gamma(\omega \to 3\pi) = \frac{f_{\rho\pi\pi}^2}{4\pi}\frac{f_{\rho\omega\pi}^2}{4\pi}\frac{(m_\omega - 3m_\pi)^4}{(m_\rho^2 - 4m_\pi^2)^2}\frac{m_\omega m_\pi^2}{\sqrt{27}} W(m_\pi), \qquad (13.62)$$

13.7 The Vector-Dominance Model

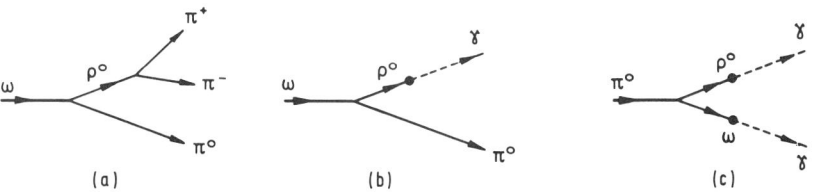

FIG. 13.6. Vector-dominance diagrams for the decays (a) $\omega \to \pi^+ + \pi^- + \pi^0$, (b) $\omega \to \pi^0 + \gamma$, (c) $\pi^0 \to \gamma + \gamma$.

$$\Gamma(\omega \to \pi^0 \gamma) = \alpha \left(\frac{\gamma_\rho^2}{4\pi}\right)^{-1} \left(\frac{f_{\rho\omega\pi}^2}{4\pi}\right) \frac{(m_\omega^2 - m_\pi^2)^3}{96 \, m_\omega^3}, \quad (13.63)$$

$$\Gamma(\pi^0 \to 2\gamma) = \alpha^2 \left(\frac{\gamma_\rho^2}{4\pi}\right)^{-1} \left(\frac{\gamma_\omega^2}{4\pi}\right)^{-1} \left(\frac{f_{\rho\omega\pi}^2}{4\pi}\right) \frac{m_\pi^3}{196}. \quad (13.64)$$

Masses are labeled by their subscripts and coupling constants f by the particles interacting. $W(m_\pi)$ is a kinematic phase-space factor equal to 3.56. Eqs. (13.62) and (13.63) show that the ratio of the decay widths for $\omega \to 3\pi$ and $\omega \to \pi\gamma$ will depend only upon $f_{\rho\pi\pi}$, γ_ρ, and known masses. The observed value of this ratio and the value of $f_{\rho\pi\pi}$ found by applying Eq. (13.57) to $\rho^0 \to \pi^+\pi^-$ gives

$$\gamma_\rho^2/4\pi = 0.9 \pm 0.1 \, .$$

Eqs. (13.63) and (13.64) show that the ratio of the decay width for $\omega \to \pi^0\gamma$ and $\pi^0 \to 2\gamma$ depends only upon γ_ω and known masses. The observed ratio gives

$$\gamma_\omega^2/4\pi = 2.5 \pm 0.5 \, .$$

The use of Eq. (13.64) depends upon the assumption that out of the three vector mesons we need only consider the pair $\rho\omega$ in the VDM diagram (Fig. 13.6c) for $\pi^0 \to 2\gamma$ decay. The $\phi\omega$ cannot be involved because the transition $\pi^0 \to \phi\omega$ does not conserve isotopic spin and the assumption depends upon the expected weakness of the $\phi\rho\pi$ coupling.

The next process of interest is the photoproduction of vector mesons:

$$\gamma + p \to V^0 + p \, .$$

The VDM suggests that this process will be dominated by a diffraction mechanism in which the photon becomes a V^0 and diffraction scatters from the proton (Ross and Stodolsky, 1966). The same arguments can be used to suggest that V^0 photoproduction from complex nuclei will be an important diffractive process. These suggestions are supported by the experimental data, which are now considerable in the case of ρ^0 photoproduction. The total cross section for ρ^0 photoproduction on hydrogen is either constant or decreasing slowly with increasing energy. Nondiffractive models would predict a decrease, e.g., s^{-2} in one-pion exchange. The angular distribution of ρ^0-mesons photoproduced on complex nuclei is what is expected from coherent production from the whole nucleus; i.e., from lead $d\sigma/dt \simeq e^{400t}$ (t in GeV^2), where the coefficient in the exponent reflects the size of the nucleus. If the mechanism were one-pion exchange, the production would be incoherent and the angular distribution the same as in hydrogen production, about e^{8t}. The decay angular distribution of the photoproduced ρ^0 is expected, in a diffraction model, to show the preservation of helicity. Thus the incoming photon has helicity ± 1 and the ρ_0 in its rest frame will have $s_z = \pm 1$ where z is along its direction of motion. The complex nuclei and hydrogen data are in agreement with this prediction.

Assuming that diffraction is the predominant process, the VDM relates

$$\gamma + p \rightarrow \rho^0 + p$$

to

$$\rho^0 + p \rightarrow \rho^0 + p$$

by

$$\frac{d\sigma}{dt}(\gamma p \rightarrow \rho^0 p) = \frac{\alpha}{4}\left(\frac{\gamma_\rho^2}{4\pi}\right)^{-1}\frac{d\sigma}{dt}(\rho^0 p \rightarrow \rho^0 p). \quad (13.65)$$

No experiments have been done on $\rho^0 p$ scattering. However, using the optical theorem [Eq. (3.16b)], and recalling that the forward-diffraction production amplitude will be entirely imaginary, we can write

$$\frac{d\sigma}{dt} = \frac{\pi}{k^2}\frac{d\sigma}{d\Omega} = \frac{\pi}{k^2}[\operatorname{Im} f(0)]^2 ;$$

hence,

$$\left.\frac{d\sigma}{dt}(\rho^0 p \rightarrow \rho^0 p)\right|_{t=0} = \frac{[\sigma_T(\rho^0 p \rightarrow \rho^0 p)]^2}{16\pi}. \quad (13.66)$$

The total $\rho^0 p$ cross section σ_T can be found from an analysis of the ρ^0 photoproduction from complex nuclei. Since diffraction is the consequence of absorption, the variation of ρ^0 photoproduction with atomic number can be used to obtain the total cross section for ρ^0 on nucleons (Drell and Trefil, 1966). The experiments indicate that σ_T is about 30 mb (Bulos et al., 1969; McClellan et al., 1969; Asbury et al., 1968). Depending on the details of the analysis, which has been oversimplified here, the result is

$$\gamma_\rho^2/4\pi = 0.5 \rightarrow 1.0 .$$

Insufficient work has been done on ϕ and ω photoproduction to obtain values for the corresponding quantities for these mesons.

Using the optical theorem applied to photons [Eq. (13.66) with γ

13.7 The Vector-Dominance Model

instead of ρ^0] and relating the forward-scattering amplitude to a sum of vector-meson photoproduction amplitudes we have

$$[\sigma_T(\gamma p)]^2 = 16\pi \left[\frac{d\sigma(\gamma p \to \gamma p)}{dt}\right]_{t=0},$$

from which we obtain

$$\sigma_T(\gamma p) = (4\pi\alpha)^{1/2}$$

$$\left[\left(\frac{\gamma_\rho^2}{4\pi}\right)^{-1}\frac{d\sigma}{dt}(\gamma p \to \rho^0 p) + \left(\frac{\gamma_\omega^2}{4\pi}\right)^{-1}\frac{d\sigma}{dt}(\gamma p \to \omega p)\right.$$

$$\left. + \left(\frac{\gamma_\phi^2}{4\pi}\right)^{-1}\frac{d\sigma}{dt}(\gamma p \to \phi p)\right]_{t=0}^{1/2}. \tag{13.67}$$

Allowing for all uncertainties this predicts $\sigma_T(\gamma p)$ in the range of about 90 to 140 μb. A result at 7.5 GeV gives 126 ± 17 μb (Ballam *et al.*, 1968) and another at $3 \to 5$ GeV gives 116 ± 17 μb (Aachen, 1968). Although this result is hardly conclusive, it does not conflict with the model.

The VDM can also be applied to relate single-pion photoproduction to vector-meson production by pions. Assuming the ρ dominates the amplitude, the model directly connects, for example,

$$\gamma + p \to \pi^+ + n \tag{13.68}$$

and

$$\rho^0 + p \to \pi^+ + n.$$

But time reversal now connects these with

$$\pi^+ + n \to \rho^0 + p,$$

and charge symmetry connects this with

$$\pi^- + p \to \rho^0 + n. \tag{13.69}$$

The final relation is

$$\frac{d\sigma}{dt}(\gamma p \to \pi^+ n) = \frac{\pi\alpha}{\gamma_\rho^2}\rho_{11}\frac{d\sigma}{dt}(\pi^- p \to \rho^0 n). \tag{13.70}$$

The ρ_{11} is the element of the ρ-meson density matrix that gives the fraction of the production that has helicity ± 1 (sometimes called transverse vector mesons). It is a function of t. Good agreement has been obtained comparing reactions (13.68) and (13.69) at 4.0 and 8.0 GeV (Diebold and Poirier, 1968; Derado and Guiragossian, 1968) using $\gamma_\rho^2/4\pi = 0.45$. Dar *et al.* (1968) have considered other relations, which take into account the ω contribution to the amplitude: this corresponds to isoscalar photon

TABLE 13.3

VALUES OF THE PHOTON-TO-VECTOR-MESON COUPLING PARAMETER

		$\frac{\gamma_\rho^2}{4\pi}$	$\frac{\gamma_\omega^2}{4\pi}$	$\frac{\gamma_\phi^2}{4\pi}$
1	Pion form factor	0.53 ± 0.04		
2	$V \to e^+e^-$	$0.52 \binom{+0.07}{-0.06}$	$3.04 \binom{+1.07}{-0.66}$	$4.69 \binom{+1.24}{-0.81}$
3	$\dfrac{\Gamma(\omega \to \pi^0 \gamma)}{\Gamma(\omega \to 3\pi)}$	0.9 ± 0.1		
4	$\dfrac{\Gamma(\omega \to \pi^0 \gamma)}{\Gamma(\pi^0 \to 2\gamma)}$		2.5 ± 0.5	
5	$\gamma + A \to \rho^0 + A$	$0.5 \to 1.0$		
6	$\dfrac{\gamma p \to \pi^+ n}{\pi p \to \rho^0 n}$	0.45		
7	$SU(3), SU(6)$ (Ratios only)	$\dfrac{1}{9}$	1	$\dfrac{1}{2}$

absorption transitions, and since there is the possibility of interference, relations have to be found in which it cancels, because the phase is not known. These involve, for example, not only the cross section for $\gamma p \to \pi^+ n$ as in Eq. (13.70) but also the cross section for $\gamma n \to \pi^- p$. The interference term has opposite sign in the cross section. These authors obtain good agreement comparing charged- and neutral-photoproduction cross-sections with corresponding vector-meson production cross-sections.

The nucleon form factors, if dominated by the vector mesons, should also give values of the coupling parameters. However, the theory of the form factors is not in a sufficiently good state to make the procedure credible. In addition, values of the vector-meson–nucleon coupling constants are required, and these are not well known.

The values we have discussed are shown in Table 13.3; in processes 1, 3, 4, 5, and 6 the essential photon-meson transition occurs with the photon on the mass shell. In process 2 the vector meson is on the mass shell. All comparisons have been made assuming the coupling constant is independent of the squared momentum transfer, which varies from zero to 0.59 GeV^2 between these extremes. Any variation would be equivalent to the existence of a form factor at the γ-ρ^0 vertex, and would introduce considerable uncertainty in the model.

Assumptions about the constancy of other coupling constants are also obvious in several of these qualitative estimates of the coupling parameters. There are various predictions about the ratios to one another of the photon coupling constants of the three vector mesons. $SU(3)$ and the ϕ-ω mixing angle given by $SU(6)$, $\cos \theta = \sqrt{\frac{2}{3}}$, gives

$$\gamma_\rho^{-2} : \gamma_\omega^{-2} : \gamma_\phi^{-2} = 9 : 1 : 2 \,.$$

Other theories are concerned with the coupling of these mesons to hadron currents. Thus the ρ^0 is coupled to the neutral component of an isotopic-spin current, the $SU(3)$ octet ω to a baryon current, and the $SU(3)$ singlet ϕ to a hypercharge current. Conservation of these currents leads to universal coupling constants. If this were correct the ρ-meson-to-nucleon and ρ-meson-to-pion coupling constants would be equal. The behavior of these currents determines the mixing and hence the values of the coupling parameters. For example, a model due to Oakes and Sakurai (1967) gives

$$\gamma_\rho^{-2} : \gamma_\omega^{-2} : \gamma_\phi^{-2} = 9 : 0.65 : 1.33 \,.$$

The present data cannot decide between these possibilities.

REFERENCES

Aachen-Berlin-Bonn-Hamburg-Heidelburg-Munich Collaboration (1968) *Phys. Lett.* B **27**, 474 (13.7).
Akerlof, C. W., Ash, W. W., Berkelman, K., and Lichtenstein, C. A. (1966). *Phys. Rev. Lett.* **16**, 147 (13.7).
Asbury, J. G. *et al.* (1968). *Phys. Rev. Lett.* **20**, 227 (13.7).
Augustin, J. E. *et al.* (1968). *Phys. Rev. Lett.* **20**, 126 (13.6).
Auslander, V. L., Budker, G. I., Pestov, Ju. N., Sidorov, V. A., Skrinsky, A. N., and Khabakhpashev, A. G. (1967). *Phys. Lett.* B **25**, 433 (13.6).
Ballam, J. *et al.* (1968). *Phys. Rev. Lett.* **21**, 1544 (13.7).
Bernstein, J. (1968). "Elementary Particle and Their Currents," Freeman, San Francisco, California (13.2).
Biedenharn, L. C. (1960). In "Nuclear Spectroscopy" (F. Ajzenberg Selove, ed.), p. 732. Academic Press, New York (13.4).
Bolon, G. C. (1967). *Phys. Rev. Lett.* **18**, 926 (13.5).
Bulos, F. *et al.* (1969). *Phys. Rev. Lett.* **22**, 490 (13.7).
Dar, A., Weisskopf, V. F., Levinson, C. A., and Lipkin, H. J. (1968). *Phys. Rev. Lett.* **20**, 1261 (13.7).
Derado, I., and Guiragossian, Z. G. T. (1968). *Phys. Rev. Lett.* **21**, 1556 (13.7).
Devons, D., and Goldfarb, L. J. B. (1957). In "Handbuch der Physik" (S. Flügge, ed.), Vol. 42, p. 362. Springer, Berlin (13.4).
de Vries, C., Hofstadter, R., and Herman, R. (1962). *Phys. Rev. Lett.* **8**, 381, 466 (E) (13.6).
Diebold, R., and Poirier, J. A. (1968). *Phys. Rev. Lett.* **20**, 1532 (13.7).
Drell, S. D., and Trefil, J. (1966). *Phys. Rev. Lett.* **16**, 552, 832 (E) (13.7).
Frauenfelder, H., and Steffen, R. M. (1966). In "Alpha, Beta and Gamma Ray Spectroscopy" (K. Siegbahn, ed.), p. 997. North-Holland Publ., Amsterdam (13.4).
Frazer, W. R., and Fulco, J. R. (1959). *Phys. Rev. Lett.* **2**, 365 (13.6).

Gatto, R. (1965). *Proc. Int. Symp. Electron Photon Interactions High Energy. Hamburg, 1965*, p. 106. Deutche Physikalische Gesellschafte V. (13.7).
Gell-Mann, M. (1962). *Phys. Rev.* **125**, 1067 (13.7).
Gell-Mann,, and Zachariasen, F. (1961). *Phys. Rev.* **124**, 953 (13.7).
Gell-Mann, M., Sharp, D. H., and Wagner, W. G. (1962). *Phys. Rev. Lett.* **8**, 261 (13.7).
Hand, L. N. (1963). *Phys. Rev.* **129**, 1834 (13.6).
Harari, H. (1967). *Proc. Int. Symp. Electron Photon Interactions High Energy. Stanford, 1967*, p. 337. International Union of Pure and Applied Physics and U.S. Atomic Energy Commission (13.5).
Höhler, G. (1965). *Proc. Easter Sch. for Physicists*, 1965, Vol. II, p. 55. CERN, Geneva (13.5).
Kroll, N. M., Lee, T. D., and Zumino, B. (1967). *Phys. Rev.* **157**, 1376 (13.7).
Littauer, R. M., Schopper, H. F., and Wilson, R. R. (1961). *Phys. Rev. Lett.* **7**, 141, 144 (13.6).
McClellan, G. *et al.* (1969). *Phys. Rev. Lett.* **22**, 377 (13.7).
Nambu, Y. (1957). *Phys. Rev.* **106**, 1366 (13.6).
Oakes, R. J., and Sakurai, J. J. (1967). *Phys. Rev. Lett.* **19**, 1266 (13.7).
Panofsky, W. K. H. (1968). *Proc. Int. Conf. High Energy Phys. 14th, Vienna, 1968*, p. 23. CERN, Geneva (13.6).
Richter, B. (1968). *Proc. Int. Conf. High Energy Phys. 14th, Vienna, 1968*, p. 3. CERN, Geneva (13.5).
Rollnik, H. (1965). *Proc. Easter Sch. for Physicists, 1965*, Vol. II, p. 19. CERN, Geneva (13.5).
Ross, M., and Stodolsky, L. (1966). *Phys. Rev.* **149**, 1172 (13.7).
Sachs, R. G. (1962). *Phys. Rev.* **126**, 2256 (13.6).
Ting, S. C. C. (1968). *Proc. Int. Conf. High Energy Phys. 14th, Vienna, 1968.*, p. 43. CERN, Geneva (13.7).
Walker, R. L. (1969). *Phys. Rev.* **182**, 1729 (13.5).
Watson, K. M. (1952). *Phys. Rev.* **85**, 852 (13.2).
Wu, T. T., and Yang, C. N. (1965). *Phys. Rev. B* **137**, 708 (13.6).

XIV

THE NEUTRAL KAONS AND *CP* CONSERVATION

14.1 Introductions

In this chapter we shall discuss the behavior of the neutral-K-meson system, initially assuming *CP* conservation and then examining the evidence and meaning of the nonconservation of *CP*, which has so far (1969) been observed only in this system. In this chapter we shall use units having $\hbar = c = 1$ and time in seconds. Then mass has the dimensions of reciprocal time, in this case seconds^{-1}. In conventional units, mass has the value given by $mc^2/\hbar$. The reason for doing this is that it has become conventional in describing neutral-kaon decays, because of the consequent simplicity of the time-development term, which has the form $\exp(-imt - \lambda t/2)$ for a state of mass m decaying with a mean life of λ^{-1} seconds.

14.2 The Time Development of Neutral-Kaon Systems

Let us consider the time development of a general state that is a superposition of K^0 and $\bar{K}^0$:

$$|\psi\rangle = a|K^0\rangle + \bar{a}|\bar{K}^0\rangle, \tag{14.1}$$

where $|\psi\rangle$, a, and $\bar{a}$ are functions of the proper time. We can represent $|\psi\rangle$ by the column matrix, ψ, where

$$\psi = \begin{bmatrix} a \\ \bar{a} \end{bmatrix}. \tag{14.2}$$

The time development is described by (Kabir, 1968)

$$i(\partial \psi/\partial t) = H\psi, \tag{14.3}$$

where H is a 2×2 matrix, which can be written, using the Pauli matrices σ,
$$H = A + \mathbf{B} \cdot \boldsymbol{\sigma}. \tag{14.4}$$
A unit matrix is understood to be included with A. Using the usual representation of these matrices we can see that $B_x\sigma_x$ and $B_y\sigma_y$ can change $K^0 \rightleftharpoons \bar{K}^0$. In Section 7.5 we made the choice of phase implied by
$$CP|K^0\rangle = |\bar{K}^0\rangle. \tag{14.5}$$
In our present notation, therefore,
$$CP \begin{bmatrix} 1 \\ 0 \end{bmatrix} = \begin{bmatrix} 0 \\ 1 \end{bmatrix} \quad \text{and} \quad CP \begin{bmatrix} 0 \\ 1 \end{bmatrix} = \begin{bmatrix} 1 \\ 0 \end{bmatrix}.$$
Representing CP by 2×2 matrix we have
$$CP = \sigma_x. \tag{14.6}$$
If the time development of our neutral-kaon beam is CP conserving, H must commute with CP and thus $B_y = B_z = 0$ and we have
$$H = A + B_x\sigma_x = \begin{bmatrix} A & B_x \\ B_x & A \end{bmatrix}. \tag{14.7}$$
The eigenvalues of this matrix are
$$A + B_x \quad \text{and} \quad A - B_x,$$
with eigenvectors
$$\phi_1 = \frac{1}{\sqrt{2}} \begin{bmatrix} 1 \\ 1 \end{bmatrix} \quad \text{and} \quad \phi_2 = \frac{1}{\sqrt{2}} \begin{bmatrix} 1 \\ -1 \end{bmatrix}, \tag{14.8}$$
respectively. We shall now see that it is convenient to express the complex eigenvalues as
$$A + B_x = m_1 - i\lambda_1/2,$$
$$A - B_x = m_2 - i\lambda_2/2.$$
It is obvious that a solution to Eq. (14.3) is
$$\psi = c_1\phi_1 \exp(-im_1t - \lambda_1t/2)$$
$$+ c_2\phi_2 \exp(-im_2t - \lambda_2t/2). \tag{14.9}$$
This equation can be interpreted in the following way. The states represented by ϕ_1, ϕ_2 are eigenstates of CP (σ_x) with eigenvalues $+1$ and -1, and each of these has a time development that contains factors $\exp(-imt)$ and $\exp(-\lambda t/2)$. The first is the proper time development of a state of total energy m, so that we associate the real part of the eigenvalues with the center-of-mass energy of the state; that is, mass. The second factor

14.2 The Time Development of Neutral Kaon Systems

represents an exponential decay of mean life λ^{-1}. Also this analysis shows that the CP eigenstates are, in bra–ket notation,

$$|K_1\rangle = \sqrt{\tfrac{1}{2}}\,\{|K^0\rangle + |\bar{K}^0\rangle\}, \qquad (14.10)$$

$$|K_2\rangle = \sqrt{\tfrac{1}{2}}\,\{|K^0\rangle - |\bar{K}^0\rangle\}. \qquad (14.11)$$

We expect these states to have different masses, since they will communicate, via weak interactions, to different virtual states. They will have different lifetimes because CP conservation decrees that they decay into different final states.

Now, let us examine the time development of a beam that is purely K^0 at the source ($t=0$). This boundary condition means that the beam at $t=0$ is represented by the ket

$$|\phi(0)\rangle = |K^0\rangle = \sqrt{\tfrac{1}{2}}\,\{|K_1\rangle + |K_2\rangle\}.$$

This follows from solving Eqs. (14.10) and (14.11). Alternatively in the matrix notation

$$\phi(0) = \begin{bmatrix} 1 \\ 0 \end{bmatrix} = \frac{1}{\sqrt{2}}\phi_1 + \frac{1}{\sqrt{2}}\phi_2.$$

Thus the constants in Eq. (14.9) are $c_1 = c_2 = 1/\sqrt{2}$, and we have

$$\phi(t) = \frac{1}{2}\begin{bmatrix} \exp[-im_1 t - \lambda_1 t/2] + \exp[-im_2 t - \lambda_2 t/2] \\ \exp[-im_1 t - \lambda_1 t/2] - \exp[-im_2 t - \lambda_2 t/2] \end{bmatrix}.$$

The elements of this column vector are the coefficients of Eq. (14.1), which are the probability amplitudes for finding K^0- and $\bar{K}^0$-mesons. Thus the probability of finding a K^0-meson is

$$P(K^0) = \frac{1}{4}\left\{\exp(-\lambda_1 t) + \exp(-\lambda_2 t) \right.$$
$$\left. + 2\exp\!\left(\frac{-\lambda_1 t - \lambda_2 t}{2}\right)\cos(m_2 - m_1)t\right\}.$$

Similarly the probability of finding a $\bar{K}^0$-meson is

$$P(\bar{K}^0) = \frac{1}{4}\left\{\exp(-\lambda_1 t) + \exp(-\lambda_2 t) \right.$$
$$\left. - 2\exp\!\left(\frac{-\lambda_1 t - \lambda_2 t}{2}\right)\cos(m_2 - m_1)t\right\}.$$

Evidently

$$P(K^0) + P(\bar{K}^0) = \tfrac{1}{2}[\exp(-\lambda_1 t) + \exp(-\lambda_2 t)].$$

This last equation expresses the fact that the probability of finding any neutral kaon decreases in the way expected from the fact that the K_1 and

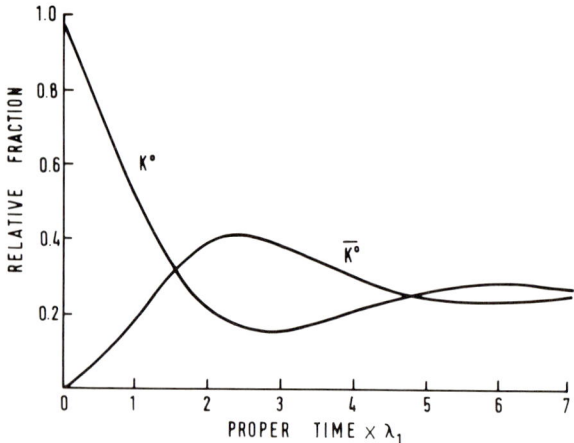

FIG. 14.1. The fraction of K^0 and $\bar{K}^0$ in a beam, which is initially K^0, as a function of proper time for $m_2 - m_1 = \lambda_1$.

K_2 decay with different lifetimes. However, the probability of finding a K^0 (or $\bar{K}^0$) has an oscillatory term that has an angular velocity $\omega = m_2 - m_1$. If there were no decay ($\lambda_1 = \lambda_2 = 0$), the neutral-kaon beam would oscillate between being 100% K^0 and 100% $\bar{K}^0$. The decay damps these oscillations. If ω is small compared to λ_1, the oscillations are very much damped and a K^0 beam smoothly becomes a 50-50% mixed K^0-$\bar{K}^0$ beam (pure K_2) by the decay of the K_1 fraction. If ω is large compared to λ_1, the oscillatory term is important and the fraction of the beam that is K^0 oscillates rapidly until damped by decay. See Fig. 14.1.

The actual situation (1969) is that $\lambda_1 = 1.1 \times 10^{10}\,\text{sec}^{-1}$, $\lambda_2 = 1.89 \times 10^7\,\text{sec}^{-1}$, and $m_2 - m_1 = 0.47\lambda_1$ ($m_2 - m_1 \simeq 7 \times 10^{-39}\,\text{g} \simeq 3 \times 10^{-6}\,\text{eV}$). Figure 14.1 shows, for example, how the fractions of K^0 and $\bar{K}^0$ vary with proper time in a beam initially K^0. Since $\bar{K}^0$-mesons can produce hyperons while K^0 cannot in normal targets at low energy, the beam will produce hyperons as a function of distance from the source in a manner that will reflect the oscillatory term. This effect was observed by Fitch *et al.* (1961) and Camerini *et al.* (1962) and can obviously be used to obtain the magnitude of the mass difference. However, the values obtained were high ($m_2 - m_1 \simeq 1.5 \to 1.9\lambda_1$) compared with the currently accepted value (about $0.5\lambda_1$). Another way of observing the strangeness oscillation is to make use of the $\Delta Q = \Delta S$ rule (Section 11.12). This allows the decays

$$K^0 \to \pi^- + l^+ + \nu_l, \qquad \bar{K}^0 \to \pi^+ + l^- + \bar{\nu}_l, \qquad (14.12)$$

but forbids

$$K^0 \to \pi^+ + l^- + \bar{\nu}_l, \qquad \bar{K}^0 \to \pi^- + l^+ + \nu_l, \qquad (14.13)$$

where l is a charged lepton of either kind. Obviously, the observation of three-body decays with a negative lepton will measure the $S = -1$ content of a beam. Thus a beam that is pure K^0 at source should have a number of $\pi^+ l^- \bar{\nu}_l$ decays as a function of proper time that follows the curve marked $\bar{K}^0$ in Fig. 14.1. Again this effect has been observed. Unfortunately the analysis is complicated by uncertainty as to the status of the $\Delta Q = \Delta S$ rule. Making no assumptions about the rule, Aubert et al. (1965) analyzed 196 K_{e3}^0 decays in a bubble chamber and found

$$|m_2 - m_1| = (0.47 \pm 0.2)\lambda_1.$$

These methods of measuring the magnitude of the mass difference have been superseded by other methods that make use of the CP-nonconserving decays (Section 14.6). Nonetheless the observed effects are evidence for the oscillatory behavior in neutral-K beams. It is worth noting that the sign of the mass difference cannot be found from this kind of experiment.

14.3 The Neutral-Kaon Mass Difference

We have seen (Section 7.5) that the K_1 decays mainly into the CP-even states $\pi^+\pi^-$ and $\pi^0\pi^0$. The K_2 decays mainly into 3π and $\pi l \nu$ states. The ratio of the phase space available in these states is comparable to the ratio of the lifetimes, as could be expected. The analysis in the previous section implies that the change

$$K^0 \rightleftharpoons \bar{K}^0,$$

is occurring in a neutral-K beam. This is a $\Delta S = 2$ transition, and if there is no direct $\Delta S = 2$ interaction it must proceed in two steps via $S = 0$ intermediate states; for example,

$$K^0 \rightleftharpoons \pi^+\pi^- \rightleftharpoons \bar{K}^0.$$

These intermediate states are reached by the $\Delta S = 1$ weak interaction, and the entire transition is then of second order in the weak interactions. The rate of this change is determined by the mass difference, which will then be given by a term that contains the weak-coupling constant squared (second-order perturbation theory). The decay rate λ_1 is also proportional to the squared coupling constant, so that we expect $\lambda_1 \simeq |m_2 - m_1|$. If there were a $\Delta S = 2$ interaction as strong as the weak interactions, then the mass difference would be very much larger. The observed value, $m_2 - m_1 \simeq 0.47\lambda_1$, means that any $\Delta S = 2$ interaction cannot have a coupling constant greater than about 10^{-6} that of the $\Delta S = 1$ weak interaction (Kabir, 1968; Okun and Pontecorvo, 1957).

14.4 The Theory of Regeneration

We have disussed (Sections 7.5) the process of regeneration, in particular of K_1-mesons from a beam of K_2-mesons traversing an absorber. Now, we can expand our description and make a distinction between two processes giving rise to K_1 mesons after the absorber. The first is due to diffraction (or scattering) regeneration. Any process that scatters neutral K-mesons with amplitudes that are different for K^0 and $\bar{K}^0$ will alter the K^0-$\bar{K}^0$ balance from that existing in the incident beam (K_2) and hence give rise to secondary K_1-mesons. The angular and energy distribution of these mesons will be typical of a normal scattering process. At high energies this distribution will have a strong forward-diffraction peak.

The second process is called coherent regeneration. Simply, it comes about because the absorber removes K^0 and $\bar{K}^0$ unequally from the incident beam, so that the transmitted beam has a disturbed K^0-$\bar{K}^0$ balance, which means K_1-mesons will be found in the transmitted beam. Their energy and angular distribution will be the same as those quantities in the incident beam.

Consider the scattering of K_2-mesons at a certain angle. The scattering process conserves strangeness, so that the amplitudes that apply are the scattering amplitudes for K^0 and $\bar{K}^0$, f and $\bar{f}$. Thus the scattering of K_2-mesons leads to a state

$$\sqrt{\tfrac{1}{2}}\,\{f\,|\,K^0\rangle - \bar{f}\,|\,\bar{K}^0\rangle\} = \tfrac{1}{2}(f - \bar{f})\,|\,K_1\rangle + \tfrac{1}{2}(f + \bar{f})\,|\,K_2\rangle. \tag{14.14}$$

Thus the amplitude for scattering a K_2 into a K_2 is $f_s = (f + \bar{f})/2$, while the regeneration amplitude, which is that for scattering K_2 into K_1, is

$$f_r = (f - \bar{f})/2\,.$$

To understand the behavior of K-mesons in condensed matter it is necessary to reanalyze the time development (Good, 1957). The matter changes the H of Eq. (14.3) and also the states that can be associated with exponential loss. To find the change we have to consider the optical behavior of a plane wave normally incident on a plane slab of transmitting material (Fig. 14.2). Consider a point P at a depth z. The incident plane wave at this point is $\exp i(kz - \omega t)$. In addition there is a contribution to the amplitude due to scattered waves from all the material of the slab. Consider a slice, thickness $\Delta z'$, at a depth z'. We can calculate its contribution to the amplitude at z by dividing it into Fresnel zones. The contribution is i/π that of the first zone, which has an area $2\pi^2(z - z')/k$. If f is the forward-scattering amplitude and n the number of scatterers per unit volume, we have a total contribution at z due to the slice of

$$\frac{i}{\pi} n \frac{2\pi^2(z - z')}{k} \Delta z' \frac{f}{z - z'} \exp i(kz - \omega t)\,.$$

14.4 The Theory of Regeneration

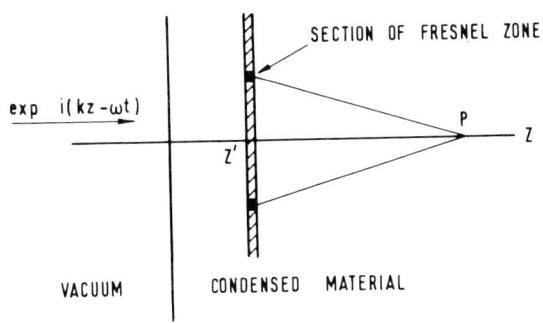

FIG. 14.2. Diagram showing a slice of thickness $\Delta z'$ in a slab of condensed material traversed by a plane wave. The slice contributes to the total amplitude at P by scattering the incident plane wave. This contribution can be calculated by dividing the slice into Fresnel zones.

Integrating from $z' = 0$ to z we find that the contribution at z due to the material is

$$\frac{2\pi z f n i}{k} \exp i(kz - \omega t).$$

Thus the total amplitude is

$$\psi(z) = \left(1 + \frac{2\pi z f n i}{k}\right) \exp i(kz - \omega t). \tag{14.15}$$

This result has been obtained in a nonrigorous fashion. It can be interpreted in terms of a refractive index (Goldberger and Watson, 1964). We can see this by considering the case when the second term in the coefficient is small compared with 1.

$$\psi(z) = \exp\left(\frac{2\pi z f n i}{k}\right) \exp i(kz - \omega t)$$

$$= \exp i\left(kz\left\{1 + \frac{2\pi n f}{k^2}\right\} - \omega t\right); \tag{14.16}$$

that is, a refractive index of $1 + 2\pi n f/k^2$. Considering now the plane wave as representing neutral K-mesons and considering the time development in the K rest frame, we have $z = \beta\gamma t$, and $i\partial\psi/\partial t$ contains an extra term $-2\pi n f \beta\gamma/k$, where β is the velocity of the K-meson through the material and γ is its Lorentz factor. The forward-scattering amplitude f and the wave number k are defined in the frame in which the scattering centers are at rest. Then $k = m_K \gamma \beta$ and the extra term is $-2\pi n f/m_K$. This term depends upon whether the kaons are K^0 or $\bar{K}^0$ (f or $\bar{f}$), so that H of Eq. (14.7) becomes

$$H' = H - \frac{2\pi n}{m_K}\begin{bmatrix} f & 0 \\ 0 & \bar{f} \end{bmatrix}. \tag{14.17}$$

In this analysis f and $\bar{f}$ are the forward-scattering amplitudes for K^0 and $\bar{K}^0$, respectively. To find the states with exponential attenuation in the absorber we must find the eigenvectors and eigenvalues of H'. These states $\mathcal{K}_1$, $\mathcal{K}_2$ are not quite K_1 and K_2, and the *regeneration parameter* ρ, which gives the deviation, is

$$\rho = \frac{2\pi n}{m_K} \frac{f_r}{[(m_1 - m_2) - i(\lambda_1 - \lambda_2)/2]}. \qquad (14.18)$$

To first order in ρ, the normalized eigenvectors $(\mathcal{K}_1, \mathcal{K}_2)$ are

$$\frac{1}{\sqrt{2}} \begin{bmatrix} 1 - \rho \\ 1 + \rho \end{bmatrix} \quad \text{and} \quad \frac{1}{\sqrt{2}} \begin{bmatrix} 1 + \rho \\ -1 + \rho \end{bmatrix}$$

with eigenvalues

and
$$\kappa_1 = m_1 - i\lambda_1/2 - (\pi n/m_K)(f + \bar{f})$$

Thus
$$\kappa_2 = m_2 - i\lambda_2/2 - (\pi n/m_K)(\bar{f} + f). \qquad (14.19)$$

and
$$|\mathcal{K}_1\rangle = |K_1\rangle - \rho |K_2\rangle$$

$$|\mathcal{K}_2\rangle = |K_2\rangle + \rho |K_1\rangle. \qquad (14.20)$$

Consider now a K_2 beam incident upon an absorber of thickness L. If the kaon velocity is β, the proper time elapsing during traversal is $\tau = L/\beta\gamma$. Thus the incident beam is (to first order in ρ)

$$|K_2\rangle = |\mathcal{K}_2\rangle - \rho |\mathcal{K}_1\rangle. \qquad (14.21)$$

As in Eq. (14.9) the solutions of Eq. (14.3) have a time development that is given by the eigenvalues, so that on leaving the regenerator the state of the beam is described by

$$\exp(-i\kappa_2\tau) |\mathcal{K}_2\rangle - \rho \exp(-i\kappa_1\tau) |\mathcal{K}_1\rangle$$
$$= \exp(-i\kappa_2\tau)[|K_2\rangle + \rho |K_1\rangle]$$
$$\quad - \rho \exp(-i\kappa_1\tau)[|K_1\rangle - \rho |K_2\rangle]. \qquad (14.22)$$

Thus the K_1 amplitude at the end of the regenerator is

$$\rho \left[\exp(-i\kappa_2\tau) - \exp(-i\kappa_1\tau)\right],$$

and the intensity of K_1 relative to that of K_2 incident is this amplitude squared,

$$\frac{16\pi^2 n^2 |f_r|^2}{m_K(1 + 4\delta^2)} \left[\exp(-\lambda_2\tau) + \exp(-\lambda_1\tau) \right.$$
$$\left. - 2\exp\{-(\lambda_1 + \lambda_2)\tau/2\} \cos(m_2 - m_1)\tau\right], \qquad (14.23)$$

where $\delta = (m_2 - m_1)/(\lambda_2 - \lambda_1)$. Note that throughout this analysis the

14.5 The Sign of the Mass Difference

amplitudes $f, \bar{f}$ are defined as forward-scattering amplitudes in the rest frame of the regenerator.

The process of regeneration was first observed by Muller *et al.* (1960) and used by Fujii *et al.* (1964) to obtain a measure of the $K_1 - K_2$ mass difference. It is now a clearly established phenomenon. It is worth noting that the regenerated intensity depends on the square of the density of scattering centers, a prediction verified by Christenson *et al.* (1965). In addition, in the course of this work the regenerator was divided into two parts so that the emerging K_1 amplitude was a sum of an amplitude due to regeneration in the two parts. If the first part was moved upbeam, the phase of its regeneration amplitude at the end changed, and thus the yield of K_1 mesons after the regenerators changed in a way depending on the gap between the regenerators, on the mass difference, and on the decay constants. This method yielded one of the more precise values of the mass difference,

$$|\delta| = 0.50 \pm 0.10.$$

14.5 The Sign of the Mass Difference

None of the methods of observing the oscillatory behavior in a neutral-kaon beam so far mentioned allows a determination of the sign of the mass difference. To find the sign it is necessary to have information about the scattering amplitudes f and $\bar{f}$. Consider a source of neutral kaons produced by strong interactions, so that it is either a K^0 or a $\bar{K}^0$-meson source (for the analysis we will assume K^0). Thus at proper time $t = 0$ the state of the beam is

$$|\phi(0)\rangle = \sqrt{\tfrac{1}{2}}\{|K_1\rangle + |K_2\rangle\}.$$

At a proper time t the beam is

$$|\phi(t)\rangle = \sqrt{\tfrac{1}{2}}\{\exp(-im_1 t - \lambda_1 t/2)|K_1\rangle \\ + \exp(-im_2 t - \lambda_2 t/2)|K_2\rangle\}.$$

Suppose it undergoes scattering at that time; the scattering amplitude for $K_1 \to K_1$ or $K_2 \to K_2$ is $f_s = (f + \bar{f})/2$ and for $K_2 \to K_1$ or $K_1 \to K_2$ is $f_r = (f - \bar{f})/2$. Thus the scattered state is

$$\sqrt{\tfrac{1}{2}}\{[f_s \exp(-im_1 t - \lambda_1 t/2) + f_r \exp(-im_2 t - \lambda_2 t/2)]|K_1\rangle \\ + [f_s \exp(-im_2 t - \lambda_2 t/2) + f_r \exp(-im_1 t - \lambda_1 t/2)]|K_2\rangle\}.$$

Therefore the intensity of K_1-mesons produced in a scattering at time t is

$$\begin{aligned}P_1(t) &= \tfrac{1}{2}|f_s \exp(-im_1 t - \lambda_1 t/2) + f_r \exp(-im_2 t - \lambda_2 t/2)|^2 \\ &= \tfrac{1}{2}\{f_s^2 \exp(-\lambda_1 t) + f_r^2 \exp(-\lambda_2 t) \\ &\quad + 2\exp[-(\lambda_1 + \lambda_2)t/2]\,\mathrm{Re}\,f_s^* f_r \exp[-i(m_2 - m_1)t]\}.\end{aligned} \quad (14.24)$$

If f_s and f_r are known, this can determine the sign of $m_2 - m_1$.

One determination used K^0-mesons produced by K^+ charge exchanging in deuterium and observed the K_1 yield after a scattering as a function of proper time between production and scatter. The scattering amplitudes used were determined from phase-shift analyses of K^+ and K^- scattering (Canter et al., 1966). A second method involved observing the regenerated K_1 yield as a function of distance between two different pieces of material comprising the regenerator (Jovanovich et al., 1966). Both methods show that the K_2 meson is heavier than the K_1. The averaged value from all methods (1969) is

$$m_2 - m_1 = (0.469 \pm 0.015)\lambda_1.$$

14.6 CP Nonconservation

The observation of the decay of the long-lived neutral kaon into two pions,

$$K_2 \to \pi^+ + \pi^-$$

(Christenson et al., 1964; Abashian et al., 1964) can be interpreted as evidence of the nonconservation of CP eigenvalues, since the K_2 has $CP = -1$ and all spin-0 two-pion states have $CP = +1$ (Table 7.4, p. 196). When these observations were announced, various other interpretations were suggested; before discussing these we shall describe subsequent observations, which reduce considerably the number to be considered seriously and give information on the possible source of CP violation.

It is necessary to make a distinction between the eigenstates of CP and the states of a neutral kaon beam that have an exponential decay. The former we continue to represent by K_1 and K_2. The latter we represent by K_S and K_L, for the short- and long-lived components, respectively.

The experimental facts can be summarized as follows.

(1) The decay

$$K_L \to \pi^+ + \pi^-$$

has been observed in a vacuum under varying conditions of energy and production with the same result for the decay rate. Defining

$$\eta_{+-} = \frac{\text{amplitude } K_L \to \pi^+\pi^-}{\text{amplitude } K_S \to \pi^+\pi^-},$$

the best value, given in a review by Steinberger (1969), is

$$|\eta_{+-}| = (1.90 \pm 0.05) \times 10^{-3}.$$

14.6 CP Nonconservation

(2) The decay

$$K_L \to \pi^0 + \pi^0,$$

has been observed (see, for example, Gaillard et al., 1967; Cence et al., 1969), with the result that if we define

$$\eta_{00} = \frac{\text{amplitude } K_L \to 2\pi^0}{\text{amplitude } K_S \to 2\pi^0},$$

then

$$|\eta_{00}| \simeq 3.6 \times 10^{-3}.$$

However, Bartlett et al. (1968) obtained $|\eta_{00}|^2 = (-2 \pm 7) \times 10^{-6}$; that is, no observable rate for this decay. At the time of writing, the situation is not clear, since these experiments are very difficult and may suffer from unrecognized errors. For the purpose of discussion we shall assume

$$|\eta_{00}| \neq |\eta_{+-}|.$$

(3) Interference has been observed between the K_L decay into $\pi^+\pi^-$ and the same decay from K_S mesons regenerated from the same beam. (Fitch et al., 1965; Alff-Steinberger et al., 1966; Bott-Bodenhausen et al., 1966).

(4) Interference has been observed between $K_S \to \pi^+\pi^-$ and $K_L \to \pi^+\pi^-$ at distances from a K^0 source where the K_S has decayed sufficiently to give a 2π-decay amplitude comparable to that from K_L (Böhm et al., 1968).

(5) K_L leptonic decay rates are not charge symmetric. Dorfan et al. (1967) observe

$$\frac{K_L^0 \to \pi^- \mu^+ \nu_\mu}{K_L^0 \to \pi^+ \mu^- \bar{\nu}_\mu} = 1.0081 \pm 0.0027,$$

and Bennett et al. (1967) observe

$$\frac{K_L^0 \to \pi^- e^+ \nu_e}{K_L^0 \to \pi^+ e^- \bar{\nu}_e} = 1.0045 \pm 0.0007.$$

We shall first examine briefly those theories that were put forward to explain the initial observation, (1), above. Cosmological explanations assume, for example, a long-range interaction that causes $K_2 \to K_1$ transitions. However, such theories predict that the transition rate $K_2 \to K_1$, and hence the apparent $K_2 \to 2\pi$ rate, depends upon the laboratory energy of the K_2; no such effect is observed. Cosmological theories also encounter other difficulties (Kabir, 1968).

There are various CP-invariant interpretations of the effect (1) on its own, which were also put forward. The simplest is that the K_2 has a decay channel

$$K_2 \to K_1 + S,$$

where S is a light particle. The K_1 then decays by a 2π mode. This can be ruled out by the observation of the interference [experiments (3) and (4)], since this proposed decay would produce K_1-mesons incoherent with those produced by regeneration, for example. Another possibility is that there is a breakdown of the exponential decay, so that at a long time there is a residual K_1 part to a beam. This hypothesis cannot explain the interference effect described in (4) above, since it says that all 2π decays are due to K_1 and therefore that there are no other amplitudes for the K_1 to 2π-decay amplitude to interfere with.

We shall see that the decay asymmetry described in (5) and the interference effects described in (4) are unequivocal demonstrations of CP-nonconservation, and we shall now proceed to analyze the neutral-kaon system in this circumstance.

14.7 CP-Noninvariant Analysis

We return to the formalism of Section 14.2 and consider the mass matrix H of Eq. (14.3); we relax the requirement that it be CP invariant, so that it can now have a $B_y \sigma_y$ term when written as in Eq. (14.4). If we include a term $B_z \sigma_z$, then we allow CPT violation. The reason for this is that the diagonal elements would then become unequal, and in the bra–ket notation that means

$$\langle K^0 | H | K^0 \rangle \neq \langle \bar{K}^0 | H | \bar{K}^0 \rangle.$$

These matrix elements are the complex mass of the K^0 and $\bar{K}^0$, respectively, and must be equal if CPT invariance holds (Section 6.9). We shall assume CPT invariance; hence, $B_z = 0$. The analysis and its consequences for CPT noninvariance has been given by Lee and Wolfenstein (1965).

H can now be written

$$H = \begin{bmatrix} \alpha & \beta/r \\ \beta r & \alpha \end{bmatrix}, \tag{14.25}$$

which has eigenvalues

$$\alpha \pm \beta,$$

with eigenvectors

$$\frac{1}{(1 + |r|^2)^{1/2}} \begin{bmatrix} 1 \\ \pm r \end{bmatrix}. \tag{14.26}$$

14.7 CP-Noninvariant Analysis

Since the CP nonconservation is small, we expect that $|r| \simeq 1$. Thus the states with well-defined exponential time variation are

$$|K_S\rangle = (1 + |r|^2)^{-1/2}\{|K^0\rangle + r|\bar{K}^0\rangle\}, \qquad (14.27)$$

$$|K_L\rangle = (1 + |r|^2)^{-1/2}\{|K^0\rangle - r|\bar{K}^0\rangle\}, \qquad (14.28)$$

and the general solution to Eq. (14.3) is, in the bra-ket notation,

$$|\psi(t)\rangle = C_S \exp[-i(\alpha + \beta)t]|K_S\rangle$$
$$+ C_L \exp[-i(\alpha - \beta)t]|K_L\rangle. \qquad (14.29)$$

In a CP-invariant theory $r = 1$. We know that η_{+-} is small, so that it follows that r is very close to 1. If we put

$$\epsilon = (1 - r)/(1 + r), \qquad (14.30)$$

then we expect ϵ to be small ($\simeq 10^{-3}$) and we can write

$$|K_S\rangle = \frac{\{(1 + \epsilon)|K^0\rangle + (1 - \epsilon)|\bar{K}^0\rangle\}}{[2(1 + |\epsilon|^2)]^{1/2}}, \qquad (14.31)$$

$$|K_L\rangle = \frac{\{(1 + \epsilon)|K^0\rangle - (1 - \epsilon)|\bar{K}^0\rangle\}}{[2(1 + |\epsilon|^2)]^{1/2}}. \qquad (14.32)$$

Comparing this with Eqs. (14.10) and (14.11) we note that we can rewrite

$$|K_L\rangle = \frac{\{(|K^0\rangle - |\bar{K}^0\rangle) + \epsilon(|K^0\rangle + |\bar{K}^0\rangle)\}}{[2(1 + |\epsilon|^2)]^{1/2}}$$
$$= \frac{\{|K_2\rangle + \epsilon|K_1\rangle\}}{(1 + |\epsilon|^2)^{1/2}}, \qquad (14.33)$$

so that K_L contains a small amount of the CP-even state K_1. Thus, even if CP were conserved in the decay, the existence of a CP-noninvariant term in the mass matrix would lead us to expect that

$$\eta_{+-} = \eta_{00} = \epsilon.$$

If it is established finally that $\eta_{+-} \neq \eta_{00}$, this is not the situation and it will be necessary to include the possibility of CP-nonconserving decays. We shall assume that is the case.

Returning to Eqs. (14.27) and (14.28) we see that the general neutral-kaon beam is described by six parameters, the real and imaginary parts of the eigenvalues and of r (or of ϵ). One of these parameters, the phase of r, depends upon the choice of relative phase between $|K^0\rangle$ and $|\bar{K}^0\rangle$, which cannot be determined. The convention followed is that suggested by Wu and Yang (1964). It rests upon the observation that the branching ratio for the decays

$$K_S \to \pi^+ + \pi^-,$$
$$K_S \to \pi^0 + \pi^0$$

is very close to that expected from the $\Delta \mathbf{t} = \frac{1}{2}$ rule (Section 11.12) and therefore that the $t = 0$ final state is dominant. Therefore, the K^0 and $\bar{K}^0$ phases are chosen so that the decay amplitudes to the $t = 0$ final state $\langle t, t_3 |$ are equal; that is,

$$\langle 0, 0 | T | K^0 \rangle = \langle 0, 0 | T | \bar{K}^0 \rangle, \quad (14.34)$$

where T is the transition operator. In addition, we can fix the phase of this transition amplitude without loss of generality. We arbitrarily put it equal to a real transition amplitude a_0 times the factor $\exp(i\delta_0)$, which is a phase shift due to the final state $t = 0$ π-π interaction (Section 6.6). Assuming conservation of angular momentum, we see from Table 7.4 that other accessible two-pion states have isotopic spin 2. We know that the final state in $K^\pm_{2\pi}$ decay has $t = 2$, so that it is reasonable (and if $\eta_{00} \neq \eta_{+-}$, necessary) to include a transition amplitude to the $t = 2$ state in $\dot{K}^0_{2\pi}$. The amplitude for K^0 decay into $t = 2$ is then

$$a_2 e^{i\delta_2} = \langle 2, 0 | T | K^0 \rangle. \quad (14.35)$$

CPT invariance then requires (Section 6.9)

$$\langle 2, 0 | T | \bar{K}^0 \rangle = a_2^* e^{i\delta_2} \quad (14.36)$$

From the discussion of Section 6.6 we know that if CP is conserved, the amplitudes a_0 and a_2 must be relatively real. Thus we are, by having a_0 real, expecting that the extent of CP nonconservation in the decay transitions will be given by the imaginary part of a_2. The parameter to be used as a measure of this nonconservation is

$$\epsilon' = \frac{\operatorname{Im} a_2}{\sqrt{2}\, a_0} \exp i\left(\frac{\pi}{2} + \delta_2 - \delta_0\right). \quad (14.37)$$

Let us now relate η_{+-} and η_{00} to ϵ and ϵ'. Decomposing the final states, we have

$$|\pi^+\pi^-\rangle = \sqrt{\tfrac{1}{6}}\{\sqrt{2}\,|0,0\rangle + \sqrt{3}\,|1,0\rangle + |2,0\rangle\},$$
$$|\pi^-\pi^+\rangle = \sqrt{\tfrac{1}{6}}\{\sqrt{2}\,|0,0\rangle - \sqrt{3}\,|1,0\rangle + |2,0\rangle\},$$
$$|\pi^0\pi^0\rangle = \sqrt{\tfrac{1}{3}}\{-|0,0\rangle + \sqrt{2}\,|2,0\rangle\}.$$

The $\pi^+\pi^-$ final state is the symmetric combination of the two states given above; that is,

$$\sqrt{\tfrac{1}{2}}\{\langle\pi^+\pi^-| + \langle\pi^-\pi^+|\},$$

14.7 CP-Noninvariant Analysis

which we abbreviate to $\langle \pi^+\pi^- |$. It follows that the total effective amplitude for decay into this state is

$$\langle \pi^+\pi^- | T | K_{S,L} \rangle = [6(1 + |\epsilon|^2)]^{-1/2} \{(1 + \epsilon)(\sqrt{2}\, a_0 e^{i\delta_0} + a_2 e^{i\delta_2})$$
$$\pm (1 - \epsilon)(\sqrt{2}\, a_0 e^{i\delta_0} + a_2^* e^{i\delta_2})\},$$

$$\langle \pi^0\pi^0 | T | K_{S,L} \rangle = [6(1 + |\epsilon|^2)]^{-1/2} \{(1 + \epsilon)(-a_0 e^{i\delta_0} + \sqrt{2}\, a_2^* e^{i\delta_2})$$
$$\pm (1 - \epsilon)(-a_0 e^{i\delta_0} + \sqrt{2}\, a_2^* e^{i\delta_2})\}.$$

The choice of sign is $+$ for K_S and $-$ for K_L. Then if we define

$$\Delta = \frac{\mathrm{Re}\, a_2}{\sqrt{2}\, a_0} \exp i(\delta_2 - \delta_0), \tag{14.38}$$

and

$$A = \frac{2a_0}{[6(1 + |\epsilon|^2)]^{1/2}},$$

we have

$$\langle \pi^+\pi^- | T | K_S \rangle = \sqrt{2}\, A\{1 + \Delta + \epsilon\epsilon'\}, \tag{14.39}$$

$$\langle \pi^+\pi^- | T | K_L \rangle = \sqrt{2}\, A\{\epsilon(1 + \Delta) + \epsilon'\}, \tag{14.40}$$

$$\langle \pi^0\pi^0 | T | K_S \rangle = A\{-1 + 2\Delta + 2\epsilon\epsilon'\}, \tag{14.41}$$

$$\langle \pi^0\pi^0 | T | L_L \rangle = A\{-\epsilon + 2\epsilon\Delta + 2\epsilon'\}. \tag{14.42}$$

Therefore,

$$\left| \frac{\langle \pi^+\pi^- | T | K_S \rangle}{\langle \pi^0\pi^0 | T | K_S \rangle} \right|^2 = 2 \left| \frac{1 + \Delta + \epsilon\epsilon'}{1 - 2\Delta - 2\epsilon\epsilon'} \right|^2.$$

This factor, when multiplied by the ratio of slightly different phase-space factors, is the branching ratio $\pi^+\pi^-/\pi^0\pi^0$ for the K_S decay. The observed value is such that the ratio of the amplitudes squared is very close to 2. Thus $\Delta + \epsilon\epsilon'$ must be small. Both ϵ and ϵ' are small (about 10^{-3}), so this observation suggests Δ is small. Hence

$$\eta_{+-} = \frac{\langle \pi^+\pi^- | T | K_L \rangle}{\langle \pi^+\pi^- | T | K_S \rangle} = \epsilon + \frac{\epsilon'}{1 + \Delta} \simeq \epsilon + \epsilon', \tag{14.43}$$

$$\eta_{00} = \frac{\langle \pi^0\pi^0 | T | K_L \rangle}{\langle \pi^0\pi^0 | T | K_S \rangle} = \epsilon - \frac{2\epsilon'}{1 - 2\Delta} \simeq \epsilon - 2\epsilon'. \tag{14.44}$$

These two equations expose the meaning of the two parameters ϵ and ϵ'. If CP violation mixes CP-even states into K_L but nothing else, $\epsilon' = 0$, and $\eta_{+-} = \eta_{00}$, as expected from a slight mixture of K_1 in K_L. If CP-violating effects show up only on decay, then $\epsilon = 0$, and $K_2 \to 2\pi$ is allowed.

We can investigate also the effect on the decay rate asymmetry

$$\mathscr{A} = \frac{R(K_L \to \pi^- l^+ \nu_l) - R(K_L \to \pi^+ l^- \bar{\nu}_l)}{R(K_L \to \pi^- l^+ \nu_l) + R(K_L \to \pi^+ l^- \bar{\nu}_l)}, \tag{14.45}$$

where l is a muon or electron and R means decay rate. The selection rules imposed by the $\Delta Q = \Delta S$ rule (Section 11.12) are given by the allowed decays of Eq. (14.12) and the forbidden decays of Eq. (14.13). Whether or not this rule is good, $\mathscr{A}$ will not be zero if $\epsilon \neq 0$ and must be zero if $\epsilon = 0$. In showing this we include the effect of a $\Delta Q = -\Delta S$ amplitude. The $\Delta Q = \Delta S$ amplitudes we take to be real (CP conserving):

$$f = \langle \pi^- l^+ \nu_l | T | K^0 \rangle = \langle \pi^+ l^- \bar{\nu}_l | T | \bar{K}^0 \rangle .$$

The $\Delta Q = -\Delta S$ amplitudes may not conserve CP, so we include the possibility that they are imaginary

$$g^* = \langle \pi^+ l^- \bar{\nu}_l | T | K^0 \rangle ,$$

and hence

$$\langle \pi^- l^+ \nu_l | T | \bar{K}^0 \rangle = g .$$

Then, using Eq. (14.32)

$$\langle \pi^- l^+ \nu_l | T | K_L \rangle = \frac{[(1+\epsilon)f - (1-\epsilon)g]}{[2(1+|\epsilon|^2)]^{1/2}},$$

$$\langle \pi^+ l^- \bar{\nu}_l | T | K_L \rangle = \frac{[(1+\epsilon)g^* - (1-\epsilon)f]}{[2(1+|\epsilon|^2)]^{1/2}}.$$

Retaining terms of lowest order in ϵ only, we find

$$\mathscr{A} = 2 \operatorname{Re} \epsilon \, \frac{1+|X|^2}{|1-X|^2}, \tag{14.46}$$

where $X = g/f$ and is the ratio of the $\Delta Q = -\Delta S$ amplitude to the $\Delta Q = \Delta S$ amplitude. The experimental results on the ratio g/f are as yet indecisive and consistent with zero. Therefore the measurement of the asymmetry $\mathscr{A}$, assuming $X = 0$, gives $\operatorname{Re} \epsilon$.

We now consider the interference between the K_L and K_S 2π-decay modes near a beam source. For example, a K^0 beam at $t = 0$ is described by

$$|K^0\rangle = \frac{[2(1+|\epsilon|^2)]^{1/2}}{2(1+\epsilon)} \{|K_S\rangle + |K_L\rangle\} .$$

K_L and K_S are the states of well-defined exponential time development [Eqs. (14.31) and (14.32)], so we put

$$\alpha + \beta = m_S - i\lambda_S/2 ,$$
$$\alpha - \beta = m_L - i\lambda_L/2 ,$$

and at time t the state of the beam is

$$|\psi(t)\rangle = N\{\exp(-im_S t - \lambda_S t/2) |K_S\rangle + \exp(-im_L t - \lambda_L t/2) |K_L\rangle\} ,$$

where $N = [2(1+|\epsilon|^2)]^{1/2}/2(1+\epsilon)$. Considering the $\pi^+ \pi^-$ decay channel

14.7 CP-Noninvariant Analysis

and using Eqs. (14.39) and (14.40) (dropping Δ and $\epsilon\epsilon'$) we have the amplitude for such a decay:

$$\langle \pi^+\pi^- | T | \phi(t) \rangle = \sqrt{\tfrac{2}{3}}\, a_0 (1+\epsilon)^{-1} \{ \exp(-im_s t - \lambda_s t/2) + (\epsilon + \epsilon') \exp(-im_L t - \lambda_L t/2) \}\, .$$

Therefore the decay intensity $I(t)$ is

$$I(t) = \frac{2|a_0|^2}{3(1+\epsilon)} (\exp(-\lambda_s t) + |\epsilon + \epsilon'|^2 \exp(-\lambda_L t) + 2|\epsilon + \epsilon'| \exp[-(\lambda_L + \lambda_s)t/2] \cos\{(m_L - m_s)t + \phi\})\,, \quad (14.47)$$

where

$$\phi = \text{Arg}(\epsilon + \epsilon')\, .$$

Considering the situation that exists, $\lambda_s \simeq 600\, \lambda_L$ and $|\epsilon + \epsilon'| \simeq 2 \times 10^{-3}$, we see that the interference term will be visible when

$$\exp(-\lambda_s t) \simeq 2|\epsilon + \epsilon'| \exp(-\lambda_s t/2)\,,$$

which happens at about $t = 12\,\lambda_s^{-1}$. The interference term changes sign if the beam is $\bar{K}^0$ at the source. The result is that the form of the interference pattern gives information on the initial state. This is shown in Fig. 14.3.

This is a very important result because it clearly distinguishes between particles and antiparticles. It is now easy to compose instructions for the inhabitants of another galaxy for a neutral-K experiment that would decide on the result of the sign of this interference term, whether the local material was particle or antiparticle by our labeling. The observation of this effect is clearly evidence of CP violation.

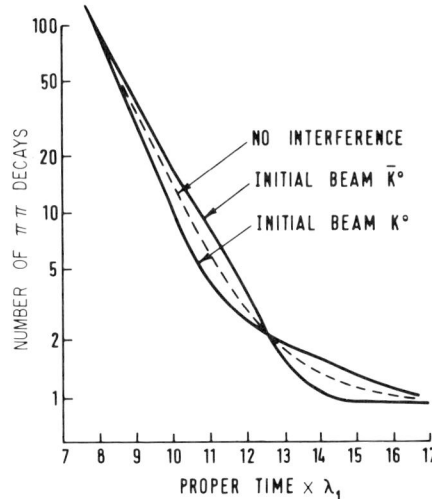

FIG. 14.3. Plot of the number of decays $K \to \pi^+\pi^-$ as a function of proper time for the case $|\eta_{+-}| = 1.9 \times 10^{-3}$, $\phi = 0.7$ radians.

14.8 The Experimental Situation

At the CERN Weak Interaction Conference in January 1969 the progress in determining the important quantities in CP violation was reviewed by Steinberger (1969). The $K_L \to \pi^+\pi^-$ branching ratio is a fairly well-established number, and the average gave

$$|\eta_{+-}| = (1.90 \pm 0.05) \times 10^{-3}.$$

The phase of η_{+-} is difficult to measure. It can be done by measuring the interference between $K_L \to \pi^+\pi^-$ and regenerated $K_S \to \pi^+\pi^-$ and making a separate determination of the phase of the regeneration amplitude by using either optical-model fits or dispersion relations. The determination of the interference between K_L and K_S into 2π near a source that is pure K^0 or $\bar{K}^0$ is very difficult, but does measure the phase directly. The average given by Steinberger was

$$\text{Arg } \eta_{+-} = 40 \pm 6°.$$

The values for the parameter $|\eta_{00}|$ are much more uncertain and range from 0 to 3.6×10^{-3}, all with considerable errors. It seems fairly certain that the process occurs, but at the time of writing it is impossible to make any firm statements about the rate. The first measurement of the phase has given $\text{Arg } \eta_{00} = 17 \pm 31°$ (Chollet et al., 1969).

If the measurements are taken at their face value and we assign $|\eta_{00}|$ the value 3.6×10^{-3} and take the weighted average of the results of Dorfan el al. (1967) and of Bennett et al. (1967) (Section 14.6) for $\text{Re } \epsilon$ (assuming $X = 0$), we find an Argand diagram for one solution of ϵ, ϵ', η_{+-}, and η_{00}, which is shown in Fig. 14.4. The difficulty is that this solution and that with $\eta_{00} \to \eta_{00}^*$ require $\delta_2 - \delta_0$ to be positive, whereas the indications of the analysis of pion–pion scattering observed in peripheral pion–nucleon interactions are that this quantity is negative. The impor-

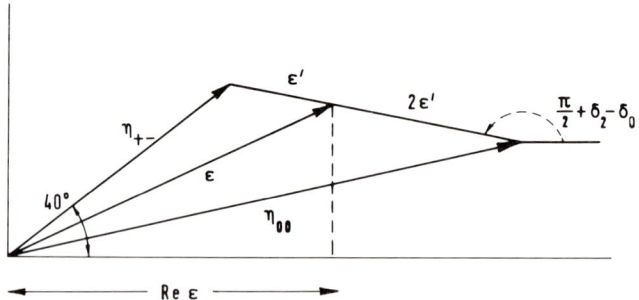

FIG. 14.4. Argand diagram showing possible values for ϵ, ϵ', η_{+-} and η_{00} (from data available in January 1969).

tance of this difficulty can be exaggerated in view of the present (1969) uncertainty in ϵ', the errors on the observed quantities, and the lack of conclusive measurements on the $\Delta Q = -\Delta S$ amplitudes. It appears quite possible that $\eta_{00} = \eta_{+-}$, in which case $\epsilon' = 0$ and no determination of $\delta_2 - \delta_0$ is possible.

14.9 The Source of CP Violation

The experimental evidence, in spite of its uncertainties, firmly indicates a breakdown of CP invariance in some elementary-particle interactions. It is first interesting to point out the relevance to the TCP theorem. If we consider TCP invariance to be established, then the observation of CP violation necessarily implies that time-reversal invariance is broken. Tests of time-reversal invariance in various interactions have been made, but with no indication of breakdown (Section 6.6). If T invariance is valid, then CP violation invariance means a breakdown of the TCP theorem.

Let us consider various interactions and how they could cause the apparent CP violation. If the interaction not conserving CP has $\Delta S = 0$ then it cannot cause the direct $K \to 2\pi$ decay, since this has $\Delta S = 1$. Its effect must therefore be combined with the assumed CP-conserving $\Delta S = 1$ part of the weak interaction, and we have a two-stage (virtual) process

$$K_2 \xrightarrow{\Delta S = 1} X \xrightarrow{\Delta S = 0} \pi^+ \pi^-,$$
$$CP = -1 \qquad = -1 \qquad = +1$$

where X is a system of strongly interacting particles having $S = 0$ and $B = 0$. The CP-nonconserving part of the transition could cause that change by failing to conserve C or P or both. The sequential nature of this process requires that the suggested interaction has a strength comparable to the electromagnetic interaction. Therefore this interaction has been considered in this role and, since it is known to conserve P, its C properties have been carefully examined (see Section 6.7). There is no conclusive evidence for C nonconservation. The case of η-decay (Section 6.7) also shows that the strong interaction does not break C invariance to any significant extent.

The next possibility is a $\Delta S = 1$ interaction. This would be equivalent to a small CP-nonconserving part (10^{-3}) to the usual weak interaction. If this is the case, CP-nonconserving effects may show up in other weak decays but will be, by their smallness compared to the CP-conserving interaction, difficult to observe.

A third possibility is that there is a scalar CP-nonconserving $\Delta S = 2$ interaction. Since this can cause $K^0 \rightleftharpoons \bar{K}^0$, it has the effect of making B_y nonzero [Eq. (14.4)] and makes ϵ finite. Since it causes this transition by first order effects, whereas weak interactions cause it by second order, and because the CP-nonconserving transition is small, this suggested interaction need only have about 10^{-8} the strength of the normal weak interaction. This "superweak" interaction would predict $\epsilon' = 0$ and hence $\eta_{+-} = \eta_{00}$. If this should be the case, then there appear to be no effects that could be observed other than those appearing in the neutral-kaon system.

This discussion has simplified matters somewhat. For example, the $\Delta S = 2$ interaction could be very much stronger if it only connected states of opposite parity. The $K^0 \rightleftharpoons \bar{K}^0$ transition would then have to be second order in this interaction, and the required strength becomes less different from that of the CP-conserving weak interaction. For a fuller discussion, readers are referred to Kabir (1968) and to the original papers proposing these interactions (for example, Wolfenstein, 1964; Sachs, 1964).

It is evident that at the time of writing we are very much in the dark about the source of CP nonconservation. If measurements show $\eta_{00} = \eta_{+-}$, then the outlook is bleak for observing CP nonconservation other than in the neutral-kaon system. If this is not the case, then there is a chance that CP-nonconserving transitions will have a detectable effect on decays such as $K^{\pm} \to \pi^{\pm}\pi^0\pi^0$ and $\pi^{\pm}\pi^+\pi^-$.

REFERENCES

Abashian, A., Abrams, R. J., Carpenter, D. W., Fisher, G. P., Mefkens, B. M. K., and Smith, J. H. (1964). *Phys. Rev. Lett.* **13**, 243 (14.6).
Alff-Steinberger, C. *et al.* (1966). *Phys. Lett.* **21**, 595 (14.6).
Aubert, B. *et al.* (1965). *Phys. Lett.* **17**, 59 (14.2).
Bartlett, D. F. *et al.* (1968). *Phys. Rev. Lett.* **21**, 558 (14.6).
Bennett, S., Nygren, D., Saal, H., Steinberger, J., and Sunderland, J. (1967). *Phys. Rev. Lett.* **19**, 993 (14.6, 14.8).
Böhm, A. *et al.* (1968). *Phys. Lett.* B **27**, 321 (14.6).
Bott-Bodenhausen, M. *et al.* (1966). *Phys. Lett.* **23**, 277 (14.6).
Camerini, U. *et al.* (1962). *Phys. Rev.* **128**, 362 (14.2).
Canter, J. *et al.* (1966). *Phys. Rev. Lett.* **17**, 942 (14.5).
Cence, R. J. *et al.* (1969). *Phys. Rev. Lett.* **22**, 1210 (14.6).
Chollet, J. *et al.* (1969). *Proc. Topical Conf. Weak Interaction*, CERN 69-7, p. 309. CERN, Geneva (14.9).
Christenson, J. H., Cronin, J. W., Fitch, V. L., and Turlay, R. (1964). *Phys. Rev. Lett.* **13**, 138 (14.6).
Christenson, J. H., Cronin, J. W., Fitch, V. L., and Turlay, R. (1965). *Phys. Rev.* B **140**, 74 (14.4).
Dorfan, D. *et al.* (1967). *Phys. Rev. Lett.* **19**, 987 (14.6, 14.8).
Fitch, V. L., Piroue, P. A., and Perkins, R. B. (1961). *Nouvo Cimento* **22**, 1160 (14.2).
Fitch, V. L., Roth, P. F., Russ, J. S., and Vernon, W. (1965). *Phys. Rev. Lett.* **15**, 73 (14.6).
Fujii, T., Jovanovich, J., Turkot, F., and Zorn, G. (1964). *Phys. Rev. Lett.* **13**, 253 (14.4).
Gaillard, J.-M. *et al.* (1967). *Phys. Rev. Lett.* **18**, 20 (14.6).

References

Goldberger, M. L., and Watson, K. M. (1964). "Collision Theory." Wiley, New York (14.4).

Good, M. L. (1957). *Phys. Rev.* **106,** 591 (14.4).

Jovanovich, J. V., Fujii, T., Turkot, F., Zorn, G. T., and Deutsch, M. (1966). *Phys. Rev. Lett.* **17,** 1075 (14.5).

Kabir, P. K. (1968). "The CP Puzzle." Academic Press, New York (14.2, 14.3, 14.6, 14.9).

Lee, T. D., and Wolfenstein, L. (1965). *Phys. Rev. B* **138,** 1490 (14.7).

Muller, F. *et al.* (1960). *Phys. Rev. Lett.* **4,** 418 (14.4).

Okun, L., and Pontecorvo, B. (1957). *Sov. Phys. JETP* **5,** 1297 (14.3).

Sachs, R. G. (1964). *Phys. Rev. Lett.* **13,** 286 (14.9).

Steinberger, J. (1969). *Proc. Topical Conf. Weak Interaction*, CERN 69-7, p. 291. CERN Geneva (14.6, 14.8).

Wolfenstein, L. (1964). *Phys. Rev. Lett.* **13,** 562 (14.9).

Wu, T. T., and Yang, C. N. (1964). *Phys. Rev. Lett.* **13,** 380 (14.7).

Appendix A
FUNCTIONS OF A COMPLEX VARIABLE

A.1 Functions and Singularities

Dispersion relations depend intimately upon the theory of complex variables, and it is worth briefly reviewing in this appendix the theorems that will be used. For a complete treatise on this subject readers are referred to Titchmarsh (1939), Copson (1935), or Phillips (1957). If a complex number $f = u + iv$ is related to another complex number $z = x + iy$ so that the value of f is given by that of z, f is said to be a function of the complex variable z. The derivative of the function $f(z)$ is defined so that df/dz at a point z is independent of the direction in which z is approached when taking the limit. In order to have such a derivative the function must satisfy the Cauchy–Riemann relations:

$$\partial u/\partial x = \partial v/\partial y \quad \text{and} \quad \partial u/\partial y = -\partial v/\partial x. \tag{A.1}$$

A function that is single valued, continuous, and has a derivative within some region of the complex plane is said to be analytic (or regular, or holomorphic) throughout that region. Every function except a constant has one or more points at which it is not analytic. Such points are called singularities. The simplest singularity is called a pole. For example, the function $1/(z-a)$ has a pole (of unit order) at the point $z = a$, for it is analytic everywhere except at this point; the function $1/[z(z-a)]$ has two poles (of unit order), one at $z = 0$, the other at $z = a$. The function $1/z^n$ has a pole at $z = 0$ of order n.

Another kind of singularity that we shall meet is called a branch point. To understand this consider the function $f(z) = z^{1/2}$. Since

$$z = r \exp[i\theta] = r \exp[i(\theta + 2\pi n)],$$

where n is an integer, then $f(z) = \sqrt{r}\, \exp[i(\theta/2 + \pi n)]$. Thus the values of $f(z)$ depend upon n, and in fact there are two, one for n even, the other for n odd. Thus $f(z)$ is a double-valued function of z and cannot, by definition, be analytic. This situation can be avoided by limiting the angular range of $\theta + 2\pi n$ to two branches. Dropping the $2\pi n$ we can define the limits by, for example,

$$0 < \theta < 2\pi,$$
$$2\pi < \theta < 4\pi.$$

This demarcation can be envisaged by a cut along the x axis from $z = 0$ to $x = \infty$, which separates the two branches. The point $z = 0$ is called the branch point. It is only a matter of convention that the cut is along the x axis. It could be a curved line going from the branch point $z = 0$ to ∞. It is evident that $f(z) = (z - a)^{1/2}$ has a branch point at $z = a$ and a cut to infinity. For a given general function, the number of branches is not limited to two, and the cuts between them can have different branch points and directions.

A.2 Integral Theorems

The integral theorems are particularly important and involve paths of integration in the variable z that are called contours. By convention, closed contours to be followed in a positive sense are to be traversed anticlockwise (i.e., in a right-handed sense looking along the third axis of the right-handed Cartesian coordinate system x, y, z). Cauchy's theorem states that if $f(z)$ is everywhere analytic on and within a closed contour c, then

$$\oint_c f(z)\, dz = 0. \tag{A.2}$$

This implies that $\int_A^B f(z)\, dz$ is independent of the path taken between A and B so long as A, B, and all paths between them, lie within the region of analyticity of $f(z)$. Under the same conditions for $f(z)$ we also have the Cauchy integral formula

$$f(z) = \frac{1}{2\pi i} \oint_c \frac{f(z')\, dz'}{z' - z}, \tag{A.3}$$

where z lies within the contour c.

Suppose $f(z)$ is analytic within a given region of the complex plane except for a pole at the point a. Then within this region, $f(z)$ can be expanded thus:

$$f(z) = a_0 + a_1(z - a) + a_2(z - a)^2 + \cdots$$
$$+ \frac{b_1}{z - a} + \frac{b_2}{(z - a)^2} + \cdots. \tag{A.4}$$

b_1 is called the residue of this pole. This expansion is known as Laurent's theorem. Consider the contour c that encloses a; then

$$\oint_c f(z) \, dz = 2\pi i b_1 . \tag{A.5}$$

This can be generalized to the case of a contour enclosing several poles by changing the right-hand side to $2\pi i$ (Σ residues).

The need for cuts between branches of a multivalued function is now clear. The integral theorems cannot apply to a contour that is closed in the z plane but in fact may start on one branch and try to close on another. If contours may not cross cuts, this situation is avoided and the integral theorems are correct.

The last point we shall discuss in this appendix is the principal value of an integral. Consider the integral

$$I = \oint_c \frac{f(z)}{z - a} \, dz$$

where $f(z) \to 0$ as $z \to \infty$ and has no singularities on or above the real axis. The contour goes from $x = -\infty$ to $x = +\infty$ and circles at infinity around the upper half-plane to $x = -\infty$. The contribution from the semicircle at infinity is zero and we have

$$I = \int_{-\infty}^{+\infty} \frac{f(x) \, dx}{x - a} . \tag{A.6}$$

The integral must be handled carefully at $x = a$, assuming a real. The theorem for doing this states that

$$I = P \int_{-\infty}^{+\infty} \frac{f(x)}{x - a} \, dx + i\pi f(a) , \tag{A.7}$$

where

$$P \int_{-\infty}^{+\infty} = \lim_{\delta \to 0} \left[\int_{-\infty}^{a-\delta} + \int_{a+\delta}^{\infty} \right] . \tag{A.8}$$

Equation (A.8) defines the principal value of an integral. Obviously Eq. (A.7) can be generalized to the case of function $F(z)$ with several singularities on the real axis. Thus

$$\int_{-\infty}^{+\infty} F(x) \, dx = P \int_{-\infty}^{+\infty} F(x) \, dx + 2\pi i \left(\sum R_1 + \tfrac{1}{2} \sum R \right) ,$$

where R_1 are the residues of poles in $F(z)$ in the upper half-plane and R are the residues of $F(z)$ on the real axis.

Consider now the application of Eq. (A.7) to

$$f(a) = \frac{1}{2\pi i} \oint_c \frac{f(z') \, dz'}{z' - a}.$$

If $|f(z')| \to 0$ as $z' \to \infty$, then we can make the contour the same as in discussion above. Then applying Eq. (A.7) we get

$$f(a) = \frac{1}{\pi i} P \int_{-\infty}^{+\infty} \frac{f(x) \, dx}{x - a}. \tag{A.9}$$

REFERENCES

Copson, E. T. (1935). "An Introduction to the Theory of Functions of a Complex Variable." Oxford Univ. Press, London and New York (A.1).

Phillips, E. G. (1957). "Functions of a Complex Variable," 8th ed. Oliver and Boyd, Edinburgh and London (A.1).

Titchmarsh, E. C. (1939). "The Theory of Functions," 2nd ed. Oxford Univ. Press, London and New York (A.1).

Appendix B
RELATIVISTIC KINEMATICS

B.1 The Properties of Four Vectors

Four-vectors were introduced in Section 1.5. The use of four-vectors and scalars (invariants) makes relativistic kinematics simple. The contravariant four-vector of momentum p^μ has components $E, \mathbf{P}$ (total energy and three momentum, respectively) and the corresponding covariant four-vector p_μ has components $E, -\mathbf{P}$. Then, the scalar product [Eq. (1.38)]

$$p^\mu p_\mu = E^2 - P^2 = m^2, \tag{B.1}$$

is an invariant; that is, it is independent of the inertial frame from which the values of E and $\mathbf{P}$ used to evaluate the scalar product are observed. This invariant quality is true of the scalar product of any two four vectors. (Note that throughout this appendix we use units having $c = 1$.)

All contravariant vectors transform [Eq. (1.37)]

$$p'^\mu = a^\mu{}_\lambda p^\lambda, \tag{B.2}$$

and the covariant vectors transform

$$p_\mu' = a_\mu{}^\lambda p_\lambda. \tag{B.3}$$

Since contravariant and covariant vectors are very simply related, we will restrict ourselves to transforming the contravariant vector; the coefficients $a^\mu{}_\lambda$ and $a_\mu{}^\lambda$ can be related using the properties of $g_{\mu\nu}$ and $g^{\mu\nu}$.

Let us consider a (contravariant) four-vector p observed from an inertial frame S. A second inertial frame S' is moving with velocity β along the z axis of S and oriented so that when the origins coincide the Cartesian axes coincide. The four-vector is p' observed from S'. Then

$$p' = Lp, \tag{B.4}$$

B.1 The Properties of Four Vectors

where the four-vectors are represented by column vectors and L is a 4×4 matrix. For this case

$$L = \begin{bmatrix} \gamma & 0 & 0 & -\gamma\beta \\ 0 & 1 & 0 & 0 \\ 0 & 0 & 1 & 0 \\ -\gamma\beta & 0 & 0 & \gamma \end{bmatrix}, \quad (B.5)$$

where $\gamma = (1 - \beta^2)^{-1/2}$. Thus

$$p' = \begin{bmatrix} E' \\ P_x' \\ P_y' \\ P_z' \end{bmatrix} = L \begin{bmatrix} E \\ P_x \\ P_y \\ P_z \end{bmatrix} = \begin{bmatrix} \gamma(E - \beta P_z) \\ P_x \\ P_y \\ \gamma(P_z - \beta E) \end{bmatrix}. \quad (B.6)$$

Conversely

$$p = L^{-1} p', \quad (B.7)$$

where L^{-1} is L with $\beta \to -\beta$. A more general case occurs if β is not along the z axis in S but in a general direction given by $\boldsymbol{\beta}$. (The frames S and S' retain the same relative orientation.) Then

$$L = \begin{bmatrix} \gamma & -\gamma\beta_x & -\gamma\beta_y & -\gamma\beta_z \\ -\gamma\beta_x & 1 + \dfrac{\gamma^2 \beta_x^2}{\gamma + 1} & \dfrac{\gamma^2 \beta_x \beta_y}{\gamma + 1} & \dfrac{\gamma^2 \beta_x \beta_z}{\gamma + 1} \\ -\gamma\beta_y & \dfrac{\gamma^2 \beta_x \beta_y}{\gamma + 1} & 1 + \dfrac{\gamma^2 \beta_y^2}{\gamma + 1} & \dfrac{\gamma^2 \beta_y \beta_z}{\gamma + 1} \\ -\gamma\beta_z & \dfrac{\gamma^2 \beta_x \beta_z}{\gamma + 1} & \dfrac{\gamma^2 \beta_y \beta_z}{\gamma + 1} & 1 + \dfrac{\gamma^2 \beta_z^2}{\gamma + 1} \end{bmatrix}. \quad (B.8)$$

An easier way to remember the transformation implied in Eq. (B.8) is

$$\mathbf{P}' = \mathbf{P} + \gamma \boldsymbol{\beta} \left[\frac{\gamma \boldsymbol{\beta} \cdot \mathbf{P}}{\gamma + 1} - E \right], \quad (B.9)$$

$$E' = \gamma E - \gamma \boldsymbol{\beta} \cdot \mathbf{P}. \quad (B.10)$$

If the coordinate axes are not parallel, then rotations must be made before applying these equations.

Apart from the energy and momentum of a particle, it is worth remembering that the four-velocity of a particle γ, $\gamma \boldsymbol{\beta}$ is also a four-vector.

In this appendix we use p to represent the (contravariant) four-momentum of a particle. $\mathbf{P}$ is the three-momentum, which has magnitude P. We avoid covariant four-vectors by writing $p^\mu p_\mu$ as $p \cdot p$ or p^2, but keep in mind that the covariant vector is implicitly present and makes

$$p \cdot p = p^2 = E^2 - P^2. \tag{B.11}$$

There is a useful trick that we can introduce at this point. Rearranging Eq. (B.1) we can put

$$(E^2/m^2) - (P^2/m^2) = 1.$$

Compare this with

$$\sec^2 \alpha - \tan^2 \alpha = 1. \tag{B.12}$$

Obviously, if we have $\alpha = \sec^{-1}(E/m)$, the following equations hold:

$$\gamma = E/m = \sec \alpha, \tag{B.13a}$$

$$\beta = P/E = \sin \alpha, \tag{B.13b}$$

$$\gamma\beta = P/m = \tan \alpha, \tag{B.13c}$$

where β is now the velocity of the particle. Thus if the energy and momentum are expressed in units of the rest mass, we can use tables of trigonometric functions to facilitate calculations of any one of these quantities from another. An alternative is to use $\cosh^2 \alpha - \sinh^2 \alpha$ instead of Eq. (B.10) and use the consequent hyperbolic function tables.

B.2 Laboratory and Center-of-Mass Coordinates

The study of nuclear and elementary-particle reactions involves Lorentz transformations between the laboratory coordinates and some defined center-of-mass coordinates (Section 1.5). Of particular interest is that center-of-mass system involved when a moving particle (1) is incident upon a second particle (2) stationary in the laboratory frame. If W is the center-of-mass energy (which is also the invariant mass of the system), then

$$W^2 = s = p \cdot p = (p_1 + p_2) \cdot (p_1 + p_2), \tag{B.14}$$

where p is the total four-momentum and p_1, p_2 the individual four-momenta involved (s is one of the Mandelstam variables, Section 12.2). Evaluating this in the laboratory system we have

$$p_1 = (E_1, \mathbf{P}_1),$$
$$p_2 = (m_2, 0).$$

Hence

$$W^2 = (m_2 + E_1)^2 - \mathbf{P}_1 \cdot \mathbf{P}_1$$
$$= m_1^2 + m_2^2 + 2m_2 E_1. \tag{B.15}$$

B.3 Particle Reactions

The velocity of the center of mass in the laboratory is, by Eq. (B.13b), its laboratory momentum divided by its center-of-mass energy,

$$\beta_c = P_1/W, \tag{B.16}$$

and the Lorentz factor, from Eq. (B.13a) is the total energy divided by the center-of-mass energy,

$$\gamma_c = (E_1 + m_2)/W. \tag{B.17}$$

We now find it convenient to distinguish a quantity as observed in the center-of-mass system by a prime, while the same quantity observed in the laboratory is unprimed. Thus $P_2 = 0$, but P_2' is obtained by noting that particle 2 has velocity β_c in the center-of-mass system, hence

$$P_2' = \beta_c \gamma_c m_2 = P_1 m_2/W, \tag{B.18}$$

and

$$E_2' = \gamma_c m_2 = (m_2^2 + m_2 E_1)/W. \tag{B.19}$$

By definition of this center of mass, $\mathbf{P}_2' = -\mathbf{P}_1'$. Using

$$W = E_1' + E_2'$$

we find that

$$E_1' = (m_1^2 + m_2 E_1)/W. \tag{B.20}$$

B.3 Particle Reactions

The first point of interest is the case where the reaction does not proceed until a certain threshold energy is reached. Suppose particle 1 is incident on particle 2 to produce particles 3, 4, ... ;

$$1 + 2 \to 3 + 4 + \cdots.$$

At the threshold let the total center-of-mass energy be W_0; then

$$W_0 = m_3 + m_4 + \cdots.$$

Let E_0 be the incident total energy of particle 1 at threshold:

$$W_0^2 = m_1^2 + m_2^2 + 2m_2 E_0.$$

Therefore

$$2m_2 E_0 = (m_3 + m_4 + \cdots)^2 - m_1^2 - m_2^2$$

and thus

$$E_0 = (-Q/2m_2)(m_1 + m_2 + m_3 + m_4 + \cdots) + m_1 \tag{B.21}$$

where $Q = m_1 + m_2 - (m_3 + m_4 + \cdots)$. This Q-value is negative for reactions having a threshold and positive for reactions that can proceed at all energies.

If Q is positive or if the reaction is occurring above threshold, then the total center-of-mass energy W is equal to the sum of the total energies in the center-of-mass of all the final particles:

$$W = E_3' + E_4' + \cdots .$$

In the case of a two-body reaction

$$1 + 2 \to 3 + 4,$$

the momentum $\mathbf{P}_3' = -\mathbf{P}_4'$ and we have

$$P_3'^2 = P_4'^2 = P'^2 = E_3'^2 - m_3^2 = E_4'^2 - m_4^2 .$$

Solving we find

$$E_4' = \frac{W^2 + m_4^2 - m_3^2}{2W}, \qquad (B.22)$$

$$E_3' = \frac{W^2 + m_3^2 - m_4^2}{2W}, \qquad (B.23)$$

$$P' = \frac{\lambda(W^2, m_3^2, m_4^2)}{2W}$$
$$= \frac{1}{2W}[W^4 + m_3^4 + m_4^4 - 2W^2 m_3^2 - 2W^2 m_4^2 - 2m_3^2 m_4^2]^{1/2}, \quad (B.24)$$

where we have (Section 3.11)

$$\lambda(x, y, z) = [x^2 + y^2 + z^2 - 2xy - 2yz - 2zx]^{1/2} . \qquad (B.25)$$

If particle 3 has polar angles (θ_3', ϕ_3') then particle 4 has polar angles $(\pi - \theta_3', \phi_3' + \pi)$ with respect to the direction $\mathbf{P}_1'$ and a suitable plane to define the azimuth. These center-of-mass quantities can be transformed to the quantities observed in the laboratory by the Lorentz transformation with the center-of-mass velocity β_c. The results are

$$\tan \theta_3 = \frac{P_3' \sin \theta_3'}{\gamma_c(\beta_c E_3' + P_3' \cos \theta_3')}, \qquad (B.26a)$$

$$\tan \theta_4 = \frac{P_4' \sin \theta_3'}{\gamma_c(\beta_c E_4' - P_4' \cos \theta_3')}, \qquad (B.26b)$$

$$\phi_3 = \phi_3' \qquad (B.27a)$$

$$\phi_4 = \phi_4' \qquad (B.27b)$$

$$E_3 = \gamma_c(E_3' + \beta_c P_3' \cos \theta_3') , \qquad (B.28a)$$

B.3 Particle Reactions

$$E_4 = \gamma_c(E_4' - \beta_c P_4' \cos\theta_3'), \quad \text{(B.28b)}$$

where θ_3 and θ_4 are with respect to the direction $\mathbf{P}_1$.

If the reaction has more than two products,

$$1 + 2 \to 3 + 4 + \cdots,$$

then it may be impossible to detect all the particles. The invariant mass of the unobserved particles is called the missing mass m. Energy and momentum conservation require

$$p_1 + p_2 = p_3 + p_4 + \cdots,$$

Suppose particles 3 to j are observed and $j + 1$ to n are missed; then

$$\begin{aligned} m^2 &= (p_{j+1} + \cdots + p_n)^2 \\ &= (p_1 + p_2 - p_3 - p_4 - \cdots - p_j)^2. \end{aligned} \quad \text{(B.29)}$$

If particle 2 is at rest in the laboratory frame, then, using laboratory quantities

$$m^2 = (E_1 + m_2 - \sum_{r=3}^{j} E_r)^2 - |\mathbf{P}_1 - \sum_{r=1}^{j} \mathbf{P}_r|^2. \quad \text{(B.30)}$$

These formulas for final-state quantities also apply to the products of the decay of a particle of mass M if we replace W by M. The velocity of the decaying particle replaces the velocity of the center of mass.

Useful relations may be obtained from the manipulation of four-vectors. For example, in the two-body reaction (2 at rest in the laboratory)

$$1 + 2 \to 3 + 4$$

we have

$$p_1 + p_2 = p_3 + p_4;$$

that is,

$$p_3^2 = (p_1 + p_2 - p_4)^4.$$

Evaluating this in the laboratory, the right-hand side contains the two three-momenta $\mathbf{P}_1$ and $\mathbf{P}_4$, so that its scalar product contains $\cos\theta_4$. No other angles are contained, and we obtain

$$\cos\theta_4 = \frac{m_3^2 - m_1^2 - m_2^2 - m_4^2 + 2(E_1 + m_2)E_4 - 2E_1 m_2}{2P_1 P_4}. \quad \text{(B.31)}$$

Thus judicious arrangement of scalar products will pick out quantities of interest.

Another useful way of using four-vectors and invariant quantities is as follows. Any quantity that can be given as a function of four-vectors is then invariant in the sense that it can be evaluated in any frame. Thus Eq. (B.14) for W is of this form:

$$W^2 = (p_1 + p_2)^2 = p_1^2 + p_2^2 + 2p_1 \cdot p_2 = m_1^2 + m_2^2 + 2p_1 \cdot p_2.$$

This can be evaluated from quantities observed in any frame and will give the correct result. Consider another example: What is the energy of particle 1 observed in the rest frame of particle 2? We can use any suitable combination of invariants formed from p_1 and p_2. In the rest frame of 2 let 1 have four-momentum $(E_1', \mathbf{P}_1')$ while 2 obviously has four-momentum $(m_2, 0)$. Evidently, evaluated in this frame, $p_1 \cdot p_2 = E_1' m_2$; hence,

$$E_1' = (p_1 \cdot p_2)/m_2.$$

This is the required equation, and it is correct in whatever frame the right-hand side is evaluated, because $p_1 \cdot p_2$ is invariant. Similarly Eq. (B.31) becomes

$$\cos\theta_4 = \frac{m_2^2 p_1 \cdot p_4 - (p_1 \cdot p_2)(p_4 \cdot p_2)}{\{[(p_1 \cdot p_2)^2 - m_1^2 m_2^2][(p_2 \cdot p_4)^2 - m_2^2 m_4^2]\}^{1/2}}.$$

B.4 Dalitz Plots

The kinematics of individual events of a final state consisting of three particles can be very conveniently represented by plotting events as points on a Dalitz plot (see also Sections 7.8, 7.9 and 8.3). If T_1, T_2, T_3 are the kinetic energies and W is the total energy of these particles in the center of mass, then

$$T_1 + T_2 + T_3 = Q = W - m_1 - m_2 - m_3.$$

An equilateral triangle has the property that the sum of the perpendicular distances from the three sides to a point within the triangle is a constant equal to the height of the triangle. Therefore if we draw an equilateral triangle of height Q, we can plot a point within the triangle whose distances from the three sides are proportional to the three kinetic energies. Conservation of momentum confines the points representing events to an area somewhat smaller than the inscribed circle. The limits of this area can be obtained by simple kinematic considerations (Williams, 1964; Hagedorn, 1963). Evidently this is not the only possible representation. Since only two of the three kinetic energies are independent, we could just as well plot the points in Cartesian coordinates T_1 and T_2, say. There is another more interesting way of plotting, which can be derived from Eq. (B.22). Suppose we have three particles 1, 2, 3; then we can consider the kinetic energy of 1, say, being given in terms of the invariant mass of 2 and 3 (m_{23}); i.e.,

$$E_1 = m_1 + T_1 = \frac{W^2 + m_1^2 - m_{23}^2}{2W},$$

where $m_{23}^2 = (p_2 + p_3)^2$. Therefore T_1 is proportional to $-m_{23}^2$ and our plot in Cartesian T_1, T_2 coordinates is equivalent to one in coordinates m_{23}^2, m_{13}^2. It can be shown that on such a plot loci of equal m_{12}^2 are straight lines. Dalitz plots of three-particle final states are frequently done in this way.

These plots have a very important property. If we consider the differential transition rate into any particular allowed part of the plot, we know it consists of two parts: a squared transition-matrix element and a density-of-states factor. This last factor is constant over the entire allowed area (Section C.2), so that the density of points representing events reflects the value of the squared matrix element.

B.5 The Mandelstam Variables

In Section 12.2 we introduced the Mandelstam variables, which are very useful in describing the kinematics of a two-body reaction

$$1 + 2 \rightarrow 3 + 4.$$

They are

$$s = (p_1 + p_2)^2,$$
$$t = (p_1 - p_3)^2,$$
$$u = (p_1 - p_4)^2,$$

for which $s + t + u = m_1^2 + m_2^2 + m_3^2 + m_4^2$. The Mandelstam plot (Section 12.2) places two-body events with respect to equilateral coordinate axes. In certain circumstances regions of the Mandelstam plot correspond to well-known Dalitz plots. Thus the plot for the weak scattering process

$$\pi^- + K^+ \rightarrow \pi^+ + \pi^-$$

has a region in which the values of s, t, and u show it corresponds to the crossed reaction

$$K^+ \rightarrow \pi^+ + \pi^+ + \pi^-,$$

which is the reaction to which the Dalitz plot technique was first applied.

B.6 Transformation of Differential Cross Sections

Consider one product of a particle reaction in which an incident particle strikes a stationary particle in the laboratory. The intensity and spectrum of such a product can be defined by

$$\frac{\partial^2 \sigma}{\partial E \, \partial \Omega}$$

where σ is a total cross section for producing that product, E is its total energy, and $d\Omega$ an element of solid angle in its direction of motion. We can define a similar quantity observed in the center of mass of the primary collision

$$\frac{\partial^2 \sigma'}{\partial E' \, \partial \Omega'}$$

Now total cross-sections are invariant:

$$\sigma = \sigma'.$$

Therefore,

$$\iint \frac{\partial^2 \sigma}{\partial E \, \partial \Omega} \, dE \, d\Omega = \iint \frac{d^2 \sigma}{\partial E' \, \partial \Omega'} \, dE' \, d\Omega'.$$

We can change the variables of integration on the left-hand side into $dE' \, d\Omega'$ by using the Jacobian $\partial(E, \Omega)/\partial(E', \Omega')$, so that

$$\iint \frac{\partial^2 \sigma}{\partial E \, \partial \Omega} \frac{\partial(E, \Omega)}{\partial(E', \Omega')} \, dE' \, d\Omega' = \iint \frac{\partial^2 \sigma}{\partial E' \, \partial \Omega'} \, dE' \, d\Omega'.$$

Hence

$$\frac{\partial^2 \sigma}{\partial E \, \partial \Omega} \frac{\partial(E, \Omega)}{\partial(E', \Omega')} = \frac{\partial^2 \sigma}{\partial E' \, \partial \Omega'}.$$

In fact, the Jacobian in this case is P'/P, and we have

$$\frac{1}{P} \frac{\partial^2 \sigma}{\partial E \, \partial \Omega} = \frac{1}{P'} \frac{\partial^2 \sigma}{\partial E' \, \partial \Omega'}. \tag{B.32}$$

If there are only two reaction products, a similar analysis yields

$$\frac{\partial \sigma}{\partial \Omega} = \frac{P^2}{\gamma_c P'(P - \beta_c E \cos \theta)} \frac{\partial \sigma'}{\partial \Omega'}. \tag{B.33}$$

Care must be observed with this formula when E is double valued at a given θ, as can happen in two-body reactions with a threshold (Dedrick, 1962).

Increasing use is being made of invariant differential cross-sections. For example, in

$$1 + 2 \rightarrow 3 + 4$$

the angular distribution of 3 with respect to 1 appears to be more meaningful if given by $d\sigma/dt$. Both σ and t are invariants, so that this quantity is invariant. If $d\sigma/d\Omega'$ is the center-of-mass differential cross section and if it is independent of azimuth (no polarization effects), we have

$$\frac{d\sigma}{d\Omega'} = \frac{1}{2\pi} \frac{d\sigma}{d \cos \theta'}. \tag{B.34}$$

Now we have

$$t = (p_1 - p_3)^2 = m_1^2 + m_3^2 - 2 p_1 \cdot p_3. \tag{B.35}$$

If 1 is the incident particle and 3 the particle whose distribution we are considering, then θ' is the angle between $\mathbf{P}_1'$ and $\mathbf{P}_3'$ and we have, evaluating in the center-of-mass frame,

$$t = m_1^2 + m_3^2 + 2(E_1'E_3' - P_1'P_3' \cos \theta').$$

The center-of-mass quantities E_1', E_3', P_1', and P_3' are all constant, so that

$$dt = -2P_1'P_3' \, d\cos\theta'$$

and

$$\frac{d\sigma}{dt} = -\frac{1}{2P_1'P_3'} \frac{d\sigma}{d\cos\theta'} = -\frac{\pi}{P_1'P_3'} \frac{d\sigma}{d\Omega'}. \tag{B.36}$$

B.7 Transformation of the Polarization Vector

The polarization vector for spin-$\tfrac{1}{2}$ particles, as introduced in Section 2.13, is a three-vector and as used is defined in the rest frame of the particle. To transform it relativistically it must be made into a four-vector and, in fact, the three-vector is already the space component of the relevant four-vector, s, which has $s^0 = 0$ so that $s = 0$, $\boldsymbol{\sigma}$ in the rest frame (Bjorken and Drell, 1964). Since $p = m, 0$ in the rest frame, we have the invariant

$$p \cdot s = 0.$$

In addition, if the polarization is $\boldsymbol{\sigma}$, we have

$$s \cdot s = -|\boldsymbol{\sigma}|^2,$$

which is also invariant. The four-vector s must transform according to Eq. (B.2). However, this is not very enlightening or necessary, since the polarization is usefully defined only in the particle rest frame. Therefore we must now consider the kind of situation in which it is necessary to recognize that the polarization might appear to be different as a result of observing it from different frames. From what we have said so far about this problem it is evident that the difference can only be one of direction, not of magnitude, of the polarization three-vector.

Consider, as an example, the decay

$$\Lambda \to p + \pi^-.$$

To an observer in the Λ rest frame the protons are polarized with the vector along their direction of motion (longitudinally). The same is true for any observer reached by a Lorentz transformation along the proton direction. If the Λ is moving in the laboratory frame, an observer in the laboratory will find that the proton can have a transverse component of the polarization. To see how this happens consider the decay. In the proton rest frame its direction of motion is well defined (apart from sign it is the direction of the Λ before decay or of the recoiling π^-). The polarization produced by the decay is a vector magnitude $+\alpha$ [Eq. (6.11)] pointing along this direction. Now consider the laboratory. In the proton rest frame the direction of motion of the laboratory is, in general, not the same as that of the Λ. As far as observations in the laboratory are concerned (for example, of a left–right scattering asymmetry) the polarization is defined with respect to the direction of the proton in the laboratory, which means with respect to the direction of the laboratory in the proton rest frame. Thus the proton is, in general, no longer longitudinally polarized to a laboratory observer. There is an apparent rotation of the polarization vector away from the longitudinal position. The angle of this rotation is the angle between the directions of the velocities of the Λ and the laboratory observed in the proton rest frame.

As expected, the effect is not so much a change of polarization vector as a change in the direction with respect to which it is defined. The sense of the apparent rotation can be found by simple consideration of the kinematics.

This discussion can be applied to all massive particles, since the polarization or alignment is referred to orthogonal axes in the rest frame. Any apparent changes due to observation from different frames correspond to rotation of these axes and can be dealt with by rotation of the density matrix, as in Section 2.16.

For a fuller discussion of this subject and of the rotation of polarization vectors in magnetic and electric fields we refer readers to Hagedorn (1963).

REFERENCES

Bjorken, J. D., and Drell, S. D. (1964). "Relativistic Quantum Mechanics." McGraw-Hill, New York (B.7).

Dedrick, K. G. (1962). *Rev. Mod. Phys.* **34,** 429 (B.6).

Hagedorn, R. (1963). "Relativistic Kinematics." Benjamin, New York (B.4, B.7).

Williams, W. S. C. (1964). In "High Energy and Nuclear Physics Data Handbook" (W. S. C. Williams and W. Galbraith, eds.), Section VII, 2nd ed. Rutherford Lab., Chilton (B.4).

Appendix C
PHASE SPACE

C.1 The Density-of-States Factor

In Section 3.11, we introduced relations between cross sections or transition rates and the matrix elements of the S-matrix or of the transition operator. An important factor in the relations is the density-of-states or phase-space factor. Obviously this factor is necessary to the calculation of rates, but in addition it can be used to make predictions about spectra or distributions in various kinematic variables if certain simplifying assumptions are made about the behavior of transition-matrix elements. Thus Fermi's statistical theory (1950) used the phase space to calculate the particle multiplicities and spectra expected in high-energy hadron collisions. We discuss the determination of this factor and some simple examples.

From Eq. (3.72) the total phase space available to q particles is

$$\rho_q = \frac{1}{(2\pi)^{3q}} \int_1 \cdots \int_q d^4p_1 d^4p_2 \cdots d^4p_q \delta^4(p_\beta - p_\alpha) \prod_{i=1}^{q} \delta(p_i^2 - m_i^2) \theta(E_i), \tag{C.1}$$

where $\theta(E_i) = 1$ for $E_i > 0$, otherwise 0. $\delta^4(p_\beta - p_\alpha)$ expresses conservation of momentum between the initial state (α) and the final state (β) of q particles. Equation (3.71) shows that

$$\rho_q = \frac{1}{(2\pi)^{3q}} \int \cdots \int \frac{d\mathbf{P}_1 d\mathbf{P}_2 \cdots d\mathbf{P}_q}{2^q E_1 E_2 \cdots E_q} \delta^4(p_\alpha - p_\beta). \tag{C.2}$$

We shall put $R_q = (2\pi)^{3q} 2^q \rho_q$ to simplify notation. Since R_q is a function of the total center-of-mass energy $p_{\alpha 0} = p_{\beta 0} = W_q$, we shall put this dependence in explicitly, using W_q to represent the total center-of-mass energy of particles $1, 2, \ldots, q$. Hence

$$R_q(W_q) = (2\pi)^{3q} 2^q \rho_q.$$

From Eq. (C.2) and the Lorentz invariance of R it follows that

$$R_q(W_q) = \int (d\mathbf{P}_q/E_q) R_{q-1}(W_{q-1}), \tag{C.3}$$

where, from Eq. (B.22) and remembering that W_{q-1} is the same as the invariant mass of $q - 1$ particles, we have

$$W_{q-1}^2 = W_q^2 + m_q^2 - 2E_q W_q.$$

E_q is the energy of particle q in the center of mass of the q particles. Note the different use of the subscript q: to W_q it gives the meaning of the center-of-mass energy of all the particles $1, 2, \ldots, q$. On all other kinematic quantities it refers to the qth particle.

Equation (C.3) allows any R to be reduced to $R_2(W_2)$, where

$$R_2(W_2) = \iint \frac{d\mathbf{P}_1 \, d\mathbf{P}_2}{E_1 E_2} \delta(W_2 - E_1 - E_2) \delta(\mathbf{P}_1 + \mathbf{P}_2). \tag{C.4}$$

The arguments of the δ-functions imply that this integral is to be evaluated in the center-of-mass of particles 1 and 2. Integrating first over $d\mathbf{P}_2$, we have that the second δ-function defines $\mathbf{P}_2$, so

$$R_2(W_2) = \int \frac{d\mathbf{P}_1}{E_1 E_2} \delta(W_2 - E_1 - E_2)$$

$$= \int \frac{P_1^2 \, dP_1 \, d\Omega_1}{E_1 E_2} \delta(W_2 - E_1 - E_2).$$

To evaluate this, the following property of the δ-function is required:

$$\int F(x) \delta(f(x)) \, dx = F(a)(\partial f/\partial x)_{x=a}^{-1},$$

assuming a is the only root to the equation $f(x) = 0$. Here

$$f = W_2 - E_1 - E_2,$$

so

$$\frac{\partial f}{\partial P_1} = P\left(\frac{1}{E_1} - \frac{1}{E_2}\right),$$

where P is the momentum of both particles in their center of mass [Eq (B.24)]:

$$P = \frac{\lambda(W_2^2, m_1^2, m_2^2)}{2W_2}$$

$$= \frac{1}{2W_2}(W_2^4 + m_1^4 + m_2^4 - 2W_2^2 m_1^2 - 2W_2^2 m_2^2 - 2m_1^2 m_2^2)^{1/2}.$$

C.1 The Density-of-States Factor

Hence

$$R_2(W_2) = \int \frac{P}{W_2} d\Omega_1 = \frac{4\pi P}{W_2}. \tag{C.5}$$

In the recurrence relation (C.3) we can put

$$d\mathbf{P}/E = P^2 \, dP \, d\Omega/E = P \, dE \, d\Omega,$$

so that

$$R_q(W_q) = \iint_{m_q}^{E_q^*} P_q \, dE_q \, d\Omega_q \iint_{m_{q-1}}^{E_{q-1}^*} P_{q-1} \, dE_{q-1} \, d\Omega_{q-1} \cdots$$
$$\times \iint_{m_3}^{E_3^*} P_3 dE_3 \, d\Omega_3 \int (P/W_2) \, d\Omega_1. \tag{C.6}$$

The upper limits on the energy integrals are set by the kinematics. So, reading from left to right, the maximum energy the qth particle may have will occur when the center-of-mass energy W_{q-1} of the remaining $q-1$ particle is just equal to the sum of their masses. Hence, by Eq. (B.22),

$$E_q^* = \frac{W_q^2 + m_q^2 - (m_1 + m_2 + \cdots + m_{q-1})^2}{2W_q}.$$

Generalizing to the rth particle we have

$$E_r^* = \frac{W_r^2 + m_r^2 - (m_1 + m_2 + \cdots + m_{r-1})^2}{2W_r}.$$

For a given $E_r < E_r^*$,

$$W_{r-1}^2 = W_r^2 + m_r^2 - 2E_r W_r.$$

It is not always the total phase space that is interesting but a particular differential with respect to one or more kinematic variables. Thus the energy spectrum of a particle depends upon the differential of the phase space with respect to that energy. To obtain that differential it is only necessary to omit the corresponding integration in Eq. (C.6), or to insert an appropriate δ-function.

Let us compute the density-of-states factor for β-decay

$$n \to p + e^- + \bar{\nu}_e.$$

From Eq. (C.6), thinking of the electron as particle 3, the neutrino as 1, and the proton as 2,

$$R_3(m_n) = \int P_e \, dE_e \, d\Omega_e \, \frac{(m_n^2 - 2E_e m_n + m_e^2 - m_p^2)}{2(m_n^2 - 2E_e m_n + m_e^2)} \, d\Omega_\nu.$$

The solid angles are uncorrelated, so if we are not interested in the direction of the neutrino in the $p\bar{\nu}$ center-of-mass, or in the overall orientation of the event, we can put

$$R_3(m_n) = \frac{(4\pi)^2}{2m_n^2} \int_{m_e}^{E_0} P_e \, dE_e \, (E_0 - E_e),$$

where we have put

$$E_0 = m_n - m_p$$

and have neglected m_e and E_e with respect to m_n. If the transition matrix element is constant, the total decay rate is proportional to R_3. In this case it is evident that the electron spectrum includes the factor

$$\frac{dR_3}{dE_e} = \frac{(4\pi)^2}{2m_n^2} P_e(E_0 - E_e).$$

The actual decay rate is this times the spin summed-and-averaged Lorentz-invariant matrix element [Eq. (3.68)]. In β-decay a factor E_r for every particle can be factored out of this matrix element to leave the correct combination of Gamow–Teller or Fermi matrix elements. Thus the decay rate per unit energy interval is proportional to

$$\frac{(4\pi)^2}{2m_n} P_e E_\nu E_e (E_0 - E_e) = \frac{(4\pi)^2}{2m_n} P_e (E_0 - E_e)^2 E_e.$$

Apart from differences in notation, this is the density-of-states factor in Eq. (11.11).

C.2 Phase Space and Dalitz Plots

Consider now the more general case of a three-body final state. A system of center-of-mass energy M changes into a system of three particles having masses m_1, m_2, and m_3. What is the distribution as a function of E_2 and E_3, where these are energies in the overall center of mass?

$$R_3(M) = \iint_{m_3}^{E_3^*} P_3 \, dE_3 \, d\Omega_3$$

$$\times \iint \frac{d\mathbf{P}_1 \, d\mathbf{P}_2}{E_1 E_2} \delta(M - E_1 - E_2 - E_3) \, \delta(\mathbf{P}_1 + \mathbf{P}_2 + \mathbf{P}_3).$$

The last part of the integral is invariant and so can be evaluated in the center-of-mass of particles 1 and 2, where we know it has the value given by Eq. (C.5). Hence, using a prime on all quantities in that frame to make a clear distinction,

$$R_3(M) = \iint P_3 \, dE_3 \, d\Omega_3 \int \frac{P'}{m_{12}} d\Omega_2',$$

C.2 Phase Space and Dalitz Plots

where $d\Omega_2'$ could represent an element of solid angle about a direction for particle 2 at an angle θ_2' to a line opposite to the direction of flight of particles 3. m_{12}, the invariant mass of particles 1 and 2, replaces W_2. Now transforming from the 1–2 center of mass to the three-body center of mass,

$$E_2 = \gamma(E_2' + P_2'\beta\cos\theta_2'),$$

where $P_2' = P'$, $\gamma = (E_1 + E_2)/m_{12} = (1 - \beta^2)^{-1/2}$, and $\beta = P_3/(E_1 + E_2)$ is the velocity of the 1–2 center of mass in the overall center of mass. It follows that

$$dE_2 = P'\gamma\beta\,d\cos\theta_2' = P'\gamma\beta\,d\Omega_2'/2\pi,$$

where we have absorbed the integration around the azimuth. Therefore

$$R_3(M) = \iint P_3\,dE_3\,d\Omega_3 \int \frac{2\pi\,dE_2}{\gamma\beta m_{12}}$$

$$= 2\pi \iiint dE_3\,dE_2\,d\Omega_3. \quad\text{(C.7)}$$

Thus, if we are not interested in the orientation of the final products $(d\Omega_3)$, we see that

$$d^2R_3(M) \propto dE_3\,dE_2.$$

This means that a Dalitz plot that uses axes E_2 and E_3 (or E_1 and E_2, etc.) will be uniformly populated if the invariant matrix element is constant. Dalitz plots are sometimes made with the axes labeled by the invariant mass of pairs of particle. By Eq. (B.22), the mass of the pair 1, 2 is given by

$$m_{12}^2 = M^2 + m_3^2 - 2E_3 M. \quad\text{(C.8)}$$

Similarly

$$m_{13}^2 = M^2 + m_2^2 - 2E_2 M. \quad\text{(C.9)}$$

Dropping the solid-angle integral from Eq. (C.7) we have

$$R_3(M) \propto \iint dE_3\,dE_2 = \iint dm_{12}^2\,dm_{13}^2\,\frac{\partial(E_2, E_3)}{\partial(m_{13}^2, m_{12}^2)},$$

where the factor is the Jacobian for the transformation

$$\frac{\partial(E_2, E_3)}{\partial(m_{13}^2, m_{12}^2)} = \begin{vmatrix} \dfrac{\partial E_2}{\partial m_{13}^2} & \dfrac{\partial E_2}{\partial m_{12}^2} \\ \dfrac{\partial E_3}{\partial m_{13}^2} & \dfrac{\partial E_3}{\partial m_{12}^2} \end{vmatrix} = \frac{1}{4M^2},$$

which is constant. It follows, as expected, that this labeling of axes also leads to a uniform plot if the invariant matrix element is constant. Similarly, using the kinetic energies also leads to the uniform phase space.

Let us now consider projections of the Dalitz plot; here the question is: What is the expected distribution in energy of one particle? From Eq. (C.6) the differential of the phase space required for the energy spectrum of particle 3 is

$$d R_3(M) = \frac{P_3 \, dE_3 \, P}{m_{12}} = \frac{P_3 \lambda(M^2, m_1^2, m_2^2) \, dE_3}{2 m_{12}^2}$$

$$= \frac{P_3 \, dE_3}{2 m_{12}^2} [M^4 + m_1^4 + m_2^4 - 2M^2 m_1^2 - 2M^2 m_2^2 - 2m_1^2 m_2^2]^{1/2}.$$

(C.10)

The indices may be cyclically permuted to obtain the phase-space factors for the energy distribution of the other two particles.

Since $2M \, dE_3 = -dm_{12}^2$, we have, dropping the nonessential -ve sign;

$$dR_3(M) = \frac{P_3 \, dm_{12}^2}{4 M m_{12}} \lambda(M^2, m_1^2, m_2^2).$$

It is evident that deviations from phase-space predictions will give information on the variation of the matrix element. The presence of a resonance in a three-body final state will show up as a concentration of events in the Dalitz plot at the corresponding invariant mass. Many resonances have been discovered in this way (Sections 7.8 and 8.3). The calculation and use of phase space is not restricted to three-body final states. The ω-meson was discovered as a deviation from phase space of the distribution of the mass of uncharged combinations of three pions from five-pion proton–antiproton annihilation (Maglic et al., 1961; Section 7.9).

Other discussions of phase space and examples are given by Hagedorn (1963), Kretzschmar (1961), and Radojicic (1964).

REFERENCES

Fermi, E. (1950). "Elementary Particles." Yale Univ. Press, New Haven, Connecticut (C.1).

Hagedorn, R. (1963). "Relativistic Kinematics." Benjamin, New York (C.2).

Kretzschmar, M. (1961). Annu. Rev. Nucl. Sci. 11, 1 (C.2).

Maglic, B. C., Alvarez, L. W., Rosenfeld, A. H., and Stevenson, M. L. (1961). Phys. Rev. Lett. 7, 178 (C.2).

Radojicic, D. (1964). "High Energy and Nuclear Physics Data Handbook." (W. S. C. Williams and W. Galbraith, eds.) Section VIII, 2nd ed. Rutherford Lab., Chilton (C.2).

Appendix D
DIRAC MATRIX ELEMENTS

D.1 The Structure of Matrix Elements for Dirac Particles

Several times we have been led to consider matrix elements between physical states of two spin-$\frac{1}{2}$ baryons of the kind $\langle 2 | j | 1 \rangle$, where j is a vector or axial-vector operator, as, for example, in weak interactions (Chapter XI, particularly Section 11.10) and in considering nucleon electromagnetic structure (Section 13.6). Because of strong interactions these matrix elements are impossible to evaluate, but on general invariance grounds can be reduced to a useful form. Consider first the vector matrix element $\langle 2 | j_V(x) | 1 \rangle$. Let p_1 and p_2 be the incoming and outgoing baryon four-momenta and put $q = p_2 - p_1$. The plane-wave solutions of the Dirac equations are $u(p_1)$, $u(p_2)$. Then

$$\langle 2 | j_V^\rho(x) | 1 \rangle = \bar{u}(p_2) [\gamma^\rho F_V(q^2) + i\sigma^{\rho\lambda} q_\lambda F_T(q^2) + F_S(q^2) q^\rho] u(p_1) e^{iq \cdot x} \quad (D.1)$$

where the F functions contain all that is unknown about the matrix element. This is evidently a vector as required, but it is not clear that this is the most general vector matrix element possible. First we know that there are no more possible vector-forming combinations of γ-matrices alone that we can use that will not reduce to γ^ρ. When we look for vector combinations of γ-matrices and four-vectors, then only two others remain:

$$\sigma^{\rho\lambda}(p_1 + p_2)_\lambda \quad \text{and} \quad (p_1 + p_2)^\rho .$$

We can show that these reduce to the terms already in Eq. (D.1) and do not have to be added. To do this consider the equation satisfied by a plane-wave spinor $u(p_1)$ [Eq. (10.22a)]

$$(\gamma^\lambda p_{1\lambda} - m_1) u(p_1) = 0 .$$

Multiply from the left by $\bar{u}(p_2)\gamma^\rho$ and we have

$$0 = \bar{u}(p_2)\gamma^\rho[\gamma^\lambda p_{1\lambda} - m_1]u(p_1)$$
$$= \bar{u}(p_2)[p_1^\rho - i\sigma^{\rho\lambda}p_{1\lambda} - m_1\gamma^\rho]u(p_1). \qquad (D.2)$$

Similarly, the adjoint spinor $\bar{u}(p_2)$ satisfies Eq. (10.22b):

$$\bar{u}(p_2)[p_{2\lambda}\gamma^\lambda - m_2] = 0.$$

Multiplying from the right by $\gamma^\rho u(p_1)$ we find

$$0 = \bar{u}(p_2)[p_{2\lambda}\gamma^\lambda - m_2]\gamma^\rho u(p_1)$$
$$= \bar{u}(p_2)[p_2^\rho + i\sigma^{\rho\lambda}p_{2\lambda} - m_2\gamma^\rho]u(p_1). \qquad (D.3)$$

Adding Eqs. (D.3) and (D.2) we find

$$\bar{u}(p_2)(p_1 + p_2)^\rho u(p_1) = \bar{u}(p_1)[i\sigma^{\rho\lambda}(p_1 - p_2)_\lambda + (m_1 + m_2)\gamma^\rho]u(p_1)$$
$$= \bar{u}(p_1)[-i\sigma^{\rho\lambda}q_\lambda + (m_1 + m_2)\gamma^\rho]u(p_1).$$

Thus, the addition of a term of the form $(p_1 + p_2)^\rho$ to Eq. (D.1) is unnecessary because it can be absorbed into those already there. Similarly by subtracting Eq. (D.3) from (D.2) it is possible to show that $\sigma^{\rho\lambda}(p_1 + p_2)_\lambda$ depends only upon q_ρ and γ^ρ and need not appear in Eq. (D.1). The factors F_V, F_T, and F_S are called form factors and will be functions of all the independent scalar variables. In the circumstances that we are considering, there is only one such variable, q^2, which explains the explicit dependence given in Eq. (D.1).

The interaction of all spin-$\frac{1}{2}$ fermions with the electromagnetic field gives matrix elements of this form. However, the electric current is conserved; that is, it satisfies a continuity equation, $\partial_\rho j_V^\rho = 0$, so that

$$\langle 2|\partial_\rho j_V^\rho|1\rangle = 0.$$

The linear operator j_V^ρ and its matrix elements are functions of the space-time coordinates, but the bra and ket vectors are not, so that we have

$$\langle 2|\partial_\rho j_V^\rho(x)|1\rangle = \partial_\rho\langle 2|j_V^\rho(x)|1\rangle.$$

The x variation of the matrix element is contained in $e^{iq\cdot x}$ and therefore

$$\langle 2|\partial_\rho j_V^\rho(x)|1\rangle = iq_\rho\langle 2|j_V^\rho(x)|1\rangle.$$

Substituting from Eq. (D.1) we find

$$\bar{u}(p_2)[F_V(q^2)\gamma^\rho q_\rho + iF_T(q^2)\sigma^{\rho\lambda}q_\lambda q_\rho + F_S(q^2)q^2]u(p_1) = 0.$$

Now $\sigma^{\rho\lambda}q_\lambda q_\rho = 0$ because of the double sum over indices and the antisymmetric property of $\sigma^{\rho\lambda}$. In addition,

$$\bar{u}(p_2)\gamma^\rho p_{2\rho} = m_2\bar{u}(p_2)$$

D.1 The Structure of Matrix Elements for Dirac Particles

and
$$\gamma^\rho p_{1\rho} u(p_1) = m_1 u(p_1)$$
so that
$$\bar{u}(p_1)[F_V(q^2)(m_2 - m_1) + F_S(q^2)q^2]u(p_1) = 0. \tag{D.4}$$

It follows that if $m_1 = m_2$, as is the case for electron or positron electromagnetic interactions,
$$F_S(q^2) = 0,$$
for all q^2. Thus the current contribution to the matrix element for the interaction of a spin-$\frac{1}{2}$ fermion with the electromagnetic field is
$$\langle 2|j_V{}^\rho|1\rangle = \bar{u}(p_2)[F_1(q^2)\gamma^\rho + i(\kappa/2m)F_2(q^2)\sigma^{\rho\lambda}q_\lambda]u(p_1)\,e^{iq\cdot x}. \tag{D.5}$$

As described in Chapter XIII, we know that e times the matrix element of Eq. (D.5) represents the observable electromagnetic interaction. The coupling strength is e at $q^2 = 0$; hence, $F_1(q^2 = 0) = 1$. The second term corresponds to an anomalous magnetic moment. If κ is put equal to the anomalous part of the magnetic moment, then $F_2(q^2 = 0) = 1$. For electrons and positrons κ is very small (Section 10.9), but for protons and neutrons it is large, presumably on account of the effect of strong interactions. At present, this cannot be calculated from the theory of strong interactions, and the observed value of κ must be used (Section 13.6).

The weak-interaction vector current will have matrix elements also limited to the same form. Since the fermion masses cannot be the same, it is possible for the scalar form factor to be nonzero. In this case it would appear that Eq. (D.4), the consequence of the CVC hypothesis, would fix the value of the induced scalar term $F_S(q^2)$. However, the CVC hypothesis is correct only in the absence of electromagnetism, and thus Eq. (D.4) cannot be expected to apply in weak transitions between particles having electromagnetic mass splittings, e.g., neutron decay. In addition, Weinberg (1958) has classified the contributions to the current according to their properties under the G-parity transformation (Section 6.8). The γ^ρ and $\sigma^{\rho\lambda}q_\lambda$ terms are even and are called first-class currents. The scalar term is odd and is called a second-class current. The only second-class current with any experimental support is that associated with F_- in K_{l3} decay [Eq. (11.50)]. No other has been observed, and thus it is usual to omit such currents in baryon leptonic decays.

The matrix element of an axial-vector current can be dealt with in the same way as the vector current. The result is that

$$\langle 2|j_A{}^\rho|1\rangle$$
$$= \bar{u}(p_2)[G_A(q^2)\gamma^\rho\gamma_5 + iG_T(q^2)\sigma^{\rho\lambda}q_\lambda\gamma_5 + G_P(q^2)q^\rho\gamma_5]u(p_1)\,e^{iq\cdot x}, \tag{D.6}$$

where the unknown parts of the interaction are contained in the form factors G. The martix element of the divergence of this current is

$$\langle 2| \partial_\rho j_A{}^\rho |1\rangle = i\bar{u}(p_2)[G_A(q^2)(m_1 + m_2) + G_P(q^2)q^2]\gamma_5 u(p_1)\, e^{iq\cdot x}. \tag{D.7}$$

and is not zero. In Eq. (D.6) the only second-class current is $\sigma^{\rho\lambda}q_\lambda\gamma_5$, and therefore it is usually assumed that this term is absent.

REFERENCE

Weinberg, S. (1958). *Phys. Rev.* **112**, 1375 (D.1).

Appendix E

TABLES OF CLEBSCH–GORDAN COEFFICIENTS AND OF ROTATION MATRIX ELEMENTS

E.1 Clebsch-Gordan Coefficients

In this section we give tables of the Clebsch–Gordan coefficients for two frequently occurring cases of the vector addition of angular momenta. The notation [see Eq. (2.44)] is $\langle j_a, j_{az}, j_b, j_{bz} | j, j_z \rangle$, where

$$|j_a - j_b| \leqslant j \leqslant j_a + j_b \quad \text{and} \quad j_{az} + j_{bz} = j_z.$$

The Tables E.1 and E.2 give the coefficients for $j_b = \tfrac{1}{2}$ and $j_b = 1$, respectively. Note that

$$\langle j_a, j_{az}, j_b, j_{bz} | j, j_z \rangle = (-1)^{j-j_a-j_b} \langle j_b, j_{bz}, j_a, j_{az} | j, j_z \rangle.$$

The Particle Data Group (1969) in their tables use the notation

$$\langle j_1, j_2, m_1, m_2 | j_1, j_2, J, M \rangle,$$

TABLE E.1

THE CLEBSCH–GORDAN COEFFICIENTS $\langle j_a, j_{az}, \tfrac{1}{2}, j_{bz} | j, j_z \rangle$

	$j_{bz} = +\tfrac{1}{2}$	$j_{bz} = -\tfrac{1}{2}$
$j = j_a + \tfrac{1}{2}$	$\left(\dfrac{j_a + j_z + \tfrac{1}{2}}{2j_a + 1}\right)^{1/2}$	$\left(\dfrac{j_a - j_z + \tfrac{1}{2}}{2j_a + 1}\right)^{1/2}$
$j = j_a - \tfrac{1}{2}$	$-\left(\dfrac{j_a - j_z + \tfrac{1}{2}}{2j_a + 1}\right)^{1/2}$	$\left(\dfrac{j_a + j_z + \tfrac{1}{2}}{2j_a + 1}\right)^{1/2}$

where

$$|j_1 - j_2| \leqslant J \leqslant j_1 + j_2 \quad \text{and} \quad m_1 + m_2 = M.$$

Tables giving values of the coefficients for larger j may be found, for example, in Condon and Shortley (1951) and in Edmonds (1960).

TABLE E.2

The Clebsch-Gordan Coefficients $\langle j_a, j_{az}, 1, j_{bz} | j, j_z \rangle$

	$j_{bz} = +1$	$j_{bz} = 0$	$j_{bz} = -1$
$j = j_a+1$	$\left(\dfrac{(j_a+j_z)(j_a+j_z+1)}{(2j_a+1)(2j_a+2)}\right)^{1/2}$	$\left(\dfrac{(j_a-j_z+1)(j_a+j_z+1)}{(2j_a+1)(j_a+1)}\right)^{1/2}$	$\left(\dfrac{(j_a-j_z)(j_a-j_z+1)}{(2j_a+1)(2j_a+2)}\right)^{1/2}$
$j = j_a$	$-\left(\dfrac{(j_a+j_z)(j_a-j_z+1)}{2j_a(j_a+1)}\right)^{1/2}$	$\dfrac{j_z}{[j_a(j_a+1)]^{1/2}}$	$\left(\dfrac{(j_a-j_z)(j_a+j_z+1)}{2j_a(j_a+1)}\right)^{1/2}$
$j = j_a-1$	$\left(\dfrac{(j_a-j_z)(j_a-j_z+1)}{2j_a(2j_a+1)}\right)^{1/2}$	$-\left(\dfrac{(j_a-j_z)(j_a+j_z)}{j_a(2j_a+1)}\right)^{1/2}$	$\left(\dfrac{(j_a+j_z+1)(j_a+j_z)}{2j_a(2j_a+1)}\right)^{1/2}$

E.2 Rotation Matrix Elements

Tables E.3 and E.4 give the elements of the rotation matrix $d^j(\beta)$ [see Eq. (2.68a)] for $j = \frac{1}{2}$ and $j = 1$, respectively. Tables of the matrix elements for higher j may be found in Edmonds (1960) and in Brink and Satchler (1962).

TABLE E.3

The Rotation Matrix Elements $d^{1/2}_{j_z j_z'}(\beta)$

	$j_z' = +\frac{1}{2}$	$j_z' = -\frac{1}{2}$
$j_z = +\frac{1}{2}$	$\cos(\beta/2)$	$-\sin(\beta/2)$
$j_z = -\frac{1}{2}$	$\sin(\beta/2)$	$\cos(\beta/2)$

TABLE E.4

The Rotation Matrix Elements $d^1_{j_z j_z'}(\beta)$

	$j_z' = +1$	$j_z' = 0$	$j_z' = -1$
$j_z = -1$	$\cos^2(\beta/2)$	$-\sqrt{\tfrac{1}{2}}\sin\beta$	$\sin^2(\beta/2)$
$j_z = 0$	$\sqrt{\tfrac{1}{2}}\sin\beta$	$\cos\beta$	$-\sqrt{\tfrac{1}{2}}\sin\beta$
$j_z = -1$	$\sin^2(\beta/2)$	$\sqrt{\tfrac{1}{2}}\sin\beta$	$\cos^2(\beta/2)$

REFERENCES

Brink, D. M. and Satchler, G. R. (1962) "Angular Momentum," Oxford University Press, London and New York (E.2).

Condon, E. U. and Shortley, G. H. (1951). "The Theory of Atomic Spectra." Cambridge University Press, London and New York (E1).

Edmonds, A. R. (1960). "Angular Momentum in Quantum Mechanics." Princeton Univ. Press, Princeton, New Jersey (E.1, E.2).

Particle Data Group (1969). *Rev. Mod. Phys.* **41**, 109 (E1).

Author Index

Numbers in italics refer to the pages on which the complete references are listed.

A

Aachen Collaboration, 459, *461*
Aamodt, R. L., 181, *216*
Abashian, B., 472, *482*
Abrams, R. J., 472, *482*
Adair, R. K., 156, 219, *175*, 219, *253*
Ademells, M., 374, *385*
Adler, S. L., 338, 360, *385*
Akerlof, C. W., 455, *461*
Albergi Quaranta, 361, *385*
Alder, K., 338, *385*
Alff-Steinberger, C., 164, *175*, 212, 213, *216*, 473, *482*
Alichanov, A. I., 349, *385*
Allaby, J. V., 406, *429*
Allen, R. A., 335, *385*
Alston, M. H., 200, 201, 247, *216*, 247, *253*
Alvarez, L. W., 210, 211, 212, *216*, 504, *504*
Amaldi, U., 414, *429*
Ambler, E., 173, *176*, 337, *387*
Anderson, H. L., 228, *253* 403, *429*,
Armenteros, R., 50, *66*
Asbury, J. G., 458, *461*
Ash, W. W., 322, *326*, 455, *461*
Ashkin, J., 362, *386*
Atac, M., 164, *175*
Aubert, B., 363, 377, *386*, 467, *482*
Augustin, J. E., 195, *216*, 452, *461*
Auril, P., 234, *253*
Auslander, V. L., 195, *216*, 452, *461*

B

Baacke, J., 404, 426, *429*
Backenstoss, G., 349, *386*
Bailey, J., 323, *326*
Baird, J. K., 174, *176*
Ballam, J., 459, *461*
Baltay, C., 167, *175*
Barber, W. C., 322, *326*
Bardon, M., 349, *386*
Bareyre, P., 123, *125*, 234, *253*
Barger, V., 427, *429*
Barkas, W. H., 329, *386*
Barnes, V. E., 253, *253*, 269, *292*
Bartlett, D. F., 473, *482*
Bauer, R. W., 338, *386*
Bazin, M. J., 167, *175*
Behrends, R. E., 256, *292*
Bellettini, G., 178, *216*
Bemporad, C., 178, *216*
Benenson, R. E., 113, *126*
Bennett, S., 480, *482*
Bergkvist, K. E., 329, *386*
Berkelman, K., 322, *326*, 455, *461*
Berman, S. M., 210, *216*, 348, *386*
Bernstein, J., 358, 368, *386*, 432, *461*
Biedenharn, L. C., 93, *110*, 123, *125*, 440, *461*
Bingham, H., 363, *386*
Bjorken, J. D., 146, *175*, 294, 303, 304, 314, 325, *326*, 497, *498*
Blankenbecler, R. B., 423, *429*
Blatt, J. M., 68, 93, *110*, 122, *125*, 333, *386*
Bock, R., 156, *175*
Bodansky, D., 164, *175*
Bogoliubov, N. N., 294, 308, *326*
Böhm, A., 473, *482*
Bolon, G. C., 444, *461*
Bott-Bodenhausen, M., 473, *482*
Boyarski, A., 281, *292*, 427, *429*

Braccini, P. L., 178, *216*
Braithwaite, W. J., 164, *175*
Bransden, B. H., 234, *253*
Breit, G., 91, *110*
Brene, N., 370, *386*
Brickman, C., 123, *125*, 234, *253*
Brink, D. M., 38, 39, 40, 48, *66*, 510, *511*
Brown, J., 190, *216*
Budker, G. I., 195, *216*, 452, *461*
Bulos, F., 458, *461*
Burcham, W. E., 335, 336, 348, *385*, *387*
Burgy, M. T., 164, *175*, 339, *386*
Butler, C. C., 136, 137, *143*
Byers, N., 248, 249, *253*

C

Cabibbo, N., 368, 369, 371, 376, *386*
Camerini, U., 466, *482*
Canter, R. J., 472, *482*
Carmony, D. D., 251, *254*
Carpenter, D. W., 472, *482*
Carruthers, P., 256, 268, 271, 282, *292*
Carter, R. E., 345, *386*
Cartwright, W. F., 178, *216*
Cassels, J. M., 181, *216*
Cence, R. J., 473, *482*
Chackett, K. F., 335, *385*
Charap, J. M., 256, *293*
Chen, J. R., 164, *175*
Chew, G. F., 198, *216*, 232, *253*, 400, 408, 423, 425, 427, *429*
Chinowsky, W., 202, *216*
Chollet, J., 480, *482*
Chrisman, B., 164, *175*
Christenson, J. H., 200, *216*, 471, 472, *482*
Chu, W. T., 362, *386*
Chupp, W. W., 141, *143*.
Clark, D. L., 180, *216*
Clegg, A. B., 417, *429*
Cline, D., 427, *429*
Coester, F. S., 173, *175*
Coleman, S., 280, 285,*293*,382,*386*,404,*429*
Condon, E. U., 38, *66*, 510, *511*
Conversi, M., 262, *386*
Cool, R. L., 289, *293*, 362, 378, *386*
Coombes, C. A., 362, *386*
Copson, E. T., 484, *487*
Cork, B., 362, 378, *386*

Costa, G., 195, 212, *216*
Courant, H., 222, *253*
Cowan, C. L., 329, 345, *387*
Crawford, F. S., 142, *142*, 154, 156, *175*, 220, *253*, 362, *386*
Cresti, M., 154, 156, *175*, 220, *216*
Cronin, J. W., 200, *216*, 348, 363, 364, 366, 367, 378, *386*, 471, 472, *482*
Cronstrom, C., 370, *386*
Cutkosky, R. E., 277, *293*

D

Dalitz, R. H., 123, 137, 142, 151, 152, *175*, 243, *253*, 289, *293*, 358, *386*, 413, *429*
Danby, G., 346, *386*
Dar, A., 459, *461*
Dashen, R. F., 284, 293
Davidon, W. C., 403, *429*
Davies, R., 345, *386*
Day, T. B., 180, *216*
Dayhoff, E. S., 318, *327*
Debrunner, P., 164, *175*
Dedrick, K. G., 496, *498*
Derado, I., 459, *461*
Derrick, M., 181, *216*
de Rujula, A., 348, *386*
de Swart, J. J., 256, 271, 273, 274, 276, 282, *293*
Deutsch, M., 338, *386*, 472, *483*
Devons, D., 440, *461*
de Vries, C., 448, 449, *461*
Diebold, R., 362, *386*, 459, *461*
Di Capua, E., 351, *386*
Di Lella, L., 362, *386*
Dirac, P. A. M., 1, 5, 23, *23*
Donnachie, A., 234, 235, *253*
Dorfan, D., 473, 480, *482*
Douglass, R. L., 142, *142*
Dreitlein, J., 256, *292*
Drell, S. D., 146, *175*, 294, 303, 304, 314, 317, 324, 325, *326*, 415, 416, *429*, 458, *461*, 497, *498*
Dress, W. B., 174, *176*
Duclos, J., 167, *175*, 348, *386*
Duke, P. J., 237, *253*
Durbin, R., 180, *216*
Durr, H. P., 417, *429*
Dyson, F., 256, 289, *293*, 318, *326*

Author Index

E

Eden, R. J., 393, 394, 395, 407, *429*
Edmonds, A. R., 39, 48, *66*, 510, *511*
Elings, U. B., 281, *293*
Eliseev, G. P., 349, *385*
Ely, R. P., 247, *253*
Engelke, C. E., 113, *126*
Engelmann, P., 164, *175*
Ericson, T. E. O., 362, *386*

F

Fackler, O., 197, *217*
Fano, U., 56, *66*
Farley, F. J. M., 348, *386*
Featherstone, F. H., 141, *143*
Feinberg, G., 174, *175*
Feld, B. T., 234, *254*
Fenster, S., 248, 249, *253*
Fermi, E., 100, *110*, 228, *253*, 331, 334, *386*, 499, *504*
Ferrari, E., 414, *429*
Ferro-Luzzi, M., 225, 244, 246, *253*, 277, *293*
Feshbach, H., 419, *430*
Fetkovich, J. G., 181, *216*
Feynman, R. P., 23, 48, *23*, *66*, 311, *326*, 343, 355, *386*
Fidecaro, G., 181, *216*
Fields, T. H., 181, *216*
Filthuth, H., 374, *387*
Fisher, G. P., 472, *482*
Fitch, V. L., 173, *175*, 200, *216*, 466, 471, 472, 473, *482*
Flatté, S. M., 212, *216*
Foa, L., 178, *216*
Fowler, W. B., 138, *143*
Franzini, P., 349, *386*
Frauenfelder, H., 164, *175*, 440, *461*
Frautshi, S. C., 422, 423, 428, *429*
Frazer, W. R., 450, *461*
Freeman, J. M., 336, 348, *387*
Freytag, D., 167, *175*
Frisch, D., 197, *217*
Froissart, M., 407, *429*
Fronsdal, C., 256, *292*
Fubini, S., 358, *386*
Fujii, T., 471, 472, *482*

Fulco, J. R., 450, *461*
Fung, S. Y., 247, *253*

G

Gaillard, J. M., 200, *216*, 473, *482*
Galaktionov, Yu. V., 349, *385*
Galbraith, W., 362, *386*
Gamow, G., 334, *387*
Garland, R., 351, *386*
Garwin, R. L., 152, *175*
Gasiorowicz, S., 124, *126*, 256, 271, *293*
Gatto, R., 321, *326*, 374, *385*, 455, *462*
Gell-Mann, M., 139, *143*, 162, *175*, 190, *216*, 255, 259, 268, 282, 288, *293*, 343, 355, 356, 358, 375, 376, *386*, 422, 423, *429*, 454, 456, *461*
Gerhardt, J. B., 335, *387*
Gidal, G., 247, *253*
Gittelman, B., 322, *326*, 338, *386*
Glashow, S. L., 280, 285, *293*, 382, *386*
Glasser, R. G., 164, *175*
Gobbi, B., 376, *387*
Goldberger, M. L., 112, 116, *126*, 360, *387*, 403, 423, *429*, 469, *483*
Goldfarb, L. J. B., 440, *461*
Goldhaber, G., 141, *143*, 202, *216*, 240, *254*
Goldhaber, M., 338, *387*
Goldhaber, S., 141, *143*, 202, *216*, 240, *253*, *254*
Good, M. L., 142, *142*, 154, 156, *175*, 220, *253*, 468, *483*
Good, R., 292, *293*
Gorodkov, Yu. V., 349, *385*
Gosham, A. T., 167, *175*
Gottfried, K., 412, 415, *429*
Gourdin, M., 276, *293*
Green, D., 376, *387*
Gribov, V. N., 427, *430*
Grodzins, L., 338, *386*, *387*
Guirigossian, Z. G. T., 459, *461*
Gursey, F., 290, 291, *293*

H

Haas, R., 156, *175*
Hadley, J., 181, *216*
Hagedorn, R., 494, 498, *498*, 504, *504*
Hague, N., 202, *216*

Hakel, W., 376, *387*
Hamermesh, M., 205, *216*, 256, *293*
Hamilton, J., 162, 163, *175*, 232, 233, *253*, *254*, 404, *429*
Hamilton, W. D., 353, *387*
Hand, L. N., *462*
Handler, R., 164, *175*
Haracz, R. D., 91, *110*
Harari, H., 444, *462*
Harmer, D. S., 345, *386*
Harris, G., 214, 215, *216*
Hartill, D. L., 322, *326*
Hayward, R. W., 173, *176*, 337, *387*
Hearn, A. C., 415, 416, *429*
Heintze, J., 167, *175*, 348, *386*
Helland, J. A., 164, *176*
Helmy, E., 141, *143*
Herman, R., 448, 449, *461*
Hofstadter, R., 448, 449, *461*
Höhler, G., 404, 426, *429*, 444, *462*
Hoppes, D. D., 173, *176*, 337, *387*
Hudson, R. P., 173, *176*, 337, *387*
Huwe, D. O., 251, *254*
Hyams, B. D., 349, *386*

I

Igi, K., 425, *429*
Iloff, E. L., 141, *143*
Ioffe, B. L., 385, *387*

J

Jackson, J. D., 123, 124, *126*, 412, 415, 427, *429*
Jacob, M., 210, *216*, 427, *430*
Jauch, J. M., 294, 308, 317, *326*
Johnson, K., 292, *293*
Jones, D. P., 156, *175*
Jones, R. B., 256, *293*
Jovanovich, J., 471, 472, *482*

K

Kabir, P. K., 463, 467, 473, *483*
Kalbflisch, G. R., 142, *142*
Källen, G., 124, *126*, 333, 334, 352, 367, *387*
Kanada, H., 407, *430*
Keck, J. C., 164, *176*

Kehoe, B., 164, *175*
Kemmer, N., 156, *175*, *175*
Kernan, A., 277, *293*
Kerth, L., 378, *386*
Khabakhpashev, A. G., 195, *216*, 452, *461*
Kim, J. K., 243, *254*, 276, 291, *293*
Kirsch, L., 164, *175*
Kirsopp, R. G., 234, 235, *253*
Kloeppel, P., 164, *175*
Knop, G., 349, *386*
Kobayashi, T., 407, *430*
Kokkedee, J. J. J., 289, *293*
Konopinski, E. J., 334, *387*
Kretzschmar, M., 504, *504*
Krohn, V. E., 164, *175*, 339, *386*
Kroll, N. M., 454, *462*
Kruse, U. E., 403, *429*
Kuehner, J. A., 181, *216*
Kunz, P. F., 200, *216*

L

Lamb, W. E., 318, *327*
Lambertson, G. R., 362, *386*
Landau, L. D., 339, *387*
Landshoff, P. V., 393, 395, *429*
Lannutti, J. E., 141, *143*
Lattes, C. M. G., 177, *216*
Lea, A. T., 234, *253*
Lebowitz, J. M., 113, *126*
Lederman, L. M., 152, *175*, 362, *386*
Lee, J., 349, *386*
Lee, T. D., 152, 153, 171, 174, 175, *175*, *176*, 220, *254*, 339, 357, 382, 383, *387*, 454, *462*, 474, *483*
Lee, W., 203, *216*, 256, *292*
Lee, Y. Y., 199, *216*
Leighton, R. B., 23, *23*
Leipuner, L. B., 156, *175*
Levinson, C. A., 459, *461*
Lichtenstein, C. A., 322, *326*, 455, *461*
Limon, P., 164, *175*
Lipkin, H. J., 256, 381, 289, *293*, 459, *461*
Littauer, R. M., 322, *326*, 448, *462*
Liu, D. C., 167, *176*
Loar, H., 180, *216*
Longo, M. J., 164, *176*
Lovelace, C., 234, 235, *253*
Low, F. E., 198, *216*, 232, *253*, 385, *387*, 408, *429*

Author Index

Lüders, G., 170, 171, *176*
Lyubimov, V. A., 349, *385*

M

McClellan, G., 458, *462*
Maglic, B. C., 195, 210, 211, 212, *216*, *217*, 504, *504*
Maksymowicz, A., 368, *386*
Mandelstam, S., 388, 400, 423, *429*
Manelli, I., 197, *217*
Margenau, H., 22, *23*
Marin, P. C., 349, *386*
Marshak, R., 342, *387*
Martin, R., 228, *253*
Massam, T., 348, *386*
Massey, H. S. W., 75, *110*
Matthews, P. T., 23, *23*, 256, 290, 292, *293*, 325, *326*, 382, *387*
Mefkens, B. M. K., 472, *482*
Melkonian, E., 113, *126*
Merrison, A. W., 181, *216*
Meshkov, S., 280, *293*
Messiah, A., 23, *23*
Meyer, J., 238, *254*
Michael, C., 123, *126*
Michel, L., *387*
Miller, D. H., 410, *430*
Miller, P. D., 174, *176*
Miller, R. H., 335, *387*
Miyazawa, H., 403, *429*
Moffet, R., 376, *387*
Montagne, J. H., 336, 348, *387*
Moorhouse, R. G., 234, *253*
Morales, A., 269, *293*
Morrison, D. R. O., 197, *216*
Morse, P. M., 419, *430*
Mott, N. F., 75, *110*
Muirhead, H., 98, *110*, 303, *326*
Muller, T., 348, *386*, 471, *483*
Munday, G. L., 335, *385*
Murphy, G. M., 22, *23*
Murphy, P. G., 156, *175*
Murray, G., 336, 348, *387*

N

Nadelhaft, I., 362, *386*
Nagle, D. E., 228, *253*
Nambu, Y., 373, *387*, 450, *462*
Nathans, R., 174, *176*
Ne'eman, Y., 255, 268, 282, *293*
Nilsson, J., 175, *176*
Nishijima, K., 139, *143*
Norton, R. E., 404, *429*
Novey, T. B., 164, *175*, 339, *386*
Nygren, D., 480, *482*

O

Oakes, R. J., 461, *462*
Occhialini, G. P. S., 177, *216*
O'Donnell, P. J., 234, *253*
Oehme, R., 171, *176*, 403, *429*
O'Halloran, T., 202, *216*
Okubo, S., 282, *293*
Okun, L. B., 333, *387*, 467, *483*
Olive, D. I., 393, 395, *429*
Olsen, S., 164, *175*
O'Neill, G. K., 322, *326*
O'Neill, P. L., 156, *175*
Orear, J., 89, *110*, 214, 215, *216*, 406, *430*
Overseth, O., 164, *176*

P

Pais, A., 138, 139, *143*, 190, 191, *216*
Pan, Y. L., 247, *253*
Panofsky, W. K. H., 181, *216*, 451, 453, *462*
Particle Data Group., 155, *176*, 177, 187, 193, 195, 196, 200, 215, *216*, 277, 287, *293*, 336, 357, 366, 369, 378, 381, *387*, 428, *430*, 509, *511*
Pauli, W., 170, *176*, 296, *326*
Perkins, D. H., 358, 384, *387*, 466, *482*
Pestov, Ju. N., 195, *216*, 452, *461*
Pevsner, A., 141, *143*, 213, *216*
Phillips, E. G., 484, *487*
Phillips, R. J. N., 426, *430*
Piccioni, O., 191, *216*
Pickup, E., 199, *217*
Pilkuhn, H., 396, 398, 402, 414, 417, *430*
Piroue, P. A., 466, *482*
Pjerrou, G., M., 251, *254*
Plano, R., 187, *217*
Poirier, J. A., 459, *461*
Polkinghorne, J. C., 156, 175, *175*, 393, 395, *429*
Pomeranchuk, I. Ya., 407, *430*
Pondrom, L., 164, *175*, 351, *386*

Pontecorvo, B., 467, *483*
Powell, C. F., 177, *216*
Powell, W. M., 247, *253*
Primakoff, H., 345, *387*
Prodell, A., 178, 187, *217*
Pursey, D. L., 156, 175, *175*

R

Radicati, L. A., 290, 291, *293*
Radojicic, D., 504, *504*
Ramsey, N. F., 174, *176*
Rarita, W., 426, *430*
Reasbeck, P., 335, *385*
Reines, F., 329, 345, *385*, *387*
Regge, T., 420, 427, 428, *430*
Rich, A., 317, *326*
Richman., C., 178, *216*
Richter, B., 322, *326*, 444, *462*
Rieseberg, H., 167, *175*
Ringo, G. R., 164, *175*, 339, *386*
Risk, W. S., 200, *216*
Ritson, D., 141, *143*
Roberts, A., 180, *216*
Roberts, W. K., 167, *176*
Robinson, D. K., 199, *217*
Robiscoe, R. T., 318, *326*
Rochester, G. D., 136, 137, *143*
Roe, B. P., 199, *216*
Rohrlich, F., 294, 308, 317, *326*
Rollnik, H., 444, *462*
Roman, P., 145, 146, *176*
Roos, M., 370, *386*
Roper, L. D., 234, *254*
Rosen, J., 376, *387*
Rosen, S. P., 345, *387*
Rosenfeld, A. H., 210, 211, 212, *216*, *217*, 375, 376, *387*, 504, *504*
Ross, M., 243, *254*, 457, *462*
Roth, P. F., 473, *482*
Roth, R., 164, *176*
Rubbia, C., 363, 367, 368, 370, *387*
Russ, J. S., 473, *482*

S

Saal, H., 480, *482*
Sachs, R. G., 156, *176*, 450, *462*, 482, *483*
Sakata, S., 255, 268, 290, *293*
Sakita, B., 290, *293*
Sakurai, J. J., 461, *462*

Salam, A., 325, *326*, 339, *387*
Salant, E. O., 199, *217*
Samaranayake, V. K., 403, *430*
Samios, N. P., 178, 187, *216*
Sands, M., 23, *23*
Satchler, G. R., 38, 39, 40, 48, *66*, 510, *511*
Schaille, H., 404, 426, *429*
Schlein, P., 203, *217*, 251, *254*
Schlupman, K., 167, *175*
Schneider, H., 164, *175*
Schopper, H. F., 156, *175*, 448, *462*
Schwartz, M., 178, 187, *216*
Schweber, S. S., 9, 10, *23*, 294, 303, 312, 324, *326*
Segré, E., 100, 112, 119, *110*, *126*
Selleri, F., 414, *429*
Sens, J. C., 361, *387*
Shafer, J. B., 251, *254*
Shapalin, E. P., 385, *387*
Sharp, D., 456, *462*
Sharp, D. H., 284, *293*
Shaw, G. L., 243, *254*
Sherr, R., 335, *387*
Shirkov, D. V., 294, 308, *326*
Shortley, G. H., 38, *66*, 510, *511*
Shreve, D. C., 164, *175*
Shull, C. G., 174, *176*
Shutt, R. P., 138, *143*
Sidorov, V. A., 195, *216*, 452, *461*
Siemann, R. H., 322, *226*
Sinclair, D., 199, *216*
Sirlin, A., 348, *386*
Skrinsky, A. N., 195, *216*, 452, *461*
Slater, W. E., 203, *217*, 240, 251, *254*
Smart, W. M., 277, *293*
Smith, J. H., 472, *482*
Smith, L. T., 203, *217*
Snow, G. A., 180, *216*, 280, *293*
Sodickson, L., 197, *217*
Soergel, V., 167, *175*, 348, *386*
Solmitz, F. T., 154, 156, *175*
Sonderegger, P., 404, 426, *429*
Soto, M. F., 318, *227*
Steck, B., 338, *385*
Steffen, R. M., 440, *461*
Steinberger, J., 178, 180, 187, *216*, *217*, 375, *387*, 472, 480, *482*, *483*
Stenger, V. J., 240, *254*
Stevenson, M. L., 142, *142*, 154, 156, *175*, 210, 211, 212, *216*, *217*, 220, *253*, 504, *504*
Stierlin, U., 349, *386*

Author Index

Stirling, A. V., 123, *125*, 234, *252*
Stodolsky, L., 457, *462*
Stork, D. H., 203, *217*, 240, 251, *254*
Storm, D. W., 164, *175*
Strauss, R., 404, 426, *429*
Strelzoff, A., 351, *386*
Sucher, J., 180, *216*
Sudarshan, G., 342, *387*
Sumi, Y., 407, *430*
Sun, C. R., 167, *175*
Sunderland, J., 480, *482*
Sunyar, A. W., 338, *386*, *387*

T

Taylor, S., 214, 215, *216*
Telegdi, V. L., 164, *175*, 339, *386*
Teller, E., 334, *387*
Thirring, W., 358, *386*
Thorndike, A. M., 138, *143*
Ticho, H. K., 142, *142*, *143*, 203, *217*, 220, 240, 251, *254*
Ting, S. C. C., 455, 456, *462*
Tinlot, J., 362, *386*
Titchmarsh, E. C., 484, *487*
Tollestrup, A. V., 164, *176*
Tornabene, S., 181, *216*
Trefil, J., 458, *461*
Treiman, S. B., 221, *254*, 292, *293*, 360, *387*, 413, *430*
Triebwasser, S., 318, *327*
Tripp, R. D., 235, 244, 246, *254*, 277, *293*
Tuan, S. F., 243, *253*
Turkot, F., 471, 472, *482*
Turlay, R., 200, *216*, 471, 472, *482*

V

Van der Velde, J. C., 199, *216*
Van Hove, L., 427, *430*
Veji, L., 370, *386*
Vernon, W., 473, *482*
Villet, G., 123, *125*, 234, *253*
Volkov, D. V., 427, *430*

W

Wagner, J. J., 345, *386*
Wagner, W. G., 456, *462*
Wahlig, M., 197, *217*
Walker, R. L., 164, *176*, 444, *462*

Watson, K. M., 112, 116, 162, 164, *126*, *176*, 244, 246, *253*, *254*, 432, *462*, 469, *483*
Webb, F. H., 141, *143*
Weinberg, S., 357, *387*, 507, *508*
Weinrich, M., 152, *175*
Weinstein, R., 322, *327*
Weisberger, W. I., 360, *387*
Weisskopf, V. F., 68, 122, *110*, 296, *326*, 333, *386*, 459, *461*
Weitkamp, W. G., 164, *175*
Wenzel, W. A. 362, 378, *386*
Werle, J., 48, 65, *66*
Wetherell, A. M., 181, *216*
Weyl, H., 256, *293*
Wheeler, P. C., 200, *216*
White, H. S., 247, *253*
White, R. E., 336, 348, *387*
Whitehead, M. N., 178, *216*
Whittemore, W. L., 138, *143*
Wick, G. C., 22, *23*, 156, *176*
Wightman, A. S., 9, *23*
Wigner, E. P., 9, *23*, 48, 116, *23*, *126*, 256, 289, *293*
Wilcox, H. A., 178, *216*
Wilkinson, D. H., 156, *176*
Williams, P. G., 256, *293*
Williams, W. S. C., 494, *498*
Wilson, R., 180, *216*
Wilson, R. R., 448, *462*
Winther, A., 338, *385*
Wolf, G., 417, *430*
Wolfenstein, L., 474, *482*, *483*
Woolcock, W. S., 404, *430*
Wormald, J. R., 181, *216*
Wright, S. C., 164, 234, *175*
Wu, C. S., 152, 153, 173, 175, *175*, *176*, 336, 337, 357, *387*
Wu, T. T., 452, *462*, 476, *483*
Wyman, M. E., 345, *386*

X

Xuong, N., 292, *293*

Y

Yang, C. N., 152, 171, *175*, 182, *217*, 220, *254*, 339, 357, 382, 383, *387*, 413, *430*, 452, *462*, 476, *483*
Yodh, G. B., 181, *216*, 280, *293*

Young, K. K., 164, *176*
Yukawa, H., 177, *217*

Z

Zachariasen, F., 422, 423, *429*, 454, *462*
Zacher, A. R., 167, *175*
Zemach, C., 203, 206, 207, 209, 210, *217*
Zichichi, A., 348, *386*
Zorn, G., 471, 472, *482*
Zumino, B., 171, *176*, 454, *462*
Zweig, G., 288, *293*

Subject Index

A

Absorption model, 415
Absorption parameter, 72
Adair spin analysis, 219
Adler–Weisberger sum rule, 360–361
Alignment, 55
Analytic functions, 484
Analyticity, 393–400
Angular momentum, 24–66
    barrier, 50, 111–112, 122, 138, 417, 443
    commuting operators of, 29
    conservation, 19, 27–28, 49
    eigenfunctions, 40–42
    eigenvalues, 25, 30–32
    of electromagnetic field, 435–441
    matrix elements, 32–33, 50–54, 509, 510
    operators, 25, 28, 29, 50–51
    orbital, 24, 25, 40–41
    partial waves and, 67–68
    rotations and, 26–28, 43–50
    spin, 28–29, *see also* Spin
    tensor, 50–54
    total, 29–30
    vector addition, 34–40
Anomalous moments, 317, 355–356, 372–373, 432, 446–447
Anomalous threshold, 404
Annihilation operators, 299, 302, 303, 306
Antibaryons, 137
Anticommuting operators, 305
Antineutrino, 328–329, *see also* Weak interactions
Antiparticle, 165
Antisymmetry, 127–129
Associated production, 138
Asymptopia, 406

Axial vector, 14–15, 150
    interaction, 323, *see also* Weak interactions
    particle, 21, 145
Axial vector current, 507, *see also* Weak interactions
    divergence of, 507–508

B

Baryon, 137, 218–253
    exchange, 410
    number, conservation of, 140, 146, 301, 410
    parity, 147, 220–222
    $SU(3)$ and, 268–269
Baryon resonances, 218, 222–226, 247–252, 269, 289, 291
    electroproduction, 453
    parity, 250–251
    spin, 247–250
    $SU(3)$ and, 276–279
Beta decay, 328–339, *see also* Weak interactions
    asymmetry, 152–153, 173, 337, 344
    charge conjugation and, 173, 330–332
    classification, 332–334
    conserved vector current and, 356
    density of states, 335, 501
    double, 345
    electron spectrum, 335, 337
    favored and unfavored transitions, 336
    Fermi transitions, 334
    Fierz interference, 335, 337
    forbidden, 333
    Gamow–Teller transitions, 334
    hyperon, 362–363, 371–374
    interactions, 331
    K-capture, 329

Kurie plot, 335
  lepton conservation and, 345
  parity nonconservation and, 152–153, 173, 331, 332, 337–339, 344
  pion, 356–357
  polarization, 338
  selection rules, 334
  time reversal invariance in, 332, 339
  V–A interactions, 339
  weak magnetism and, 356
Born approximation, 160, 401
Born term model, 414–416
Bose–Einstein statistics, 128
Boson, 128, 177–215, 268, 291
  symmetry, 128
Bra–ket notation, 1–5
Branch point and cut, 394, 404, 484
Breit–Wigner formula, 118–124
Bremsstrahlung, 313
Byers–Fenster method, 248–252

C

Cabibbo angle, 369–374
Cascade hyperons Ξ, 138, 218
  decay, 363, 372, 374, 381
  parity, 220–221
  spin, 220
Casimir operators, 267–268
Cauchy principal value, 486–487
Cauchy–Reimann relations, 484
Cauchy theorem, 485
Causality, 393
Center-of-mass system, 22–23, 490–491
Channel, 93–94, 390
Charge conjugation, 165–167, 189, *see also* Weak interactions
  eigenvalue, 165–167
  invariance under, 166–167, 393
  selection rules, 169
  violation, 166–167, 332, 338
Charge conservation, 140, 301, 410
Charge independence, 131, 134–135, *see also* Isotopic spin
Charge parity, *see* Charge conjugation
Charge renormalization, 316
Charge symmetry, 131
Chew–Frautschi plot, 427
Chew–Low plot, 410
Chew–Low theory, 116, 232–233

Chirality invariance, 342
Clebsch–Gordan coefficients, 35–40, 271–274
  recursion relation, 37
  symmetry relation, 509
Clebsch–Gordan series, 263–266, 271, *see also* Angular momentum, vector addition of
Coherence, 70
Coleman–Glashow relation, 285
Commutation relations, *see* appropriate operators
Complex variables, 484
Conservation, 20, *see also* appropriate observable
Conserved vector current, 301, 353–361, 431, 446, 506–507, *see also* Weak interactions, Electromagnetic interactions
Constants of motion, 29
Continuity equation, 295, 301, 506
Contravariant notation, 15–16, 296, 297
Coordinate transformation, 12
Coulomb scattering, 75, 89, 243, 310–311, 316–317
Coupling constant, *see* Strong interactions, Weak interactions, Electromagnetic interactions, Vector dominance
Covariant notation, 15–16, 296–297
CP, 164, 174, 189–190, 463–482
  conservation of, 174, 344, 374, 463–467
  two-component neutrino and, 343–344
  violation of, 374, 472–482
CPT theorem, 170–174, 393, 481
Creation operators, 299, 302, 303, 306
Crossing symmetry, 392, 393
Current, 301, 507, *see also* Weak Interactions
  conservation, 301, 431, 507
  electromagnetic, 431
  first and second class, 357, 507–508
  matrix elements of, 505–508

D

D/F ratio, 276, 291, 371–374
Dalitz plot, 203–211, 494–495, 502–504
Decay, *see* particle of interest, Beta decay, Weak interactions

Subject Index

of mixed states, 61–65
of pure states, 49–50
transition rate for, 98–99
Decuplet, 269, 285, 291
Density matrix, 56–61
   one-particle exchange model and, 412
   rotation of, 66
Density of states, 98–99, *see also* Phase space
Detailed balance, 108–110
Diffraction scattering, 73, 406, 424
Dirac equation, 15, 295, 303–308, 505–508
Dirac form factors, 447
Dispersion relations, 395–405
Divergences, 316–318
Duality, 427

E

Effective range formula, 112–116, 119, 233, 240–243
Eigenfunctions, 40
Eigenstates, 5
Eigenvalues, 4, 5
Eigenvectors, 4, 5
Eightfold way, 268–270, *see also* $SU(3)$
Elasticity, 120
Electric charge, 296, 325, 431–432, 437
Electric dipole moment, 174
Electromagnetic field, 302–303
   angular momentum, 435–441
   multipole, 439–441
   polarization, 436–437, 440
   vector potential, 295, 302
Electromagnetic form factors, 444–453
   dipole fit, 451
   isotopic scalar and vector, 448–449
   neutron, 448
   pion, 452, 454
   proton, 448–449
   vector dominance model and, 454, 460
Electromagnetic interaction, 310–311, 431–435
   angular distribution and, 440, 442–443
   charge conjugation invariance of, 166–167
   coupling constant, 325
   CP violation and, 481
   current, 431–432, 445

current conservation, 446, 506
decays, 168–170, 213–214, 328, 432, 455–457
of hadrons, 431–461
minimal, 432
parity conservation and, 156
radiative corrections and, 314–318, 435
selection rules, 168–170, 441
$SU(3)$ properties, 280–281, 285–287
time reversal invariance of, 164
Electromagnetic mass splitting, 131, 284–285, 435
Electron, 295, 306
   beta decay and, 328–329
   charge renormalization, 316
   magnetic moment, 317
   mass renormalization, 315–316
   propagator, 314, 321–322
   scattering, 316–317, 444–447, 453
Electron–muon universality, 319, 349–350, 351, 363, 367–368, 453
Electron–positron colliding beams, 455–456
Electron–positron field, 306–308
Electroproduction, 453
Energy, 10
   conservation, 19–20, 300–301
Energy–momentum four-vector, 16, 301, 488
Eta meson, 167, 169, 193, 212–214, 268, 283, 284
Euler angles, 43
Exclusion principle, 127, 306–307
Expectation value, 5

F

f-meson, 193, 196–200, 215
Fermi–Dirac statistics, 127
Fermi transitions, *see* Weak interactions
Fermions, 127
   antisymmetry, 91–93, 128–129
   parity of, 21, 146–147
Feynman diagrams, 311–314, 432
Field theory, 294–326
Field variables, 297
Fierz interference, *see* Weak interactions
Fine structure constant, 318
Form factors, *see* Electromagnetic form factors, Weak interactions, One par-

ticle exchange, Quantum electrodynamics
Formation of resonances, 194–195, 222–223, 441
Four-vectors, 14, 488–490
Four-velocity, 489

G

G-parity, 167–170
  selection rules, 169–170
Gamma matrices, 304
Gauge transformation, 257, 301, 431
Gell-Mann–Okubo mass formula, 282–283
Generators, 19–20, 257
  of isotopic spin, 133
  of rotations, 26–28, 258–259
  of $SU(3)$, 259–260
Goldberger–Treiman relation, *see* Weak interactions
Groups, 256–268
  fundamental representation of, 260–261
  order of, 258
  products of representations, 263–266
  rank of, 258
  representations of, 257, 258–268, 270–275
  unitary and special unitary, 257
  weight diagrams, 260–261

H

Hadrons, 255, 268–270
Hamiltonian operator, 10–11, 100, 300
  commutation properties, 19–20, 149–150
  time reversal and, 157, 160
Heisenberg equation of motion, 10, 299
Helicity, 341
Hermitian analyticity, 398
Hilbert space, 2
Hypercharge, 136–140, 238–239, 267–268, *see also* particle of interest
Hyperfragment, 188
Hyperons, 137, *see also* Cascade, Sigma hyperons, Lambda hyperons
  decay, *see* Weak interactions, baryon decay
  mass, 282–283
  parity, 220–222

I

Impact parameter, 74
Induced interactions, *see* Weak interactions
Inertial frame, 12
Infrared divergence, 317
Integral theorems, 485
Interactions, *see* Electromagnetic interactions, Strong interactions, Weak interactions
Invariance, 393, *see also* specific conservation law
Isoscalar factors, 272–273
Isotopic spin, 130–136, 255, 258–259, 268, 324, *see also* specific particle or particle systems
  conservation of, 133, 142, 310, 393
  eigenfunctions, 131–132
  eigenvalues, 131
  generators, 258–259
  multiplets, 131, 146–147, 255, 268
  operators, 131
  rotations, 133, 167–168
  selection rules, 169–170, 364–366, 375–382
  strangeness and, 139–140
  $SU(2)$ and, 258–259
  vector addition of, 135
  weak interactions and, *see* Weak interactions, $\Delta t = \frac{1}{2}$ rule

K

Kaon(s), 138–142, 187–193, *see also* Kaon–nucleon scattering
  decay
    CP invariance and, 167, 463, 472–474
    $\Delta S = \Delta Q$ rule and, 363–364, 466–467, 477
    $\Delta t = \frac{1}{2}$ rule and, 376–377
    leptonic, 187, 189, 363–369
    nonleptonic, 187, 189, 374–377
    parity nonconservation and, 151–152, 362–363, 374

Subject Index

weak interactions and, 369–370, 374, 377
hypercharge, 136–142, 268
isotopic spin, 187
mass, 187, 283
mean life, 187, 190, 466
neutral, 188–193, 463–482
neutral mass difference, 466–467, 470–471
neutral regeneration, 468–471
neutral time development, 463–467, 474–479
parity, 187, 188
photoproduction, 441
resonances, 200–202
spin, 187–188
$SU(3)$ and, 268
strangeness, 136–142
tau–theta puzzle, 151–152
Kaon–nucleon scattering, 239–246
isotopic spin, 223, 238–239
resonances, 222–224, 243–246
regeneration and, 468
S-wave, 115–116, 239–243

L

Laboratory system, 22, 490–491
Lagrangian formalism, 275–276, 298–302, 309–311
Lambda hyperon, 137–140
decay, 153–156, 363, 372–378, see also Weak interactions
mass, 282–285
parity, 220
spin, 219–220
strangeness, 138–139
$SU(3)$ and, 268
Laurent's theorem, 485–486
Legendre polynomials, 40–41
Leptonic decays, see Weak interactions
Leptons, 319, 329, 346–347
conservation of, 329, 345–347
Linear operators, 1–5
Lorentz condition, 303
Lorentz covariance, 12
Lorentz invariance, 393, see also Weak interactions
Lorentz transformations, 12–17, 300–301, 488

M

Magnetic moment, 285–287, 291–292
Magnetic multipole transitions, 439–441
Majorana neutrino theory, 345
Mandelstam representation, 400
Mandelstam variables, 197–198, 388–391, 495
Mass, see particle of interest, $SU(3)$
Mass renormalization, 315–316
Mass shell, 321
Matrix element, 3, 32–33
Maxwell's equations, 12
Meson(s), see Pions, Kaons
Meson resonances, 193–215, 268
Michel parameter, 348
Minami ambiguity, 87–90, 236–237, 245
Missing mass technique, 195
Mixed states, 54–59
Momentum transfer, 321–322, 389, 448
Multipole transitions, 439–441
Muon, 349, 361–362, see also Weak interactions
capture, 349, 360–362
electron universality, 319, 349–350, 363, 367–368, 453
magnetic moment, 323
neutrino, 329, 346
parity violation in decay, 152, 347
scattering, 453

N

Neutrino, 328–329, see also Weak interactions
absorption, 330, 346
beta decay and, 328
conservation of leptons and, 329, 346–347
electron, 329
helicity, 341–342
lepton number, 329, 346
Majorana, 345
mass, 329
spin, 329
two-component theory, 339–344
V–A theory and, 343
Neutron, 130, 131
decay, 328, 353–354, 372
electronic dipole moment, 174

526 Subject Index

electromagnetic form factor, 448–449
isotopic spin, 131
magnetic moment, 286–287, 291–292
mass, 285
parity, 21, 146–147, 220
spin, 218–219
$SU(3)$ and, 218–219, 268
time reversal invariance and, 174, 339
Neutron–proton scattering, 112–115, 133–134
Nonleptonic decays, see Weak interactions
Normal threshold, 394
Normalization, 2, 4
Nucleons, 129–131, see also Neutron, Proton

O

Observables, 5–6
Octet, 268, 288, 291
Omega hyperon, 252–253, 269
Omega meson, 193, 212, 440, 450, 456–457
  mixing with phi meson, 283–284, 291
  photoproduction, 407, 458–459
One particle exchange, 408–417
  absorption and, 415–416
  Born term model, 414–415
  form factors and, 414
  K-matrix model, 416
Operators, 2–5
  adjoint, 3
  generators of, 19, 257, 259–260
  Hermitian, 3, 5
  linear, 2
  unitary, 3, 8–9, 256–258
Optical theorem, 74, 103, 396, 402, 423, 458–459
Orthogonal property, 4
Orthonormal property, 4

P

Pair production, 313, 321–323
Panofsky ratio, 181
Parity, 20–22, 144–147, 270
  baryon, 146–147, 220, 247–251
  conservation, 21, 147–151, 156, 393
  electromagnetic transitions and, 439–441

fermion, 21–22, 146–147
intrinsic, 21–22, 145
invariance, 20–21, 145–146
nonconservation, 14, 20–21, 151–156, see also Weak interactions
of spherical harmonics, 41
transformation, 14, 20–21, 144–146, 181–183
Partially conserved axial vector current, 358
Partial wave analysis, 67–80, 230–238
  ambiguities, 87–90
  amplitudes, 70
  dispersion relations and, 404
  of kaon–nucleon scattering, 244–246
  phase shifts, 69, 79
  of pion-nucleon scattering, 75, 79–80, 230–238
  Wigner condition and, 116–118
Pauli exclusion principle, 127, 306–317
Pauli form factors, 447–449
Pauli–Lüders theorem, 170–174
Pauli spin matrices, 42–43, 258
Peripheral model, 197–200, 409, see also One particle exchange
Perturbation theory, 100, 107, 310–312, 325
Phase space, 98–100, 111–112, 499–503
Phi meson, 193, 202–203, 215, 283–284, 291, 407, 453, 457–461
Photoelectric effect, 313
Photon, 432–435
  exchange, 434
  polarization, 185–187, 436–437
  propagator, 314, 320
  vector mesons and, 453–461
  virtual, 432–435, 437
Photoproduction, 280–281, 330–331, 441–444, 459
  vector dominance and, 457–459
Pion, 177–187, 324, see also Pion–nucleon interaction, Pion–nucleon scattering
  absorption, 180–182
  charge conjugation parity, 165–166
  Dalitz decay, 178
  decay of charged, 178, 346, 356–357, 359
  of neutral, 178
  exchange, 408–410

Subject Index

G-parity, 167–169
isotopic spin, 135
mass, 177–178, 283
mean life, 178
parity, 21, 179, 181–182, 184–187
photoproduction, 111, 441–444
spin, 178, 179–180
$SU(3)$ and, 268
Pion–nucleon interaction, 182, 324, *see also* Pion-nucleon scattering
   charge independence of, 226–229, 324, 404
   Chew–Low theory of, 116, 232–233
   coupling constant, 233, 275–276, 325–326, 403–404, 408–409
   isotopic spin, 226–228
Pion–nucleon scattering, 226–238
   charge exchange, 227, 230, 404
   constants of motion, 75
   dispersion relations, 400–405
   Minami ambiguity, 236–237
   nucleon polarization in, 81, 235
   partial wave analysis of, 75, 79–80, 230–238
   radiative scattering, 227
   resonances in, 232, 237–238
   scattering lengths, 233, 403–404
Pion–pion scattering, 409, 424–426
Plane wave, 68
Polarization, 54–56, *see also* Weak interactions
   alignment, 55
   density matrix applied to, 58–61
   double scattering and, 84
   light, 7, 436
   neutrino, 341
   photon, 436
   in photoproduction, 442–443
   in spin-0–spin-½ scattering, 60, 80–87
   spin-½ particles, 60–61
   vector, 54, 497–498
Poles, 394–395, 408, 484–485
Pomeranchuk theorem, 407–408, 423
Pomeron, 423–425
Positron, 306–308
Positronium, 166, 322
Poynting vector, 439–440
Primakoff effect, 434
Primitive divergence, 315–316
Principal value, 486

Probability, 7
   amplitude, 6–7
Production of resonances, 194–195, 224–226, 247–252
Propagators, 314, 321–322
Proton, 129–133, 218–219
   beta decay, 328
   electromagnetic form factors, 448–452
   isotopic spin, 131
   magnetic moment, 286–287
   mass, 282, 287
   parity, 21, 146–147, 218–219
   spin, 218–219
Proton–proton scattering, 133–134
Pseudoscalar, 14, 17, 308
   interaction, 323–324, 401–402, *see also* Weak interactions
   induced, *see* Weak interactions
Pseudotensor, 15
Pseudovector, *see* Axial vector

Q

Quantization, 294, 296
Quantum electrodynamics, 310–318
Quantum mechanics, 5–9
   postulates of, 5
Quarks, 288–289

R

Radiative corrections, 314–318, 435
Rationalized units, 325–326
Reactions, 67, 96–98
   cross section, 72, 79, 93–95, 116, 119–120
Reciprocity, 103–106, 393
Regeneration, 468–471
Regge poles, 418–427
   applications, 426–427
   conspiracy, 426–427
   cuts and, 426–427
   daughters, 427
   evasion, 427
   pomeron, 423–424
   recurrences, 428–429
   rho, 425–426
   signature, 422
   vacuum, 425
Relativity, 11–12
Renormalization, 315–318

Representations, 258–261, 270–275, *see also* Groups, $SU(3)$
  matrix, 42–43, 258–259
Residue, 485–486
Resonances, 118–124
  Breit–Wigner formula and, 119–121, 421
  Jackson formula and, 124
Rho meson, 193, 195–200, 268
  charge, 196
  decay, 322, 455–456
  mass, 193
  photoproduction, 407, 457–459
  production, 195–199, 409–410, 414
  Regge pole, 425–426
  vector dominance and, 454–460
  width, 193
Rotations, 26–28, 43–48, 183
  of density matrix, 66
  generators of, 28
  invariance under, 28, 48
  matrix elements of, 44–48, 510
  of spin-$\frac{1}{2}$ states, 47–48

S

S-matrix, *see* Scattering matrix
Sachs' form factors, 449–452
Scalar, 14, 275
  interactions, 323, *see also* Weak interactions
    induced, 367, 507
  invariant, 15, 488
  particle, 21, 297–302, 323
  product, 16
Scattering, *see* Partial wave analysis, particle of interest
  alpha-alpha, 92
  amplitude, 70
  diffraction, 73, 406–407
  effective range, 112–114
  elastic, 72–73, 224, 406
  of identical particles, 91–93
  inelastic, 224
  lengths, 112–114, 233, 241, 402
  partial wave analysis, 67–75
  S-wave, 112–116
  from virtual particles, 408–410
Scattering cross section, *see* Partial wave analysis, Regge poles
  Breit–Wigner, 118–121
  differential, 69, 86, 92, 95, 102–103
  Lorentz transformation of, 495–497
  total, 72, 80, 103, 114–115, 120, 223
Scattering matrix, 93–96, 98–103
Schrödinger equation, 10–11, 294
Schrödinger wavefunction, 7, 11
Second class current, 357, 507–508
Selection rules, *see also* Conservation, appropriate observable
  in beta decay, 333–334, 376
  $\Delta S = \Delta Q$, 363–364, 466, 477–478
  $\Delta S = 1$, 363
  $\Delta t = \frac{1}{2}$, 364–366, 376–382
  in electromagnetic transitions, 168–170, 435, 437, 440
  isotopic spin, 169, 435
  strangeness, 138, 363–364
Self conjugate, 165–166
Shift operators, 261–262
Sigma hyperons, 138–140
  charge, 139
  decay, 363, 372, 379–381
  isotopic spin, 139
  magnetic moment, 286–287
  mass, 282, 285
  parity, 220, 221, 245–246
  spin, 220
Signature, 422
Singularities, 484–485
  scattering amplitudes, 394–395, 404–405
Sommerfeld–Watson transformation, 418–419
Space–time coordinates, 13
Spherical harmonics, 41, 129–130
Spin, 28–29, 303–304, *see also* particles of interest
  eigenvalues, 28
  matrix representation, 42–43, 258–259
  operators, 29
  states, 71, 92–93, 129
Spin–orbit force, 29, 75
Spinors, 15, 150–151, 161, 305
State vector, 1–5
States, 6
  base, 8
  mixed, 54–59
  pure, 15, 49, 54

Subject Index

Strange particles, 136–142, *see also* particles of interest
 decay of, 138–139, 362–382
Strangeness, 136–142
 conservation of, 138–140
 nonconservation of, 141, 363–364
Strong interactions, 323–326, 388–429
 charge conjugation invariance of, 167
 charge independence of, 131, 324, 393
 coupling constant, 275–276, 323–326, 401–402, 408
 effects in electromagnetism, 432–435
 parity conservation in, 156
 $SU(3)$ and, 275–278
 time reversal invariance of, 164
Superposition principle, 2, 5–7
Superselection rules, 146
Superweak theory, 482
$SU(2)$, 258–259, 262
$SU(3)$, 258–287
 baryon resonance decays and, 276–277
 broken, 281–287
 Clebsch–Gordon coefficients for, 271–274
 commutation relations, 258
 decuplet, 269, 276, 291
 F and D coupling, 276, 292, 371–374
 isoscalar factors, 272
 magnetic moments and, 285–287
 mass and, 282–285
 matrices, 258–259
 matrix elements of, 270
 octets, 268, 288, 291
 omega baryon, 269
 operators, 257, 260
 phi–omega mixing, 283–284, 291
 photoproduction and, 444
 Regge recurrences and, 429
 representations of, 259–260, 270–275
 scattering amplitudes and, 277–280
 strong interactions and, 275–276
 structure constants of, 258–260
 U-spin, 278–281, 444
 V-spin, 281
 vector dominance model and, 461
 weight diagrams, 260–269
$SU(4)$, 289
$SU(6)$, 290–292, 461
Symmetry, 127–129, *see also* Conservation, Invariance

T

Tensor, 15, 205
 angular momentum, 307
 energy, 300
 force, 29, 91
 interaction, 331, 367
 operators, 50–54, 274
Time reversal, 103–105, 156–165
 invariance, 164, 331, 339
 reciprocity and, 105
 tests of, 164
 violation, 481
Transformations, 17–20, *see also* specific transformation
 antiunitary, 158
 similarity, 18
 unitary, 8–9, 256–257
Transition amplitude, 241, 392–395
Transition matrix, 98–103
Transition rate, 98
Treiman–Yang test, 413

U

U-Spin, *see* $SU(3)$
Unitarity, 124–125, 393–394
Unitary circle, 73–74, 122–125
Unitary group, 257
Unitary symmetry, *see* $SU(n)$

V

V-spin *see* $SU(3)$
Vacuum polarization, 316
Vector, 1–5, 14
 adjoint, 2
 base, 6
 covariant and contravariant, 15, 488
 current, *see* Conserved vector current
 field, 302–303
 four-, 14
 particle, 21, 146
Vector dominance, 453–461
 coupling constant, 453–454, 460
 form factors and, 450, 454
 particle decays and, 454–457
 photoproduction and, 457–460
 $SU(3)$ and, 460–461
Vector mesons, 146, 215, 268, 283–284,

288, 291, *see also* Rho mesons, Omega mesons, Phi mesons
  decays of, 454–457
  photoproduction of, 457–458
Vector potential, 302, 439
Vector space, 1–2, 256
Vector spherical harmonics, 437–439
Vertex, 311, 314, 319–320
Virtual particles, 320, 432

W

Watson theorem, 162–164
Wave equations, *see* Dirac equation, Klein–Gordon equation, Schrödinger Equations
Wave functions, 7
Weak interactions, 328–385, *see also* Beta decay
  Adler–Weisberger sum rule, 360–361
  axial vector current, 338, 359, 370
  axial vector interaction, 331, 334, 338–339, 343, 366
  baryon decay, 363, 368–369, 371–374
  Cabibbo angle, 369–374
  charge conjugation and, 171, 332, 338, 343–344, 374
  conserved vector current and, 353–361
  coupling constant, 336, 371–374
  CP conservation, 344, 481–482
  decay asymmetry in, 337, 344, 348
  $\Delta S = \Delta Q$ rule, 363–364, 466–467, 477
  $\Delta t = \frac{1}{2}$ rule, 364–366, 375–382
  F to D ratio, 371, 374
  favored and unfavored transition, 336
  Fermi transitions, 334, 335–336
  Fierz interference, 335, 337
  form factors, 358, 366–367, 384
  Gamow–Teller transitions, 334, 335, 336
  Goldberger–Treiman relation, 360
  intermediate vector boson, 382–384
  K-capture, 329
  kaon decays, 363, 365–368, 374–377, 467, 472, 481–482
  lepton conservation, 329, 345, 346–347
  leptonic decays, 362–374
  Lorentz invariance, 329–330
  muon capture, 349, 360–362
  muon decays, 347–349
  neutral currents, 353
  neutrinos, 328–329, 343, 345–347, 384–385
  nonleptonic decays, 374–382
  parity nonconservation in, 151–156, 337–339, 343–345, 346–348, 362, 378–381
  partially conserved axial vector current, 358–359
  pion decay, 346, 356–357, 359
  polarization effects, 338, 344, 368
  pseudoscalar interaction, 331, 334, 351
    induced, 359–360, 362, 374
  scalar interaction, 331, 334, 338, 367
  selection rules, 334, 355, *see also* $\Delta S = \Delta Q$, $\Delta t = \frac{1}{2}$ rule
  tensor interaction, 331–332, 334, 335, 338
  time reversal invariance in, 164, 332, 339, 368
  time scale, 328
  universal, 352–353
  V–A interaction, 339, 343, 366
  vector interaction, 331, 334, 367, 370
  weak magnetism, 356, 373
Weak magnetism, *see* Weak interactions
Weight diagrams, *see* SU(3)
Wigner condition, 116–118
Wigner–Eckart theorem, 53, 274, 369
Wigner 3-j symbol, 39

X

X^0-meson, 215, 268, 284

Y

Yang ambiguity, 88–89
Yukawa meson, 177

PURE AND APPLIED PHYSICS

A Series of Monographs and Textbooks

Consulting Editors

H. S. W. Massey
University College, London, England

Keith A. Brueckner
University of California, San Diego
La Jolla, California

1. F. H. Field and J. L. Franklin, Electron Impact Phenomena and the Properties of Gaseous Ions. (Revised edition, 1970.)
2. H. Kopfermann, Nuclear Moments, English Version Prepared from the Second German Edition by E. E. Schneider.
3. Walter E. Thirring, Principles of Quantum Electrodynamics. Translated from the German by J. Bernstein. With Corrections and Additions by Walter E. Thirring.
4. U. Fano and G. Racah, Irreducible Tensorial Sets.
5. E. P. Wigner, Group Theory and Its Application to the Quantum Mechanics of Atomic Spectra. Expanded and Improved Edition. Translated from the German by J. J. Griffin.
6. J. Irving and N. Mullineux, Mathematics in Physics and Engineering.
7. Karl F. Herzfeld and Theodore A. Litovitz, Absorption and Dispersion of Ultrasonic Waves.
8. Leon Brillouin, Wave Propagation and Group Velocity.
9. Fay Ajzenberg-Selove (ed.), Nuclear Spectroscopy. Parts A and B.
10. D. R. Bates (ed.), Quantum Theory. In three volumes.
11. D. J. Thouless, The Quantum Mechanics of Many-Body Systems.
12. W. S. C. Williams, An Introduction to Elementary Particles. (Second edition, 1971.)
13. D. R. Bates (ed.), Atomic and Molecular Processes.
14. Amos de-Shalit and Igal Talmi, Nuclear Shell Theory.
15. Walter H. Barkas. Nuclear Research Emulsions. Part I.
    Nuclear Research Emulsions. Part II. *In preparation*
16. Joseph Callaway, Energy Band Theory.
17. John M. Blatt, Theory of Superconductivity.
18. F. A. Kaempffer, Concepts in Quantum Mechanics.
19. R. E. Burgess (ed.), Fluctuation Phenomena in Solids.
20. J. M. Daniels, Oriented Nuclei: Polarized Targets and Beams.
21. R. H. Huddlestone and S. L. Leonard (eds.), Plasma Diagnostic Techniques.
22. Amnon Katz, Classical Mechanics, Quantum Mechanics, Field Theory.
23. Warren P. Mason, Crystal Physics in Interaction Processes.
24. F. A. Berezin, The Method of Second Quantization.
25. E. H. S. Burhop (ed.), High Energy Physics. In four volumes.

26. L. S. Rodberg and R. M. Thaler, Introduction to the Quantum Theory of Scattering.
27. R. P. Shutt (ed.), Bubble and Spark Chambers. In two volumes.
28. Geoffrey V. Marr, Photoionization Processes in Gases.
29. J. P. Davidson, Collective Models of the Nucleus.
30. Sydney Geltman, Topics in Atomic Collision Theory.
31. Eugene Feenberg, Theory of Quantum Fluids.
32. Robert T. Beyer and Stephen V. Letcher, Physical Ultrasonics.
33. S. Sugano, Y. Tanabe, and H. Kamimura, Multiplets of Transition-Metal Ions in Crystals.
34. Walter T. Grandy, Jr., Introduction to Electrodynamics and Radiation.

In preparation

J. Killingbeck and G. H. A. Cole, Physical Applications of Mathematical Techniques.

Herbert Uberall, Electron Scattering from Complex Nuclei. Parts A and B.